T0206493

Internal Phosphorus Loading in Lakes

Causes, Case Studies, and Management

Edited by
Alan D. Steinman
Bryan M. Spears

ISBN-13: 978-1-60427-144-7

Printed and bound in the U.S.A. Printed on acid-free paper.

10 9 8 7 6 5 4 3 2 1

For Library of Congress Cataloging-in-Publication Data, please visit the WAV section of the publisher's website at www.jrosspub.com/wav.

Front cover photo of Lake Okeechobee, Florida, is courtesy of Nicholas Aumen, USGS.

Phone: (954) 727-9333
Fax: (561) 892-0700
Web: www.jrosspub.com

DEDICATION

We dedicate this book to the memory of Dr. Karl Havens, most recently the Director of the Florida Sea Grant College Program and Professor in the Department of Fisheries and Aquatic Sciences at the University of Florida School of Forest Resources and Conservation. Al had the privilege of working with Karl from 1993 to 2001 at the South Florida Water Management District. Karl was an incredibly focused, insightful, and respected scientist. His clarity of thought was a wonder to behold, and truly delightful to see in action. The aquatic science community suffers a huge loss with his passing. May his memory serve as a blessing.

CONTENTS

Section II—Case Studies from Around the World

Chapter 6: Observed and Modeled Internal Phosphorus Loads in Stratified and Polymictic Basins of a Mesotrophic Lake in Canada

Gertrud Nürnberg

Chapter 7: Internal Phosphorus Loads in Subtropical Shallow Lakes: Two Florida Lakes as Case Examples

K. Ramesh Reddy, Todd Z. Osborne, Dean R. Dobberfuhl, and Laura K. Reynolds

Chapter 8: Alum Treatment Did Not Improve Water Quality in Hypereutrophic Grand Lake St. Mary's, Ohio 147

Geraldine Nogaro, Amy J. Burgin, Astrea Taylor, and Chad R. Hammerschmidt

Chapter 9: Internal Pools and Fluxes of Phosphorus in Dimictic Lake Arendsee, Northeastern Germany 169

Michael Hupfer, Andreas Kleeberg, and Jörg Lewandowski

Chapter 10: Studies of Legacy Internal Phosphorus Load in Lake Peipsi (Estonia/Russia) 187

Olga Tammeorg, Jukka Horppila, Tõnu Möls, Marina Haldna,
Reet Laugaste, and Juha Niemistö

Chapter 11: Phosphorus Dynamics and Its Relationship with Cyanobacterial Blooms in Lake Taihu, China 211

Liqiang Xie, Xiaomei Su, and Hai Xu

Chapter 14: Internal Phosphorus Loading in Esthwaite Water, United Kingdom: Considering the Role of Weather and Climate 263

Eleanor B. Mackay and Ian D. Jones

Chapter 15: Lake Søbygaard, Denmark: Phosphorus Dynamics During the First 35 Years After an External Loading Reduction 285

Martin Søndergaard and Erik Jeppesen

Chapter 16: Accumulation of Legacy Sediment Phosphorus in Lake Hjälmaren, Sweden: Consequences For Lake Restoration 301

Brian J. Huser, Mikael Malmaeus, Ernst Witter, Anders Wilander, and Emil Rydin

Chapter 17: Limited Role of Internal Loading in a Formerly Hypertrophic Shallow Lake in the Netherlands 323

Ruurd Noordhuis, Gerlinde Roskam, and Leonard Osté

Chapter 18: A Review of Internal Phosphorus Loading Evidence in Säkylän Pyhäjärvi, Finland. 345

Anne-Mari Ventelä, Petri Ekholm, Teija Kirkkala, Jouni Lehtoranta,
Gertrud Nürnberg, Marjo Tarvainen, and Jouko Sarvala

Chapter 19: Internal Loading of Phosphorus to Lake Erie: Significance, Measurement Methods, and Available Data. 359

Eliza M. Kaltenberg and Gerald Matisoff

Section III—Integration and Synthesis 431

Chapter 23: Factors Influencing Internal Phosphorus Loading: A Meta-Analysis. 433
Emily Kindervater, Nicole Hahn, and Alan D. Steinman

Chapter 24: Synthesis, Implications, and Recommendations 445
Bryan M. Spears and Alan D. Steinman

FOREWORD

Macronutrients, such as phosphorus, nitrogen, and potassium, are essential for life on this planet. Yet, in too high a quantity, they can cause profound environmental problems. Phosphorus, in particular, presents a unique case. It is a critical component of the high-energy compounds ATP and ADP, as well as nucleic acids, several essential co-enzymes, and cell membranes. Yet, its bioavailability under natural conditions is low, resulting in phosphorus limitation of plant growth in most freshwater ecosystems. Humans have altered these systems, however. Excess application of fertilizer, beyond the natural assimilative capacity of our ecosystems, has resulted in the eutrophication of the planet's lakes, wetlands, and streams, resulting in the proliferation of algal blooms, and subsequent conditions of depleted dissolved oxygen in the water column. Over decades, if not centuries, this phosphorus has accumulated in lake sediments (as well as catchment soils), creating a "legacy" of phosphorus. This accumulated reservoir of phosphorus is being released into the water column in lakes throughout the world (*internal* phosphorus loading, as opposed to phosphorus entering from the catchment, which is *external* phosphorus loading), resulting in noxious algal blooms and threatening the water supply of many millions of people. In addition, because internal phosphorus loading can persist for long periods of time, it can counterbalance the anticipated benefits of control measures taken to reduce phosphorus inputs from the catchment. In total, the impacts of internal phosphorus loading can result in impaired water bodies, economic impacts to local communities due to health issues, loss of tourism, and depressed civic pride.

There is a pressing need for better understanding of internal phosphorus loading on a global basis. The content provided in this book aligns with the Global Partnership on Nutrient Management (GPNM), which was launched during the 17th session of the UN Commission on Sustainable Development in 2009 as a global partnership of governments, policy makers, industry, the scientific community, civil society organizations, and UN agencies with UNEP providing the Secretariat. More recently, the 4th session of the UN Environment Assembly in March 2019 adopted a landmark resolution on Sustainable Nitrogen Management, which was followed by the Colombo Declaration in October 2019 marking the launch of a UN Global Campaign on Sustainable Nitrogen Management. It is my hope that the information contained in *Internal Phosphorus Loading in Lakes: Causes, Case Studies, and Management* will help raise the profile of phosphorus and lead to wiser management strategies of this very important element and, in turn, result in cleaner fresh waters, healthier people, and invigorated economies.

Habib N. El-Habr, PhD

Coordinator
Global Programme of Action for the Protection of the
    Marine Environment from Land-Based Activities (GPA)
Global Partnership on Nutrient Management (GPNM)
Ecosystems Division, United Nations Environment Programme (UNEP)

LIST OF CONTRIBUTING AUTHORS

Yaron Beeri-Shlevin, Israel Oceanographic and Limnological Research, The Yigal Allon Kinneret Limnological Laboratory, Migdal, Israel

Harvey A. Bootsma, University of Wisconsin-Milwaukee, School of Freshwater Sciences, Milwaukee, Wisconsin, USA

Amy J. Burgin, Department of Ecology and Evolutionary Biology, The University of Kansas, Lawrence, Kansas, USA

Dean R. Dobberfuhl, Saint Johns River Water Management District, Palatka, Florida, USA

Grant Douglas, CSIRO Land and Water, Centre for Environment and Life Sciences, Floreat, Western Australia, Australia

Werner Eckert, Israel Oceanographic and Limnological Research, The Yigal Allon Kinneret Limnological Laboratory, Migdal, Israel

Petri Ekholm, Finnish Environment Institute, Helsinki, Finland

Björn Grüneberg, Department of Freshwater Conservation, Brandenburg University of Technology, Cottbus-Senftenberg, Bad Saarow, Germany

Nicole Hahn, Annis Water Resources Institute, Grand Valley State University, Muskegon, Michigan, USA

Marina Haldna, Centre for Limnology, Estonian University of Life Sciences, Tartumaa, Estonia

Chad R. Hammerschmidt, Department of Earth & Environmental Sciences, Wright State University, Dayton, Ohio, USA

Robert E. Hecky, Large Lakes Observatory, University of Minnesota-Duluth, Duluth, Minnesota, USA

Jukka Horppila, Department of Environmental Sciences, University of Helsinki, Helsinki, Finland

Michael Hupfer, Department of Chemical Analytics and Biogeochemistry, Leibniz Institute of Freshwater Ecology and Inland Fisheries, Berlin, Germany

Brian J. Huser, Department of Aquatic Sciences and Assessment, Swedish University of Agricultural Sciences, Uppsala, Sweden

Vera Huszar, Museu Nacional, Federal University of Rio de Janeiro, Rio de Janeiro, Brazil

Stephan C. Ives, Freshwater Restoration and Sustainability Group, UK Centre for Ecology & Hydrology, Penicuik, Midlothian, United Kingdom

Erik Jeppesen, Department of Bioscience, Aarhus University, Aarhus, Denmark

Ian D. Jones, Earth Observation and Geotechnologies, Stirling University, Stirling, United Kingdom

Eliza M. Kaltenberg, Battelle, Norwell, Massachusetts, USA

Andrea Kelly, The Broad's Authority, Norwich, Norfolk, United Kingdom

Emily Kindervater, Annis Water Resources Institute, Grand Valley State University, Muskegon, Michigan, USA

Teija Kirkkala, Pyhäjärvi Institute, Kauttua, Finland

Andreas Kleeberg, Department Geology, Soil, Waste, State Laboratory, Berlin-Brandenburg, Germany

Reet Laugaste, Centre for Limnology, Estonian University of Life Sciences, Tartumaa, Estonia

Jouni Lehtoranta, Finnish Environment Institute, Helsinki, Finland

Jörg Lewandowski, Department of Ecohydrology, Leibniz Institute of Freshwater Ecology and Inland Fisheries and Geography Department Humboldt-University of Berlin, Berlin, Germany

Miquel Lürling, Department of Environmental Sciences, Wageningen University and Department of Aquatic Ecology, Netherlands Institute of Ecology (NIOO-KNAW), Wageningen, The Netherlands

Eleanor B. Mackay, Lake Ecosystems Group, UK Centre for Ecology & Hydrology, Bailrigg, Lancaster, United Kingdom

Leonardo de Magalhães, Laboratory of Ecology and Physiology of Phytoplankton, Department of Plant Biology, University of Rio de Janeiro State, Rio de Janeiro, Brazil

Mikael Malmaeus, IVL Svenska Miljöinstitutet AB, Stockholm, Sweden

Marcelo Manzi Marinho, Laboratory of Ecology and Physiology of Phytoplankton, Department of Plant Biology, University of Rio de Janeiro State, Rio de Janeiro, Brazil

Gerald Matisoff, Case Western Reserve University, Cleveland, Ohio, USA

Linda May, Freshwater Restoration and Sustainability Group, UK Centre for Ecology & Hydrology, Penicuik, Midlothian, United Kingdom

Jônatas de Souza Mercedes, Laboratory of Ecology and Physiology of Phytoplankton, Department of Plant Biology, University of Rio de Janeiro State, Rio de Janeiro, Brazil

Tõnu Möls, Centre for Limnology, Estonian University of Life Sciences, Tartumaa, Estonia

Juha Niemistö Department of Environmental Sciences, University of Helsinki, Helsinki, Finland

Aminadav Nishri, Israel Oceanographic and Limnological Research, The Yigal Allon Kinneret Limnological Laboratory, Migdal, Israel

Geraldine Nogaro, EDF Research and Development, National Hydraulic and Environment Laboratory (LNHE), Chatou, France

Ruurd Noordhuis, Deltares, Utrecht, The Netherlands

Natália Pessoa Noyma, Laboratory of Ecology and Physiology of Phytoplankton, Department of Plant Biology, University of Rio de Janeiro State, Rio de Janeiro, Brazil

Gertrud Nürnberg, Freshwater Research, Baysville, Ontario, Canada

Todd Z. Osborne, Wetland Biogeochemistry Laboratory, Soil and Water Sciences Department, Institute of Food and Agricultural Sciences, University of Florida and Whitney Laboratory of Marine Sciences, University of Florida, Gainesville, Florida, USA

Leonard Osté, Deltares, Utrecht, The Netherlands

Jo-Anne Pitt, Environment Agency, Bristol, United Kingdom

Geoff Phillips, Biological and Environmental Sciences, University of Stirling, Stirling, United Kingdom

K. Ramesh Reddy, Wetland Biogeochemistry Laboratory, Soil and Water Sciences Department, Institute of Food and Agricultural Sciences, University of Florida, Gainesville, Florida, USA

Kasper Reitzel, Department of Biology, University of Southern Denmark, Odense, Denmark

Laura K. Reynolds, Wetland Biogeochemistry Laboratory, Soil and Water Sciences Department, Institute of Food and Agricultural Sciences, University of Florida and Coastal Ecology Laboratory, Soil and Water Sciences Department, Institute of Food and Agricultural Sciences, University of Florida, Gainesville, Florida, USA

Gerlinde Roskam, Deltares, Utrecht, The Netherlands

Emil Rydinb, Naturvatten i Roslagen AB, Norra Malmavägen, Norrtälje, Sweden

Jouko Sarvalab, Department of Biology, University of Turku, Turku, Finland

Alfons J.P. Smoldersb, Department of Aquatic Ecology and Environmental Biology, Radboud University Nijmegen, Heyendaalseweg and B-WARE Research Centre, Radboud University Nijmegen, Nijmegen, The Netherlands

Martin Søndergaardb, Department of Bioscience, Aarhus University, Aarhus, Denmark

Bryan M. Spears, Freshwater Restoration and Sustainability Group, UK Centre for Ecology & Hydrology, Penicuik, Midlothian, United Kingdom

Alan D. Steinman, Annis Water Resources Institute, Grand Valley State University, Muskegon, Michigan, USA

Xiaomei Su, State Key Laboratory of Lake Science and Environment, Nanjing Institute of Geography and Limnology, Chinese Academy of Sciences and Jiangsu Provincial Key Laboratory of Environmental Engineering, Jiangsu Provincial Academy of Environmental Sciences, Nanjing, China

Marjo Tarvainen, Southwest Finland's Centre for Economic Development, Transport and the Environment, Turku, Finland

Olga Tammeorg, Department of Environmental Sciences, University of Helsinki, Helsinki, Finland and Centre for Limnology, Estonian University of Life Sciences, Tartumaa, Estonia

Astrea Taylor, Department of Earth & Environmental Sciences, Wright State University, Dayton, Ohio, USA

Anne-Mari Ventelä, Pyhäjärvi Institute, Kauttua, Finland

Anders Wilander, Department of Aquatic Sciences and Assessment, Swedish University of Agricultural Sciences, Uppsala, Sweden

Ernst Witter, County Administrative Board of Örebro, Örebro, Sweden

Liqiang Xie, State Key Laboratory of Lake Science and Environment, Nanjing Institute of Geography and Limnology, Chinese Academy of Sciences, Nanjing, China

Hai Xu, State Key Laboratory of Lake Science and Environment, Nanjing Institute of Geography and Limnology, Chinese Academy of Sciences, Nanjing, China

ACKNOWLEDGMENTS

The authors gratefully acknowledge the support of our home institutions, Grand Valley State University (Michigan, USA) and the the UK Centre for Ecology & Hydrology. Al expresses special gratitude to Allen and Helen Hunting, whose Research and Innovation Fund helped support his efforts on this book and to Bryan Spears, who hosted him at the the UK Centre for Ecology & Hydrology while working on the book. He is also grateful for the efforts of Emily Kindervater, whose reviewing and editing skills are second to none. Bryan is especially grateful to his PhD students Stephen Ives and Kate Waters for their creative contributions, and to his wife, Cheryl, and daughters, Poppy and Catherine, for their unwavering patience and support during production. Finally, both Al and Bryan are deeply grateful to the authors who contributed to this book, and to Gwen Eyeington from J. Ross Publishing, whose inspiration and guidance helped bring this book to fruition.

ABOUT THE AUTHORS

ALAN D. STEINMAN, Ph.D.

Alan (Al) Steinman is the Director of Grand Valley State University's Annis Water Resources Institute, a position he has held since 2001. Previously, he was Director of the Lake Okeechobee Restoration Program at the South Florida Water Management District. Steinman has published over 175 scientific articles, book chapters, and books; has been awarded over $55 million in grants for scientific and engineering projects; and has testified before the U.S. Congress and the Michigan and Florida state legislatures.

Among his awards are Phi Beta Kappa; the 2017 Award of Excellence from the National Garden Clubs; the U.S. Army Corps of Engineers Outstanding Planning Achievement Award; the Joan Hodges Queneau Palladium Medal from the National Audubon Society; Paul Harris Fellow; Keiser Distinguished Lecturer in Life Sciences from Ohio Northern University; and the Patricia B. Johnson Award for Leadership and Innovative Grantmaking from the Community Foundation for Muskegon County.

Al is a member of science advisory boards for the U.S. EPA, the International Joint Commission, Michigan DEQ, Sea Grant, Healing our Waters, University of Michigan's Water Center, and the Cooperative Institute for Great Lakes Research. He currently serves as Associate Editor for the journal *Freshwater Biology*. He also has served on the State of Michigan's Groundwater Conservation Advisory Council and Phosphorus Advisory Committees. Steinman's research interests include aquatic ecosystem restoration, harmful algal blooms, phosphorus cycling, and water policy.

His current community service includes serving on the Board of Directors of Goodwill International of West Michigan, the Community Foundation for Muskegon County, and the West Michigan Symphony. Prior board service included the Michigan Chapter of The Nature Conservancy, the Land Conservancy of West Michigan, and Congregation B'Nai Israel.

Dr. Steinman holds a Postdoctoral Research Fellowship from Oak Ridge National Laboratory, a Ph.D. in Botany/Aquatic Ecology from Oregon State University, an M.S. in Botany from the University of Rhode Island and a B.S. in Botany from the University of Vermont.

BRYAN M. SPEARS, Ph.D.

Bryan Spears is a Principal Scientific Officer with the Freshwater Restoration and Sustainability Group at the UK Centre for Ecology & Hydrology (CEH), Edinburgh, part of the Natural Environment Research Council. He has worked at CEH since 2007 in which time he has published over 100 scientific articles and research reports. Much of Bryan's work has focused on identifying ecological responses to human pressures in inland and coastal waters. These pressures include climate change, industrial pollution, and nutrient enrichment and their interactions. He has managed national scale surveys of biogeochemical cycling in rivers and estuaries in Scotland and initiated a series of whole lake experiments to examine geoengineering in lakes for internal phosphorus loading control in the UK, coordinating international efforts in this field through networking initiatives.

In recognition of his contribution to teaching and research, Bryan serves as an Honorary Fellow at the University of Edinburgh, Department of Geosciences, UK. He has supervised 12 Ph.D. students, mostly in the field of lake restoration and biogeochemical cycling in aquatic ecosystems.

Bryan serves as an Associate Editor for the journals *Inland Waters* and the *Journal of Environmental Quality* and has served as a Guest Editor for the journals *Water Research* and *Hydrobiologia*. He is a member of the United Nations Environment Programme's Global Phosphorus Task Team, reporting to the Global Partnership for Nutrient Management, under the The Global Programme of Action for the Protection of the Marine Environment from Land-Based Activities (GPA), and he contributes to the UNEP World Water Quality Alliance. This enables him to raise awareness of the wide spread problems caused by phosphorus enrichment of lakes, including the importance of internal loading and its management.

Bryan's work on lake restoration has a strong focus on delivering benefits to local communities in the UK and internationally. Bryan, in collaboration with Miquel Lürling, Wageningen University, the Netherlands, co-founded the International Society of Limnology Working Group on Lake Restoration which works collectively to deliver knowledge and expertise to those countries that need it most.

Dr. Spears holds a Ph.D. in Limnology from St Andrews University, UK, an M.Sc. in Aquatic Ecology from Simon Fraser University, Canada, and a B.Sc. (hons) In Environmental Sciences from Robert Gordon University, Aberdeen, UK.

Section I

Introduction to and Overview of Internal Phosphorus Loading

CHAPTER 1

WHAT IS INTERNAL PHOSPHORUS LOADING AND WHY DOES IT OCCUR?

Alan D. Steinman[1] and Bryan M. Spears[2]

Abstract

Lake eutrophication is a global problem that is being exacerbated by climate change, excess nutrient runoff, and land-use alterations. While nutrient inputs to lakes from surrounding watersheds (external loading) have historically received considerable attention, phosphorus inputs (along with other elements) that are generated from within the lake have received far less attention until recently. But there is growing recognition and evidence that impairments that are created from phosphorus sources within lakes are a global phenomenon. Despite this awareness, there is still uncertainty regarding some of its most fundamental characteristics, including: (1) the definition of internal phosphorus loading; (2) the most appropriate way to measure it; (3) how to predict where, when, and how long it will occur; and (4) how to control or manage it. In this chapter, we briefly introduce the concept of internal phosphorus loading, provide an overview of various causes for this phenomenon, and set the stage for the remaining chapters of this book.

We have divided this book into three main parts: Part 1 is an overview of the internal phosphorus loading concept; Part 2 includes case studies from iconic lakes throughout the world; and Part 3 explains the integration and synthesis of the information that has been generated. Our ultimate goals for the book are to increase awareness of internal loading, compare and contrast internal loading from lakes around the world, and identify emerging themes regarding what drives internal loading along with which measures are best suited to manage or limit its impacts.

Keywords: Internal phosphorus loading; lake eutrophication; case studies; integration, and synthesis.

---

[1] Annis Water Resources Institute, Grand Valley State University, 740 West Shoreline Drive, Muskegon, MI 49441, USA. E-mail: steinmaa@gvsu.edu.

[2] Centre for Ecology and Hydrology, Edinburgh, Bush Estate, Penicuik, Midlothian, EH26 0QB. E-mail: spear@ceh.ac.uk.

1.1   INTRODUCTION

1.1.1   Definitions

Internal phosphorus (P) loading can be generically considered as all physical, chemical, and biological processes by which P is mobilized and translocated from the benthic environment. Other definitions exist for internal P loading; for example, Hupfer and Reitzel (see Chapter 2) explain why the term *internal loading* should be used only in cases where sediments are a net source of P at time scales of one or more years. Orihel et al. (2017) recognized that operational definitions of internal loading have not been consistent, which has resulted in confusion and ambiguity as to what is meant by the term. They qualified their definition, restricting it to P leaving the sediment that reaches the overlying water column, given the management concern regarding the influence of P on algal blooms and also excluding groundwater-driven P moving through the sediment matrix. Our approach in this book is to be less prescriptive, recognizing that users will define internal P loading based on their needs and objectives (see Chapter 2); however, it is important to recognize that internal loading defies one universal definition (cf. Orihel et al. 2017). Hence, it is critical that when authors use the term, they define their explicit intent.

1.1.2   History

Our understanding of internal P loading is grounded in a hypothesis proposed over 75 years ago by Mortimer (1941), who described the redox-mediated exchange of dissolved substances across the sediment-water interface in Esthwaite Water (UK). As oxygen and other electron donors become depleted, compounds that bind P (predominantly iron, i.e., Fe-P) become chemically reduced, releasing P and allowing it to diffuse into the overlying water. This hypothesis has been modified over the years as researchers examined the mechanisms associated with Mortimer's original ideas. Since the 1980s, a few seminal papers have refined our understanding of the physical, chemical, and biological processes driving Mortimer's central hypothesis. For example, Gächter et al. (1988), Boström et al. (1988), and Golterman et al. (2001) showed the pivotal role of the microbial community in driving phosphorus remineralization and subsequent redox chemistry in bed sediments.

The late 1980s can be considered the springboard of contemporary internal loading research, and we review a collection of seminal works below, including some significant recent contributions. Sas (1989; 826 citations as of 25 July 2019) produced the first comprehensive collection of case studies, from which he proposed a set of general principles governing internal loading in lakes, especially those recovering from catchment nutrient loading. Three distinct phases of recovery were defined relative to catchment nutrient load reduction: pre-management, a transient recovery phase, and a new steady-state. Using case studies with long-term monitoring data, differences were demonstrated in the functioning of shallow versus deep lakes through these phases. In shallow lakes, internal loading was initiated generally following a reduction in catchment loading; the length of the transient period was several years or longer in lakes where the average sediment P concentration in the upper 15 cm exceeded 1 mg g^{-1} dw. In these shallow lakes, catchment nutrient load reduction triggered a change in functioning where sediments became a source of phosphorus, with net annual sediment P release being common. However, in deep lakes where sediment P concentrations were generally low, net annual sediment P retention was common regardless of catchment load. This detailed analysis of long-term changes in sediment processes and the mass balance modeling approach produced from these studies, demonstrated the critical role that internal loading can play in driving ecological structure and function at the ecosystem scale—and confirmed the problem to be globally relevant.

Sas' analyses were further developed by Nürnberg (1984 and 1988; 679 collective citations as of 25 July 2019) in two important papers relating internal P load to sediment P content and composition, providing a novel and simple predictive approach and characterizing increasing sediment P fluxes with increasing total and reductant-soluble sediment P concentrations. Across 82 North American and European lakes, sediment P flux ranged from <1 mg P m^{-2} d^{-1} for oligotrophic sediments and up to 50 mg P m^{-2} d^{-1} for hypertrophic sediments.

Boström et al. (1988; 845 citations as of 25 July 2019) produced a complementary seminal work that outlined the complex pathways through which P was cycled between the sediment and the overlying water. This paper filled the gaps in process understanding and proposed key hypotheses that have shaped the research field. Specifically, the role of the microbial community in bed sediments was highlighted as a critical pathway for P, a hypothesis that even after 30 years, we are only just starting to address comprehensively given the development of powerful chemical and molecular analytical approaches. Pettersen et al. (1988; 192 citations as of 25 July 2019) produced a comprehensive description of the chemical pathways and constraints on P cycling, demonstrating the power of previously proposed techniques that were designed to operationally define sediment P (such as Psenner et al. 1988; 304 citations as of 25 July 2019). With modifications (e.g., Hupfer et al. 1995; 336 citations as of 25 July 2019), this fractionation approach is still in use today.

In the early 2000s, research moved toward testing the hypotheses of the 1980s. Søndergaard et al. (2003; 1144 citations as of 25 July 2019) demonstrated the power of long-term monitoring data in providing general understanding of internal loading and its drivers using data from Danish lakes. This work characterized the typical bell curve pattern in lakes dominated by internal loading where P is released to the water column during periods of low catchment loading. This work was followed by a global scale analysis of whole lake responses to reduced catchment P loading across 35 lakes with long-term data, demonstrating that internal loading could prolong recovery for years—or even decades—especially in shallow lakes (Jeppesen et al. 2005; 968 citations as of 25 July 2019). This work also highlighted the importance of climate change in future regulation of internal loading, which remains a knowledge gap in the field. More recently, the concept of legacy phosphorus has been developed, allowing the effects of internal loading to be placed into the context of catchment phosphorus recovery times, potentially reaching centuries or possibly millennia (Sharpley et al. 2013; 421 citations as of 25 July 2019). Collectively, the 10 papers that were previously cited represent an essential reading list for any researcher who may consider entering the field. They have amassed nearly 5000 citations and continue to influence the research field. It has not escaped our attention that most of these studies focus primarily on data from lakes in North America and Europe, highlighting the need to expand the study of internal P loading to lakes that are located on other parts of the globe.

This body of work advanced Mortimer's seminal geochemical P pump hypothesis and confirmed that the liberation of P from bed sediments to bottom waters could drive ecosystem scale responses, and is governed by a complex mosaic of physical, biological, and chemical processes. It also suggested that this process is globally relevant. However, the extent to which these processes respond to environmental change—including anthropogenic changes in land use, invasive species, and climate, as well as the influence of latitude and longitude—remains unclear. Muddying the waters further, we still lack robust operational classifications for internal loading and its processes, even though the field has made impressive advances in detection (see Chapter 2), modeling (see Chapter 3), prediction (see Chapter 4), and control (see Chapter 5) in recent years.

1.1.3 Why Phosphorus?

Our focus on P is driven by a number of factors. First, historically, P has been considered the primary nutrient limiting autotrophic production in freshwater ecosystems, given its limited bioavailability in nature (Schindler 1977; Hecky and Kilham 1988; Hudson et al. 2000; but see upcoming text). Second, P concentrations in healthy plants are relatively low, usually ranging from 0.1 to 0.8% of dry mass (Raven et al. 1981), although P is essential for growth. Some of the more important functions played by P in plants and animals include being a structural component of *high-energy phosphate* compounds (e.g., ADP and ATP), nucleic acids, several essential coenzymes, and cell membrane constituents (phospholipids), as well as being involved in the phosphorylation of sugars. Third, most of the sediment P that is the source of internal loading ultimately comes from the watershed, so the question of how best to manage P—in the watershed or in the lake—is a fertile area of debate with significant economic, societal, and ecological implications (Sharpley et al. 2013; Osgood 2017; Steinman et al. 2018a). Despite concerns of a global phosphorus shortage (Cordell and White 2011), the mass of P stockpiled in freshwater ecosystems as a result of anthropogenic activities continues to grow at a rate of about 5.0 Tg P yr^{-1} (Beusen et al. 2016).

While many lakes certainly are limited by phosphorus alone (cf. Paterson et al. 2011; Schindler et al. 2016), that paradigm has come into question in recent years, as nitrogen (N) has been found to be either the limiting or co-limiting (with P) nutrient in some lakes (Elser et al. 1990; Leavitt et al. 2006; Conley et al. 2009; Paerl and Scott 2010; Paerl et al. 2016; Steinman et al. 2016). Excess internal P loading can lead to N limitation in lakes (Ding et al. 2018); this type of secondary N limitation may be mitigated by controlling internal P loading. Internal nitrogen loading has received relatively little attention compared to P, although release of ammonia has certainly been documented (Beutel 2006). Internal processes of N and P cycling in lakes should not be considered decoupled, although the mechanisms by which they interact are not yet fully understood. Indeed, nitrate is an important precursor to Fe in the redox series, and nitrate losses in bottom waters are expected to occur more rapidly through denitrification in warmer lakes, leading to a higher likelihood of internal loading (Weyhenmeyer et al. 2007).

1.2 CAUSES OF INTERNAL P LOADING

Internal P loading is measurable only when sediment phosphorus release exceeds sediment phosphorus retention. Both release and retention can occur simultaneously in lakes, but P accumulation in the water column along with the attendant management concerns, emerge only when release exceeds retention. The factors driving internal P loading can be partitioned into biological, chemical, and physical mechanisms (see Figure 1.1)—although nature rarely behaves so simplistically. Hence, although we use these three categories, in reality they often interact, resulting in outcomes that may not align with predictions or preconceptions.

1.2.1 Biological Causes

Bioturbation is perhaps the best known biological mechanism for P release from sediments (cf. Mermillod-Blondin and Rosenberg 2006; Roskosch et al. 2012; Hölker et al. 2015; Nogaro et al. 2016). The most common bioturbators in eutrophic lakes are usually chironomid larvae and tubificid oligochaetes. Chironomid larvae can form u-shaped tubes in the sediment; pumping of water at the sediment-water interface can flush nutrients out of the tubes and into the overlying water column (Hansen et al. 1997), although the amount of P that is released is influenced by sediment properties

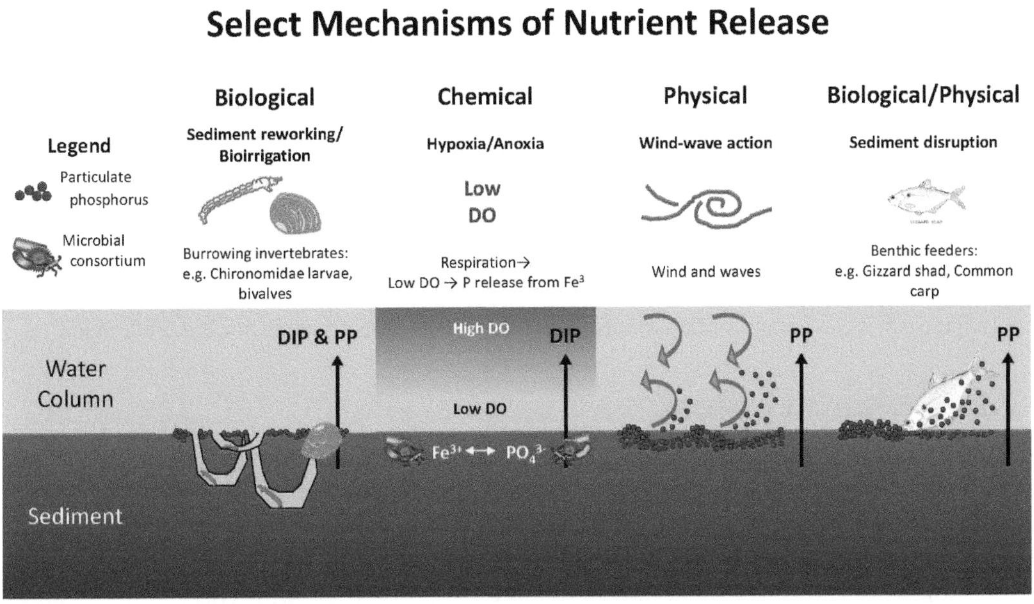

Select Mechanisms of Nutrient Release

Figure 1.1 Schematic diagram of different mechanisms responsible for internal P loading in lakes. Figure credit: Emily Kindervater.

(Nogaro and Steinman 2014). In contrast, oligochaetes ingest sediment at depth and egest fecal pellets at the sediment surface; hence, while they can stimulate solute exchange between sediment and water via their constructed galleries, their bioirrigation activity is limited in comparison with chironomid larvae (Svensson et al. 2001). Certain species of benthic fish, such as ruffe and gizzard shad (Kelly et al. 2018), as well as crayfish (Ottolenghi et al. 2002) and mussels (Nogaro and Steinman 2014; Chen et al. 2016a), also are known to disturb sediments and result in P movement from sediments to the water column, although some nutrients may derive from fish excretion and not from internal loading, *sensu lato* (Vanni 2002; Tarvainen et al. 2005; Schaus et al. 2010).

Another biologically mediated mechanism by which P can be moved from the sediment to the water column is vertical transport. Tang et al. (2017) showed that *Chaoborus* larvae, through both oxygen demand from sediment and the water column as well as nutrient excretion, enhance internal nutrient loading in lakes. In addition, Xie et al. (2003a, b) found that *Microcystis* blooms can be responsible for internal P loading, through either mineralization of decaying cells or by inducing a massive release of P from the sediment, perhaps due to either seasonal migration or high pH caused by intense algal photosynthesis, revealing the tight linkage between biology and chemistry in driving internal P loading (cf. Katsev 2017). Benthic algae and macrophytes also can play important roles in the movement of P; both groups of autotrophs can release P as they senesce and mineralize (Paalme et al. 2002; Higgins et al. 2008; Gao et al. 2013; Zhu et al. 2013). Macrophytes can also help prevent sediment resuspension by serving as a physical barrier to diffuse wind-wave action (Horppila and Nurminen 2003), although under dense canopies, internal loading can be enhanced via both anaerobic and aerobic diffusive flux pathways (Frodge and Pauley 1991). For benthivorous fish, where most data are available for carp, densities in excess of a 200 kg ha^{-1} to 700 kg ha^{-1} threshold result in increased turbidity and internal loading in shallow lakes (Williams and Moss 2003). Several chapters in this book highlight the fact that in certain lakes, sediment resuspension accounts for the majority of internal P loading (see Chapters 7 and 18).

Although inferred in earlier models as a major conduit of organic-P turnover in sediments that is ultimately driving internal loading in lakes, the role of the microbial community has remained a *black box*, until recently. We now know that microbial communities are capable of performing a range of functions that are designed to access P from inorganic and organic P compounds; this access can fuel both microbial production and the remineralization of relatively refractory and complex P compounds, thereby forming an important link with the well-described inorganic P cycle (Vila-Costa et al. 2013). Although the lake bed has been demonstrated as a major site of these biochemical pathways at the whole-lake scale (Reitzel et al. 2012), much remains to be learned of the environmental cues driving underlying processes, of the importance of microbial community succession for functional performance, and of the sensitivity of these processes to environmental change.

1.2.2   Chemical Causes

The best known chemically driven mechanism for P release derives from redox reactions that release P from iron hydoxides (Mortimer 1941). However, as thoroughly reviewed by Orihel et al. (2017) and Katsev (2017), there are many other mechanisms besides iron cycling to account for P diffusion from sediment porewater, including desorption, dissolution, mineralization, exudate excretion, and dissociation. These processes are dealt with in more detail in the following chapters of this book, as well as in the two prior citations.

1.2.3   Physical Causes

Disturbance is related to resuspension of sediment particles, which can be due to either physical forces such as wind-wave action or biological activity (Havens 1991; Steinman et al. 2006; Thomas and Schallenberg 2008; Tammeorg et al. 2013; Chen et al. 2016b; Chao et al. 2017; Matisoff et al. 2017). Bioturbation is often included as a physical process, but it also can be attributed to biotic activity (see Section 1.2.1), given that both sediment-dwelling organisms and sediment-surface feeders are responsible for solute or particle transport into the water column (cf. Vanni 2002; Mermillod-Blondin and Rosenberg 2006).

These complex biogeochemical processes interact and combine to govern the connection between the benthos and water column, the net effects of which can result in hysteresis in ecosystem scale responses to catchment and internal loading variation. These interactions make up the building blocks of ecological resilience theory in shallow lakes (Scheffer et al. 2004) and have been used to produce process models capable of predicting large-scale ecological responses to nutrient loading and climate change in some of the world's largest lakes (e.g., Taihu, China; Janssen et al. 2017).

Carey and Rydin (2011), based on a meta-analysis of sediment burial patterns of P in lakes around the world, found that the sediment total P (TP) concentrations changed with depth and that the shape of these distributions varied with lake trophic state; this suggested that lake sediment TP profiles may be indicative of nascent eutrophication. Using their database and adding data from lake studies conducted by the authors, we examined relations between water column and sediment P. We did not focus on the role of sediment depth in our analysis, which was one of the key findings in Carey and Rydin (2011). We used the same approach as Carey and Rydin for water column TP, averaging as many samples as available in the year that sediments were sampled for TP and separating lakes into three trophic states based on water column TP concentration (oligotrophic: < 10 µg/L; mesotrophic: 10–30 µg/L; and eutrophic: > 30 µg/L). There was high variance in the relationships in all trophic levels, resulting in low predictive ability (see Figure 1.2). It is likely that variance would be reduced if we used mobile sediment P instead of sediment TP because the more stable sediment P fractions

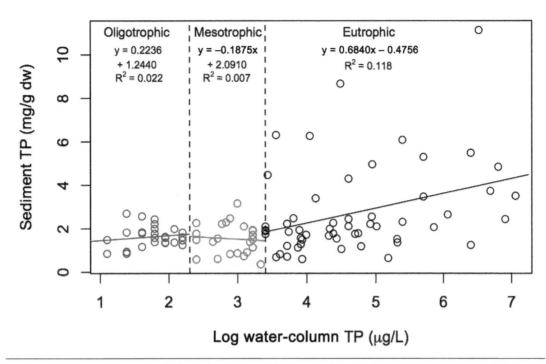

Figure 1.2 Surface sediment TP (mg/g dry weight) relationship with the log of water-column TP (μg/L) over three trophic levels. The linear model of each trophic level was calculated separately. No outliers were removed for the purposes of these linear models, but therefore the residuals of log (water column TP) for the mesotrophic and eutrophic lakes were not normally distributed. The oligotrophic to meso-trophic boundary was set at 10 μg/L and the mesotrophic to eutrophic boundary was set at 30 μg/L (vertical dotted lines). Data were obtained from Carey and Rydin (2011) and various west Michigan lakes (Brennan et al. 2016; Steinman and Ogdahl 2008, 2012; Steinman et al. 2004, 2018b).

will not be contributing to water column TP (see Chapter 9). Nonetheless, the relationships varied among trophic state, revealing a slightly positive slope between sediment TP and log water-column TP for oligotrophic lakes, a slightly negative slope for mesotrophic lakes, and a stronger positive slope for eutrophic lakes. From a management perspective, water depth should also be taken into consideration since recent studies have shown that the influence of water depth on lake water quality will vary based on trophic state: oligotrophic lakes get clearer when lake levels decline but more turbid when lake levels rise, with the opposite pattern for eutrophic lakes (Ji and Havens 2019; Lisi and Hein 2019).

1.3   CONTROL AND MANAGEMENT

Cataloging the mechanisms of internal P loading is more than an academic exercise because effective control and management is absolutely dependent on knowing the source. One theme that emerges from the case studies that are described in the following chapters is the need for lake-specific analyses to determine the best management strategy; a one-size-fits-all approach is doomed to failure (see Chapter 5). Biomanipulation involving the removal of benthivorous fish will be an expensive and ineffective strategy if the P source is diffusive flux from the sediment; conversely, a chemical inactivation treatment will have limited benefit if the main P source from the watershed is not addressed.

The influence of internal P loading is likely to become more important in the future for several reasons. First, a warming climate is resulting in warming lake temperatures (O'Reilly et al. 2015).

As lakes warm, stratification intensifies resulting in a greater chance for hypoxia/anoxia to form in the hypolimnion, thereby driving P desorption from Fe hydroxides and more P diffusion from sediments. Second, continued population growth is creating pressure on agricultural production. This intensification is resulting in greater P runoff around the world (Macintosh et al. 2018), which ultimately finds its way to lake sediments, setting the stage for future internal loading.

Like much of limnology, most studies of internal loading have occurred in North America and Europe. However, it is likely that the problem is of a global nature, and we have attempted to demonstrate that with select case studies. As we identify best practices for measuring (see Chapters 2 and 3), understanding (see Chapter 4), and managing (see Chapter 5) internal phosphorus loading, we also explore the most effective societal and scientific approaches to address this phenomenon. Our objective is to present a comprehensive account of the research field, focussing on identifying drivers of variation in internal loading over the longer term and synthesising evidence across the peer-reviewed literature and from some of the world's most iconic long-term monitoring programs that are centred on lakes, their ecology, and their vital role in society.

1.4   ACKNOWLEDGMENTS

We are grateful to the Allen and Helen Hunting Research and Innovation Fund, which provided support to author ADS during the writing of this chapter, as well as Cayalen Carey and Emil Rydin for sharing their data, which were used in creating Figure 1.2.

1.5   REFERENCES

Beusen, AH; Bouwman, AF; Van Beek, LP; Mogollón, JM; and Middelburg, JJ. 2016. Global riverine N and P transport to ocean increased during the 20th century despite increased retention along the aquatic continuum. Biogeosciences. 13:2441–2451.

Beutel, MW. 2006. Inhibition of ammonia release from anoxic profundal sediments in lakes using hypolimnetic oxygenation. Ecol Eng. 28: 271–279.

Boström, B; Andersen, JM; Fleischer, S; and Jansson, M. 1988. Exchange of phosphorus across the sediment-water interface. Hydrobiologia. 170:229–244.

Brennan, AK; Hoard, CJ; Duris, JW; Ogdahl, ME; and Steinman, AD. 2016. Water quality and hydrology of Silver Lake, Oceana County, Michigan, with emphasis on lake response to nutrient loading, 2012–14. U.S. Geological Survey Scientific Investigations Report. 2015–5158, p. 75.

Carey, CC and Rydin, E. 2011. Lake trophic status can be determined by the depth distribution of sediment phosphorus. Limnol Oceanogr. 56:2051–2063.

Chao, JY; Zhang, YM; Kong, M; Zhuang, W; Wang, LM; ShA o, KQ; and GA o G. 2017. Long-term moderate wind-induced sediment resuspension meeting phosphorus demand of phytoplankton in the large shallow eutrophic Lake Taihu. PloS One. 12: e0173477.

Chen, M; Ding, S; Liu, L; Xu, D; Gong, M; Tang, H; and Zhang, C. 2016a. Kinetics of phosphorus release from sediments and its relationship with iron speciation influenced by the mussel (*Corbicula fluminea*) bioturbation. Sci Total Environ. 542:833–840.

Chen, N; Liu, L; Chen, M; Li, Y; Xing, X; and Lv, Y. 2016b. Effects of benthic bioturbation on phytoplankton in eutrophic water: a laboratory experiment. Fund Appl Limnol. 188:25–39.

Conley, DJ; Paerl, HW; Howarth, RW; Boesch, DF; Seitzinger, SP; Havens, KE; Lancelot, C; and Likens, GE. 2009. Controlling eutrophication: nitrogen and phosphorus. Science. 323:1014–1015.

Cordell, D and White, S. 2011. Peak phosphorus: clarifying the issues of a vigorous debate about long-term phosphorus security. Sustainability. 3:2027–49.

Ding, S; Chen, M; Gong, M; Fan, X; Qin, B; Xu, H; Gao, S; Jin, Z; Tsang, DC; and Zhang, C. 2018. Internal phosphorus loading from sediments causes seasonal nitrogen limitation for harmful algal blooms. Sci Total Environ. 625:872–884.

Elser, JJ; Marzolf, ER; and Goldman, CR. 1990. Phosphorus and nitrogen limitation of phytoplankton growth in the freshwaters of North America: a review and critique of experimental enrichments. Can J Fish Aquat Sci. 47:1468–1477.

Frodge, JD; Thomas, GL; and Pauley, GB. 1991. Sediment phosphorus loading beneath dense canopies of aquatic macrophytes. Lake Reserv Manage. 7:61–71.

Gächter, R; Meyer, JS; Mares, A. 1988. Contribution of bacteria to release and fixation of phosphorus in lake sediments. Limnol Oceanogr. 33:1542–1558.

Gao, L; Zhang, L; Hou, J; Wei, Q; Fu, F; and Shao, H. 2013. Decomposition of macroalgal blooms influences phosphorus release from the sediments and implications for coastal restoration in Swan Lake, Shandong, China. Ecol Eng. 60:19–28.

Golterman, HL. 2001. Phosphate release from anoxic sediments or 'What did Mortimer really write?' Hydrobiologia. 450:99–106.

Hansen, K; Mouridsen, S; and Kristensen, E. 1997. The impact of Chironomus plumosus larvae on organic matter decay and nutrient (N, P) exchange in a shallow eutrophic lake sediment following a phytoplankton sedimentation. Hydrobiologia. 364:65–74.

Havens, KE. 1991. Fish-induced sediment resuspension: effects on phytoplankton biomass and community structure in a shallow hypereutrophic lake. J Plankton Res. 13:1163–1176.

Hecky, RE and Kilham, P. 1988. Nutrient limitation of phytoplankton in freshwater and marine environments: a review of recent evidence on the effects of enrichment 1. Limnol Oceanogr. 33:796–822.

Higgins, SN; Malkin, SY; Howell, ET; Guildford, SJ; Campbell, L; Hiriart-Baer, V; and Hecky, RE. 2008. An ecological review of Cladophora glomerata (CHLOROPHYTA) in the Laurentian Great Lakes 1. J Phycol. 44:839–854.

Hölker, F; Vanni, MJ; Kuiper, JJ; Meile, C; Grossart, HP; Stief, P; Adrian, R; Lorke, A; Dellwig, O; Brand, A; et al. 2015. Tube-dwelling invertebrates: tiny ecosystem engineers have large effects in lake ecosystems. Ecol Monogr. 85:333–351.

Horppila, J and Nurminen, L. 2003. Effects of submerged macrophytes on sediment resuspension and internal phosphorus loading in Lake Hiidenvesi (southern Finland). Water Res. 37:4468–4474.

Hudson, JJ; Taylor, WD; and Schindler, DW. 2000. Phosphate concentrations in lakes. Nature. 406:54–56.

Hupfer, M; Gachter, R; Giovanoli R. 1995. Transformation of phosphorus species in settling seston during early sediment diagenesis. Aquat Sci. 57:305–324.

Janssen, ABG; de Jager, VCL; Janse, JH; Kong, X; Liu, S; Ye, Q; and Mooij, WM. 2017. Spatial identification of critical nutrient loads of large shallow lakes: implications for Lake Taihu. Water Res. 119:276–287.

Jeppesen, E; Søndergaard, M; Jensen, JP; Havens, K; Anneville, O; Carvalho, L; Coveney, MF; Deneke, R; Dokulil, MT; Foy, B; et al. 2005. Lake responses to reduced nutrient loading—an analysis of contemporary long-term data from 35 case studies. Freshwat Biol. 50:1747–1771.

Ji, G and Havens, KE. 2019. Periods of extreme shallow depth hinder but do not stop long-term improvements in water quality in Lake Apopka, Florida (USA). Water. 11:538.

Katsev, S. 2017. Phosphorus effluxes from lake sediments. In: Lal, R and Stewart, BA, editors. Soil P phosphorus. Boca Raton (FL): CRC Press. pp. 115–131.

Kelly, PT; González, MJ; Renwick, WH; and Vanni, MJ. 2018. Increased light availability and nutrient cycling by fish provide resilience against reversing eutrophication in an agriculturally impacted reservoir. Limnol Oceanogr. 63:2647–60.

Leavitt, PR; Brock, CS; Ebel, C; and Patoine, A. 2006. Landscape-scale effects of urban nitrogen on a chain of freshwater lakes in central North America. Limnol Oceanogr. 51:2262–2277.

Lisi, PJ and Hein, CL. 2019. Eutrophication drives divergent water clarity responses to decadal variation in lake level. Limnol Oceanogr. 64(S1):S49–59.

Macintosh, KA; Mayer, BK; McDowell, RW; Powers, SM; Baker, LA; Boyer, TH; and Rittmann, BE. 2018. Managing diffuse phosphorus at the sources versus at the sink. Environ Sci Technol. In Press.

Matisoff, G; Watson, SB; Guo, J; Duewige, A; and Steely, R. 2017. Sediment and nutrient distribution and resuspension in Lake Winnipeg. Sci Total Environ. 575:173–186.

Mermillod-Blondin, F and Rosenberg, R. 2006. Ecosystem engineering: the impact of bioturbation on biogeochemical processes in marine and freshwater benthic habitats. Aquat Sci. 68:434–442.

Mortimer, CH. 1941. The exchange of dissolved substances between mud and water in lakes I. J Ecol. 29: 280–329.

Nogaro, G and Steinman, AD. 2014. Influence of ecosystem engineers on ecosystem processes is mediated by lake sediment properties. Oikos. 123:500–512.

Nogaro, G; Harris, AM; and Steinman, AD. 2016. Alum application, invertebrate bioturbation, and sediment characteristics interact to affect phosphorus exchange in eutrophic ecosystems. Freshwater Science. 35:597–610.

Nürnberg, GK. 1984. Prediction of internal phosphorus load in lakes with anoxic hypolimnia. Limnol Oceanogr. 29:111–124.

Nürnberg, GK. 1988. Prediction of phosphorus release rates from total and reductant-soluble phosphorus in anoxic sediments. Can J Fish Aquat Sci. 45:453–462.

O'Reilly, CM; Sharma, S; Gray, DK; Hampton, SE; Read, JS; Rowley, RJ; Schneider, P; Lenters, JD; McIntyre, PB; Kraemer, BM; et al. 2015. Rapid and highly variable warming of lake surface waters around the globe. Geophys Res Lett. 42:10–773.

Orihel, DM; Baulch, HM; Casson, NJ; North, RL; Parsons, CT; Seckar, DC; and Venkiteswaran, JJ. 2017. Internal phosphorus loading in Canadian fresh waters: a critical review and data analysis. Can J Fish Aquat Sci. 74:2005–2029.

Osgood, RA. 2017. Inadequacy of best management practices for restoring eutrophic lakes in the United States: Guidance for policy and practice. Inland Waters. 7:401–407.

Ottolenghi, F; Qin, JG; and Mittiga, L. 2002. Enhancement of phosphorus release from lake sediments by aeration and crayfish activity. J Freshw Ecol. 17:635–640.

Paalme, T; Kukk, H; Kotta, J; and Orav, H. 2002. 'In vitro' and 'in situ' decomposition of nuisance macroalgae *Cladophora glomerata* and *Pilayella littoralis*. Hydrobiologia. 475/476:469–476.

Paerl, HW and Scott, JT. 2010. Throwing fuel on the fire: synergistic effects of excessive nitrogen inputs and global warming on harmful algal blooms. Environ Sci Tech. 44:7756–7758.

Paerl, HW; Scott, JT; McCarthy, MJ; Newell, SE; Gardner, WS; Havens, KE; and Hoffman, DK; Wilhelm, SW; and Wurtsbaugh, WA. 2016. It takes two to tango: When and where dual nutrient (N & P) reductions are needed to protect lakes and downstream ecosystems. Environ Sci Tech. 50:10805–10813.

Paterson, MJ; Schindler, DW; Hecky, RE; Findlay, DL; Rondeau, KJ. 2011. Comment: Lake 227 shows clearly that controlling inputs of nitrogen will not reduce or prevent eutrophication of lakes. Limnol Oceanogr. 56:1545–1547.

Pettersson, K; Boström, B; and Jacobsen O-S. 1988. Phosphorus in sediments: speciation and analysis. Hydrobiologia. 170:91–101.

Psenner, R; Boström, B; Dinka, M; Petterssen, K; and Puckso, R. 1988. Fractionation of phosphorus in suspended matter and sediment. Verh Internat Verein Limnol. 22:219–228.

Raven, PH; Evert, RF; and Curtis, H. 1981. Biology of plants. 3rd ed. New York (NY): Worth Publishers.

Reitzel, K; Ahlgren, J; Rydin, E; Egemose, S; Turner, BL; and Hupfer, M. 2012. Diagenesis of settling seston: identity and transformations of organic phosphorus. J Environ Monit. 14:1098–1106.

Roskosch, A; Hette, N; Hupfer, M; and Lewandowski, J. 2012. Alteration of *Chironomus plumosus* ventilation activity and bioirrigation-mediated benthic fluxes by changes in temperature, oxygen concentration, and seasonal variations. Freshw Sci. 31:269–281.

Sas, H. 1989. Lake restoration by reduction of nutrient loading. Expectations, experiences, extrapolation. St. Augustin: Academic Verlag.

Schaus, MH; Godwin, W; Battoe, L; Coveney, M; Lowe, E; Roth, R; Hawkins, C; Vindigni, M; Weinberg, C; and Zimmerman, A. 2010. Impact of the removal of gizzard shad (*Dorosoma cepedianum*) on nutrient cycles in Lake Apopka, Florida. Freshw Biol. 55:2401–2413.

Scheffer, M. 2004. Ecology of shallow lakes. Dordrecht (The Netherlands): Kluwer Academic Publishers.

Schindler, DW. 1977. Evolution of phosphorus limitation in lakes. Science. 195:260–262.

Schindler, DW; Carpenter, SR; Chapra, SC; Hecky, RE; and Orihel, DM. 2016. Reducing phosphorus to curb lake eutrophication is a success. Environ Sci Tech. 50:8923–8929.

Sharpley, A; Jarvie, HP; Buda, A; May, L; Spears, B; and Kleinman, P. 2013. Phosphorus legacy: overcoming the effects of past management practices to mitigate future water quality impairment. J Environ Qual. 42:1308–1326.

Søndergaard, M; Jensen, JP; and Jeppesen, E. 2003. Role of sediment in and internal loading of phosphorus in shallow lakes. Hydrobiologia. 506:135–145.

Steinman, A; Abdimalik, M; Ogdahl, ME; and Oudsema, M. 2016. Understanding planktonic vs. benthic algal response to manipulation of nutrients and light in a eutrophic lake. Lake Reserv Manage. 32:402–409.

Steinman, AD; Hassett, MC; and Oudsema, M. 2018a. Effectivness of best management practices to reduce phosphorus loading to a highly eutrophic lake. Int J Environ Res Pub Health. 15: 2111; doi:10.3390/ijerph15102111.

Steinman, AD; Hassett, MC; Oudsema, M; and Rediske, R. 2018b. Alum efficacy 11 years following treatment: phosphorus and macroinvertebrates. Lake Reserv Manage. 34:167–181.

Steinman, AD; Nemeth, L; Nemeth, E; and Rediske, R. 2006. Factors influencing internal phosphorus loading in a west-Michigan, drowned river mouth lake. J N Am Benthol Soc. 25:304–312.

Steinman, AD and Ogdahl, ME. 2008. Ecological effects after an alum treatment in Spring Lake, Michigan. J Environ Qual. 37:22–29.

Steinman, AD and Ogdahl, ME. 2012. Macroinvertebrate response and internal phosphorus loading in a Michigan Lake after alum treatment. J Environ Qual. 41:1540–1548.

Steinman, AD; Rediske, R; and Reddy, KR. 2004. The reduction of internal phosphorus loading using alum in Spring Lake, Michigan. J Environ Qual. 33:2040–2048.

Svensson, JM; Enrich-Prast, A; and Leonardson L. 2001. Nitrification and denitrification in a eutrophic lake sediment bioturbated by oligochaetes. Aquat Microb Ecol. 23:177–186.

Tammeorg, O; Niemistö, J; Möls, T; Laugaste, R; Panksep, K; and Kangur, K. 2013. Wind-induced sediment resuspension as a potential factor sustaining eutrophication in large and shallow Lake Peipsi. Aquat Sci. 75:559–570.

Tang, KW; Flury, S; Grossart, H-P; and McGinnis, DF. 2017. The *Chaoborus* pump: migrating phantom midge larvae sustain hypolimnetic oxygen deficiency and nutrient internal loading in lakes. Water Res. 122:36–41.

Tarvainen, M; Ventelä, AM; Helminen, H; and Sarvala, J. 2005. Nutrient release and resuspension generated by ruffe (*Gymnocephalus cernuus*) and chironomids. Freshw Biol. 50:447–458.

Thomas, DB and Schallenberg, M. 2008. Benthic shear stress gradient defines three mutually exclusive modes of non-biological internal nutrient loading in shallow lakes. Hydrobiologia. 610:1–11.

Vanni, MJ. 2002. Nutrient cycling by animals in freshwater ecosystems. Annu Rev Ecol Syst. 33:341–370.

Vila-Costa, M; Sharma, S; Moran, MA; and Casamayor, EO. 2013. Diel gene expression profiles of a phosphorus limited mountain lake using metatranscriptomics. Environ Microbiol. 15:1190–1203.

Weyhenmeyer, GA; Jeppesen, E; Adrian, R; Arvola, L; Blenckner, T; Jankowski, T; Jennings, E; Noges, P; Noges, T; and Straile D. 2007. Nitrate-depleted conditions on the increase in shallow northern European lakes. Limnol Oceanogr. 52:1346–1353.

Williams, AE and Moss, B. 2003. Effects of different fish species and biomass on plankton interactions in a shallow lake. Hydrobiologia. 491:331–346.

Xie, LQ; Xie, P; and Tang, HJ. 2003a. Enhancement of dissolved phosphorus release from sediment to lake water by Microcystis blooms—an enclosure experiment in a hyper-eutrophic, subtropical Chinese lake. Environ Pollut. 122:391–399.

Xie, L; Xie, P; Li, S; Tang, H; and Liu, H. 2003b. The low TN:TP ratio, a cause or a result of *Microcystis* blooms? Water Res. 37:2073–2080.

Zhu, M; Zhu, G; Zhao, L; Yao, X; Zhang, Y; Gao, G; and Qin, B. 2013. Influence of algal bloom degradation on nutrient release at the sediment–water interface in Lake Taihu, China. Environ Sci Poll Res. 20:1803–1811.

CHAPTER 2

METHODS FOR MEASURING INTERNAL PHOSPHORUS LOADING

Michael Hupfer[1], Kasper Reitzel[2], and Björn Grüneberg[3]

Abstract

This chapter gives an overview of the principles and the practical application of methods for the determination of internal lake fluxes and pools of phosphorus (P). The methods for the quantification of vertical P fluxes include mass balance calculations, *in situ* measurements, and laboratory experiments with sediment cores. The application of sequential P fractionation techniques and the determination of total P in different sediment layers allow the differentiation and quantification of temporary and permanent P pools in sediments. The advantages and limitations of the respective methods and the process of data evaluation and interpretation are presented. Isolated consideration of P flux from sediment can lead to misinterpretation of the function of sediments as P sinks or sources. Therefore, for the selection of suitable water quality management measures and for the prediction of future water quality status, it is necessary to determine the internal pools and balance variables in relation to P release from sediments. Under steady state conditions, sediments act as a sink for P and simple mobilization of particulate P cannot be regarded as an additional P source. Due to the time delay between P deposition and P release, sediments in polymictic lakes can act as temporary net sources of P in summer and cause corresponding ecological effects. Sediments can also become long-term P sources after external load reduction or can exhibit reduced P retention. Through simultaneous determination of pools and fluxes, it is possible to apply balance models to better understand the exchange processes involved with P.

Keywords: lake sediments, phosphorus release, mass balance, phosphorus fractionation, potentially mobile phosphorus

[1] Department of Chemical Analytics and Biogeochemistry, Leibniz Institute of Freshwater Ecology and Inland Fisheries; Berlin, D. E-mail for correspondence: hupfer@igb-berlin.de

[2] Department of Biology, University of Southern Denmark, Odense, DK

[3] Department of Freshwater Conservation, Brandenburg University of Technology Cottbus-Senftenberg, Bad Saarow, D.

2.1   INTRODUCTION: ANALYSIS OF INTERNAL PHOSPHORUS INVENTORIES AND FLUXES

The prediction of phosphorus (P) concentrations under changed land use and climate conditions, and/or the selection of appropriate management measures to address eutrophication, require careful analysis of internal P fluxes and pools.

Phosphorus is introduced into lakes in dissolved, colloidal, or particulate form—from point as well as nonpoint sources. This P input is dependent on land-use practices, as well as on the hydrological and geological conditions of the catchment. In lake water, dissolved P forms can be transferred to particulate forms by both biotic (mainly autotrophic organisms) and physico-chemical processes (e.g., sorption at surfaces). Intensive cycling between dissolved and particulate P takes place in the trophogenic zone of a lake (Hupfer and Lewandowski 2008). Some allochthonous and autochthonous particulate P is transported to the sediment by sedimentation processes (see Figure 2.1). During and after sedimentation, some of this settled P is re-released by various biological, chemical, and physical processes (see Chapter 1). In stratified lakes, physical decoupling of the hypolimnion from the epilimnion causes the released P to accumulate in the hypolimnion. Compared to the epilimnion, the nonclosed P cycle is dominated by redissolution of P in the hypolimnion and the sediment. In shallow lakes, released P is then immediately available for further primary production.

From a mass-balance point of view, all P in a lake system is initially introduced by external sources. Under steady state conditions, the average P pool in the water body of a lake (P_{lake}) is determined by the P input (P_{in}), which is in equilibrium with P losses stemming from the sum of P output (P_{out}) and permanent P retention in the sediment (P_{ret}) [see Equation 2.1].

$$P_{in} = P_{out} + P_{ret} \qquad \text{[Eq. 2.1]}$$

Under steady state conditions, a lake ecosystem is a sink for P ($P_{ret} > 0$) because there is always some deposition of material containing P and other elements at the bottom of a lake. P_{ret} is controlled by downward flux from the water column by P sedimentation (P_{sed}) and upward flux (P release) (P_{rel}) from the sediment [see Equation 2.2].

$$P_{ret} = P_{sed} - P_{rel} \qquad \text{[Eq. 2.2]}$$

Additionally, the function of sediment as either a sink ($P_{ret} > 0$) or a source ($P_{ret} < 0$) can be calculated from the difference between P_{in} and P_{out} and the changes of P mass in the lake (ΔP_{lake}) [see Equation 2.3].

$$P_{ret} = P_{in} - P_{out} - \Delta P_{lake} \qquad \text{[Eq. 2.3]}$$

Under steady state conditions, P_{sed} is higher than P_{rel}, which leads to P accumulation and its net retention in the sediment. Because immobilization and mobilization processes during P diagenesis after sedimentation need time, two principal P pools in the sediment can be distinguished: a temporary (mobile) P pool, which is the source for further P release, and a permanent (immobile) P pool, which determines P_{ret} (see Figure 2.1). Hence, the source for P_{rel} is the temporary P pool (which includes the legacy P) complemented by the continuous supply of mobilizable settled P. Synonymous terms in the literature for P_{ret} rate are the *net sedimentation rate* and the *burial P rate* (see e.g., Imboden and Gächter 1978; Sas 1989). The P_{rel} is often termed *gross P release* or *gross benthic P flux* (Orihel et al. 2017).

A net P release (P_{ret} < 0) is possible under conditions where a short- or long-term delay exists between P_{sed} and P_{rel}, and two possible cases have been illustrated in Figure 2.2. The first case (see Figure 2.2a) shows seasonal changes of the sink and source function of the sediment, which is often described for shallow lakes (Søndergaard et al. 2013). Except in summer, the sediments act as a sink—then during the summer months the sediments release more P than is deposited during

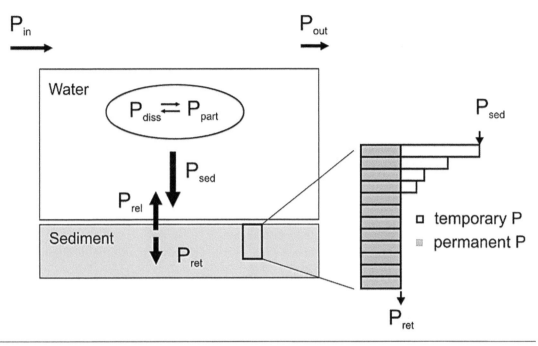

Figure 2.1 Phosphorus mass balance in a lake with the relevant P fluxes (arrows) and P pools, and the resulting portions of temporary and permanent P forms in the sediment (modified from Hupfer and Hilt 2008).

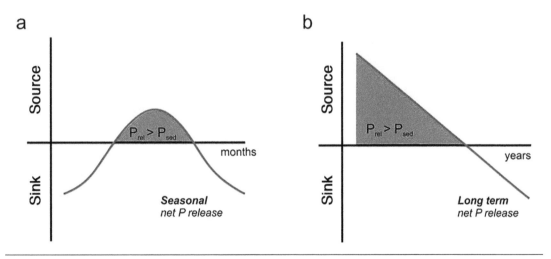

Figure 2.2 Lake sediments as P sources in the P balance: (a) seasonal changes between net release and net retention of phosphorus; (b) long-term net release of legacy phosphorus from sediments after external load reduction on an annual basis.

the same period. This is explained by well-known mobilization processes, such as mineralization of organic detritus and iron reduction, which are temperature-controlled. In the second case (see Figure 2.2b), the situation represents non-steady state conditions, whereby a distinct external load reduction in shallow lakes can lead to longer periods when the release of P from the mobile P pool is the major controlling factor for the P concentration in water. Under these circumstances, the net P release from the sediment can delay the expected positive effects of external P reduction for decades (e.g., Jeppesen et al. 2005). The long-term impact of sediment P release is mainly controlled by changes in the mobile P pool in the sediment, diagenetic release rates, and water residence time of the system.

In many publications, the P_{rel} (gross P release) from sediment into water is termed as *internal P load*. This term is misleading, because P release from the sediment is in fact not an additional load to the lake. However, the amount of mobile P in the sediment from former loadings could reduce the difference between P_{sed} and P_{rel}, so that P_{ret} is lowered but is not less than zero. The term *internal load* is therefore appropriate only in cases where sediments are actually a net source of P over time. The temporal resolution for seasonal internal load may be a month or several months (seasons), while for long-term (legacy) internal load, net sedimentation should be negative for at least one year (see Figure 2.2). Since mass balance calculations and models usually cover a period of at least one year, Orihel et al. (2017) used the term *internal P loading* only for the net annual P release. The synonymous usage of the terms *P release* and *internal P load* leads to confusion—and the meanings of the terms should be clearly differentiated (see Chapter 1).

P availability in lake water can be decreased ($\Delta P_{lake} < 0$) by reducing external P input (P_{in}), by increasing P retention (P_{ret}) in the sediment, or by increasing P output (P_{out}) [see Equation 2.4].

$$\Delta P_{lake} = P_{in} - P_{out} - P_{ret} \qquad \text{[Eq. 2.4]}$$

Controlling P_{in} is the most appropriate measure to act on for sustainable reduction of P in lakes (Goyette et al. 2018; Wang et al. 2018). P_{out} can be influenced by deep-water withdrawal, artificial circulation, sediment removal, or water treatment in external plants (see Chapter 5). The P_{ret} is often increased by the addition of P-binding chemicals and by oxidation measures. Therefore, analysis of P pools and fluxes is a prerequisite for using balance models, as well as for the selection of adequate management measures (Hupfer et al. 2016). The following sections describe the available methods for determining internal P fluxes and pools, the advantages and limitations of each method, and the corresponding data evaluation, and interpretation process.

2.2   DETERMINATION OF P RELEASE RATES FROM SEDIMENTS

2.2.1   Mass Balance Approach: Hypolimnetic P Accumulation

2.2.1.1   Principle

In monomictic and dimictic lakes, the rate of gross P release can be calculated from an increase in the P mass in the hypolimnion during thermal stratification. Hypolimnetic P accumulation (HPA) is mainly caused by an increase in soluble reactive P (SRP). The HPA rate includes both the sediment P release rate (SRR) and the water release rate (WRR); the latter is defined as P mobilized in the hypolimnion during sedimentation. Preconditions require that the hypolimnion be sufficiently separated from the epilimnion and from any exchange with groundwater. However, transport between epilimnion and hypolimnion cannot be completely excluded because of the transport of particulate P from the epilimnion to the hypolimnion by sedimentation, and from loss of hypolimnetic SRP across the thermocline due to turbulent diffusion. Hypolimnetic P accumulation integrates the spatial heterogeneity of the hypolimnetic area with its P release during and after sedimentation. Repeated measurements of P profiles enable the calculation of P release rates for different periods. Determining HPA requires data on the depth-volume relationship of the lake.

2.2.1.2 Method Description

Vertical P concentration (cP) profiles for a lake's water column are often included in routine water quality monitoring programs, and serve multiple purposes in stratified lakes; these include the determination of mean cP_{lake} and data provision for lake modelling, among others. In addition, information on P accumulation in the hypolimnion during the stagnant period can be used to calculate hypolimnetic sediment P release rates. Monitoring of the P mass at different water depths should be conducted multiple times during the stagnant period, in order to get sufficient information of temporal development of the P release rates. P concentration is measured at different depth layers, with a depth resolution adequate to describe its gradient in the water column. P content is calculated by multiplying the mean P concentration at each depth layer by the volume of the respective layer (see Figure 2.3 and Equation 2.5). The P contents of different hypolimnetic layers are then summed, to calculate the total P content in the hypolimnion at a certain time.

$$HPA = \frac{\sum_{z=1}^{i} (cP_z \times V_z)_{t2} - \sum_{z=1}^{i} (cP_z \times V_z)_{t1}}{A \times (t_2 - t_1)} \qquad \text{[Eq. 2.5]}$$

where:

    HPA = hypolimnetic P accumulation rate (mg m^{-2} d^{-1})
    cPz   = mean P concentration in a layer (mg m^{-3})
    Vz   = volume of a layer (m^3)
    A   = hypolimnetic area (m^2)
    $t_2 - t_1$ = period between two defined times (d)

The annual HPA is estimated by extrapolating the HPA of the sampling period (between two or more sampling times during stratification) to the whole stratification period, normalized for the area of the entire lake.

A P_{lake} concentration measured preferably shortly before the onset of stratification and a P profile measurement during stratification are the minimum requirements for HPA calculation, although multiple P profile measurements during summer stratification are recommended. Profile measurements may be substituted by depth-weighted sampling, where a mixed sample for the whole hypolimnion is produced by adjusting the sample volume for each layer according to its volume. Calculation of P release for different water layers is an additional option.

2.2.1.3 Advantages and Limitations

This method is the most commonly used technique to determine P release in stratified lakes (e.g., Grüneberg et al. 2011; see Chapter 3). Its biggest advantage, compared to the other methods, is that the comparison is independent of the spatial heterogeneity of the sediment and short-term P release variability. However, precision can be impaired when unstable stratification (e.g., due to periodic strong wind events) leads to P transport between epilimnion and hypolimnion. The determination in the hypolimnion via mass balance could underestimate the actual gross P release since part of the

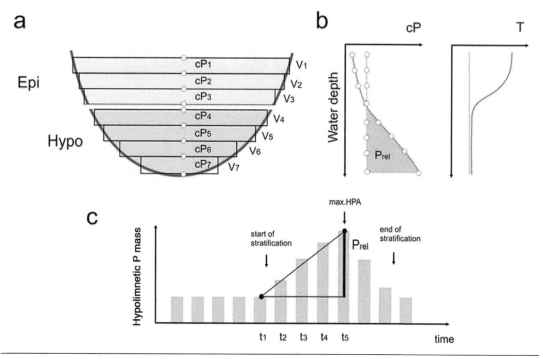

Figure 2.3   Determination of hypolimnetic P accumulation (HPA) in a lake: (a) calculation of P content of different water layers by multiplication of the respective P concentrations and volumes; (b) P and temperature profiles at the beginning (blue lines) and the end of summer stratification (red lines); (c) changes of the P content in the hypolimnion are used for the calculation of the P_{rel}. The rates are calculated by using the respective time interval and the hypolimnetic area.

released P could be re-adsorbed to, for example, oxidized iron (Fe) at the oxycline, which could, in turn, lead to re-precipitation and subsequent sedimentation of particulate P. Additionally, water column profiles in monitoring programs may not have the necessary spatial resolution to fully describe the P release.

It is often incorrect to assume that the P concentration at one depth is horizontally valid for the entire depth interval. In reality, that is not the case because the P concentrations (and concentrations of other solutes) depend on the distance from the sediment—and only in lakes with a relatively large hypolimnion is this effect negligible. In lakes with a smaller hypolimnion, this effect should be explored by taking additional measurements at sites with different water depths at the mid or end of summer stratification. Water-level fluctuation and internal seiches must also be taken into account, but overall, the most important drawbacks for this method are that the calculation provides reliable data only for the stratification period, rather than for a whole year—and only for the hypolimnion, not for the whole lake. These disadvantages can be overcome only by well-reasoned assumptions or by additional experiments, such as core analyses that use sediments from hypolimnetic and epilimnetic sites during both stratification and mixing periods.

2.2.1.4   Data Evaluation and Interpretation

Vertical P data in stratified lakes are indispensable for the evaluation of internal processes and as input data for models. The vertical profiles are used to quantify both the total P content in the lake and the average (volume weighted) P concentration. Profiles collected at different times during the year provide information about seasonal changes in mean P concentration in a lake and are the basis for calculating an annual average P concentration. Analysis of the distribution of P and its changes over a year allows additional evaluations. First, P loss from the epilimnion is a measure for P sedimentation in cases where the P supply by external load during summer is low, in relation to losses by sedimentation. Second, based on the distribution of P in the lake, the stratification factor (ß) can be calculated as the ratio between mean cP in the epilimnion and mean cP in the lake.

This stratification factor is used in mass balance calculations to describe the deviation of mean cP_{out} from the mean cP_{lake} (see Sas 1989). The higher the hypolimnetic P accumulation, the lower the P output, relative to the mean cP_{lake}. Therefore, some internal management measures aim at increasing the ß, by using artificial mixing or hypolimnetic withdrawal. The mobile P pool (see Section 2.3) in the sediment, divided by the gross P release rate (determined as hypolimnetic P accumulation), gives an estimate of the time range over which the mobile P stock in the sediment will sustain this release without additional delivery of P through sedimentation. In cases where the stock in the sediment supports P release only for a few months, the release is driven mainly by sedimentation, whereas a longer theoretical temporal range is an indication that the sediments could act as a long-term P source (case B in Figure 2.2).

The vertical distribution of P in the hypolimnion can also be used to discriminate P mobilization during sedimentation from P release from the sediment, by using the approach of Livingstone and Imboden (1996) for oxygen. The cP increases that were observed in different hypolimnetic water bodies are related to the area to volume ratio (α) in the same layer. Assuming that the sediments from all hypolimnetic water layers have similar P release rates, and contribute equally to hypolimnetic P accumulation, a linear relationship is formed between α and P increase in the respective layers. The intersection with the y axis (where theoretically the sediment area is zero) represents the mean P WRR (see Figure 2.4). The SRR per area is the difference between the total hypolimnetic accumulation and the WRR, per area (see Chapter 9, Figure 9.4).

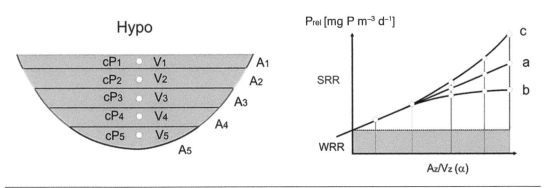

Figure 2.4 Left: Hypolimnetic P accumulation rates (mg m⁻³ d⁻¹) in different water layers are influenced by the area to volume ratio. Right: the intersection with the y axis is the P increase in the water, assuming that no sediment has contributed to the P increase in the water (adapted from Livingstone and Imboden 1996). (a) P release rates in all hypolimnetic sediments are the same; (b) P release rates of sediments from deeper areas are lower compared to other areas (e.g., due to reduced mineralization under anoxic conditions); (c) P release rates of sediments from deeper areas are higher compared to other areas (e.g., due to reductive dissolution of iron).

2.2.2   *In Situ* Measurements of Dissolved P Gradients in the Sediment-Water Interface

2.2.2.1   *Principle*

Flux and turnover rates for dissolved P can be calculated from the SRP concentration gradients in the sediment and at the sediment–water interface (Berner 1980; Berg et al. 1998). Prerequisites for using this method are that the system is under a steady state condition and that molecular diffusion is the main transport mechanism. Dialysis pore water samplers based on the *peeper* principle (Hesslein 1976) are mostly used for the fine-scale and minimally invasive recording of the concentrations of dissolved ions in the sediment. The peeper technique is based on the establishment of an equilibrium between the ion concentration of the surrounding water and the small chambers filled with demineralized water before incubation. Subsequent sampling of the chambers provides a snapshot of the ion concentrations at the sediment-water interface, reflecting the water concentrations within the last two to three days. Gel samplers are special forms of dialysis samplers and are categorized as either DET (diffusional equilibrium in thin films; Davison et al. 1991; Krom et al. 1994) or DGT (diffusive gradients in thin films) types (see Section 2.3.3). Small amounts of pore water can also be obtained for analysis by suction through a horizontally inserted, porous polymer tube with a pore size of 0.1 μm (Rhizon) after sampling of a sediment core (Seeberg-Elverfeldt et al. 2005; Steinsberger et al. 2017).

2.2.2.2   *Method Description*

The base plate of the peeper, often made of Plexiglas (acrylic), consists of chambers with a spatial resolution of 0.5 or 1.0 cm, a membrane (e.g., polysulfone, 0.2 μm), and a cover plate for fixing the membrane (see Figure 2.5). The chambers are filled with oxygen-free, demineralized water and the cover plate is screwed in place. It is recommended that the peeper is stored for 24 hours in a water-filled container under anoxic conditions (e.g., by constant gassing with N_2) before it is deployed into the lake sediment. The peeper is installed in a frame on site and then lowered to the bottom with a rope. The penetration depth of the peeper into the sediment can be controlled by the frame. Alternatively, dialysis samplers can be installed by divers. Small peepers can also be deployed in undisturbed sediment columns (diameter 9 to 10 cm) after sampling in the lab.

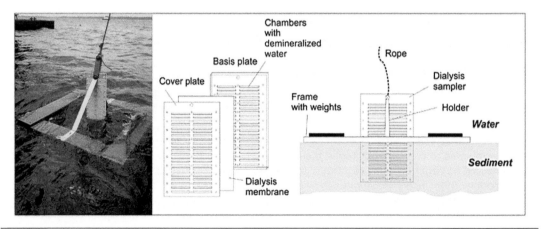

Figure 2.5   Dialysis sampler (peeper) for fine-scale measurements of dissolved substances in the sediment-water interface. Left: single parts of the sampler. Right: exposition of the dialysis sampler in the sediment with a frame.

The exposure time of peepers under field conditions is usually 10 to 14 days longer than the time necessary to reach the equilibrium. The extended time is used to compensate for disturbances in the sedimentary environment. After removing the peepers from the sediment, the dialysis membrane is punctured with a syringe or a pipette, and a sample is removed. Determination of SRP is then carried out, using standard methods. Due to the small sample quantities and splitting of the sample into several aliquots, miniaturized photometric methods can be used (Laskov et al. 2007). The diffusive P release rate is calculated using Fick's first law, independent of the method used for pore water sampling [see Equation 2.6]:

$$J_i = \frac{\varphi}{\theta^2} * Di * \frac{dC_i}{dz}$$ [Eq. 2.6]

where:

J_i = diffusion flux of ion i (mg m^{-2} d^{-1})
φ = porosity (dimensionless)
θ = tortuosity (dimensionless)
Di = molecular diffusion coefficient of the ion (m^2 d^{-1}) (H$_2$PO$_4^-$ = 7.31·10^{-5} m^2 d^{-1} at 25°C)
dC_i / dz = concentration gradient of the ion (mg L^{-1} cm^{-1})

The pH must also be taken into account when calculating the flux of P since different molecular diffusion coefficients apply to the different dissociation stages of phosphoric acid (H$_3$PO$_4$) (Li and Gregory 1974). The Stokes-Einstein relationship is used to convert the diffusion coefficients at 25°C to the diffusion coefficients at the *in situ* temperature. A detailed description of this calculation (including porosity and tortuosity) is given in Lewandowski et al. (2002).

The program Profile V 1.0 by Berg et al. (1998) can be used to calculate the turnover rates (see Figure 2.6).

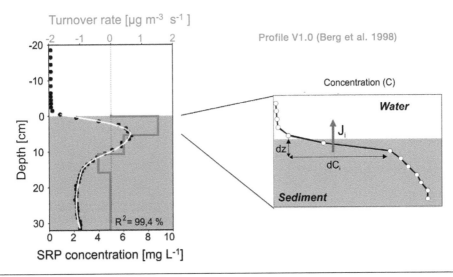

Figure 2.6 Left: SRP concentration profile between sediment and water (black dots). The white line shows the best fit of the points generated with the program profile V 1.0 (Berg et al. 1998) as the basis for the calculation of turnover rates in different layers (red bars). Right: illustration of the calculation of the diffusion flux (J$_i$) by the concentration difference (dC) per distance (dz) at the sediment water interface (according to Equation 2.6).

2.2.2.3   Data Evaluation and Interpretations

Calculation of the P release rates using SRP gradients aims at recording the gross flux from the sediment into the water. Since other transport mechanisms, such as bioirrigation, bioturbation, turbulence, and outgassing (see Chapter 1), can also play a crucial role under natural conditions, P release calculated on the basis of molecular diffusion represents a minimum of the actual P flux (Berner 1980). Additionally, the calculated diffusive flux could be underestimated because real gradients are steeper than the resolution of the dialysis chambers (see Section 2.2.3).

Determination of redox parameters (NO_3^-/NH_4^+, Fe^{2+}, Mn^{2+}, SO_4^{2-}, HS^-), allows a process-oriented assessment of P mobilization. For example, pore water profiles provide information about the penetration depth of nitrate, which is an indication of the redox state of the sediment surface. In addition, the significance of the mineralization of organic matter for the release of P can be estimated using stoichiometric ratios. Therefore, pore water profiles allow determination as to whether P mobilization is controlled by either the reductive dissolution of metals or by organic matter mineralization (Urban et al. 1997). The evaluation of pore water profiles also provides information on which sediment horizons contribute to P mobilization along with whether—and to what extent—deeper layers act as P sinks or sources.

2.2.2.4   Advantages and Limitations

The main advantage of peepers compared to other pore water separation methods, such as centrifugation or pressure filtration, is that the sampling takes place *in situ*, which prevents introduction of artifacts due to oxidation, temperature, and/or pressure changes. The method is simple and is suitable for routine measurements because it does not require special laboratory equipment. Measurement accuracy can be increased by use of gel peepers, because a higher resolution can be selected when cutting the gel (with a maximum resolution of 1 mm). The program profile V.10 considers not only diffusion as an essential transport process, but also bioturbation and bioirrigation, by using coefficients for these transport processes.

Phosphorus release calculations using single pore water gradients are often not representative over long periods of time, as well as for the entire sediment surface; that is, temporal and spatial variabilities are high (cf. Lewandowski et al. 2002). SRP concentration gradients in pore water allow only the quantification of the diffuse flux as a transport mechanism. Thus, the calculated release rates are lower than those achieved by other methods described in this chapter. The peeper method may be especially suitable for deep, stratified lakes, with limited small-scale sediment heterogeneity, with diffusion as the major release process. However, the sole quantification of the diffuse flux (which differs from other processes) is also an advantage of the method, and is applicable to all kinds of lakes.

2.2.3   Benthic Chambers

2.2.3.1   Principle

Benthic chambers allow determination of *in situ* fluxes directly at the sediment surface, without disturbance to the sediment or alteration of physical or chemical conditions, which occurs if sediment samples for flux studies are brought to the surface. Benthic chambers are closed containers that isolate an area of sediment surface from the water and allow sampling at specific time intervals (see Figure 2.7). From concentration changes within the chamber over time, a release rate (increasing concentration) or consumption rate (decreasing concentration) can be calculated. Benthic chambers have long been used in deep sea oceanography in combination with landers (Tengberg et al. 1995). They are also increasingly being used in lakes, as they allow a wide range of applications, such

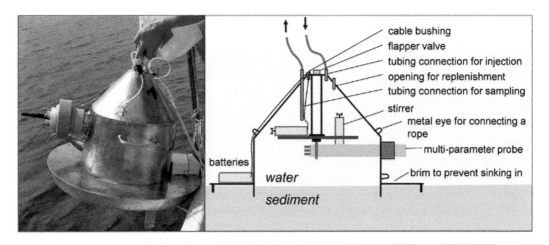

Figure 2.7 Picture and drawing of a metal benthic chamber with 0.126 m² area and 39 L volume, designed for flux measurements and chemical manipulations by injections

as measuring oxygen depletion rates in combination with nitrate and sulfate consumption, as well as the flux of carbon species (dissolved organic carbon: DOC; dissolved inorganic carbon: DIC), nutrients (SRP, NH_4^+, NO_3^-; Burger et al. 2007; Clavero et al. 2000; Petranich et al. 2018), heavy metals (Emili et al. 2016), and gases (H_2S, CH_4, N_2O; L'Helguen et al. 2014). Additionally, benthic chambers allow manipulation of physicochemical conditions and are therefore used for process studies, e.g., to test nitrate influence on P release, temperature and redox influences on nutrient fluxes, or CO_2 influence on benthic biota.

2.2.3.2 Method Description

Benthic chamber designs differ depending on the aim of the study and the application environment. Flux chambers for limnological studies have been constructed in box-shaped or cylindrical forms. They are made from metal (stainless steel), or from transparent or opaque plastics (e.g., PVC, acrylic) to allow light penetration. Plastic chambers are lighter and less expensive than metal chambers, but have thicker chamber walls, and may therefore cause more disturbances when entering the sediment. The volume of flux chambers that are used in fresh waters varies between 3.4 and 150 L, with enclosed surface areas between 50 and 2500 cm². Larger chambers, and designs with multiple chambers, are used in marine studies. A large surface area reduces the variability caused by the small-scale sediment heterogeneity, while large volumes minimize the influence of the sampling.

Flux chambers used in fresh waters are either deployed by divers or lowered to the bottom from a boat using a wire. As the chamber is lowered to the lake bottom and onto the sediment, water must be allowed to escape easily, either by automatic flap valves, or by lids. Great care must be taken when installing the chamber to prevent surface disturbance or resuspension by increased hydrostatic pressure or turbulence. Sensors (turbidity, electric conductivity) assist in minimizing sediment disturbance.

The penetration depth of the chamber, which depends on the softness of the sediment, is a critical technical problem because it determines the chamber volume required for correct flux calculations. Penetration of the chamber too deeply into the sediment can be prevented by a rim on the outside of the chamber, which for heavy chambers, can be supported by a float, to increase its buoyancy.

The isolation of a small sediment area from the natural bottom currents would result in development of abnormally steep chemical gradients in the overlying water (similar to the phenomenon in core experiments). All chambers are therefore equipped with a water stirring device to mimic natural hydrodynamic conditions. The stirring is performed by stirrers, rotating discs, or submersible pumps. Despite the immense influence of stirring on nutrient fluxes, only a few authors have made the effort to estimate hydrodynamics and boundary layer thicknesses in chambers, for example, by using microelectrodes. When measuring nutrient fluxes, physicochemical conditions should be studied in parallel. Many chambers are therefore equipped with probes connected to data loggers to measure oxygen, electric conductivity, and/or pH.

Flux chambers are left on the sediment surface for time periods varying from a few hours up to several days, and samples are taken every 0.3 to 24 hours. Deployment time and sampling intervals depend on biochemical conditions, flux intensities, and the research questions being investigated. Oxygen is normally depleted within a few hours in small chambers, so nutrient flux measurements are time-limited since they are largely influenced by redox conditions, except if redox changes are part of the experimental design. A further limitation derives from the build-up of high solute concentrations over time inside the chamber, which lowers the concentration gradient and suppresses the diffusive flux.

Sampling of chamber water is done manually, with syringes by divers, or via tubes which run to the surface. Sophisticated lander-chambers are equipped with automatic samplers. Sample volumes are usually below 1% of chamber volume, to minimize sampling artifacts. Chambers must have a small opening that allows replenishment of the sample volume with lake water. Measured ion concentrations must be corrected for the dilution associated with sample removal. Dilution can be assessed by injecting a tracer (bromine) at the start of each deployment, and measuring changes in Br concentration at specific time intervals.

Fluxes (J) across the sediment-water interface can be calculated for each time step with Equation 2.7:

$$J_t = \frac{V_t \cdot (C_t - C_{t-1})}{A \cdot t} \qquad \text{[Eq. 2.7]}$$

where:

J_t = flux of an ion (mg m^{-2} d^{-1})
V = volume of the chamber (m^3)
C_t, C_{t-1} = dissolved ion concentrations in the water at times t and t−1 (mg L^{-1})
A = surface area of the sediment enclosed by the flux chamber (m^2)

The flux for a certain period can also be calculated from the slope of the linear regression of chamber nutrient concentrations over time, divided by volume and chamber area (see Figure 2.8; Burger et al. 2007).

2.2.3.3   Advantages and Limitations

Benthic chambers allow the quantification of the most realistic, *in situ* fluxes, under ambient conditions (temperature, pH), with ideally minor disturbance of sediments. These are also characterized by the absence of artifacts occurring when the samples are subjected to changes in hydrostatic pressure and temperature as they are brought up to the surface. They provide integrated mean fluxes for a relatively large sediment surface, diminishing the problem of small-scale sediment heterogeneity that is often encountered with sediment core or pore water studies. Fluxes into benthic chambers represent the sum of surface (upper few millimeters) and subsurface mineralization, and consider

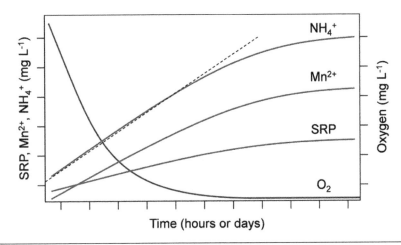

Figure 2.8 Typical course of concentrations in a benthic chamber experiment—with the slope of linear regression (dotted line as example) used for flux calculation.

not only molecular diffusion but also bioirrigation as transport processes. Seasonal flux variations (with repeated measurements), as well as spatial heterogeneity (with sets of chambers), can be covered. Benthic chambers can be applied to sandy littoral or riverine systems, as well as to soft sediments at all water depths. Benthic flux studies are often combined with high resolution pore water sampling and analysis to link surface fluxes with turnover rates in the sediment. Finally, they allow manipulations under *in situ* conditions.

A major drawback of benthic chamber application is the isolation of the sediment from resupply of fresh particles, so that the *real in situ* flux, largely driven by mobilization from the fresh uppermost sediment layer, can be measured for only a short time. Measurements are quite laborious and time consuming because they require permanent (over several hours) or temporal (daily) presence for sampling—and for these reasons, measurement verification by double determinations is uncommon. There are drawbacks in practical handling of the chambers. Insertion of chambers by wires without sediment disturbance is difficult to handle and control. Moreover, water stirring is a critical factor; too little stirring might result in underestimation due to steep concentration gradients, while excessive stirring might cause high fluxes due to advective transport or even resuspension.

Besides the standard design that was previously described, there are several highly complex and expensive chamber-lander systems in use in both oceanography and freshwater science. They are developed to work autonomously for several weeks—they automatically take water and sediment samples, are equipped with internal microelectrodes and cameras, and some have a system to maintain oxygen and pH levels constant inside the chamber. However, benthic chambers can be kept very simple and cheap (e.g., a pot with an opening for sampling), especially for short-term measurements in shallow fresh waters.

2.2.3.4 *Data evaluation and Interpretations*

In situ fluxes measured with benthic chambers are gross fluxes. These rates can be 5 to 10 times higher than the diffusive fluxes that are calculated from standard, low resolution (1 cm) pore water profiles. This has been described as *flux enhancement* (Clavero et al. 2000; Grüneberg et al. 2015), and has been attributed to macrofaunal irrigation (Callender and Hammond 1982), and preferential P release from fresh surface sediment due to rapid mineralization (Urban et al. 1997; Grüneberg et al. 2015). Thus, the methods presented in this chapter differ with respect to the specific P release processes measured (e.g., transport processes for dissolved P from pore water to overlying water). While the dialysis sampler method considers only molecular diffusion, benthic chambers and column experiments also account for natural macroinvertebrate activity (pore water pumping, irrigation) or advection. Urban et al. (1997) have pointed to the sediment surface as the site of intense diagenetic activity. Mineralization of up to 85% of the organic matter contributed from gross sedimentation occurs within the upper few millimeters of sediment. Thus, methods such as benthic chambers, which consider the sum of subsurface and surface mineralization, give more realistic (high) rates than conventional peepers, which usually do not have sufficient resolution to measure the steep gradients at the surface (Urban et al. 1997).

The relatively high rates of P release measured with benthic chambers have prompted some authors to interpret benthic fluxes as major (internal) loads for lakes. However, great care must be taken in reaching such conclusions, as these gross rates are influenced by the constant resupply of organic matter by sedimentation.

2.2.4 Laboratory Release Experiments with Sediment Cores

2.2.4.1 Principle

Periodic P release into the pore water and overlying water may lead to internal loading, when sediment P release is higher than P sedimentation. After P sedimentation, the fate of P depends upon complex interactions between physical, chemical, and biological processes, which are controlled by various environmental parameters (see Chapter 1). The combined effects of these interactions may result in either a net uptake of P by the sediment or a net release of P from the sediment. One way to measure this P exchange over the sediment-water interface is to use intact sediment cores to study P release over time.

2.2.4.2 Method Description

To measure sediment P release, undisturbed sediment cores are collected using a Kajak gravity corer or similar device, and then incubated under laboratory conditions. Five sediment cores are often used as replicates (e.g., Hansen et al. 2003; Dithmer et al. 2016); however, this depends on sediment heterogeneity. When sediment incubation is performed immediately after sample collection, a 24-hour incubation period under temperature and redox regimes mimicking *in situ* conditions will give a snapshot of the actual P release at the time of sampling. This approach gives rates similar to benthic chamber fluxes. During such short-term incubation, the sediment cores can be kept separately, with their own isolated water column above. Redox conditions can be controlled by gently adding nitrogen or oxygen to the overlying water. However, there is a risk that the P gradient above the sediment surface during incubation will decrease. This could lead to an underestimation of natural diffusive flux, which depends on the concentration gradient (see Section 2.2.2), due to lower diffusion of P from the sediment to the water. Therefore, it is important to ensure vertical mixing of the water column in the cores. This can be done by including internal magnets and placing the individual sediment cores around a larger central magnet, which will ensure rotation of the internal magnet. Vertical mixing can also be achieved by gently bubbling air (Steinman et al. 2009) or by cooling the surface water, which will lead to density-driven water column mixing (Hupfer and Uhlmann 1991).

Subsamples of the overlying water are collected by, for example, a 10 ml syringe, onto which a filter can be added to collect samples for SRP analysis. It is also recommended that glass tubes are used to take a depth-integrated sample if gradients occur. If long-term experiments are required, it is advisable to incubate the replicate sediment cores in containers where they can be flooded with

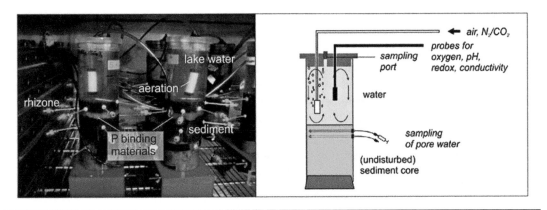

Figure 2.9   P release experiment with sediment cores for testing of different P-binding chemicals. Rhizons served for the sampling of pore water at different times.

the lake water. This allows the sediment to be in contact with a larger water volume, and thereby reduces the risk of too high P accumulation in individual cores. P release (or uptake) from the sediment is calculated by subtracting the P concentration at time $t+1$ from that at time t, multiplied by the volume of overlying water, and divided by the surface area of sediment. Rates can be calculated either step by step, or for each time step, relative to the starting time. The P release rate is often estimated in mg P m^{-2} d^{-1}. Additionally, column experiments can be conducted with a flow of water through the column, as an open system. The advantage of this approach is that the concentration in the overlying water remains relatively constant, and the outflow acts as an infinite sink. For this approach, the release rate is calculated as the difference between inflow and outflow loads, normalized to column surface area.

2.2.4.3   Advantages and Limitations

Short-term P release experiments can provide valuable information on the ability of the sediment to release P under natural conditions thus helping to identify hotspots for P release in lakes and can also be used to compare P release rates among lakes. Long-term incubation of sediment cores that are used to follow P release over time can also be used to assess the effects of different manipulations (such as changing pH, temperature, or redox conditions) and the effects of bioturbation on P-binding materials (Nogaro et al. 2016). However, long-term incubations are especially prone to experimental artifacts. Long-term experiments exclude the role of sedimentation. Thus, the input of potentially mobile P from settling particles is absent, which leads to underestimations of the P release with time. Furthermore, compared to natural conditions, lower input of organic matter as an electron donor could reduce redox-driven P mobilization processes. Long-term incubations also result in less exchange of water above the sediment surface than is observed under natural conditions. This may also lead to underestimates of P release, due to lower diffusion gradients across the sediment-water interface, or altered redox conditions. In addition, P release experiments do not provide any information on the size of the mobile P pool; hence, similar P release rates can be observed from sediments with very different mobile P pools (see Figure 2.10).

2.2.4.4   Data Evaluation and Interpretations

P release experiments have been successfully used to evaluate the effects of many different manipulations. Hence, Dithmer et al. (2016) used short-term P release experiments to evaluate the effect of

Phoslock treatments in 10 European lakes. Similarly, the effects of aluminium treatments in seven Danish lakes were evaluated using sediment P release experiments (Jensen et al. 2015). Hansen et al. (2003) tested the effect of different P-reducing manipulations on P release from intact sediment cores (see Figure 2.10), whereas Jensen and Andersen (1992) and Grüneberg et al. (2015) used sediment P release experiments to demonstrate how temperature and nitrate impacted sediment P release rates in lakes. Finally, the importance of the sediment Fe:P ratio on P release from aerobic sediments in 15 Danish lakes was demonstrated by Jensen et al. (1992).

2.3 QUANTIFICATION OF POTENTIALLY MOBILE PHOSPHORUS IN SEDIMENTS

2.3.1 Gradient Method

2.3.1.1 Principle

During P diagenesis, sediment loses P through many complex processes (see Chapter 1). In undisturbed sediments, this loss is reflected by a decrease in total P with increasing sediment depth and age. Calculation of a P release potential is based on the assumption that stationary conditions prevailed in the period represented by the sediment. Stationary conditions mean that the trophic level was constant leading to constant annual P sedimentation and similar diagenetic processes. According to the diagenesis model, P is then released over time, within a few years, until a new constant total P content is achieved. The minimal (background) sediment total phosphorus (TP) content is defined as the endpoint of early diagenesis and represents the permanent P pool (Hupfer and Lewandowski 2005; Grüneberg et al. 2015).

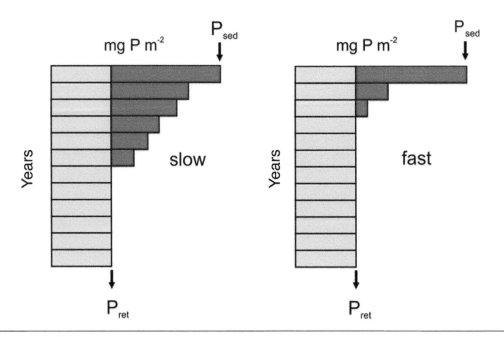

Figure 2.10 Illustration of P diagenesis in lakes sediments. Left: slow diagenesis leads to an accumulation of mobile P. Right: fast diagenesis prevents an accumulation of mobile P. Both types of sediments have the same P release (P_{rel}) because the differences between P_{sed} flux and P_{ret} flux are the same in a and b. Red: mobile (temporary) P. Gray: permanent P.

2.3.1.2 Method Description

Depending on the size and morphometry of the lake, undisturbed sediment cores are collected from different representative sites, with the aim to cover the dominant lake areas, based on water depth. In small lakes or lakes with a simple morphometry, at least three sediment cores should be sampled from the deepest point. Each sediment core is sliced into 1 cm layers. Replicate sediment samples from similar depths can be combined into one pooled sample, or alternatively, treated individually, which allows for heterogeneity evaluation at that station. The dry mass (DM), loss on ignition (LOI), and TP are then determined in the laboratory, for each layer.

The mass of the mobile P pool in each sediment layer can be calculated by the difference between the measured TP and minimal P background content (see Figure 2.11). The differences of mobile P for each layer are summed down to a depth where no mobile P is present, in order to yield the mobile P mass per unit area, according to Equation 2.8:

$$P_{mobil} = \sum_{i=1}^{n} (TP_i - TP_{min}) * x_i * (1 - \varphi_i) * \rho_{dry,i} \qquad \text{[Eq. 2.8]}$$

where:

P_{mobil} = mobile P (mg P cm^{-2})
TP_i = TP content of a layer i (mg P g dw^{-1})
TP_{min} = TP content of permanently fixed P (minimal background content)
x_i = thickness of layer i (cm)
φ_i = porosity of layer i (dimensionless)
$\rho_{dry,i}$ = density of dry mass, measured or calculated as mixture of mineral (2.65 g cm^3) and organic material (= LOI) (1.4 g cm^3).

2.3.1.3 Advantages and Limitations

These measurements enable determination of P loss during diagenesis under natural environmental conditions. If the age of a single sediment depth is known, the contribution of each layer to the P release and the duration for complete P mobilization can be determined. One problem is the upscaling of results to cover the entire lake due to possible sediment heterogeneity issues caused by focused or differential sedimentation. It is therefore recommended that sediment cores are taken from different stations, in order to cover the different lake areas. Even under steady conditions, P content shows seasonal variations (e.g., dilution by calcite precipitation), so that the idealized course of diagenesis could be disturbed. An increased TP content in deeper layers is an indication that in former periods, the loading conditions and state of the lake were different from more recent periods. The gradient technique can be used in cases only where the TP profiles principally follow the theoretical curve of declining TP contents with depth in the upper sediment layer. Besides the seasonal variability of total P content in settling material, the stratigraphy could be influenced by sediment resuspension and redeposition, as well as by biological activity (bioturbation).

2.3.1.4 Data Evaluation and Interpretation

The calculated mobile P pool provides important information for predicting the long-term impact of this pool on the observed P release. It also provides information on the period of time expected for exhaustion of the mobile P pool, assuming no new P sedimentation. Clearly, quantification of this P pool is important when selecting an appropriate restoration measure, or the optimal dose of P-binding chemicals. However, it is important to note that sediment capping and dredging will not have a significant impact on the lake water quality in cases where the mobile P pool is low compared to either P pools in the water or the external P load (see Chapter 5). Particularly under nonstationary conditions—which are often extant prior to lake restoration—great care must be taken with interpreting whether or not P surface accumulation is in the form of mobile P, and it should be verified by other methods, such as P fractionation (see Section 2.3.2).

2.3.2   Fractionation Method

2.3.2.1   Principle

As it is impossible to describe the wide range of individual P species found in lake sediments, numerous operationally defined, sequential P extraction procedures have been developed in the past (e.g., Hieltjes and Lijklema 1980; Golterman and Booman 1988; Psenner and Pucsko 1988; Paludan and Jensen 1995; Puttonen et al. 2014). Sequential P extraction procedures serve as important tools for identifying different binding forms of P in lake sediments and to distinguish between mobile and permanent P forms. Sequential extraction procedures take advantage of the fact that various solid P phases show different reactions toward the different solutions employed in the sequential procedures. This allows partition of the sediment P into operationally defined P pools, according to specific chemical properties, such as pH and redox conditions. Generally, potentially mobile P forms can be defined as the sum of immediately bioavailable P (dissolved phosphate), or P that can be made bioavailable after transformation into dissolved P forms by naturally occurring processes in the sediment. These processes could include redox-dependent dissolution of P from Fe-P, or mineralization of organic P compounds.

2.3.2.2   Method Description

Undisturbed sediment cores are collected, with a Kajak gravity corer or similar equipment, to determine the mobile P pool using the sequential extraction method. By slicing and dividing the sediment into discrete intervals (normally 1 cm depth intervals) prior to sequential P extraction, valuable information on depth-integrated changes in sediment P pools can be obtained. In many recent studies, the scheme proposed by Psenner and Pucsko (1988) (see Table 2.1), or modifications thereof, have been used to determine potentially mobile P, but at present no standardized procedure exists (see Chapter 24). However, a common approach for all of the procedures is that fresh sediment is extracted sequentially, using a series of extractants, each chosen to selectively dissolve specific P phases from the sediment matrix. The sequence of extractants in the sequential P extraction procedures is designed to remove the most reactive phases first; harsher extractants gradually remove more resistant P phases in successive steps (see Table 2.1).

The mobile P pool is calculated as the sum of P forms potentially contributing to P release, that is, loosely adsorbed P (water-P/NH_4Cl-P), redox-sensitive P (BD-P), and organic-bound P (nonreactive P = NRP, in the first three extraction steps) (Rydin 2000; Reitzel et al. 2005). Normally the mobile P pool is calculated to 5 or 10 cm thick surface layer, but this depth should be selected based on the declining contents of the mobile P pool with increasing depth (see Figure 2.11).

2.3.2.3   Advantages and Limitations

In general, sequential sediment extractions do not require any special equipment and are relatively inexpensive. Sequential P extractions have also provided background data for many lake restoration projects, by providing information on the potentially mobile sediment P pools, even in situations where the gradient method would not work. It is important, however, to keep in mind that not all of the potentially mobile P forms may be mobile under natural conditions (Puttonen et al. 2014). Reitzel et al. (2007) used ^{31}P NMR spectroscopy to demonstrate that only 50% of the NRP was mobile in the sediment of Lake Erken. Hence, the sequential P extraction procedure might overestimate the mobile P pool, in comparison with the TP gradient method (Hupfer et al. 2016). It is also difficult to determine a standard sediment depth at which the potentially mobile P pool will contribute to P release; however, depths of 4–10 cm are often used with sequential P extraction procedures to estimate P capping material dosages (e.g., Reitzel et al. 2005; Meis et al. 2012).

Some methodological artifacts also exist. Besides incomplete extraction of the individual P phases by each extract, there is potential to form new P phases during the extractions, such as the formation of apatite P during NaOH extraction, leading to underestimation of Al-P, and overestimation of Ca-P in Al enriched sediments (Hupfer et al. 2009), or re-adsorption of extracted P phases onto existing P-binding sites in the sediment (Ruttenberg 1992). Unless detailed information on the Al-P and Ca-P forms is required, the potential overestimation of Ca-P will not be problematic, since both Al-P and Ca-P are considered to be permanent P forms. Finally, it is important to re-emphasize that sediment heterogeneity in general requires that multiple sediment cores are sampled covering the whole lake morphometry, so that information from sequential extraction procedures can be adequately extrapolated to represent the entire lake.

Table 2.1 Example of a modified Psenner extraction procedure (Psenner et al. 1988; Paludan and Jensen 1995)

Pools and Extractants	Target P Pools	Mode of Natural Release
Water-P **Deionized water**	Pore water P or loosely adsorbed P	**Mobile P.** Diffusion driven release
BD-P **Bicarbonate-buffered dithionite**	Reducible P species. Mainly Fe-P	**Mobile P.** Anoxic conditions. Often caused by high mineralization rates and stratification during summer.
NaOH-NRP **Non-reactive P in the water, BD, and NaOH extraction**	Mainly organic P and condensed inorganic P (polyphosphate, pyrophosphate)	**Mobile P.** Mineralization processes in the sediment. Stimulated by temperature and quality of electron acceptors.
NaOH-SRP **NaOH extraction after removal of nrP and humic-P**	Al-P and P in not reducible Fe oxides	**Permanent P.** Released under periods with high pH. For example, resuspension of sediment P into high productivity water.
Humic-P **Acidified NaOH extract**	P in organic compounds (mainly humic acids) precipitation under low pH conditions	**Permanent P.** Refractory P compounds that will only be very slowly mineralized.
HCl-P **Hydrochloric acid extraction**	Mineral P species; mainly apatite P	**Permanent P.** Mineral P should not be released under natural conditions.
Residual-P **Total P in the remaining sediment**	All P forms not extracted in previous steps; organic and inorganic P	**Permanent P.** This mixture of P forms, which has resisted the harsh extractions, should not be released under natural conditions.

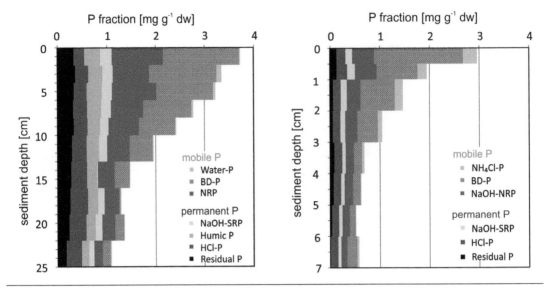

Figure 2.11 Left: sequential P fractions from the Danish Lake Sønderby. Red fractions are considered mobile (redrawn from Reitzel et al. 2005). The calculated potentially mobile P mass in 0–10 cm was 15.86 mg P m^{-2}. Right: sequential P fractions from the pre-alpine Swiss Lake Sempach (Hupfer et al. 1995). The hypolimnetic aeration of the lake led to higher iron bound P at the sediment surface but not to an increase of permanently bound P. The potentially mobile P mass in 0–5 cm was 4.55 mg P m^{-2}.

2.3.2.4 Data Evaluation and Interpretation

In general, the first extraction step (e.g., water/NH$_4$Cl extraction) is considered highly mobile and is mainly a combination of pore water P and loosely adsorbed P released due to diffusion or desorption. However, these pools generally constitute only a few percent of the sediment TP. The BD-P pool reflects P that will be mobilized under reducing conditions, which would be expected during summer blooms. Jensen and Thamdrup (1993) found a highly significant correlation between P and Fe extracted in the BD extract. The NRP pool is also considered mobile, as it may consist of easily degradable organic and condensed inorganic P, and this P pool was shown to decline with depth in the Danish Lake Sønderby and in the Swiss Lake Sempach (see Figure 2.12), indicating the mobile nature of the NRP pool (Reitzel et al. 2005; Hupfer et al. 1995). As NRP made up 20–30% of the sediment TP in a series of Danish and Swedish lakes (cf. Reitzel et al. 2005), it is likely that NRP constitutes an important P pool in many types of sediments (e.g., Rydin 2000), and should be included in sequential sediment extractions.

2.3.3 Diffusive Gradients in Thin Films (DGT)

2.3.3.1 Principle

Diffusive gradients in thin films (DGT) is a passive sampling technique, where micro-niches of different analytes in the sediment, such as P, can be studied. Unlike dialysis systems (peepers) where equilibrium concentrations are measured, DGT measures the flux of a dissolved element from the pore water to the binding gel. Hence, when DGT is deployed in the sediment, it equilibrates with the pore water, P diffuses through a filter membrane and a diffusive gel, which is then immobilized in a binding gel (see Figure 2.12) specifically designed to bind the relevant analytes (Zhang et al. 1998).

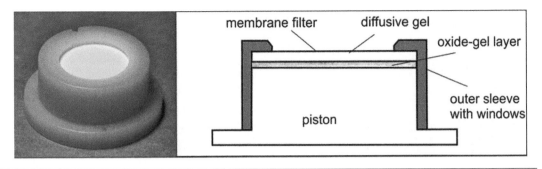

Figure 2.12 Picture and cross section of a DGT device (modified from Zhang et al. 1998).

The technique relies on a steady state concentration gradient from the solution (e.g., pore water) to the binding gel. This binding gel can be designed to simultaneously bind multiple analytes, such as P, Fe, and sulfide (Wu and Wang 2017). After deployment in the sediment, the P that is bound in the binding gel can be eluted with an acid. By relating the eluted P concentration to the surface area of the binding gel and the time of deployment, a time-averaged DGT P flux can be estimated, which is usually expressed as μmol P cm^{-2} s^{-1}. This P flux can be converted to an average P concentration at the surface of the DGT device at the time of deployment by using Fick's first law of diffusion. This P concentration will be similar to the pore water concentration, as long as P resupply from the sediment matrix can secure available P in the pore water. In sediment zones with high P resupply to the pore water, the DGT-determined pore water concentration will be higher than in layers where resupply is lower.

2.3.3.2   Method Description

The DGT probe is deployed in the sediment, either *in situ* or in intact sediment cores. The period of deployment depends on the P uptake, and it is important that the binding gel is not saturated during deployment. This can be tested by initial P adsorption studies to determine the P-binding capacity of the binding gel. After deployment, the DGT is sacrificed to isolate the P-binding gel. This gel is subsequently eluted in acid and P is analyzed. In addition, improvements of the DGT technology allow the capture of 2D, high resolution DGT data by using laser ablation, inductively coupled, plasma mass spectrometry (Stockdale et al. 2008).

2.3.3.3   Advantages and Limitations

The DGT process offers a powerful technique to study P mobilization processes in sediments, due to its high resolution and the lack of disturbance compared with sediment coring and pore water extractions. However, the DGT technique should not be used for assessing P release from the sediment to the water, since the rate of P accumulation by the DGT process differs from that in the sediment and the overlying water. The DGT technique acts like a local sink in the sediment, where it depletes the SRP pore water concentration and induces an unknown degree of resupply from the solid phase— which may not be able to keep up with P diffusion into the DGT binding gel—while there is constant resupply available from the overlying water due to water movement (leading to the replenishment of the DGT accumulated P). To use the DGT-measured pore water P concentration as a measure for the SWI flux ignores this fact, resulting in P flux estimates that are likely to be incorrect. Therefore, to measure sediment P release, incubation of sediment cores (see Section 2.2.4) or deployment of *in situ* incubation chamber systems (see Section 2.2.3) are more appropriate.

2.3.3.4 Data Evaluation and Interpretations

Zhang et al. (1998) used DGT to determine P concentrations in the lake water and pore water of a small eutrophic pond, and demonstrated that the DGT technique gave results similar to direct photometric SRP analysis, and for the first time demonstrated that DGT measurements could be used to measure sediment P concentrations at very high resolution (millimeter scale). Later, Stockdale et al. (2008) demonstrated high resolution imaging of P, using DGT in freshwater sediment where they showed tight coupling between organic matter decomposition by sulfate-reducing bacteria and P depletion due to enhanced P uptake by the bacteria during their growth period. In sediment zones with high P resupply to the pore water, the DGT-determined pore water concentration will be higher than in layers where resupply is lower. Therefore, DGT analysis of P can be used to locate microzones in the sediment for P mobilization or retention.

2.4 SEDIMENT AS SINK OR SOURCE OF P

2.4.1 Determination of P Retention Using Dated Sediment Cores

2.4.1.1 Principle

Deposition of P and other substances is reflected in the sediment. Sedimentary P retention (P_{sed} – P_{rel}) and its historical changes can be quantified using datable sediment cores (Dillon and Evans 1993; Hambright et al. 2004). Radionuclide dating of undisturbed cores using Cs-137 (Wieland et al. 1993) or Pb-210 (Appleby and Oldfield 1992) is the most common method for recent (~ 100 years) sediments, while other methods, using C-14 or visual varve counting (Zolitschka et al. 2015), coupled with knowledge of recent events (e.g., material transport by floods or snow melting), can also be used for longer time scales. In addition, anthropogenic time markers, such as polychlorinated biphenyls (PCBs), spheroidal carbonaceous particles, and Hg or other heavy metals are increasingly used for dating (Moros et al. 2017). The recent use of P-binding chemicals for lake restoration can also serve as a time marker (Hupfer et al. 2016). The simultaneous determination of total P content in individual sediment layers has made it possible to assign a P mass to specific time periods.

2.4.1.2 Method Description

Undisturbed sediment cores are taken from the deepest part of the lake, using a gravity corer, and are then sliced into 0.5 or 1 cm layers. The dating method determines how much material is needed and how the sediment should be dried. At a minimum, the dry mass, the organic content as LOI, and the TP content have to be analyzed. The area-related P accumulation rate can be determined using the following equation [see Equation 2.9]:

$$P_{ret} = \sum_{i=1}^{n} (TP_i * (1 - \varphi_i) * i_x * \rho_{dry,i}) / t \qquad \text{[Eq. 2.9]}$$

where:

P_{ret} = P retention [mg P m^{-2} yr^{-1}]
TP_i = TP content of a layer i [mg g dw^{-1}]
i_x = thickness of layer i [cm]
φ_i = porosity of the layer i (dimensionless)

$\rho_{dry,i}$ = density of dry mass, measured or calculated as mixture of mineral (2.65 g cm^{-3}) and
   organic material (= LOI) (1.4 g cm^{-3})

t = time between two time markers [years]

2.4.1.3 Advantages and Limitations

The method based on sediment chronology offers the possibility of measuring P retention directly by integrating time periods longer than one year. Additionally, a comparison with historical P deposition is possible. The most important prerequisite is that the chronology of the core is not disturbed. It is recommended that at least two independent dating methods are used. One problem for quantification is that the P diagenesis is not completely finished, so that the existing mobile P in younger sediment layers can lead to overestimation of P retention. In that case, value correction can be achieved by subtracting the mobile P pool estimate, based on the TP gradients (see Section 2.3.1), from the TP baseline. Insufficient resolution of sediment layers in relation to the annual deposition rate could also cause inaccuracies.

The main problem for the application of the method is the assumption that sediment deposition is homogenous within a lake. Sediment focusing may cause severe overestimation of whole lake accumulation rates, although radionuclide data can also be used to quantify focusing (Crusius and Anderson 1995). Determination of whole lake accumulation rates normally requires programs involving multiple core observations (Dillon and Evans 1993; Brezonik and Engstrom 1998) combined with bio-stratigraphic or clay markers (Rippey et al. 2008) to correlate sediment horizons. It is therefore recommended that preliminary studies determine whether the lake's deepest point is representative of the whole lake bed or whether more sampling points should be included.

2.4.1.4 Data Evaluation and Interpretations

In studies of anthropogenic lake eutrophication and P dynamics, historic sedimentary P accumulation can be used as an indicator for anthropogenically increased P loading and to support P mass balance calculations (e.g., Dillon and Evans 1993; Findlay et al. 1998). P accumulation in the sediment is a crucial term in the P mass balance. With the help of this loss term, the external P load can be estimated by converting Equation 2.4, with the prerequisite that the system is either in equilibrium or that the change of P_{lake} over time and the P_{out} are known. P_{out} is often simply calculated by multiplying outflow discharge with the mean epilimnetic P concentration. Further information on geochemical conditions in the lake or on changes in the catchment affecting P accumulation can be obtained by parallel studies of the content and accumulation rates of various elements.

Figure 2.13 shows an example for the determination of P retention since 1950, using dated sediment cores for a German, pre-alpine lake with a history of eutrophication and re-oligotrophication. Age-dating of core sediments was done by analysis of Cs-137, a radioisotope released as a by-product of nuclear weapons testing or from nuclear accidents. These measurements resulted in the identification of three points in time:

1. Cs-137 first occurred in the atmosphere in about 1951
2. Cs-137 peaked during 1963–64 with maximum atmospheric bomb testing
3. Cs-137 was detected after radioactive fallout from the Chernobyl nuclear reactor accident in April 1986

The example demonstrates that both the dry matter accumulation rate and the P accumulation rate showed no significant changes during the eutrophication period between the beginning of the 1950s and 1960s.

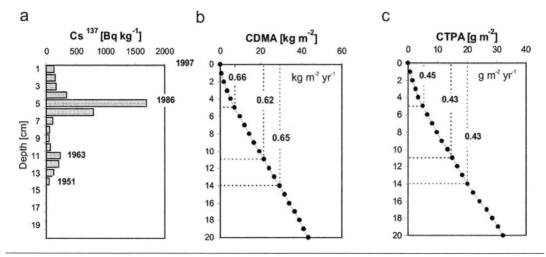

Figure.2.13 Sediment core analysis in the pre-alpine Lake Schliersee in Bavaria (Germany) taken in 1997. (a) Dating with cesium-137 enables the determination of three points in time (see text). The dating is used to calculate (b) dry mass accumulation rates based on cumulative dry mass per area (CDMA) and (c) the P accumulation based on the cumulative total P mass per area (CTPA) for each three time interval.

2.4.2   Difference between Gross Sedimentation (Trap Measurements) and Gross P Release

2.4.2.1   Principle

In order to assess whether the measured P release rate is to be classified as internal P loading, it can be related to the P gross sedimentation from traps representing the transport process in the opposite direction. Transport of P toward the sediment is mainly achieved by sedimentation of particles under the influence of gravity, and this downward vertical transport can be measured using sediment traps. Cylindrical traps are mostly used to quantify the sedimentation rate in lakes (Bloesch and Burns 1980). The exposure of traps to different water depths makes it possible to determine which changes occur during settling of particles through the water body (Hupfer et al. 1995). A comparison of the composition of the settling particles with that of the sediment surface also makes it possible to identify changes in the P forms that are due to early diagenesis (Eckert et al. 2003; Kleeberg 2002).

2.4.2.2   Method Description

The cylinder traps used to measure sedimentation rate should have diameters of at least 4 cm and a height-to-diameter ratio within the range of 3 to 10. The cylinder traps are equipped with a removable trap vessel. In order to prevent microbial degradation processes and zooplankton grazing during exposure, 0.01–0.25 M formalin can be introduced into the traps (Lee et al. 1992). The traps are attached with rope and are anchored (see Figure 2.14) with vertical adjustment achieved by an underwater buoy close to the water surface. Using multiple traps, the time regime for filling trap vessels can be automatically controlled, so that a longer exposure without intermittent recovery is possible (see Figure 2.14). After trap recovery, column water is carefully drained, pumping by tube or through openings in the wall. Dead zooplankton are separated with a sieve (0.5 mm). The dry mass is determined from the aliquot of the mixed suspension before the water is removed, by centrifugation or decantation. After drying, the solids are analyzed for TP and other parameters, using

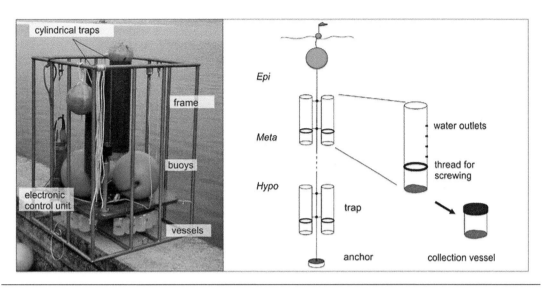

Figure 2.14   Left: a multi-trap system with two parallel cylindrical columns for parallel sample collection. Right: a schematic view of an installation for conventional sedimentation traps in a lake.

standard methods. The gross P sedimentation rate per area is determined from the dry mass rate multiplied by TP content.

2.4.2.3   Advantages and Limitations

Collecting particles using traps usually provides reliable values for the determination of downward P fluxes and their seasonal course. At least one annual cycle is required to determine the sedimentation rate in a lake for mass balance approaches and for qualitative comparison with the sediment. Trap material can also be used for the determination of P forms (e.g., Hupfer and Lewandowski 2005; Kleeberg 2002) and to predict and observe the mobility of the settling P after sedimentation, recalling that above ground inflows can lead to both heterogeneity and to longitudinally different sedimentation behavior in a lake or reservoir. Additionally, further processes such as focusing and resuspension determine local sediment composition, which makes direct comparison between settling particles and sediment more difficult.

2.4.2.4   Data Evaluation and Interpretations

Comparison between P retention rate by datable sediment cores and the P sedimentation rate allows calculation of the released P portion, after sedimentation (Hupfer et al. 1995). The method also makes it possible to consider whether and how the seasonality of sedimentation is synchronized with P sedimentation and whether it is therefore to be regarded as the actual P sedimentation trigger. If additional parameters such as iron, manganese, calcium, carbon, sulfur, or phytoplankton production are determined, important conclusions can be made about the factors controlling P sedimentation.

2.5   CONCLUSIONS

Although it has been known for a long time that lake sediments play an important role in the P balance of a lake, the exact determination of the internal P load is still a major challenge. There are theoretical and methodological reasons for this. Firstly, the key terms *internal load, P release, net sedimentation*, and *retention* are not used consistently in the literature. Under steady state conditions, sediments acts as a P sink. Since mass balances for lakes and mass balance models usually cover a period of at least one year, we recommend using the term *internal P loading* only for the net P release on an annual basis. Only under these conditions can the importance of internal P load as an extra load be directly related to the external P load, which is very important information when making decisions on appropriate management measures.

Besides the lack of consensus on key terminology, P release from sediments is often determined without measuring additional internal pools and fluxes, and this is a further source of interpretation problems. Review of available investigation methods shows that the P release rate could be independent of the inventory of the potentially mobile P pool in sediment. High P release rates could be the result of the fast transformation of settled particulate P to dissolved P, at the sediment surface. According to this concept, a high P release rate could be sustained by a permanent downward P flux as sedimentation of organic and inorganic P. Therefore, it is necessary to determine the sink or source function of sediments, either by mass balances or by comparison of vertical fluxes in both directions.

Mass balance calculations involving the measurement P input, P output, and changes of P in the lake are complex, but allow reliable determination of the function of the sediments in relation to the P budget in a lake. Determining P release, by means of benthic chambers or laboratory experiments, is not representative for the whole lake or the whole year. Such experiments are, however, particularly suitable for process studies such as investigating the impact of environmental factors or internal lake changes on P release. The methods presented for determining fluxes and pools are largely compatible with the terms and required input data for mass balance models (e.g., one box model).

2.6 ACKNOWLEDGMENTS

We are very grateful to Henning Jensen (University of Southern Denmark, DK), Andreas Kleeberg (State Laboratory Berlin-Brandenburg, D), Jörg Lewandowski (Leibniz Institute of Freshwater Ecology and Inland Fisheries, D), and René Gächter (Eawag, CH) for inspiring discussions and for sharing their experiences.

2.7 REFERENCES

Appleby, PG and Oldfield, F. 1992. Applications of ^{210}Pb to sedimentation studies. In: Ivanovich, M and Harmon RS, eds.Uranium-series Disequilibrium. Appl to Earth, Marine and Environ Sci. Oxford Science, pp. 731–778.

Berg, P; Risgaard-Petersen, N; and Rysgaard, S. 1998. Interpretation of measured concentration profiles in sediment pore water. Limnol Oceanogr. 43:1500–1510.

Berner, RA. 1980. Early diagenesis: A theoretical approach. Princeton: Princeton University Press. p. 241.

Bloesch. J and Burns, NM. 1980. A critical review of sedimentation trap technique. Schweiz Z Hydrol. 42(1):15–55.

Brezonik, PL and Engstrom, DR. 1998. Modern and historic accumulation rates of phosphorus in Lake Okeechobee, Florida. J Paleolimnol. 20:31–46.

Burger, D; Hamilton, DP; Pilditch, CA; and Gibbs, MM. 2007. Benthic nutrient fluxes in a eutrophic, polymictic lake. Hydrobiologia. 584:13–25.

Callender, E and Hammond, DE. 1982. Nutrient exchange across the sediment-water interface in the Potomac River estuary. Estuar Coast Shelf Sci. 15 (4):395–413.

Clavero, V; Izquierdo, JJ; Fernandez, JA; and Niell, FX. 2000. Seasonal fluxes of phosphate and ammonium across the sediment-water interface in a shallow small estuary (Palmones River, southern Spain). Mar Ecol Prog Ser. 198:51–60.

Crusius, J and Anderson, RF. 1995. Sediment focusing in six small lakes inferred from radionuclide profiles. J Paleolimnol. 13:143–155.

Davison, W; Grime, GW; Morgan, JAW; and Clarke, K. 1991. Distribution of dissolved iron in sediment pore waters at submillimeter resolution. Nature. 352:323–325.

Dillon, PJ and Evans, HE. 1993. A comparison of phosphorus retention in lakes determined from mass balance and sediment core calculations. Water Res. 27:659–668.

Dithmer, L; Nielsen, UG; Lurling, M; Spears, BM; Yasseri, S; Lundberg, D; Moore, A; Jensen, ND; and Reitzel, K. 2016. Responses in sediment phosphorus and lanthanum concentrations and composition across 10 lakes following applications of lanthanum modified bentonite. Water Res. 97:101–110.

Eckert, W; Didenko, J; Uri, E; and Eldar, D. 2003. Spatial and temporal variability of particulate phosphorus fractions in seston and sediments of Lake Kinneret under changing loading scenarios. Hydrobiologia. 494:223–229.

Emili, A; Acquavita, A; Covelli, S; Spada, L; Di Leo, A; Giandomenico, S; and Cardellicchio, N. 2016. Mobility of heavy metals from polluted sediments of a semi-enclosed basin: in situ benthic chamber experiments in Taranto's Mar Piccolo (Ionian Sea, Southern Italy). Environ Sci Pollut Res. 23:12582–12595.

Findlay, DL; Kling, HJ; Rönicke, H; and Findlay, WJ. 1998. A paleolimnological study of eutrophied Lake Arendsee (Germany). J Paleolimnol. 19:41–54.

Golterman, H and Booman, A. 1988. Sequential extraction of iron-phosphates and calcium-phosphate from sediment by chelating agents. Verh Internat Verein Limnol. 23:904–909.

Goyette, J-O; Bennett, EM; and Maranger, R. 2018. Low buffering capacity and slow recovery of anthropogenic phosphorus pollution in watersheds. Nat Geosci. 11:921–925.

Grüneberg, B; Dadi, T; Lindim, C; and Fischer, H. 2015. Effects of nitrogen and phosphorus load reduction on benthic phosphorus release in a riverine lake. Biogeochemistry. 123:185–202.

Grüneberg, B; Rücker, J; Nixdorf, B; and Behrendt, H. 2011. Dilemma of non-steady state in lakes-development and predictability of in-lake P concentration in dimictic Lake Scharmützelsee (Germany) after abrupt load reduction. Internat Rev Hydrobiol. 96:599–621.

Hambright, KD; Eckert, W; Leavitt, PR; and Schelske, CL. 2004. Effects of historical lake level and land use on sediment and phosphorus accumulation rates in Lake Kinneret. Environ Sci Technol. 38:6460–6467.

Hansen, J; Reitzel, K; Jensen, HS; and Andersen, FO. 2003. Effects of aluminum, iron, oxygen and nitrate additions on phosphorus release from the sediment of a Danish softwater lake. Hydrobiologia. 492(1–3):139–149.

Hesslein, RH. 1976. An in situ sampler for close interval pore water studies. Limnol Oceanogr. 22: 912–914.

Hieltjes, A and Lijklema, L. 1980. Fractionation of inorganic phosphates in calcareous sediments. J Envir Qual. 9:405–407.

Hupfer, M; Gächter, R; and Giovanoli, R. 1995. Transformation of phosphorus species in settling seston and during early sediment diagenesis. Aquat Sci. 57(4):305–324.

Hupfer, M and Hilt, S. 2008. Ecological Engineering: Lake Restoration. In: Jorgensen SE, Fath B, editors. Encyclopedia of Ecology. Oxford: Elsevier. pp. 2080–2093.

Hupfer, M and Lewandowski, J. 2005. Retention and early diagenetic transformation of phosphorus in Lake Arendsee (Germany)—consequences for management strategies. Arch Hydrobiol. 164(2):143–167.

Hupfer, M and Lewandowski, J. 2008. Oxygen controls the phosphorus release from lake sediments—a long lasting paradigm in limnology. Internat Rev Hydrobiol. 93:415–432.

Hupfer, M; Reitzel, K; Kleeberg, A; and Lewandowski, J. 2016. Long-term efficiency of lake restoration by chemical phosphorus precipitation: scenario analysis with a phosphorus balance model. Water Res. 97:153–161.

Hupfer, M and Uhlmann, D. 1991. Microbially mediated phosphorus exchange across the mud-water-interface. Verh Int Verein Limnol. 24(3):2999–3003.

Hupfer, M; Zak, D; Rossberg, R; Herzog, C; and Pöthig, R. 2009. Evaluation of a well-established sequential phosphorus fractionation technique for use in calcite-rich lake sediments: identification and prevention of artefacts due to apatite formation. Limnol Oceanogr Meth. 7:399–410.

Imboden, D; and Gächter, R. 1978. A dynamic model for trophic state prediction. J Ecol Modell. 4:77–98.

Jensen, HS and, Andersen, FO. 1992. Importance of temperature, nitrate, and pH for phosphate release from aerobic sediments of 4 shallow, eutrophic lakes. Limnol Oceanogr. 37(3):577–589.

Jensen, HS; Kristensen, P; Jeppesen, E; and Skytthe, A. 1992. Iron-phosphorus ratio in surface sediment as an indicator of phosphate release from aerobic sediments in shallow lakes. Hydrobiologia. 235:731–743.

Jensen, HS; Reitzel, K; and Egemose, S. 2015. Evaluation of aluminum treatment efficiency on water quality and internal phosphorus cycling in six Danish lakes. Hydrobiologia. 751(1):189–199.

Jensen, HS and Thamdrup, B. 1993. Iron-bound phosphorus in marine sediments as measured by bicarbonate-dithionite extraction. Hydrobiologia. 253(1–3):47–59.

Jeppesen, E; Søndergaard, M; Jensen, JP; Havens, KE; Anneville, O; Carvalho, L; Coveney, MF; Deneke, R; Dokulil, MT; Foy, B; et al. 2005. Lake responses to reduced nutrient loading—an analysis of contemporary long-term data from 35 case studies. Freshw Biol. 50:1747–1771.

Kleeberg, A. 2002. Phosphorus sedimentation in seasonal anoxic Lake Scharmützel, NE Germany. Hydrobiologia. 472(1–3): 53–65.

Krom, MD; Davison, P; Zhang, H; and Davison, W. 1994. High-resolution pore water sampling with a gel sampler. Limnol Oceanogr. 39(8):967–1972.

Laskov, C; Herzog, C; Lewandowski, J; and Hupfer, M. 2007. Miniaturised photometrical methods for the rapid analysis of phosphate, ammonium, ferrous iron and sulfate in pore water of aquatic sediments. Limnol Oceanogr Meth. 5:63–71.

Lee, C; Hedges, JI; Wakeham, SG; and Zhu, N. 1992. Effectiveness of various treatments in retarding microbial activity in sediment trap material and their effects on the collection of swimmers. Limnol Oceanogr. 37:117–130.

Lewandowski, J; Rüter, K; and Hupfer, M. 2002. Two-dimensional small-scale variability of pore water phosphate in freshwater lakes: results from a novel dialysis sampler. Environ Sci Technol. 36: 2039–2047.

L'Helguen, SP; Chauvaud, L; Cuet, P; Frouin, P; Maguer, JF; and Clavier, J. 2014. A novel approach using the 15N tracer technique and benthic chambers to determine ammonium fluxes at the sediment-water interface and its application in a back-reef zone on Reunion Island (Indian Ocean). J Exp Mar Biol Ecol. 452:143–151.

Li, VH and Gregory, S. 1974. Diffusion of ions in sea water and in deep sea sediments. Geochim Cosmochim Acta. 38:703–714.

Livingstone, DM and Imboden, DM. 1996. The prediction of hypolimnetic profiles: a plea for a deductive approach. Can J Fish Aquat Sci. 53(4):924–932.

Meis, S; Spears, BM; Maberly, SC; O'Malley, MB; and Perkins, RG. 2012. Sediment amendment with Phoslock® in Clatto Reservoir (Dundee, UK): investigating changes in sediment elemental composition and phosphorus fractionation. J Environ Manage. 93(1):185–93.

Moros, M; Andersen, TJ; Schulz-Bull, D; Hausler, K; Bunke, D; Snowball, I; Kotilainen, A; Zillen, L; Jensen, JB; Kabel, K; et al. 2017. Towards an event stratigraphy for Baltic Sea sediments deposited since AD 1900: approaches and challenges. Boreas. 46(1):129–142.

Nogaro, G; Harris, A; and Steinman, AD. 2016. Alum application, invertebrate bioturbation and sediment characteristics interact to affect nutrient exchanges in eutrophic ecosystems. Freshw Sci. 35:597–610.

Orihel, DM; Baulch, HM; Casson, NJ; North, RL; Parsons, CT; Seckar, DCM; and Venkiteswaran, JJ. 2017. Internal phosphorus loading in Canadian fresh waters: a critical review and data analysis. Can J Fish Aquat Sci. 74:2005–2029.

Paludan, C and Jensen, HS. 1995. Sequential extraction of phosphorus in freshwater wetland and lake sediment: significance of humic acids. Wetlands. 15(4):365–373.

Petranich, E; Covelli, S; Acquavita, A; De Vittor, C; Faganeli, J; and Contin, M. 2018. Benthic nutrient cycling at the sediment-water interface in a lagoon fish farming system (northern Adriatic Sea, Italy). Sci Total Environ. 644:137–149.

Psenner, R and Pucsko, R. 1988. Phosphorus fractionation: advantages and limits of the method for the study of sediment P origins and interactions. Arch Hydrobiol Beih. 30:43–59.

Puttonen, I; Mattila, J; Jonsson, P; Karlsson, OM; Kohonen, T; Kotilainen, A; Lukkari, K; Malmaeus, JM; and Rydin, E. 2014. Distribution and estimated release of sediment phosphorus in the northern Baltic Sea archipelagos. Estuar Coast Mar Sci. 145:9–21.

Reitzel, K; Ahlgren, J; DeBrabandere, H; Waldeback, M; Gogoll, A; Tranvik, L; and Rydin, E. 2007. Degradation rates of organic phosphorus in lake sediment. Biogeochemistry. 82(1):15–28.

Reitzel, K; Hansen, J; Andersen, FO; Hansen, KS; and Jensen, HS. 2005. Lake restoration by dosing aluminum relative to mobile phosphorus in the sediment. Environ Sci Technol. 39(11):4134–4140.

Rippey, B; Anderson, N; Renberg, I; and Korsman, T. 2008. The accuracy of methods used to estimate the whole-lake accumulation rate of organic carbon, major cations, phosphorus and heavy metals in sediment. J Paleolimnol. 39:83–99.

Ruttenberg, KC. 1992. Development of a sequential extraction method for different forms of phosphorus in marine sediments. Limnol Oceanogr. 37(7):1460–1482.

Rydin, E. 2000. Potentially mobile phosphorus in Lake Erken sediment. Water Res. 34(7):2037–2042.

Sas, H. 1989. Lake restoration by reduction of nutrient loading: expectations, experiences, extrapolations. Academia Verlag, St. Augustin, p. 497.

Seeberg-Elverfeldt, J; Schlüter, M; Feseker, T; and Kölling, M. 2005. Rhizon sampling of pore waters near the sediment-water interface of aquatic systems. Limnol Oceanogr Meth. 3:361–371.

Søndergaard, M; Bjerring, R; and Jeppesen, E. 2013. Persistent internal phosphorus loading during summer in shallow eutrophic lakes. Hydrobiologia. 710(1):95–107.

Steinman, A; Chu, X; and Ogdahl, M. 2009. Spatial and temporal variability of internal and external phosphorus loads in Mona Lake, Michigan. Aquat Ecol. 43(1):1–8.

Steinsberger, T; Schmid, M; Wüest, A; Schwefel, R; Wehrli, B; and Müller, B. 2017. Organic carbon mass accumulation rate regulates the flux of reduced substances from the sediments of deep lakes. Biogeosciences. 14:3275–3285.

Stockdale, A; Davison, W; and Zhang H. 2008. High-resolution two-dimensional quantitative analysis of phosphorus, vanadium and arsenic, and qualitative analysis of sulfide, in a freshwater sediment. Environ Chem. 5:143–149.

Tengberg, A; De Bovee, F; Hall, P; Berelson, W; Chadwick, D; Ciceri, G; Crassous, P; Devol, A; Emerson, S; Gage, J; et al. 1995. Benthic chamber and profiling landers in oceanography—A review of design, technical solutions and functioning. Prog Oceanogr. 35:253–294.

Urban, NR; Dinkel, C; and Wehrli, B. 1997. Solute transfer across the sediment surface of a eutrophic lake: I. Porewater profiles from dialysis samplers. Aquat Sci. 59:1–2.

Wang, YP; Hu, WP; Peng, ZL; Zeng, Y; and Rinke, K. 2018. Predicting lake eutrophication responses to multiple scenarios of lake restoration: a three-dimensional modeling approach. Water. 10(8): 994.

Wieland, E; Santschi, PH; Höhener, P; and Sturm, M. 1993. Scavenging of Chernobyl ^{137}Cs and natural ^{210}Pb in Lake Sempach, Switzerland. Geochim Cosmochim Acta. 57(13): 2959–2979.

Wu, Z and Wang, S. 2017. Release mechanisms and kinetic exchange of phosphorus (P) in lake sediment characterized by diffusive gradients in thin films (DGT). J Hazard Mater. 331: 36–44.

Zhang, H; Davison, W; Gadi, R; and Kobayashi, T. 1998. In situ measurement of dissolved phosphorus in natural waters using DGT. Anal Chim Acta. 370(1):29–38.

Zolitschka, B; Francus, P; Ojala, AEK; and Schimmelmann, A. 2015. Varves in lake sediments—a review. Quat Sci Rev. 117:1–41.

CHAPTER 3

INTERNAL PHOSPHORUS LOADING MODELS: A CRITICAL REVIEW

Gertrud Nürnberg[1]

Abstract

Internal phosphorus (P) load as phosphate released from lake-bottom sediments is an important nutrient source that can trigger cyanobacteria blooms. Its quantification can be complicated and is affected by the thermal stratification status of the lake; therefore, modeling is often the approach to incorporate internal load into water quality predictions. This review critically evaluates two basically different model classes: (1) detailed, process-based models (*mechanistic models*) often developed for individual lakes and (2) empirical, cross-system models developed from the data of many lakes over a range of conditions. Performance, testability, and other model characteristics are described for a wide range of published models. Simpler, empirical models predict internal load from sediment P release rates (RRs) and the area and time involved in release, while complex models often incorporate seasonal P fluxes. Steady state mass balance models just distinguish between annual P fluxes to and from the sediments to assess internal load, while more complex models can incorporate time and spatial dynamics. The main application of internal load estimates is the prediction of lake total P concentration for different periods. Internal load is becoming more important because of synergistic effects of temperature increases on P RR and area, as well as changes in thermal stratification, all of which is predicted in many climate change models. Model choice depends on data availability, financial and technical resources, and the application history of specific models, while testability and verifiability combined with the principle of parsimony are considered most important.

Key words: Internal phosphorus load; mechanistic and empirical models; model testing; verification; lake sediment

[1] Freshwater Research, 3421 Hwy 117, RR.1, Baysville, Ontario, P0B 1A0 Canada. E-mail: gkn@fwr.ca

3.1   INTRODUCTION

Acknowledging the need for internal P load determination is the first step toward its modeling and quantification (Nürnberg 2009). The quantification of an internal P flux, which is directed upward from the sediment into the water, is rendered difficult by opposing fluxes such as settling and sedimentation. These downward fluxes are often high, for example, more than 57% of annual external inputs were on average retained in stratified lakes with no apparent hypolimnetic anoxia and an average water load of less than 16.8 m yr^{-1} (Nürnberg 1984). In oligo- and mesotrophic lakes downward fluxes far exceed the upward internal fluxes, which is probably the reason that internal loading has not always been considered in lake models. However, internal P loading is recognized as part of the P cycle, especially in deteriorated systems (Søndergaard et al. 2012).

Many different models that include internal P load exist, and reviews for modeling aquatic ecosystems include Mooij (2010) and Robson (2014). Two basically different model approaches are: (1) detailed, process-based models (*mechanistic models*) often developed for individual lakes and (2) empirical, cross-system models developed from the data of many lakes over a wide range of conditions. While process-based models are *data-hungry* (Robson 2014), the empirical approaches follow the principle of parsimony; they are powerful because they can predict in situations with limited data availability and hence limited effort and costs. While the application of cross-system models to individual situations (specific lake with system-specific data) has statistical limitations, they can be overcome by using a Bayesian hierarchical framework (Cheng et al. 2010).

Process-based models that are developed for individual systems where sufficient interest and funds are available provide more detailed predictions. This is especially relevant where degraded systems are large and complicated, as in multiple basins (e.g., run-of-the-river reservoirs; Komatsu et al. 2006), the Laurentian Great Lakes (Kim et al. 2013), and Lake Taihu, China (Hu 2016)), or when testing hypotheses for specific conditions. However, such process-based models are not always reliable because the lack of long-term observations needed for calibration and the large number of required variables can prevent model validation (Arhonditsis and Brett 2004). Hence, such models can be misleading when used to provide management suggestions (Brett et al. 2016; Brett and Arhonditsis 2016).

To reduce the need for local data, many complex, process-based models use empirical, cross-system sub-models. All models require tests and statistical methods including measures of variation, comparison with observations, and sensitivity analysis for model validation and verification, and evaluating model robustness (see Chapter 6).

3.2   RESULTS AND DISCUSSION

3.2.1   Sediment Models (RRs and Spatially and Temporally Active Sediment)

Models that predict internal loading from sediment characteristics need to consider both a RR and the extent of the releasable sediment expanse in time and space. I have repeatedly used an areal RR, defined as P release per actively releasing (anoxic) sediment area per day (mg m^{-2} d^{-1}, Section 3.2.1.1), and the observed anoxic factor (AF) or the predicted active area factor (AA) to represent time and spatial extent (d yr^{-1} or d period^{-1}, Section 3.2.1.2; Nürnberg 2009). The product of RR and AA (or AF) then yields an annual or seasonal gross internal load (see Equation 3.1, where L$_{int}$, mg m^{-2} of lake surface area per year or per growing period) (see Figure 3.1).

$$L_{int} = RR \times AA$$
[Eq. 3.1]

A. Internal P load determination in a stratified lake

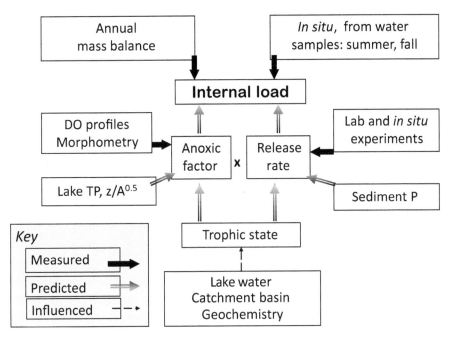

B. Internal P load determination in a shallow mixed lake

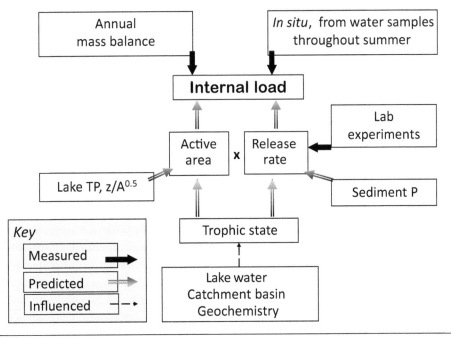

Figure 3.1   Schematic of possible modeling of internal P load, (A) for stratified lakes and (B) for shallow mixed lakes. (Based on models summarized in Nürnberg 2009.)

Similar descriptions of RR and area have been used by others including some more detailed and complex descriptions that may include seasonal variability as a function of physical and chemical determinants or simple simulation and calibration with available data (see Sections 3.2.1.1 and 3.2.1.2).

3.2.1.1 Sediment P RR

The internally derived P load is released from lake-bottom sediments. Consequently, sediment characteristics, including specific P fractions, have been used repeatedly in empirical as well as mechanistic models of internal load.

A theoretically plausible approach to predicting RR is the use of diffusion models (e.g., Fick's law; Chapra 1997) to compute a diffusion rate (similar to RR) from the phosphate pore water concentration. This approach commonly yields lower RR estimates than other approaches, including the experimental determination of RR from sediment cores (Carter and Dzialowski 2012; James 2017), and growing period P increases (Paytan et al. 2017) because the depth resolution is often too crude (see Chapter 2). This creates an inherent bias for lower calculated RRs because the apparent concentration gradients are unrealistically low. Typical pore water measurements for 1 cm intervals can be considered too large for adequate determinations of the P concentration gradient (Davison 1982), but micro-measurements using microelectrodes or diffusional gradients in thin films (DGT) with a resolution of 1 mm or better yield more realistic diffusion rates (Davison et al. 1991; Ding et al. 2011, 2018; see Chapter 2).

Laboratory or *in situ* studies can determine P release in sediment cores in the lab, in barrels, or in sophisticated chambers deployed on the sediment (Nürnberg 1988; Loh et al. 2013; Doig et al. 2016; James 2017). Such experimental treatment provides the opportunity of testing various influences on release, including temperature, pH, redox potential, and oxygen and P concentration of the overlying water. Knowledge gained from these experiments can be (carefully) extrapolated to growing period or annual conditions for a specific year and lake to provide variability in internal load estimates. For example, temperature differences between years were incorporated into the prediction of annual RRs by using experimentally derived relationships and into annual AAs (see Section 3.2.1.2) by using theoretical temperature dependencies (Nürnberg et al. 2012; Nürnberg and LaZerte 2016a).

Sediment total P (TP) and specific fractions have been considered to affect P release with varying success (see Figure 3.1). TP and redox-dependent, iron-bound P fractions were significantly correlated with experimentally determined RRs (Nürnberg 1988), but others did not find such correlations (Boström et al. 1988; Loh et al. 2013). Regression models were statistically significant but differed for deep location sediment TP of eutrophic lakes and oligo-mesotrophic lakes (see Table 3.1).

Such simple empirical models of deep sediment RRs can yield representative whole-lake daily areal P RRs and can be explained as follows: lake conditions and characteristics are integrated in the lake sediments at the deepest sites after historical accumulations of external inputs as influenced by water column productivity and sediment focusing. But these models are based on many world-wide lakes with variable geochemistry using variable methods (see Table 3.1, world-wide equation) and should be verified for any specific lake (see Section 3.2.4).

To determine the dosage in chemical lake restoration, P fractions are analyzed that are most likely involved in sediment P release as *mobile P* (see Table 3.2). They include three fractions of the Psenner type fractionation method (Psenner and Pucsko 1988) and then yield the necessary dose to decrease or cease P release from a specified release area and an arbitrarily considered sediment depth (Reitzel et al. 2005). Such *mobile P* concentration is likely correlated with RR, and I suggest further studies to determine possible regression models.

Table 3.1 Some regression models for predicting sediment P release rates (RR) (mg P/m^2/day) (Nürnberg 1988). Sediment TP and reductant soluble P fraction (BD-P, according to Psenner et al. 1984, releasable P fraction from iron oxy-hydroxides) concentrations (mg/g dry weight) and loss on ignition (LOI) as measure of organic content (%).

Model	R^2	*P*-value	N, Application	Sediment Layer
Log(RR) = 0.80 + 0.76 log(TP)	0.21	<0.001	63, worldwide eutrophic	0–5 cm
RR = −4.18 + 3.77 TP	0.63	<0.0001	14, North American oligo-mesotrophic	0–5 cm
RR = 4.78 + 2.75 TP − 0.177 LOI	0.85	<0.0001	14, North American oligo-mesotrophic	0–10 cm
RR = −0.58 + 13.72 BD-P	0.71	<0.0001	14, North American oligo-mesotrophic	0–5 cm

Table 3.2 Predictors and models of P RRs based on sediment characteristics.

Predictors	Approach	Specifics	Evaluation (advantage/ disadvantage)	Reference
Sediment TP, organic content (LOI), reductant-soluble fraction (BD-P)	Regression	At deepest location in lake, 0–5 cm or 0–10 cm sediment layer	Varying parameterizations can cause erroneous applications: • literature data set applies to eutrophic lakes • study data set applies to oligo-, mesotrophic lakes	Nürnberg 1988 (Table 3.1)
Mobile P (specific fractions of extracted P)	Mechanistic	Dosage calculation for chemical treatment yields a *total* releasable quantity	The releasable P mass is not necessarily related to rates; arbitrarily selected depths	Reitzel et al. 2005; James and Bischoff 2015
TP, recycling depth	Mechanistic with empirical sub-models	Applied to deposition area when DO <1.5 mg/L (Table 3.5)	Needs input to specific lake, includes sediment compartment, useful for recovery predictions	Chapra and Canale 1991

RRs were correlated with lake trophic state (as determined from growing season average TP, chlorophyll, and/or Secchi transparency (Nürnberg 1996)) in lakes (Nürnberg 1994) and reservoirs (Carter and Dzialowski 2012), as well as with watershed characteristics (Carter and Dzialowski 2012). While these relationships generally have only limited predictive power (see Table 3.3), they are a beginning point for the determination of RRs where sediment data are scarce and can serve as a *reality check* for more complicated models.

Table 3.3   Predictors and models of P RRs based on lake and catchment characteristics.

Predictors	Approach	Specifics	Evaluation (advantage/ disadvantage)	Reference
Trophic state (mixed layer growing period TP, chlorophyll, Secchi average)	Regression	U.S. Plains reservoirs	mesotrophic 6.1 mg/m²/d, eutrophic 10.8 mg/m²/d, hyper-eutrophic 25.6 mg/m²/d.	Carter and Dzialowski 2012
Trophic state (mixed layer growing period TP, chlorophyll, Secchi average)	Regression	North American and European lakes	mesotrophic 4.9 mg/m²/d, eutrophic 12.5 mg/m²/d, hyper-eutrophic 21.6 mg/m²/d.	Nürnberg 1994, unpubl. data
Lake TP	Regression	U.S. Plains reservoirs, North American and European lakes	$R^2 = 0.37$ $R^2 = 0.25$	Carter and Dzialowski 2011, Nürnberg 1986
Secchi	Regression	U.S. Plains reservoirs	$R^2 = 0.24$	Carter and Dzialowski 2012
Watershed use	Regression	Percent cropland in U.S. Plains reservoirs	$R^2 = 0.44$	Carter and Dzialowski 2012

3.2.1.2   Duration and Area of P Release

Because P RRs from oxic sediment surfaces are typically much smaller than those from anoxic surfaces, most models predicting internal P load from sediments consider variation in oxygen state at the sediment-water interface explicitly or implicitly (season, temperature).

In stratified systems that are relatively deep compared to their surface area, monitored dissolved oxygen (DO) concentration profiles are used if available, to determine the anoxic ($<1–2$ mg L^{-1} DO) sediment area and period (see Equation 3.2, AF = anoxic factor) (Nürnberg 1995, 2019).

$$AF = \sum_{i=1}^{n} \frac{t_i \cdot a_i}{A_o} \qquad \text{[Eq. 3.2]}$$

where t_i is the period of anoxia (days), a_i is the corresponding lake area that is hypoxic (m²), A_o is the lake surface area (m²), and n is the number of periods with different depths for the specific hypoxic level considered (e.g., $<1–2$ mg L^{-1} DO for anoxia at the sediment surface).

Based on such observations, multiple regression models were developed to predict AA, which is the extent and duration of an anoxic hypolimnetic area in stratified lakes or the duration and extent of actively releasing sediment surfaces in polymictic lakes (Nürnberg 2002). Predictors include growing period (summer and early fall) or annual average lake water nutrient concentration with morphometric variables and external TP load (Nürnberg 1995, 1996) (see Table 3.4, Table 3.5, and Figure 3.1).

An equivalent to the active area was modeled as depositional area with DO <1.5 mg L^{-1} using areal hypolimnetic oxygen deficit (AHOD) predicted from lake TP and hypolimnion depth (Chapra and Canale 1991). These models have been applied to numerous lakes to quantify anoxia (Chapra

Table 3.4 Regression models for predicting the actively releasing area (AA, days/summer) in stratified lakes with anoxic hypolimnia and in shallow mixed lakes (Nürnberg 1995, 1996, 2004). Standard errors of the regression coefficients are given in parentheses—z, mean depth (m); A_o, lake surface area (km²); TP_{annual}, annual water-column average TP (µg/L); TP_{summer}, epilimnetic growing period average TP (µg/L); TN_{summer}, epilimnetic growing period average TN (µg/L); L_{ext}, external TP load (mg/m²/yr).

Model	R^2	*P*-value	N
-36.2 (5.2) + 50.1 (4.4) log (TP_{summer}) + 0.762 (0.196) $z/A_o^{0.5}$	0.67	<0.001	70
-35.4 (5.1) + 44.2 (4.3) log (TP_{annual}) + 0.950 (0.187) $z/A_o^{0.5}$	0.65	<0.001	73
-39.9 (9.7) + 27.0 (4.0) log (L_{ext})	0.76	<0.001	17
-173 (22) + 73.0 (8.6) log (TN_{summer}) + 0.925 (0.272) $z/A_o^{0.5}$	0.62	<0.001	54

Table 3.5 Models to predict sediment release activity or spread and duration of anoxia with areal hypolimnetic oxygen deficit (AHOD) and sediment oxygen demand (SOD).

Predictors	Approach	Specifics	Applicability	Reference
Lake TP, z, A TN, TP load	Regression	Anoxic factor, AF, and active area factor, AA: Spatial and temporal spread as AF or AA in stratified and as AA in polymictic lakes.	Any stratified or polymictic lake, possibly reservoirs.	Equation 3.2, Table 3.4
Lake TP, hypolimnion depth	AHOD = f(TP)	Needs initial DO, stratification onset.	Many lakes, but needs input to specific lake.	Chapra and Canale 1991
Chlorophyll	Regression	Proportion of anoxic samples.	Subgroups of lakes (Secchi, lake elevation, morphometry) yield different relationships.	Yuan and Pollard 2014
	Mechanistic Mass balance	Sed2K, based on particulate organic matter.		Chapra et al. 2014
Many	Simulation	Calibrated on available DO profiles.	Specific to study lake.	Luo et al. 2017
Many	Process-based	Lots of data input (e.g., SOD).	Sub-models used in study lake.	Yuan and Pollard 2014

1997; Moore et al. 2012; Nürnberg et al. 2013b) and internal load (see Section 3.2.1.3). Statistically significant regressions were found between chlorophyll concentration and the proportion of anoxic samples as an estimate of lake anoxia (Yuan and Pollard 2014), which could also serve to model the extent and duration of anoxia in a specific lake (see Table 3.5).

The more detailed, vertically segmented, mechanistic mass balance model Sed2K (see Table 3.5) predicts DO concentration from particulate organic matter diagenesis in lake sediments (Chapra et al. 2014). It was used in combination with other models to describe spread of hypolimnetic anoxia and support management related to eutrophication (Gelda et al. 2013).

Numerous simulation and mechanistic models predict temperature and oxygen profiles in lakes and reservoirs (reviewed in Luo et al. 2017) but the data input and number of calibrated parameters are large compared to simple regression models. Computer methods were developed recently that simultaneously calibrate the parameters of such models. For example, the nine key variables of a

DYRESM-CAEDYM model were simultaneously calibrated to satisfactorily predict DO profiles in a polymictic lake (Luo et al. 2017). CAEDYM (Computational Aquatic Ecosystem DYnamics Model) is a process-based library of water quality, biological and geochemical sub-models driven either by the DYnamic REservoir Simulation Model (DYRESM) or the Estuary and Lake COMputer model (ELCOM) to account for transport and mixing (Mooij et al. 2010).

CE-QUAL-W2 is a 2-D, hydrodynamic and water quality model, widely used in U.S. riverine systems and reservoirs. It can require close to 100 input parameters or variables for its calibration and the extraction of internal loading-related results is difficult. At least one application with respect to hypolimnetic oxygen response to loading variability over time was controversial (Brett et al. 2016; Brett and Arhonditsis 2016; Wells and Berger 2016).

3.2.1.3 P Release from the Active Sediment Area in Complex, Process-Based Ecosystem Models

When RR and AA values or their equivalents are available, their simple multiplication yields an internal load estimate (see Equation 3.1 and Figure 3.1). Most established ecosystem models do not provide separate RR and AA estimates or internal load, but include this information implicitly in the model formulation (see Table 3.6).

Internal fluxes can be divided between influences from macrophytes and macrobenthos—for example, dreissenid mussels, resuspension, and redox-related release—but not all P compartments and contributing fluxes are known or can be experimentally determined (Kim et al. 2013). In fact, Mooij et al. (2010) concluded in their extensive review of ecosystem models that a lack of available empirical data prevents the detailed dynamic modeling of sediment processes and exchange with the water column. Nonetheless, there are numerous mechanistic approaches of sediment diagenesis modeling in both marine and freshwater systems as reviewed in detail by Paraska et al. (2014).

In applications to eutrophic lakes and restoration scenarios, sediments are sometimes included in a sub-model that specifies P release that is dependent on redox or oxygen state (Wang et al. 2003) (see Table 3.6). Detailed sediment diagenesis models that predict changes in RRs depending on various downward fluxes, burial, and chemical kinetics render such models very complex (Jørgensen et al. 1982; Boudreau 1997) (see Table 3.6). Therefore, it is challenging to obtain necessary data for testing, modification, calibration, and validation of such process-driven models (Mooij et al. 2010).

Models in combination with monitoring data can help separate fluxes. For example, using the simulation model STELLA (see Table 3.6), observed increases in orthophosphate during the late summer months were determined to be caused by processes other than wind-driven sediment resuspension, most likely redox-related sediment P release (Lee 2015).

Approaches that link empirical information with conceptual (mechanistic) relationships include *structural equation modeling* (see SEM in Table 3.6), which is a multivariate statistical method that can evaluate a network of relationships (Arhonditsis et al. 2006). SEM determined a significant correlation between epilimnetic depth and phosphate concentration in Lake Mendota (Wisconsin), which was explained by entrainment of internal load from the anoxic hypolimnion. Studies on shallow eutrophic lakes determined empirical relationships between variables involved in sediment-based reduction processes (organic matter, proteins, carbohydrates, and lipids mediated by extracellular enzymes) and P release from ferrous hydroxides using SEM (Li et al. 2016). Such studies could lead to explicit internal loading models and may also help testing or even parameterizing the sediment diagenesis models (Paraska et al. 2014).

Table 3.6 Examples of complex, process-based ecosystem models

Approach	Specifics	Evaluation (advantage/disadvantage)	Reference
3D, dynamic	3 sediment P components oxic/anoxic sediment layers.	Simple mechanistic, sediment-water exchange.	Wang 2003
Diagenesis	Thermodynamic, kinetics.	Complex mechanistic sediment-water.	Jørgensen et al. 1982, Boudreau 1997
SPIEL	Amount and composition settling particles, DO, NO3. Sediment-water interface temperature. Benthic boundary layer thickness and sediment properties.	Deep stratified lakes only. To evaluate restoration effect. Many complicated formulas.	Schauser et al. 2004
SWITCH	4 (transient) sediment layers (oxic to anoxic, bioturbation), e.g., 3 phosphate forms.	Many complicated formulas, differential equations, >27 pre-set variables.	Smits and van der Molen 1993
PCLake	Simulation, sediment top layer (10 cm), Models switch between *clear* and *turbid* state (Scheffer and Nes 2007).	Eutrophic shallow lakes (Dutch) restoration.	Mooij et al. 2010
MyLake	1-D simulation model that includes snow conditions.	Boreal lakes, includes uncertainty analysis.	(Saloranta and Andersen 2007).
STELLA	Simulation model, determines increased SRP from sediment P release rather than wave action.	Requires extensive calibration.	Lee 2015
CAEDYM	Sediment DO, redox, P flux	Process-based sub-model.	Luo et al. 2017
Delft 3D-ECO	Sediment/water exchange; includes FLOW model.	Complex; includes sub-models.	Mooij et al. 2010
CE-QUAL-W2	2-D, hydrodynamic and water quality model, especially in reservoirs.	Controversial application to a reservoir study.	Brett et al. 2016
SEM	Internal load entrainment in Lake Mendota.	Combines process understanding with statistical error evaluation.	Arhonditsis et al. 2006
SEM	Sediment/water interactions in 29 eutrophic shallow lakes.	Combines process understanding with empirical information.	Li et al. 2016

3.2.2 Whole Lake Models and P Retention

Internal loading occurs from the bottom sediment directly into the overlaying water and sediment-derived P is distributed throughout the lake immediately or after varying delays. Shallow lakes that mix on a daily or weekly basis (polymictic or oligomictic) often show influences of internal load such as continuous increases in water column P concentration and phytoplankton biomass throughout the summer and fall (Nürnberg 2009) and can simply be considered a *continuously stirred tank reactor* or a *one-box model* (Chapra 1997). But deeper lakes deviate from the simple one-box model because they stratify during part of the growing period. This means that lake TP settles during different periods in stratified and mixed lakes depending on their morphometry (see Section 3.2.3.1).

In all lakes, a settling process occurs that counteracts the flux out of the sediment. Therefore, approaches that separate upward (internal load) from downward fluxes of TP seem to be the most promising for the modeling of internal load (see Table 3.7).

Retention models have been used since Vollenweider's steady state mass balance modeling efforts (Vollenweider 1975) and are some of the most studied subjects concerning lake models, as

Table 3.7 P retention and mass balance models (examples)

Predictors	Approach	Specifics	Evaluation (advantage/ disadvantage)	Reference
Retention, R (proportion per year)				
tau (yr), q_s (m/yr)		Mass in and out without consideration of hypolimnetic oxygen state. R_{net} = (in-out)/in	Can overestimate P in lakes without internal load and underestimate in lakes with internal load.	*Reviewed in*: Nürnberg 1984, Brett and Benjamin 2008
q_s (m/yr)		Specifically parameterized on lakes with oxic hypolimnia. R_{pred} = 15/(18+q_s)	May underestimates R in reservoirs; more R models available.	Nürnberg 1984, 2009
Internal load				
		Net L_{int} = L_{ext} x (R_{pred} − R_{net}) Gross L_{int} = L_{ext} x (R_{pred} − R_{net})/(1 −R_{pred})	Needs estimates of P mass balance components.	Nürnberg 2009
Complex mass balance				
Many	LakeMab	In- and outflow, sedimentation, resuspension, diffusion, bio-uptake, retention.	Lots of calibration and input needed.	Håkanson 2006, reviewed in Mooij et al 2010
	LEEDS	Lake eutrophication, effect, dose, sensitivity.	Same as above.	Håkanson 2000
		Time dynamic, also includes macrophytes, dreissenids.	Bay of Quinte of Lake Ontario.	Kim et al 2013
		Fish catch for biomanipulation, annual temperature variability.	Finnish shallow lake.	Nürnberg et al. 2012
	Bayesian	Hierarchical strategy to relax the assumption of cross-system information applications to individual lakes.	Statistically defensible	Shimoda and Arhonditsis 2015
	Stratification factor	Ratio of the annual mean outflow versus annual mean lake TP concentration.	Needs information of lake TP, which is to be predicted, and outflow TP.	Hupfer et al. 2017

previously reviewed (Nürnberg 1984; Brett and Benjamin 2008). Retention (R) is typically expressed as a proportion (or percentage) of the total external input that is retained (R_{net} = (in-out)/in), which is a net quantity. Sometimes retention is expressed in units of mass (cf. Hupfer et al. 2017), which is equivalent to the multiplication of a proportional R with external load.

The classic Vollenweider model and many subsequent steady-state mass balance models were developed from lakes with and without sediment P release and hence predict net retention (R_{net}) where both upward and downward P fluxes are considered. Therefore, their use in internal load modeling (only upward flux) is questionable (see Table 3.7). Instead, a retention model was developed on lakes without internal load with oxic hypolimnia, R_{pred}, which represents only the downward or settling flux (see Table 3.7). The mass balance is typically based on an entire year so that the difference between the predicted and the computed retention is the upward flux or net internal loading for that year (see Table 3.7). Assuming that R_{pred} applies to internal load in a similar fashion as to external load (Nürnberg 2009), the net internal load can be turned into a gross annual load (see Table 3.7) so that it is comparable to internal load estimates of Section 3.2.1.3 and external load.

Some more complex process-based mass balance models incorporate changes with time and space in dynamic models. For example, many models were developed by Håkanson, including Lake-Mab, a dynamic mass balance model (using differential equations) and the more comprehensive Lake Eutrophication Effect Dose Sensitivity (LEEDS)-model to predict lake eutrophication, effect, dose, and sensitivity (Håkanson 2000) (see Table 3.7).

Also, many process-based ecosystem models adhere to the constraints of mass balance and conservation of mass. But because of their complexity, they rarely explicitly distinguish internal sources from in-lake processes and external input. One advantage of complex ecosystem models is that they are often time dynamic so that they can predict climate change effects or evidence and permanence of restoration efforts (Lewis et al. 2007). They are also typically multidimensional and can address complex systems, including large lakes and reservoirs (Mooij et al. 2010; Robson 2014). They are usually developed for specific lakes, conditions, or problems (Malmaeus and Rydin 2006; Lewis et al. 2007). For example, a model PCLake, developed for shallow, highly eutrophic Dutch lakes (Mooij et al. 2010) does not include a depth dimension and therefore, is probably not applicable to deep, stratified lakes, while the model SPIEL was developed for deeper stratified lakes (Schauser et al. 2004) and hence would not apply to shallow polymictic lakes, and the model MyLake accounts for cold and snow conditions and is especially useful in northern latitudes (Saloranta and Andersen 2007) (see Table 3.6).

Restricted applicability can yield increased model fit to specific conditions and lake types. Usually, initial and boundary conditions have to be provided and the model must be calibrated. Further, dynamic simulation and partially mechanistic models typically require a large amount of data input that may not be available and has to be modeled empirically itself. This requires a lot of computer activity, which is hard to follow so that error detectability is low. Because of these disadvantageous properties of most complex, process-based ecological models, their testing and verification are especially important, but not done routinely (Arhonditsis and Brett 2004) (see Section 3.2.4).

3.2.3 Applications

The comparison of the quantity of internal with external P loads provides guidance in lake management decisions concerning total maximum daily loads in the United States (Steinman and Ogdahl 2015) and similar activities elsewhere (Grüneberg et al. 2011; Copetti et al. 2017). But most applications involve the modeling of lake P concentration as discussed next.

3.2.3.1   Prediction of Lake TP Concentration and Water Quality Variables for Management and Restoration Options

The most direct use for internal P load predictions is its application in a mass balance to model lake TP concentration. The predicted TP concentrations are then often used as input into models such as TP-algal biomass and TP-Secchi transparency regression models (Nürnberg 1996).

When predicting summer or growing-period mixed surface layer TP, the lake stratification status has to be considered because of the differing access of sediment-derived P to the upper water column (see Section 3.2.2). Assuming an exposure to settling—expressed as R_{pred}—of internal load to a various degree, TP concentrations can be predicted for different periods in stratified and mixed lakes (see Table 3.8). In contrast, Formula A (as shown in Table 3.8) is not a model but just the mass balance equation that expresses the annual average TP concentration.

The modeled TP values have been used in various lake studies (Nürnberg 2009). These models envision different influences from external and internal P loads at different times in lakes with differing stratification. For example, gross estimates of the sum of external and internal load divided by annual water load to represent an annual average input concentration and reduced by the proportion that settles (R_{pred}) predicts the annual water column average in all lakes (see Formula B in Table 3.8). While the mixed-layer summer average, which is the variable of most concern, is

Table 3.8   Predictive equations for TP averages for different periods in stratified and mixed lakes (Nürnberg 1998, 2009). External (L_{ext}) and internal (L_{int}) load values are gross estimates (mg/m²/yr). $L_{int\_in\_situ}$ is determined from in situ TP increases throughout the growing season and is a partially net estimate. $R_{net} = (in-out)/in$, $R_{pred} = 15/(18+q_s)$ with q_s, annual areal water load (m/yr) (see Table 3.7). Formula A is not a model but just the mass balance equation.

Formula	Predicted TP averages for different periods and water layers	
	Stratified lake	**Polymictic (mixed) lake**
Internal load not explicitly incorporated: A $TP = \dfrac{L_{ext}}{q_s} \times (1 - R_{net})$	Annual water column	Annual, water column, probably also annual mixed surface layer
Internal load as gross estimate: B $TP = \dfrac{L_{ext} + L_{int}}{q_s} \times (1 - R_{pred})$	Annual water column, Close to summer epilimnetic	Annual water column; minimum of growing period
C $TP = \dfrac{L_{ext}}{q_s} \times (1 - R_{pred}) + \dfrac{L_{int}}{q_s}$	Maximum fall	Maximum growing period TP, close to fall average.
Internal load as partially net estimate (in situ increases): D $TP = \dfrac{L_{ext}}{q_s} \times (1 - R_{pred}) + \dfrac{L_{int\_in\_situ}}{q_s}$	Summer epilimnetic	Growing period
Internal load not considered: E $TP = \dfrac{L_{ext}}{q_s} \times (1 - R_{pred})$	Minimum summer epilimnetic, close to early summer	Not applicable, underestimate

closely predicted in stratified lakes (n = 33, paired t-test was not significant at p = 0.383 and the slope was not significantly different from 1.0) (Nürnberg 2009), it is usually underestimated in polymictic lakes using Formula B because of the different timing of exchange between the surface and bottom layers.

In the model for a theoretically possible maximum internal load influence in either lake type, internal load is not exposed to any settling (see Formula C in Table 3.8). This maximum can theoretically be reached at fall turnover in a mono- or dimictic lake and late during the growing period in a polymictic lake, but usually Formula C overestimates observations (Nürnberg 1998). If the estimate of internal load already includes settling and is therefore reduced (partially net estimate) when determined from *in situ* increases throughout the growing season, the epilimnetic summer concentration in stratified lakes and the growing period concentration in polymictic lakes are adequately predicted (see Formula D in Table 3.8). When internal loading is not considered in the mass balance model (see Formula E in Table 3.8), the early summer minimum TP concentration is often closely predicted in stratified lakes, but TP is underestimated in polymictic lakes (Nürnberg 1998).

In contrast to the models that include internal load as presented here, the often-cited Vollenweider Model (Vollenweider 1976) consistently underestimated observed summer epilimnetic TP averages in stratified lakes where internal load is large (n = 44, paired t-test was significant at p < 0.05 and the slope was 0.697 with SE = 0.07 instead of 1.0 for a perfect prediction (Nürnberg 2009).

Other approaches to predict lake TP concentration in lakes with elevated hypolimnetic TP consider a *stratification factor*, that is, the quotient of the annual mean outflow and annual mean P concentration of the lake (Hupfer et al. 2017), while internal load is not explicitly determined.

When assessing remediation options in P-replete systems, typically individual P sources are quantified and their comparable effects on lake water quality determined. Effects of internal P loading are often underestimated in comparison to external loading effects because a net estimate (see Section 3.2.2) is compared to the gross estimate of external load, and because its chemical form is phosphate, which was found to be more than 80% biologically available when hypolimnetic samples were added to surface samples (Nürnberg and Peters 1984). This high bioavailability is probably the reason for its triggering effect on cyanobacteria blooms as determined with increasing frequency in studies on cyanobacteria proliferation (Smith et al. 2011). Therefore, going beyond the predictions of seasonal lake TP of Table 3.8, P concentrations from internal and external sources can be estimated separately (Nürnberg and LaZerte 2016b) and can be used to evaluate their specific effects on the probability of cyanobacteria blooms.

Often a prediction of treatment success or of the duration of treatment benefits is the goal for modeling internal load. Some of the more complex ecosystem models try to forecast such phases (see Table 3.6), but results are rarely verified (see Section 3.2.4). Further, the usefulness of complex models in predicting seasonal TP concentration variability can be questioned. For example, a model comparison study for the U.S. Everglades wetlands revealed that a multicompartment, mechanistic biogeochemical model did not significantly improve simulated time series TP concentrations compared to a simpler two-compartment model (Min et al. 2011).

3.2.3.2   Modeling of Climate Effects on Internal Load

Warming of surface waters has led to larger temperature gradients and longer and more stable summer stratification, which means larger internal loads in three alpine Austrian lakes (Ficker et al. 2017). Similar climate change effects on internal loading have been hypothesized in many studies (Planas and Paquet 2016). But longer stratification can also reduce the P transport from the hypolimnion, which has led to an apparent oligotrophication of deep Lake Zürich, Switzerland (Yankova et al. 2017). Therefore, internal load modeling has to include various potential climate effects to be able to predict water quality in future.

A combined physical lake (PROBE) and mechanistic phosphorus model (LEEDS) was applied to several temperature scenarios to examine the difference in responses of Swedish lakes with different physical characteristics (Malmaeus et al. 2006). The lake with the longer residence time and hence more influential internal load was predicted to become eutrophic faster in response to climate change-related temperature increases than the lakes with less internal loading effects.

Even relatively simple, steady state, mass balance models can determine lake P response to climate change, as long as they incorporate a dependency for lake water temperature on internal load as in Nürnberg et al. (2012). Models that incorporate physical changes in stratification duration and anoxia (e.g., using AA) in addition to temperature effects on RR, permit the prediction of internal load changes due to climate changes.

3.2.4   Model Verification

The most direct way of verifying a model is the comparison of predicted with observed data and subsequent statistical analysis using accepted measures of data fit (Arhonditsis and Brett 2004). Observed internal P load estimates can be determined from *in situ* hypolimnetic P increases during hypolimnetic anoxia in deep lakes (see Chapters 1, 2, and 6), even though such estimates should be considered underestimates because some sedimentation would have occurred (Nürnberg 2009). However, internal loading cannot easily be quantified from seasonal increases in polymictic lakes. Instead of observed *in situ* internal load estimates, lake TP concentration is the variable often used in internal load modeling verification. When using observed TP concentration for model verification, the variability of mixed layer P concentrations throughout the growing period has to be considered (e.g., Table 3.8) and external inputs have to be accounted for.

The assumption that cross-system information (empirical models such as in Table 3.1, Table 3.4, Table 3.7, and Table 3.8) applies to an individual lake has been challenged repeatedly (Shimoda and Arhonditsis 2015). The applicability of models developed on other lakes to a specific lake has to be verified by using statistically robust comparisons of observed with modeled data. For example, the sub-model to predict AA in a stratified lake can be compared to AF determined from dissolved oxygen profiles (see Section 3.2.1.2). If re-parameterization is needed after calibration, further testing has to be added (e.g., split years, etc., Nürnberg 2009). Other ways for supporting simple empirical models include the adherence to the model assumptions, range of important model input variables consistent with the original lake data set, and consideration of the original statistics concerning predictability (e.g., regression parameter, R^2, level of significance, and confidence intervals) (see Table 3.4).

Bayesian hierarchical strategy is another strategy to relax the assumption of cross-system information applicability to individual lakes (Shimoda and Arhonditsis 2015) and has been used in various types of predictive models in many studies (Arhonditsis et al. 2006; Cheng et al. 2010; Kim et al. 2013). This approach requires extensive effort and has been used mainly in large, complex, and societally important systems.

Another way of improving the confidence in internal load values is the use of several different model approaches to obtain independent estimates. In this way a range of possible results is obtained and, if all estimates are reasonably similar, one can be more confident in their validity. This principle led to various approaches used in stratified and shallow lakes (Nürnberg 2009; Nürnberg et al. 2012, 2013a, 2019; Nürnberg and LaZerte 2016a) and marine systems (Klump and Martens 1987).

Model robustness can be evaluated by including sensitivity and uncertainty analysis, for example, as in a simulation model for northern lakes (Saloranta and Andersen 2007). In all modeling exercises, the need for testing stresses the importance of adequate monitoring data, which can be summarized in different ways (Brett et al. 2016) or may not be available at all. This leads to suggestions for better integration between data collection and modeling efforts (Min et al. 2011).

3.3   CONCLUSIONS

The choice of a specific model for predicting internal P load depends on data availability, financial and technical resources to support such tasks, and the application history of specific models, among other constraints. In my view, the most important part is the possibility of model verification considering available resources and obeying the principle of parsimony. It makes sense to use a more robust but less detailed model to outline the most likely predictions. Subsequently, more dynamic and complex modeling may be useful. Such modeling may support specific input into the simpler model, for example, the determination of settling velocity in a calcium-precipitating reservoir informed the retention model (Nürnberg 2009). Of the many possible applications, lake management and climate change effects on water quality, including cyanobacteria blooms, may be the most frequent reasons for modeling internal P load.

3.4   ACKNOWLEDGMENTS

I am grateful to Bruce LaZerte and editor Alan Steinman for useful discussions and comments.

3.5   REFERENCES

Arhonditsis, GB and Brett, MT. 2004. Evaluation of the current state of mechanistic aquatic biogeochemical modeling. Mar Ecol Prog Ser. 271:13–26.

Arhonditsis, GB; Stow, CA; Steinberg, LJ; Kenney, MA; Lathrop, RC; McBride, SJ; and Reckhow, KH. 2006. Exploring ecological patterns with structural equation modeling and Bayesian analysis. Ecol Model. 192:385–409.

Boström, B; Andersen, JM; Fleischer, S; and Jansson, M. 1988. Exchange of phosphorus across the sediment-water interface. Hydrobiologia. 170:229–244.

Boudreau, BP. 1997. Diagenetic models and their implementation: modelling transport and reactions in aquatic sediments. Berlin, Heidelberg: Springer Berlin Heidelberg.

Brett, MT; Ahopelto, SK; Brown, HK; Brynestad, BE; Butcher, TW; Coba, EE; Curtis, CA; Dara, JT; Doeden, KB; Evans, KR; et al. 2016. The modeled and observed response of Lake Spokane hypolimnetic dissolved oxygen concentrations to phosphorus inputs. Lake Reserv Manag. 32:243–255.

Brett, MT and Arhonditsis, GB. 2016. Modeling the dissolved oxygen response to phosphorus inputs in Lake Spokane: the fallacy of using complex over-parameterized models as the basis for TMDL decisions. Lake Reserv Manag. 32:280–287.

Brett, MT and Benjamin, MM. 2008. A review and reassessment of lake phosphorus retention and the nutrient loading concept. Freshw Biol. 53:194–211.

Carter, LD and Dzialowski, AR. 2012. Predicting sediment phosphorus release rates using landuse and water-quality data. Freshw Sci. 31:1214–1222.

Chapra, SC. 1997. Surface water-quality modeling. Boston: McGraw-Hill Companies, Inc.

Chapra, SC and, Canale, RP. 1991. Long-term phenomenological model of phosphorus and oxygen for stratified lakes. Water Res. 25:707–715.

Chapra, SC; Gawde, RK; Auer, MT; Gelda, RK; and Urban, NR. 2014. Sed2K: modeling lake sediment diagenesis in a management context. J Environ Eng. 141:04014070.

Cheng, V; Arhonditsis, G; and Brett M. 2010. A revaluation of lake-phosphorus loading models using a Bayesian hierarchical framework. Ecol Res. 25:59–76.

Copetti, D; Salerno, F; Valsecchi, L; Viviano, G; Buzzi, F; Agostinelli, C; Formenti, R; Marieri, A; and Tartari G. 2017. Restoring lakes through external phosphorus load reduction: the case of Lake Pusiano (Southern Alps). Inland Waters. 7:100–108.

Davison, W. 1982. Transport of iron and manganese in relation to the shapes of their concentration-depth profiles. Hydrobiologia. 92:463–471.

Davison, W; Grime, GW; Morgan, JAW; and Clarke, K. 1991. Distribution of dissolved iron in sediment pore waters at submillimetre resolution. Nature. 352:323–325.

Ding, S; Chen, M; Gong, M; Fan, X; Qin, B; Xu, H; GA o, S; Jin, Z; Tsang, D; and Zhang, C. 2018. Internal phosphorus loading from sediments causes seasonal nitrogen limitation for harmful algal blooms. Sci Total Environ. 625:872–884.

Ding, S; Jia, F; Xu, D; Sun, Q; Zhang, L; Fan, C; and Zhang, C. 2011. High-resolution, two-dimensional measurement of dissolved reactive phosphorus in sediments using the diffusive gradients in thin films technique in combination with a routine procedure. Env Sci Technol. 45:9680–9686.

Doig, LE; North, RL; Hudson, JJ; Hewlett, C; Lindenschmidt, K-E; and Liber, K. 2016. Phosphorus release from sediments in a river-valley reservoir in the northern Great Plains of North America. Hydrobiologia.

Ficker, H; Luger, M; and Gassner, H. 2017. From dimictic to monomictic: Empirical evidence of thermal regime transitions in three deep alpine lakes in Austria induced by climate change. Freshw Biol. 62:1335–1345.

Gelda, RK; Owens, EM; Matthews, DA; Effler, SW; Chapra, SC; Auer, MT; and Gawde, RK. 2013. Modeling effects of sediment diagenesis on recovery of hypolimnetic oxygen. J Environ Eng. 139:44–53.

Grüneberg, B; Rücker, J; Nixdorf, B; and Behrendt, H. 2011. Dilemma of nonsteady state in lakes—development and predictability of in-lake P concentration in dimictic Lake Scharmützelsee (Germany) after abrupt load reduction. Int Rev Hydrobiol. 96:599–621.

Håkanson, L. 2000. The role of characteristic coefficients of variation in uncertainty and sensitivity analyses, with examples related to the structuring of lake eutrophication models. Ecol Model. 131:1–20.

Hu, W. 2016. A review of the models for Lake Taihu and their application in lake environmental management. Ecol Model, 40th Anniversary of Ecological Modelling Journal. 319:9–20.

Hupfer, M; Reitzel, K; Kleeberg, A; and Lewandowski, J. 2017. Long-term efficiency of lake restoration by chemical phosphorus precipitation: scenario analysis with a phosphorus balance model. Water Res. 97:153–161.

James, WF. 2017. Diffusive phosphorus fluxes in relation to the sediment phosphorus profile in Big Traverse Bay, Lake of the Woods. Lake Reserv Manag. 1–9.

Jørgensen, SE; Kamp-Nielsen, L; and Mejer, HF. 1982. Comparison of a simple and a complex sediment phosphorus model. Ecol Model. 16 99–124.

Kim, D-K; Dong-Kyun Zhang, W; RA o, YR; Watson, S; Mugalingam S; Labencki T; Dittrich M; Morley A; Arhonditsis, GB. 2013. Improving the representation of internal nutrient recycling with phosphorus mass balance models: A case study in the Bay of Quinte, Ontario, Canada. Ecol Model. 256:53–68.

Klump, JV, and Martens. CS. 1987. Biogeochemical cycling in an organic-rich coastal marine basin. 5. Sedimentary nitrogen and phosphorus budgets based upon kinetic models, mass balances, and the stoichiometry of nutrient regeneration. Geochim Cosmochim Acta. 51:1161–1173.

Komatsu, E; Fukushima, T; and Shiraishi, H. 2006. Modeling of P-dynamics and algal growth in a stratified reservoir—mechanisms of P-cycle in water and interaction between overlying water and sediment. Ecol Model. 197:331–349.

Lee, TA. 2015. Influences of the environment and plankton community interactions on toxic cyanobacterial blooms in Vancouver Lake, Washington, a temperate shallow freshwater system (Ph.D., Washington State University, USA).

Lewis, GN; Auer, MT; Xiang, X; and Penn, MR. 2007. Modeling phosphorus flux in the sediments of Onondaga Lake: Insights on the timing of lake response and recovery. Ecol Model. 209:121–135.

Li, H; Song, C-L; CA o, X-Y; and Zhou, Y-Y. 2016. The phosphorus release pathways and their mechanisms driven by organic carbon and nitrogen in sediments of eutrophic shallow lakes. Sci Total Environ. 572:280–288.

Loh, PS; Molot, LA; Nürnberg, GK; Watson, SB; and Ginn, B. 2013. Evaluating relationships between sediment chemistry and anoxic phosphorus and iron release across three different water bodies. Inland Waters. 3:105–117.

Luo, L; Hamilton, D; Lan, J; McBride, C; and Trolle, D. 2017. Auto-calibration of a one-dimensional hydro-dynamic-ecological model (DYRESM 4.0-CAEDYM 3.1) using a Monte Carlo approach: simulations of hypoxic events in a polymictic lake. Geosci Model Dev Discuss. 1–26.

Malmaeus, JM; Blenckner, T; Markensten, H; and Persson, I. 2006. Lake phosphorus dynamics and climate warming: a mechanistic model approach. Ecol Model. 190:1–14.

Malmaeus, JM; and Rydin, E. 2006. A time-dynamic phosphorus model for the profundal sediments of Lake Erken, Sweden. Aquat Sci. 68:16–27.

Min, J-H; Paudel, R; and Jawitz, JW. 2011. Mechanistic biogeochemical model applications for Everglades restoration: a review of case studies and suggestions for future modeling needs. Crit Rev Environ Sci Technol. 41:489–516.

Mooij, WM; Trolle, D; Jeppesen, E; Arhonditsis, G; Belolipetsky, PV; Chitamwebwa, DBR; Degermendzhy, AG; DeAngelis, DL; Domis, LNDS; Downing, AS; et al. 2010. Challenges and opportunities for integrating lake ecosystem modelling approaches. Aquat Ecol. 44:633–667.

Moore, BC; Cross, BK; Beutel, M; Dent, S; Preece, E; and Swanson, M. 2012. Newman Lake restoration: a case study Part III. Hypolimnetic oxygenation. Lake Reserv Manage. 28:311–327.

Nürnberg, GK. 1984. The prediction of internal phosphorus load in lakes with anoxic hypolimnia. Limnol Oceanogr. 29:111–124.

Nürnberg, GK. 1988. Prediction of phosphorus release rates from total and reductant-soluble phosphorus in anoxic lake sediments. Can J Fish Aquat Sci. 45:453–462.

Nürnberg, GK. 1994. Phosphorus release from anoxic sediments: what we know and how we can deal with it. Limnetica. 10:1–4.

Nürnberg, GK. 1995. Quantifying anoxia in lakes. Limnol Oceanogr. 40:1100–1111.

Nürnberg, GK. 1996. Trophic state of clear and colored, soft- and hardwater lakes with special consideration of nutrients, anoxia, phytoplankton and fish. Lake Reserv Manage. 12:432–447.

Nürnberg, GK. 1998. Prediction of annual and seasonal phosphorus concentrations in stratified and polymictic lakes. Limnol Oceanogr. 43:1544–1552.

Nürnberg, GK. 2002. Quantification of oxygen depletion in lakes and reservoirs with the hypoxic factor. Lake Res Manage. 18(4):299–306.

Nürnberg, GK. 2004. Quantified hypoxia and anoxia in lakes and reservoirs. The Scientific World. 4:42–54.

Nürnberg, GK. 2009. Assessing internal phosphorus load – problems to be solved. Lake Reserv Manage. 25:419–432.

Nürnberg, GK. 2019. Quantification of anoxia and hypoxia in water bodies (2). In: Water Encyclopedia. John Wiley & Sons, Inc.

Nürnberg, GK; Howell, T; and Palmer, M. 2019. Long-term impact of Central Basin hypoxia and internal phosphorus loading on north shore water quality in Lake Erie. Inland Waters. 9:362–373.

Nürnberg, GK, LaZerte, BD. 2016a. More than 20 years of estimated internal phosphorus loading in polymictic, eutrophic Lake Winnipeg, Manitoba. J Gt Lakes Res. 42:18–27.

Nürnberg, GK, LaZerte, BD. 2016b. Trophic state decrease after lanthanum-modified bentonite (Phoslock) application to a hyper-eutrophic polymictic urban lake frequented by Canada geese (*Branta canadensis*). Lake Reserv Manage. 32:74–88.

Nürnberg, GK; LaZerte, BD; Loh, PS; Molot, LA. 2013a. Quantification of internal phosphorus load in large, partially polymictic and mesotrophic Lake Simcoe, Ontario. J Gt Lakes Res. 39:271–279.

Nürnberg, GK; Molot, LA; O'Connor, E; Jarjanazi, H; Winter, JG; and Young, JD. 2013b. Evidence for internal phosphorus loading, hypoxia and effects on phytoplankton in partially polymictic Lake Simcoe, Ontario. J Gt Lakes Res. 39:259–270.

Nürnberg, GK and Peters, RH. 1984. Biological availability of soluble reactive phosphorus in anoxic and oxic freshwater. Can J Fish Aquat Sci. 41:757–765.

Nürnberg, GK; Tarvainen, M; Ventelä, A-M; and Sarvala, J. 2012. Internal phosphorus load estimation during biomanipulation in a large polymictic and mesotrophic lake. Inland Waters. 2:147–162.

Paraska, DW; Hipsey, MR; and Salmon, SU. 2014. Sediment diagenesis models: Review of approaches, challenges and opportunities. Environ Model Softw. 61:297–325.

Paytan, A; Roberts, K; Watson, S; Peek, S; Chuang, P-C; Defforey, D; and Kendall, C. 2017. Internal loading of phosphate in Lake Erie Central Basin. Sci Total Environ. 579:1356–1365.

Planas, D and Paquet, S. 2016. Importance of climate change-physical forcing on the increase of cyanobacterial blooms in a small, stratified lake. J Limnol. 75.

Psenner, R and Pucsko, R. 1988. Phosphorus fractionation: advantages and limits of the method for the study of sediment P origins and interactions. Arch Hydrobiol Suppl. 30:43–59.

Reitzel, K; Hansen, J; Andersen, FØ; and Jensen, HS. 2005. Lake restoration by dosing aluminum relative to mobile phosphorus in the sediment. Env Sci Technol. 39:4134–4140.

Robson, BJ. 2014. State of the art in modelling of phosphorus in aquatic systems: review, criticisms and commentary. Environ Model Softw. 61:339–359.

Saloranta, TM and Andersen, T. 2007. MyLake—A multi-year lake simulation model code suitable for uncertainty and sensitivity analysis simulations. Ecol Model. 207:45–60.

Schauser, I; Hupfer, M; and Brüggemann, R. 2004. SPIEL—a model for phosphorus diagenesis and its application to lake restoration. Ecol Model. 176:389–407.

Scheffer, M and Nes, EH. 2007. Shallow lakes theory revisited: various alternative regimes driven by climate, nutrients, depth and lake size. Shallow Lakes Chang World, Developments in Hydrobiology. 584:455–466.

Shimoda, Y and Arhonditsis, GB. 2015. Integrating hierarchical Bayes with phosphorus loading modelling. Ecol Inform. 29:77–91.

Smith, L; Watzin, MC; and Druschel, G. 2011. Relating sediment phosphorus mobility to seasonal and diel redox fluctuations at the sediment-water interface in a eutrophic freshwater lake. Limnol Oceanogr. 56:2251–2264.

Søndergaard, M; Bjerring, R; and Jeppesen, E. 2012. Persistent internal phosphorus loading during summer in shallow eutrophic lakes. Hydrobiologia. 1–13 %U http://link.springer.com/article/10.1007/s10750-012-1091–1093.

Steinman, AD and Ogdahl, ME. 2015. TMDL reevaluation: reconciling internal phosphorus load reductions in a eutrophic lake. Lake Reserv Manag. 31:115–126.

Vollenweider, RA. 1975. Input-output models. Schweiz Z Für Hydrol. 37:53–84.

Vollenweider, RA. 1976. Advances in defining critical loading levels for phosphorus in lake eutrophication. Mem Ist Ital Idrobiol. 33:53–83.

Wang, H; Appan, A; and Gulliver, JS. 2003. Modeling of phosphorus dynamics in aquatic sediments: I—model development. Water Res. 37:3928–3938.

Wells, SA and Berger, CJ. 2016. Modeling the response of dissolved oxygen to phosphorus loading in Lake Spokane. Lake Reserv Manag. 32:270–279.

Yankova, Y; Neuenschwander, S; Köster, O; and Posch, T. 2017. Abrupt stop of deep water turnover with lake warming: drastic consequences for algal primary producers. Sci Rep. 7.

Yuan, LL and Pollard, AI. 2014. Classifying lakes to quantify relationships between epilimnetic chlorophyll a and hypoxia. Environ Manage. 55:578–587.

CHAPTER 4

UNDERSTANDING THE DRIVERS OF INTERNAL PHOSPHORUS LOADING IN LAKES

Martin Søndergaard and Erik Jeppesen[1]

Abstract

Internal loading or release of phosphorus (P) from lake sediments can be an important factor determining lake water quality in both deep and shallow lakes. In deep lakes, anoxic conditions increase the accumulation of phosphate in the hypolimnion during stratification, which eventually may reach the epilimnion during mixing. In shallow and mixed lakes with oxic water throughout the column, internal P loading can also be significant when the P retention capacity in the uppermost thin and oxidized part of the sediment is exceeded. Generally, internal P loading is a highly complex process that involves a number of biogeochemical mechanisms operating on different spatial and temporal scales. Phosphorus may be released from the surface sediment due to metabolic processes occurring during diagenesis or from deeper parts of the sediment. Here, we provide an overview of the key drivers of internal P loading in lakes. Overall, the importance of internal loading depends on a number of interacting mechanisms, including external P loading, the types and amounts of P accumulated in the sediment, the mobilization from particulate to dissolved P forms, and the transport mechanisms of P from the sediment to the water column.

Key words: Lake, sediment, phosphorus, oxygen, redox potential

4.1 INTRODUCTION

Internal loading of P may be an important and, in shallow lakes, dominant source of P during most of the growing season. As P is often regarded as a key limiting nutrient for primary producers in lakes, internal P loading can have a very significant impact on the overall function of lakes, including

---

[1] Department of Bioscience, Aarhus University, Denmark. E-mail: ms@bios.au.dk

the food web structure, and thereby substantially affect their ecological state and water quality. The importance of internal P loading has been documented in multiple studies conducted in both deep and shallow lakes (for examples, see Chapters 6–22 in this book).

Internal P loading is considered particularly important in lakes where the external P loading has been recently reduced, resulting in diminished P retention in the sediment; in such cases the sediment may even be a net source of P (Søndergaard et al. 2013). This internal loading can delay lake recovery for typically 10 to 15 years, but sometimes longer (Jeppesen et al. 2005; see also Chapter 15 about Lake Søbygaard). In eutrophic shallow lakes, internal P loading may persist, particularly in summer, even though the lake acts as a sink on a yearly scale (Søndergaard et al. 2013). It is, however, important to remember that internal P loading, unless fed via sediment by groundwater, does not represent release of *new* P to the lake. It is a recycling of P having entered the lake earlier. Ultimately, therefore, the external loading of P defines the amount of available P in a lake over the long term.

The significance of internal P loading in eutrophic shallow lakes is illustrated on the left side of Figure 4.1. As summer proceeds, the total P increases steadily and in late summer concentrations that are as much as two to three times higher than in winter are reached. This constant release of P from the sediment feeds directly into the euphotic layer and thus further stimulates phytoplankton growth and primary production. The actual P increase in the lake water due to internal P loading depends on a number of factors, such as the hydraulic retention time—in rapidly flushed lakes, P concentrations in the lake water will mainly reflect the inlet concentrations. The magnitude of the seasonal variation in internal P loading also depends on the level of eutrophication and trophic structure. In meso-oligotrophic shallow lakes, for example, internal P release does not cause elevated total phosphorus (TP) levels during summer, possibly because the oxidized sediment surface receives enough light to stimulate benthic primary production and/or growth of and P uptake by submerged macrophytes (see Figure 4.1, upper left). Moreover, high grazing from zooplankton helps to prevent an excessive growth of phytoplankton (Jeppesen et al. 2005).

In deep and summer-stratified lakes in temperate and subtropical regions, the seasonal changes in lake water P differ from the pattern in shallow lakes (see the right side of Figure 4.1). Here, P decreases over summer as particulate P, often in the form of phytoplankton—then dies and settles, thereby transporting P from the epilimnion to the hypolimnion/sediment. During stratification, some of this P, when mineralized and released from the sediment, accumulates in the hypolimnion as orthophosphate and is then transported to the surface water in autumn when the surface water temperature declines and the lake water becomes fully mixed. However, even in dimictic lakes with permanent summer stratification, there may be some interaction between the hypolimnion and the epilimnion during thermocline erosion and periodic upwelling of hypolimnetic water (Weinke and Biddanda 2018).

The effect of internal P loading also depends on morphological features. Even though lakes stratify during summer, a substantial part of the lake bottom may still be above the thermocline, and these shallow parts of the lake will exhibit the usual characteristics of a shallow lake with a significant benthic-pelagic coupling and potential high internal P loading. Medium-deep lakes may stratify temporarily when a thermocline establishes for short periods—days or weeks—during calm and warm weather, and even some shallow lakes in subtropical/tropical climate may temporarily stratify on a 24-hour period. In eutrophic lakes, a relatively short period of stratification may suffice to induce development of anoxic conditions at the sediment-water interface or even in the lower water column (Wilhelm and Adrian 2008). Anoxic conditions increase P release from the sediment, implying that temporarily stratified lakes may be even more significantly affected by internal loading than both polymictic and dimictic lakes.

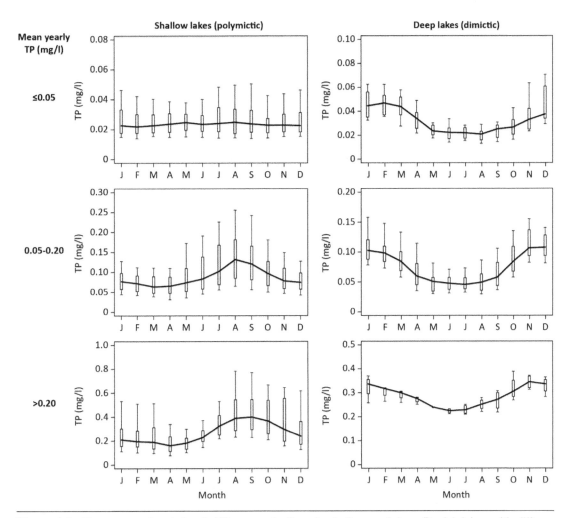

Figure 4.1 The seasonal variation in TP concentrations in shallow and deep Danish lakes with different mean yearly TP concentrations. The shallow and polymictic lakes include 14 lakes (mean depth = 0.8 to 4.6 m, total lake months = 4104). The deep and dimictic lakes include four lakes (mean depth = 3.6 to 16.5 m, total lake months = 945). Deep lakes with TP > 200 µg L^{-1} are represented by only 48 lake months (Lake Furesøen 1991–1994). Boxes represent 10, 25, 75, and 90% fractiles, median values being connected. Data were collected from 1989 to 2015. See Søndergaard et al. (2017) for description of methods and monitoring program. Note different scales on the y-axis.

4.2 INTERNAL P LOADING

The drivers of internal P loading have been the subject of multiple studies during the past decades and several reviews are available (Boström et al. 1988; Søndergaard et al. 2003; Orihel et al. 2017). Overall, internal P loading is regarded as a highly complex process involving a great variety of interacting physical and biogeochemical mechanisms operating on different spatial and temporal scales. New techniques have revealed significant sub-millimeter-scale heterogeneity in the spatial pattern of labile P, illustrating the complexity of biogeochemical processes also at the very fine scale (Meng et al. 2018; see Chapter 2).

4.2.1   Gross Versus Net Flux of P

As addressed in Chapter 1, the term *internal P loading* is not used consistently in the literature and it is important to keep this in mind when describing the drivers behind internal P loading. One of the main reasons for this inconsistency is probably that the P exchange between sediment and water is a dual process involving P release from the sediment and P sedimentation. The difference between these two processes represents a net flux of P (which can be negative if the sediment retains P), while the P release from the sediment represents a gross release of P. A large gross release of P may be counterbalanced by a similar high sedimentation rate and, if this is the case, there will be no net flux of P from the sediment.

Gross P release rates are often estimated in laboratory experiments involving incubation of sediment cores under different experimental conditions for a number of days or weeks where the release or uptake of P is measured. The net release of P can be estimated by mass balance calculations of the total inlet and outlet of P from the lake during, for example, one year. The difference between the gross release and the net flux of P and the resulting effects in a shallow lake during a two-week period exhibiting low phytoplankton biomass and thereby low sedimentation rates is illustrated in Figure 4.2.

4.2.2   Origin of Internal P Loading

The P released from the sediment of a lake originates from different sources, one of which is the mineralization of organic-bound P at the sediment surface; this P is then released as phosphate that may be transported back to the water column. Mineralization includes multiple biogeochemical mechanisms and is a process that may influence the patterns of other release mechanisms, and it will often change over the season depending on, among other things, temperature and the production and sedimentation of fresh organic matter from the water column above. The classical theory of sediment-water interactions states that an oxidized microzone can prevent or reduce the release of P because P is bound to oxidized iron hydroxides. However, the thickness of this oxidized microzone may diminish during summer at high mineralization rates (and use of electron acceptors as oxygen) and increase during winter at low mineralization rates (oxygen and other oxidizers, such as nitrate, can penetrate deeper into the sediment before they are used, thereby keeping the sediment oxidized). In this way, the sediment surface may demonstrate a seasonal retention capacity pattern where less P is retained (or net release occurs) during summer compared with the winter situation (see Figure 4.3).

P released from the sediment also may originate from deeper parts where it is transported from a pool accumulated in the lake's past—often in periods when the external nutrient loading and the retention of P were higher (Carey and Rydin 2011). Lakes with high external loading of P and iron have a particularly high capacity to retain P. The release of legacy P is steady and long-term and is usually the reason why a lake, also on a yearly basis, suffers from internal P loading for a number of years. So, the P released from sediment at a given time may originate from different sources and parts in the sediment (see Figure 4.3), where short-term release mainly depends on seasonal factors and long-term release mainly on the amount of P transported from deeper parts of the sediment. Phosphorus originating from deeper parts of the sediment may be trapped in the oxidized surface layer if there is capacity to take up this P; otherwise, the P continues upward and ends in the water column. Usually, at least the upper 10 cm of the sediment is regarded as interacting with the water above, but some studies have shown that P may be transported upward from as far down as 20 to 25 cm depth (see Chapter 15).

The active part of the sediment interacting with the water above is highly lake specific depending on sediment composition, loading history, and morphological conditions such as water depth.

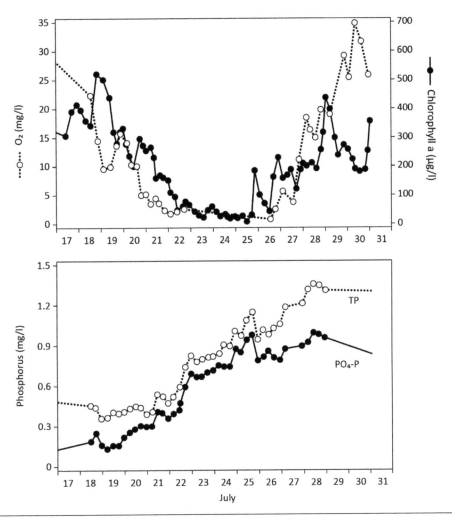

Figure 4.2 Illustration of changes in hypertrophic and shallow Lake Søbygaard during a phytoplankton collapse (see Chapter 15 for more data and details on Lake Søbygaard). As the phytoplankton collapse occurs, concentrations of O_2 and chlorophyll a decrease in the water column (top panel). Simultaneously with the reduction in DO, there is an increase in both total P and phosphate in the water column (bottom panel). After a few days, the DO and chlorophyll a in Lake Søbygaard returned to pre-collapse concentrations and the P concentrations stabilized. Under normal conditions, a large gross release of P will be counteracted by a similar-sized large sedimentation rate of particulate P bound in phytoplankton, and lake water P concentrations will remain more or less constant. However, following a phytoplankton collapse, the downward transport (sedimentation) will be more or less eliminated and a continuously high gross release of P will give rise to high net internal loading and an increase in lake water P concentrations. Thus, during just one week in Lake Søbygaard, TP concentrations increased from 0.5 mg L^{-1} to 1.2 mg L^{-1}, corresponding to a net increase of about 100 mg m^{-2} day^{-1}. Based on Søndergaard et al. (1990).

Furthermore, lake sediments can be highly heterogeneous—not just vertically but also horizontally, which must be considered when evaluating the importance of internal loading in lakes. It is also important to remember that any sediment surface eventually will be covered by newly settled material. This will change the environment over time toward more reduced conditions and consequently influence the original adsorption and desorption mechanisms of P.

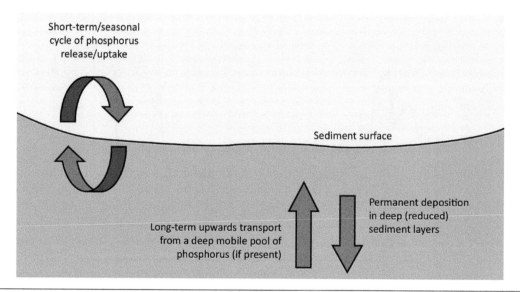

Figure 4.3   Illustration of short- and long-term release of phosphorus originating from different sources and caused by different mechanisms in the sediment.

4.3   DRIVERS OF INTERNAL P LOADING

There are many potential and diverse drivers of P release from the lake sediment, and these may function in different and often interacting ways. Overall, the internal loading of P and its magnitude depend on three main aspects and their interactions: (1) *mobile P*—a releasable or mobile pool of P must be present; (2) *mobilization*—P from the mobile pool must be mobilized; and (3) *transport*—mobilized P must be transported from the sediment from where it is mobilized to the water above (see Figure 4.4). All three aspects are influenced by a number of interacting and constantly changing physical and biogeochemical mechanisms.

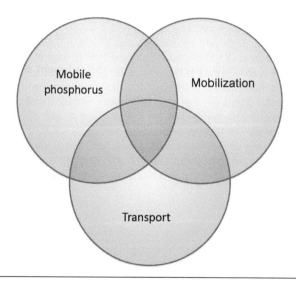

Figure 4.4   Conceptual model of the three main aspects driving internal P loading.

The concept of including these three aspects can be exemplified with the classical iron-P release mechanisms (see later): 1) mobile P in the form of iron-bound P must be present; 2) P must be mobilized from iron complexes, for example via redox-dependent release when oxidizers (oxygen, nitrate) disappear; and 3) upward transport must occur, for example by diffusion, as pore water with high phosphate concentrations migrates toward pore water with lower phosphate concentrations at the sediment-water interface.

4.3.1   P Forms—Mobile P in the Sediment

The total amount of P in the sediment is normally much higher than in the water column. Even if considering only the upper 10 to 15 cm of the sediment, the total amount here will often be 100 to 1000 times larger than that in the lake water, and also much higher than the total yearly external loading of P to a lake. Typically, the highest P concentrations are found in the surface sediment. This is due to the larger presence of non-refractory organic-bound P and the higher sorption capacity caused by the more oxidized conditions near the surface. In some lakes, it may also be due to a recently higher external loading than in the past, resulting in greater retention in the upper part of the sediment.

A number of different chemical fractionation schemes have been developed, by which the different types of P present in the sediment are separated into a number of inorganic and organic components (Psenner et al. 1984; Jensen and Thamdrup 1993; Lukkari et al. 2007). Usually, the relative amount of organic-bound P decreases with increasing depth of the sediment since most of the organic matter originally present in the surface sediment is eventually mineralized. Organic-bound P in deeper parts of the sediment thereby represents mainly refractory (residual) P that is permanently deposited in contrast to the more labile organic-bound P in surface sediment, which may be mineralized and released. The inorganic fractions are often separated into four to six different types of P that are extracted by using different chemical reagents. Based on such fractionation, a mobile pool of inorganic P can be defined according to how easily the fractions are extracted. Usually, this pool comprises a very labile fraction of loosely bound P (including pore water P) and a fraction of redox-sensitive P (bound to iron or manganese). P can also be present as the stable iron (II) phosphate mineral vivianite ($Fe_3(PO_4)_2*8H_2O$, which regularly occurs in the presence of organic remains in iron-rich sediments (Rothe et al. 2016).

Overall, the different forms of P identified by extraction are defined only by the chemical reagents and extraction methods. The mobile pool of P defined by the extractions can be used to predict the total amount of P potentially releasable from the sediment at a given depth—and with this, the influence and/or magnitude of internal P loading can be estimated. However, because internal P loading is not only a question of mobile P but also of P mobilization and transport, there may not necessarily be a direct relationship between the mobile pool of P and internal loading. In lake modeling, the mobile pool of P may be one of the driving forces used to predict future P concentrations in lake water.

The amount of iron (Fe) in the sediment relative to the amount of P (the Fe:P ratio) is another measure used as a guide to determine under which conditions sediment can be expected to have an additional capacity for binding P under oxic conditions. A Fe:P ratio above 15 (by weight) has been suggested to control P release from oxic sediments in shallow lakes (Jensen et al. 1992). A maximum capacity to bind P implies that P release may occur from iron-rich sediment even under aerobic conditions at the sediment-water interface if the maximum capacity is exceeded.

4.3.2 Mobilization of P

4.3.2.1 Mineralization

When particulate P settles to the sediment, it may be in the form of inorganic P, originating, for example, from river inlets; or as organic-bound P, originating either from primary producers in the lake such as phytoplankton or transported to the lake via inlets. When reaching the sediment, decomposition of the remaining non-refractory organic matter starts by microorganisms and macro-invertebrates living in the sediment through which phosphate is released. Depending on the amount and type of settled organic matter, mineralization can be an important process driving the release of P. The rate of organic matter degradation increases with increased oxygen supply and temperature or due to the presence of other oxidizing agents such as nitrate, producing higher phosphate concentrations in the pore water.

During mineralization, some of the P in the sediment may be accumulated by microorganisms as polyphosphate in different quantities depending on the availability of oxygen (Gächter et al. 1988; Hupfer et al. 2004). Changing environmental conditions may cause release of P from this bacterial pool; polyphosphate metabolism may accordingly contribute to the P release.

4.3.2.2 Redox Potential and Oxygen

The classical mobilization mechanism transforming particulate-bound P to dissolved P as phosphate—first described by Einsele (1936) and Mortimer (1941)—is the redox-sensitive release of P from iron compounds. This occurs when oxidized iron oxy-hydroxides (iron III) adsorbing P are chemically reduced at low oxygen concentrations to the dissolved form of iron (iron II), which does not sorb P. Consequently, phosphate is released to the pore water. Reduction of iron oxides may also be a part of dissimilatory reduction caused by metabolism by anaerobic bacteria. A further description and critical review of this long-lasting paradigm in limnology and the importance of oxygen concentrations are given, for example, by Hupfer and Lewandowski (2008).

Reduction of iron III to iron II takes place when the redox potential decreases to a level below approximately 200 mV. Through a comparable mechanism, manganese-bound P may be released via reduction of manganese IV to manganese II. The redox potential describes the overall capacity of a solution (or sediment) to receive or donate electrons (Søndergaard 2009). The redox potential depends strongly on oxygen concentrations and 200 mV corresponds to an oxygen concentration of about 0.1 mg L^{-1} or about 1% oxygen saturation. Therefore, even low oxygen concentrations will prevent redox-dependent P mobilization. That the redox potential is highly dependent on oxygen means that the redox-dependent release and uptake of P fluctuate according to changes in oxygen concentrations. In dimictic lakes, accumulation of phosphate in the hypolimnion begins when all the oxygen has been consumed in respiration processes. This is a process lasting days, weeks, or months depending on the volume of the hypolimnion, the level of eutrophication, and the amount of easily degradable organic matter in the sediment. The presence of other electron acceptors such as nitrate can also keep iron in its oxidized form (Beutel et al. 2016).

Shallow lakes experience large fluctuations in oxygen concentrations at the sediment-water interface, and this also affects the retention of P. These fluctuations can take place on a seasonal (Jensen et al. 1992) or even a diurnal scale by light-dark variations (Gerhardt et al. 2005). Seasonally, lower temperatures and lower sedimentation rates of organic matter during winter decrease the oxygen consumption in the sediment, allowing oxygen and other oxidizers to penetrate deeper into the sediment, with a consequent increase in the overall redox-sensitive P sorption. In clear shallow lakes with a significant benthic primary production, comparable changes in oxygen concentrations at the sediment-water interface may be seen on a diurnal scale, with even hourly fluctuations. In the presence of light during daytime, primary production enables oxygen to penetrate several millimeters into the sediment, while during night the extent of the oxidized zone decreases (Wetzel 2001). Macroinvertebrates such as chironomid larvae and tubificid worms living in the surface sediment may augment the influx of oxygen and other oxidizing substances to the sediment, thereby increasing the redox potential and improving the redox-sensitive binding of P (Zhang et al. 2010). Macroinvertebrates may, however, also increase the mineralization rate and thereby the potential release of P.

If a shallow lake changes from a turbid to a clear water state, for example due to successful biomanipulation involving removal of zooplanktivorous and benthivorous fish, the system will shift from one with benthic rather than pelagic dominance and a higher P retention capacity in the surface sediment. There are several examples demonstrating how this fundamentally changes P dynamics and markedly increases the overall retention of P, also on a yearly scale (Søndergaard et al. 2002; Jensen et al. 2017; Søndergaard et al. 2017). In shallow lakes, dense submerged macrophyte vegetation can prevent water mixing, which may create stratification and a mini-scale oxygen profile similar to that in deep, temperature-stratified lakes. Depending on macrophyte density, such small-scale patchiness may produce a chemical micro-scale structure with increased P levels not far below the lake surface (Barker et al. 2010).

4.3.2.3 Sulfate Reduction

Sulfate reduction by sulfate-reducing bacteria using sulfate as terminal electron acceptor in the mineralization of organic matter usually takes place a few centimeters below the sediment-water interface where oxygen and nitrate are depleted and sulfate is still present. A product of sulfate reduction is hydrogen sulfide (H_2S), which may react with soluble iron(II) to precipitate as iron(II) sulfide (FeS) that can convert into pyrite (FeS_2) during further diagenesis. Both iron(II) sulfide and pyrite represent terminal sinks for iron with low solubility, resulting in low concentrations of dissolved iron(II). This reduces the upward transport of iron(II) and disables or reduces the formation of iron oxy-hydroxides under more oxidized conditions, implying a diminished potential of P binding to iron and an enhanced risk of P release from the sediment.

The impact of sulfate reduction challenges the classical coupling of carbon, iron, and P cycles. As a consequence, the permanent redox-dependent P retention in stratified lakes does not necessarily depend only on the oxygen supply to the sediment surface but also on the balance between the sedimentation of organic matter, P, and iron and sulfide production in the sediment (Gächter and Müller 2003). If iron concentrations are high and sulfate concentrations low, the P release might be low, but if sulfate reduction is high, this may decrease the permanent deposition of P in the sediment and induce enhanced internal P loading.

4.3.2.4 pH

Changes in pH affect the sorption and desorption of phosphate to mineral surfaces, but the overall effect of changed pH on P mobilization is complex and potentially involves many factors and types of minerals. Increased pH may reduce the sorption of inorganic P because increased concentrations of hydroxyl ions compete with phosphate ions and, due to ligand exchange reactions, replace phosphate ions on the surface of iron minerals (Andersen 1975). Consequently, increased pH caused by high photosynthetic activity by phytoplankton may create a positive feedback loop between high phytoplankton biomass and augmented P release from the sediment.

Changes in pH also affect the dissolution or precipitation of carbonates, which may again impact the sorption or binding of P (Golterman 2001). A lower pH can be induced by mineralization processes in sediments as a result of CO_2 production. This may lead to dissolution of calcite and apatite and their sorption of P.

4.3.2.5 Temperature

Temperature influences most chemical and, in particular, biological processes, including those mentioned in this chapter. Accordingly, temperature also indirectly affects the drivers behind many of the mechanisms that are important for internal P loading. Experimentally, it has been demonstrated that higher temperatures reduce the thickness of the oxidized zone in the surface sediment of shallow lakes and increase the release of P (Jensen et al. 1992). This is likely due to increased respiration by heterotrophs.

The significance of temperature for a number of the drivers of internal P loading means that the predicted future climate change with higher temperatures is expected to speed up the release mechanisms. There also may be indirect effects of climate change on the internal nutrient loading through alterations in trophic structure, including for example the fish community with cascading top-down and bottom-up effects (Jeppesen et al. 2009; Jeppesen et al. 2012).

4.3.3 Transport of P

P from the sediment is transported in particulate and dissolved form to the water column. Particulate P is mainly transported via bioturbation and resuspension by wind, whereas dissolved P is transported by chemical diffusion, and this may be enhanced by an eruption of gas produced in deeper parts of the sediment and perhaps also by macroinvertebrate mixing of the upper sediment. The two types of transport mechanisms interact—wind-induced resuspension may, for example, influence the concentrations of phosphate in the pore water.

4.3.3.1 Diffusion

Once phosphate is released from organic or inorganic-bound forms in the sediment to the pore water, it becomes labile and subject to chemical diffusion. The resultant phosphate concentrations in the pore water depend on numerous chemical and biological reactions between dissolved and particulate forms of P and sediment mixing/disturbance. With constantly changing physical and chemical environmental conditions, a true chemical equilibrium between dissolved and particulate forms of P may never be reached.

Chemical diffusion implies direct transport of dissolved phosphate along a concentration gradient. Normally, phosphate in the pore water constitutes only a small fraction of the total P present in the sediment (less than 1%). However, phosphate concentrations are usually much higher in the pore water than in the water column above, and this drives an upward molecular diffusion and transport of P. The importance of diffusion for the overall P release from the sediment varies between seasons and between lakes. The diffusive flux, assuming steady state conditions, can be calculated based on Fick's first law using the gradient in phosphate concentrations and information on the diffusion coefficient, but due to sediment disturbance, the actual release rates are often much higher (see the upcoming paragraphs).

4.3.3.2   Gas Ebullition

Gas ebullition is a mechanism that increases the upward transport rates of P in the sediment. This process is particularly important in organic-rich sediments where non-refractory organic matter is still present in the chemically reduced part and where fermentation processes producing CO_2 and methane (CH_4) may give rise to the formation of gas bubbles. Gas bubbles eventually travel to the surface of the sediment, while simultaneously pulling particles and phosphate in the pore water toward the sediment surface. The effect of gas ebullition on the overall P transport is difficult to quantify, but studies on the emission of methane from lake sediments suggest that gas ebullition may constitute an important part of the whole gas flux in eutrophic lakes, particularly at increasing temperatures and high nutrient levels (Davidson et al. 2018).

4.3.3.3   Bioturbation

Bioturbation is created by macroinvertebrates living in the upper sediment or by benthivorous fish, such as carp, when feeding on invertebrates in the sediment. Bioturbation and resuspension of sediment may also be the result of plant-feeding waterfowl rooting up submerged macrophytes.

Bioturbation affects the transport of both particulate and dissolved P in many ways (Hölker et al. 2015; see Chapter 1). Phosphate may be released from the pore water, and bioturbation by benthic organisms may indirectly have a significant influence on the release mechanism by burrow construction, which induces redox oscillations due to the cyclic redox patterns (Aller 1994). Resuspended particles may release P to, or take up phosphate from, the water depending on equilibrium conditions. Increased turbidity caused by bioturbation may have secondary effects on the internal P loading due to reduced light availability at the sediment surface (mentioned earlier). The presence of microphytobenthos, algae living in the surface sediment layers, may attenuate the effects on P release of bioturbating fauna in shallow lakes (Benelli et al. 2017).

4.3.3.4   Resuspension by Wind

In shallow, exposed lakes, the sediment is often influenced by wind-induced resuspension, creating more turbid water (Kristensen et al. 1992; Qin et al. 2006). In deep lakes, seiche-induced resuspension may occur. Resuspension increases the contact between sediment particles and the water above, and P-binding particles will be exposed to an environment very different from that in the sediment. This may induce further mineralization of organic matter in the usually oxidized water column as well as create changes in the equilibrium conditions between the P bound to inorganic particles. P may then be released or trapped depending on the phosphate concentration in the water column and the equilibrium conditions (Søndergaard et al. 1992). In some lakes and during parts of the year, resuspension by wind is thought to be of key importance for the P cycling because it is many times greater than the diffusive release (Tammeorg et al. 2015).

Resuspension may also act as a pump and thus increase the transport of P within the sediment by enhancing the diffusive flux and the transport of phosphate from the pore water. In this way, resuspension fluxes can exceed the diffusive flux of P and be the main P transport mechanism in the upper part of the sediment (Reddy et al. 1996).

4.4   CONCLUSIONS

Internal P loading is a complex but very important process influencing the P availability in most lakes. Release of P from the sediment depends on a number of drivers and involves multiple interacting biogeochemical processes that may differ from lake to lake. An overview of important drivers and factors affecting the internal P release is presented in Table 4.1.

Table 4.1 Overview of drivers and important factors influencing internal P loading in lakes

Factors	Mechanisms Involved	Main Effects/Examples
Overall factors	• External P loading	• Defines the overall P availability. After reduced external loading, internal loading from the sediment can be a major net P source.
	• Lake morphology	• Water depth and lake size define mixing or stratification with significant effects on P dynamics.
	• Hydraulic retention time	• Influence on the relative importance of internal P loading.
P forms	• Depth distribution	• Legacy P may exist as a pool accumulated in deeper parts of the sediment when P loading was higher.
	• Size of P pool	• More P → higher potential P release.
	• Type of P	• Large redox sensitive P pool → higher potential P release.
P mobilization	• Mineralization	• Release of organic-bound P to pore water as phosphate.
	• Redox conditions	• Reduction of iron(III) and release of P due to anoxic conditions.
	• Sulfate reduction	• Immobilization of Fe due to production of H_2S during sulfate reduction.
	• pH	• Higher P release from iron-bound P due to increased pH.
	• Temperature	• Increased mineralization rate and oxygen demand. Enhances all biological processes.
P transport	• Diffusion	• Higher phosphate in pore water → higher release rates.
	• Gas ebullition	• Increased upward flux in the sediment.
	• Bioturbation	• Can have both negative and positive effects on internal P loading.
	• Wind resuspension	• Increased contact between sediment particles and water.

4.5 ACKNOWLEDGMENTS

The project was supported by the EU project MARS (Managing Aquatic Ecosystems and Water Resources under Multiple Stress) funded under the 7th EU Framework Programme. Anne Mette Poulsen and Juana Jacobsen are acknowledged for editorial and layout assistance.

4.6 REFERENCES

Aller, RC. 1994. Bioturbation and remobilization of sedimentary organic matter: effects of redox oscillation. Chem Geol. 114:331–345.
Andersen, JM. 1975. Influence of pH on release of phosphorus from lake sediments Arch Hydrobiol. 76:411–419.

Barker, T; Irfanullah, H; and Moss B. 2010. Micro-scale structure in the chemistry and biology of a shallow lake. Freshw Biol. 55:1145–1163.

Benelli, S; Bartoli, M; Zilius, M; Vybernaite-Lubiene, I; Ruginis, T; Petkuviene, J; and Fano, EA. 2017. Microphytobenthos and chironomid larvae attenuate nutrient recycling in shallow-water sediments. Freshw Biol. 63:187–201.

Beutel, MW; Duvil, R; Cubas, FJ; Matthews, DA; Grizzard, TJ; Wilhelm, FM; Austin, D; Horne, AJ; and Gebremariam, SY. 2016. A review of managed nitrate addition to improve surface water quality. Crit Rev Environ Sci Technol. 46:673–700.

Boström, B; Andersen, JM; Fleischer, S; and Jansson, M. 1988. Exchange of phosphorus across the sediment-water interface. Hydrobiologia 170:229–244.

Carey, CC and Rydin, E. 2011. Lake trophic status can be determined by the depth distribution of sediment phosphorus. Limnol Oceanogr. 56:2051–2063.

Davidson, TA; Audet, J; Jeppesen, E; Landkildehus, F; Lauridsen, TL; Søndergaard, M; and Syväranta, J. 2018. Synergy between nutrients and warming enhances methane ebullition from experimental lakes. Nat Clim Change 8:156–160.

Einsele, W. 1936. Über die Beziehungen des Eisenkreislaufs zum Phosphatkreislauf im eutrophen See. Arch Hydrobiol. 29(6):664–686.

Gerhardt, S; Brune, A; Schink, B. 2005. Dynamics of redox changes of iron caused by light–dark variations in littoral sediment of a freshwater lake. Biogeochemistry 74(3):323–339.

Golterman, HL. 2001. Phosphate release from anoxic sediments or 'What did Mortimer really write?' Hydrobiologia 450: 99–106.

Gächter, R; Meyer, JS; and Mares, A. 1988. Contribution of bacteria to release and fixation of phosphorus in lake sediments. Limnol Oceanogr. 33:1542–1558.

Gächter, R and Müller B. 2003. Why the phosphorus retention of lakes does not necessarily depend on the oxygen supply to their sediment surface. Limnol Oceanogr. 48(2):929–933.

Hölker, F; Vanni, MJ; Kuiper, JJ; Meile, C; Grossart, HP; Stief, P; Adrian, R; Lorke, A; Dellwig, O; Brand, A; Hupfer, M; Mooij, WM; Nützmann, G; and Lewandowski, J. 2015. Tube-dwelling invertebrates: tiny ecosystem engineers have large effects in lake ecosystems. Ecol Monogr. 85:333–351.

Hupfer, M and Lewandowski, J. 2008. Oxygen controls the phosphorus release from lake sediments—a long-lasting paradigm in limnology. Int Rev Hydrobiol. 93(4–5):415–432.

Hupfer, M; Rübe, B; and Schmieder, P. 2004. Origin and diagenesis of polyphosphate in lake sediments: a 31P-NMR study. Limnol Oceanogr. 49:1–10.

Jensen, HS; Kristensen, P; Jeppesen, E; and Skytthe, A. 1992. Iron:phosphorus ratio in surface sediment as an indicator of phosphate release from aerobic sediments in shallow lakes. Hydrobiologia 235/236(1):731–743.

Jensen, HS and Thamdrup, B. 1993. Iron-bound phosphorus in marine sediments as measured by bicarbonate-dithionite extraction. Hydrobiologia 253:47–59.

Jensen, M; Liu, ZW; Zhang, XF; Reitzel, K; and Jensen, HS. 2017. The effect of biomanipulation on phosphorus exchange between sediment and water in shallow, tropical Huizhou West Lake, China. Limnologica 63:65–73.

Jeppesen, E; Kronvang, B; Meerhoff, M; Søndergaard, M; Hansen, KM; Andersen, HE; Lauridsen, TL; Liboriussen, L; Beklioglu, M; and Özen, A, et al. 2009. Climate change effects on runoff, catchment phosphorus loading and lake ecological state, and potential adaptations. J Environ Qual. 38:1930–1941.

Jeppesen, E; Søndergaard, M; Jensen, JP; Havens, KE; Anneville, O; and Carvalho, L. 2005. Lake responses to reduced nutrient loading—an analysis of contemporary long-term data from 35 case studies. Freshw Biol. 50(10):1747–1771.

Jeppesen, E; Søndergaard, M; Lauridsen, TL; Davidson, TA; Liu, Z; Mazzeo, N; Trochine, C; Özkan, K; Jensen, HS; and Trolle, D, et al. 2012. Biomanipulation as a restoration tool to combat eutrophication: recent advances and future challenges. Adv. Ecol. Res. 47:411–487.

Kristensen, P; Søndergaard, M; and Jeppesen, E. 1992. Resuspension in a shallow hypertrophic lake. Hydrobiologia 228:101–109.

Lukkari, K; Hartikainen, H; and Leivuori, M. 2007. Fractionation of sediment phosphorus revisited. I: Fractionation steps and their biogeochemical basis. Limnol Oceanogr.: Methods 5: 433–444.

Meng, Y; Ding, S; Gong, M; Chen, M; Wang, Y; Fan, X; Shi, L; and Zhang, C. 2018. Submillimeter-scale heterogeneity of labile phosphorus in sediments characterized by diffusive gradients in thin films and spatial analysis. Chemosphere 194:614–621.

Mortimer, CH. 1941. The exchange of dissolved substances between mud and water in lakes. J Ecol. 29(2):280–329.

Orihel, DM; Baulch, HM; Casson, NJ; North, RL; Parsons, CT; Seckar, DCM; and Venkiteswaran, JJ. 2017. Internal phosphorus loading in Canadian fresh waters: a critical review and data analysis. Can J Fish Aquat Sci. 74:2005–2029.

Psenner, R; Pucsko, R; and Sager, M. 1984. Die Fraktionierung organischer und anorganischer Phosphorverbindungen von Sedimenten. Versuch einer Definition ökologisch wichtiger Fraktionen. (Fractionation of organic and inorganic phosphorus compounds in lake sediments: an attempt to characterize ecologically important fractions). Arch Hydrobiol. 70:111–155.

Qin, BQ; Zhu, GW; Luo, LC; GA o, G; and Gu, BH. 2006. Estimation of internal nutrient release in large shallow Lake Taihu, China. Sci China Ser. D 49:38–50.

Rothe, M; Kleeberg, A; and Hupfer, M. 2016. The occurrence, identification and environmental relevance of vivianite in waterlogged soils and aquatic sediments. Earth-Sci Rev. 158:51–64.

Søndergaard, M. 2009. Redox potential. Encyclopedia of Inland Waters. Oxford: Pergamon Press. p. 852–859.

Søndergaard, M; Bjerring, R; and Jeppesen, E. 2013. Persistent internal phosphorus loading during summer in shallow eutrophic lakes. Hydrobiologia 710(1):95–107.

Søndergaard, M; Jensen, JP; Jeppesen, E; and Hald Møller, P. 2002. Seasonal dynamics in the concentrations and retention of phosphorus in shallow Danish lakes after reduced loading. Aquat Ecosys Health Manage. 5:19–29.

Søndergaard, M; Jensen, JP; and Jeppesen, E. 2003. Role of sediment and internal loading of phosphorus in shallow lakes. Hydrobiologia. 506/509(1–3):135–145.

Søndergaard, M; Jeppesen, E; Kristensen, P; and Sortkjær, O. 1990. Interactions between sediment and water in a shallow and hypertrophic lake: a study on phytoplankton collapses in Lake Søbygård, Denmark. Hydrobiologia 191:139–148.

Søndergaard, M; Kristensen, P; and Jeppesen E. 1992. Phosphorus release from resuspended sediment in the shallow and wind exposed Lake Arresø, Denmark. Hydrobiologia. 228:91–99.

Søndergaard, M; Lauridsen, TL; Johansson, LS; and Jeppesen, E. 2017. Repeated fish removal to restore lakes: case study of Lake Væng, Denmark—two biomanipulations during 30 years of monitoring. Water 9(1).

Tammeorg, O; Niemisto, J; Mols, T; Laugaste, R; Panksep, K; and Kangur, K. 2013. Wind-induced sediment resuspension as a potential factor sustaining eutrophication in large and shallow Lake Peipsi. Aquat Sci. 75:559–570.

Weinke, AD and Biddanda, BA. 2018. From bacteria to fish: ecological consequences of seasonal hypoxia in a Great Lakes estuary. Ecosystems 21:426–442.

Wetzel, RG. 2001. Limnology. 3rd Ed. Lake and river ecosystems. San Diego, Calif: Academic Press.

Wilhelm, S and Adrian, R. 2008. Impact of summer warming on the thermal characteristics of a polymictic lake and consequences for oxygen, nutrients and phytoplankton. Freshw Biol. 53:226–237.

Zhang, L; Gu, X; Fanl, C; Shang, J; Shen, Q; Wang, Z; and Shen, J. 2010. Impact of different benthic animals on phosphorus dynamics across the sediment-water interface. J Environ Sci. 22(11):1674–1682.

CHAPTER 5

METHODS FOR THE MANAGEMENT OF INTERNAL PHOSPHORUS LOADING IN LAKES

Miquel Lürling[1,2], Alfons J.P. Smolders[3,4], and Grant Douglas[5]

Abstract

An overview of the most common categories of internal load management is provided. Curtailment of internal recycling of phosphorus (P) is centered on the reduction of excess nutrients in the water column and the control of phosphate release from sediments. This chapter introduces key techniques and approaches for achieving P management objectives—including their advantages, disadvantages, and possible side effects. The first section deals with sediment removal via excavation and dredging and an overview of potential environmental impacts along with cases studies of sediment removal to improve water quality. The next sections address aeration, oxygenation, and hypolimnetic withdrawal to reduce internal loading. Finally, we examine the use of chemical additions to immobilize P with a focus on those materials used to oxidize sediments, liquids to bind P, and solid-phase P-binders.

5.1 INTRODUCTION

Eutrophication is the most important water quality issue worldwide (Downing 2014). Despite eutrophication and its nuisance symptoms having been recognized for many decades, problems related to eutrophication are still a growing concern. There is general consensus that key symptoms of

[1] Department of Environmental Sciences, Wageningen University, P.O. Box 47, 6700 AA Wageningen, The Netherlands. E-mail: miquel.lurling@wur.nl

[2] Department of Aquatic Ecology, Netherlands Institute of Ecology (NIOO-KNAW), P.O. Box 50, 6700 AB Wageningen, The Netherlands

[3] Department of Aquatic Ecology and Environmental Biology, Radboud University Nijmegen, Heyendaalseweg 135, 6525 AJ Nijmegen, The Netherlands. E-mail: a.smolders@science.ru.nl

[4] B-WARE Research Centre, Radboud University Nijmegen, Toernooiveld 1, 6525 ED Nijmegen, The Netherlands

[5] CSIRO Land and Water, Centre for Environment and Life Sciences, Floreat, WA, Australia. E-mail: grant.douglas@csiro.au

eutrophication, including harmful phytoplankton blooms, are exhibiting a worldwide expansion (Heisler et al. 2008; O'Neil et al. 2012; Paerl and Paul 2012). These blooms occur not only in developing countries and countries in transition, where sewerage coverage and sewage treatment is in general less than 10% (Sato et al. 2013; van Loosdrecht and Brdjanovic 2014), but also in countries with strong external load control and state-of-the-art waste water treatment. For example, in The Netherlands, harmful algal blooms are prevalent despite 99.4% of the population being connected to sewerage systems, ranking it number one in the world for sanitation in this respect (OECD 2018). Not all blooms are the result of eutrophication; cyanobacterial blooms and surface scums are, for instance, a regular occurrence in oligotrophic North Patagonian lakes (Nimptsch et al. 2016). Hence, the symptom—high cyanobacterial biomass—may not always be directly linked to external nutrient inputs but may also be caused by legacy internal nutrient load or by the mere physical process of an over-buoyant surface accumulation. The management of eutrophication problems should be based on site-specific diagnoses: a lake system analysis (LSA). In lakes with legacy internal P load or seasonal internal P load as the main driver of eutrophication-related problems, as well as in lakes where diffuse loading is transporting nutrients via ground water, in-lake measures are essential to address the symptoms of eutrophication.

The application of in-lake measures has been met with skepticism in the past by some. Effective in-lake actions are sometimes viewed as being *a last resort* (Ibelings et al. 2016) or have been advised against while prescribing broad-scale external nutrient load control as the silver bullet in eutrophication management (e.g., Paerl 2014; Hamilton et al. 2016; Paerl et al. 2016). Curtailment of external loading is considered a prerequisite for eutrophication control and without it, in-lake measures aimed to reduce internal loading would not be worth considering (Schindler 2006). Restricting external load is generally agreed to be the most obvious action in tackling eutrophication (Cooke et al. 2005). However, it is becoming increasingly clear that eutrophication may not always be effectively addressed by decreasing external nutrient loading alone (Van Liere et al. 1990; Carpenter et al. 1999; Cooke et al. 2005). In fact, in many U.S. lakes that suffer from nonpoint source nutrient pollution, watershed management practices often fail to reduce P inputs to levels that would improve water quality (Osgood 2017). To determine if external P load reduction is the first logical step in eutrophication control, if additional interventions are needed, or if only in-lake measures are the best option to mitigate eutrophication and control eutrophication nuisance, an LSA is required. Nonetheless, such site-specific diagnostics are rare, despite strong advocacy for their need (Cooke et al. 2005; Lürling et al. 2016; Stroom and Kardinaal 2016).

The consequence of a *catchment only* prescription for addressing eutrophication in lakes (Hamilton et al. 2016; Paerl et al. 2016) is that it fails to provide scope for lake managers and water authorities for adequate treatment of site-specific problems (Eviner and Hawkes 2008). Of course, a eutrophication problem is typically an imbalance between a lake's nutrient inputs and outputs; however, not all cases are the result of point-source pollution as previously indicated. Decades of minor diffuse P inputs may result in an internal P load that is orders of magnitude larger than the P inflows. By curtailing the external inflows in this case, it will take multiple decades until the eutrophication issues will be solved, as a result of internal P load. Similarly, in lakes where external load has been reduced effectively, recycling of legacy P between the sediment and water may maintain eutrophication problems (Ryding and Forsberg 1976; Sharpley et al. 2013). The eutrophic state can be further maintained due to the dramatic ecosystem alterations that took place during the process of eutrophication, such as the increase in general phytoplankton biomass, the loss of submerged plants, accumulation of organic matter and nutrients in the sediment, changes in zooplankton community structure to small bodied grazers, and an increase in planktivorous and benthivorous fish biomass (Moss 2010). Inevitably, more actions are needed in combination with external nutrient load control to achieve rapid relief from eutrophication problems. Without additional in-lake measures, recovery may not only take decades to centuries (Søndergaard et al. 1999; Carpenter 2005; Cooke et al. 2005), but in certain ecosystems, such as those in urbanized areas, no other option than in-lake measures may be feasible (Huser et al. 2016). Hence, to meet society's demands, including legislative ones such as the European Water Framework Directive (EU 2000), water authorities and lake managers need a tool box filled with effective measures, both catchment and in-lake in scope, that encompass the collection of unique characteristics of each lake, grounded in a comprehensive LSA. Recent advances and demonstration studies have delivered an impressive evidence base on in-lake measures and we review here those designed for internal P load control. We address briefly the advantages, disadvantages, and possible side effects of the most commonly applied in-lake measures for the control of internal P cycling. The most common strategies of internal load management have been divided into categories which are described in the following order: sediment removal, aeration/oxygenation, hypolimnetic withdrawal, and chemical additions.

5.2   SEDIMENT REMOVAL

There are two main methods for removal of sediment: excavation and dredging (Cooke et al. 2005). *Excavation* requires first that the overlying water is completely removed and subsequently the upper layer of the lake bed is scraped off. *Dredging* is the removal of sediment without drawdown. Sediment removal is one of the standard operating techniques to bring a waterway back to its desired profile, usually for navigational, recreational, or hydrological reasons, but it is also applied to improve chemical and/or ecological water quality, including internal loading control.

5.2.1   Excavation

Excavation is possible in shallow waterways, such as canals or ditches, but also in rivers, in which segments can be isolated using sheet piles to create impoundments facilitating drawdown. In small lakes and ponds, a full drawdown can be achieved (see Figure 5.1). An advantage is that the targeted sediments can be adequately removed and that the excavated sediment has less volume because of the dewatering. Dewatering, however, is also a limiting factor because of both cost and ensuring that the degree of dewatering is sufficient to allow excavation machines to operate efficiently (Cooke et al. 2005). There may also be a need to treat removed water, resulting in added expense (Oldenborg and Steinman 2019). In general, standard excavation equipment is used, including backhoes, bulldozers, and excavators. Insufficient dewatering may result in heavy machines sinking into the muck (see Figure 5.1).

Excavation examples include Campus Lake (3.7 ha) and College Lake (1.4 ha) in the United States, where after drawdown, 24000 m^3 and 4000 m^3 of sediment, respectively, were removed by dragline dredge and moved using trucks (Knaus and Malone 1984). In lakes that experience natural drying out in warm, dry periods, sediment can more easily be removed. For instance, the tropical shallow lakes Udai Sagar and Fateh Sagar (India) experienced massive sediment removal in May/June 2001, 2003, and 2004 when the lakes dried up, which reduced internal P supply and improved water quality (Pandey and Yaduvanshi 2005).

Figure 5.1   Excavation of an urban pond in Heesch, The Netherlands, and in a ditch near Alblasserdam (insert bottom right panel, picture from Alblasserdamsnieuws.nl, Peter Stam).

5.2.2   Dredging

Dredging is the wet removal of sediment, which can be done using mechanical or hydraulic dredges. Common mechanical dredges include the bucket ladder dredge, the grab or clamshell dredge, and the backhoe and front shovel dredge, whereas hydraulic dredges include the plain suction dredge, the cutter suction dredge, and trailing suction hopper dredge (Vlasblom 2003).

Of the mechanical dredges, cranes or excavators are most frequently used. These excavators can be operated from a pontoon or from the water's edge. Cranes are used as grab dredges, as backhoes, and as front shovels (Vlasblom 2003). Dredging by cranes is intermittent as single grabs or buckets are taken that are then emptied at the surface in a truck, on a barge, or in a hopper. Backhoe and front shovels can reach depths of 15 m with bucket sizes varying from 1 to 20 m^3, whereas grabs can reach deeper, with grab sizes reaching up to 200 m^3 (Vlasblom 2003). Standard buckets include the dredging bucket, which is a digging bucket with holes allowing water to escape while retaining the bulk of excavated sediment, the open or closed clamshell, and the visor bucket that is equipped with a revolving valve—the visor—used to close the top of the bucket. The closed clamshell and the visor limit the mixing of dredged materials with water when they are elevated to the surface. The bucket ladder dredge is a floating dredger that uses a chain of buckets to excavate sediment. Filled buckets (30–1200 L) are emptied on top of the ladder and sediment is deposited in a barge (Vlasblom 2003).

Figure 5.2 Two dredging events in an urban pond in Sint-Oedenrode, The Netherlands, using a silt pusher to push the sediment to a collection site from where it is moved into trucks. The left picture shows a manned silt pusher near an excavator bringing the sediment in a truck, while the right picture shows an unmanned silt pusher in operation.

In shallow waters, often in urban areas, silt pushers are used (see Figure 5.2). A silt pusher is a kind of underwater bulldozer equipped with modifiable dozer blades that push sediment toward the collection point, where an excavator then takes out the sediment (see Figure 5.2).

Hydraulic dredges use pumps to transport the sediment slurry from the suction head via pipes either to a nearby collection area or, if direct pumping is not possible, the pumped sediment is collected in barges or hoppers (Vlasblom 2003). Hoppers are usually equipped with valves that can be opened to dump the collected sediment elsewhere, for example, in the sea. Suction dredges can also be used in shallow urban waters (see Figure 5.3). This sediment can either be used in construction, or when polluted, will be dumped in a deep sand excavation (https://www.bodemplus.nl/onderwerpen/wet-regelgeving/bbk/grond-bagger/).

5.2.3   Environmental Impacts

Removing sediment inevitably comes with environmental impacts, such as effects on benthic invertebrate species, water column oxygen depletion, release of pollutants, resuspension of sediments, and increased turbidity (Peterson 1982; Knott et al. 2009; Manap and Voulvoulis 2015). Drawdown and sediment removal in the 2.2 ha Štěpánek fishpond (Czech Republic), for example, strongly decreased diversity (30% of taxa) and abundance (90% of individuals) of littoral invertebrates that only gradually increased over subsequent years (Sychra and Adámek 2011). Veiga et al. (2013) reported on a planned dredging project in an 8 ha lake in Brazil; the removal of 120,000 m³ of sludge was stopped because resuspended matter from dredging caused fish kills and promoted an algal bloom.

The aforementioned dredging in an urban pond (see Figure 5.2) is an example of where inadequate planning and execution of the dredging activities led to undesirable side effects. The use of

Figure 5.3 Example of suction dredging in pond Kienehoef (Sint-Oedenrode, The Netherlands) where sediment and water are directly pumped via pipes into a nearby disposal site for dewatering. The pictures show the suction dredges (red boats) and the black transport pipe equipped with yellow floaters.

a silt pusher and standard excavator to remove the sediment—in total 850 m^3 *in situ* sediment was removed (pers. comm. R. van Otterlo, Water Authority De Dommel, November 6, 2014)—increased water turbidity drastically due to spillage and resuspension, which immediately led to anoxia and fish kills (see Figure 5.4). Municipality and Water Authority officials had forgotten to inform and include the local angling society, who owned the legal fishing rights for the pond, and volunteers tried to rescue as many fish as possible during the dredging. Nonetheless, dozens of fish died (see Figure 5.4).

Negative impacts can, however, be reduced by good planning, including all stakeholders, such as fishing associations, in the preparation phase and application of technical solutions to minimize impacts (e.g., silt curtains, flocculants). Large-scale resuspension of sediments should be managed carefully (Knott et al. 2009), considering both concentration and duration of elevated suspended sediment load (Newcombe and MacDonald 1991).

5.2.4 Pond Molenwiel Case Study

The 2013 pond Molenwiel (The Netherlands) dredging event was closely monitored. Authorities discovered that not all sediment had been removed during the first dredging (March 4–6, 2013) and performed a second dredging (April 3–4, 2013). Both events caused a sharp decline in water column

Figure 5.4   The dredging method in urban pond De Molenwiel (The Netherlands, March 2013) came with (A) spillage and (B) anoxia, (C) prompting volunteers to rescue as many fish as they could, along with (D) high turbidity, and (E) dozens of dead fish.

oxygen concentration and saturation (see Figure 5.5A). After dredging, oxygen conditions gradually improved (first event) or more rapidly improved (second event) reaching supersaturating conditions (see Figure 5.5A). Turbidity strongly increased from 4 NTU before to almost 1100 NTU during the first dredging and from 10 NTU to 500 NTU during the second event, after which turbidity gradually decreased (see Figure 5.5B). Likewise, suspended solid concentrations increased from 7 mg L^{-1} to 708 mg L^{-1} and from 14 to 325 mg L^{-1} during the first and second dredging event, respectively, and then steadily declined again (see Figure 5.5B). Secchi depth, which was more than 50 cm before dredging, declined to about 1 cm during dredging.

The dredging in this pond was conducted to improve water quality and to eliminate nuisance cyanobacterial blooms. In 2013, a green algal bloom, consisting primarily of *Chlamydomonas sp.*, occurred after dredging, while cyanobacteria biomass remained low during summer (see Figure 5.6). In 2014, cyanobacteria bloomed again in the pond, but from 2015 to 2017 they only occasionally reached high densities and their annual mean biomass remained lower than before dredging (see Figure 5.6). The overall mean cyanobacterial chlorophyll-*a* concentration before dredging (from 2006 to 2012) was 47 ($\pm$76) µg L^{-1} and after (from 2013 to 2017) it was 14 ($\pm$39) µg L^{-1}.

Removal of the 850 m^3 of sediment (where 1150 m^3 had been estimated) reduced the sediment P release considerably, but did not have a strong impact on the water column total phosphorus (TP) concentrations (see Figure 5.7). Sediment P release was determined by measuring the phosphate concentration in water overlying the sediment in core incubations. Before dredging, five cores (60 cm long, 6 cm diameter) were collected with an Uwitech core sampler. The water was deoxygenated by bubbling with N$_2$ gas, after which the cores were closed with a rubber stopper. After dredging, seven cores were collected. Four cores were made anoxic and closed with a rubber stopper, while the remaining three remained open to air. The cores were placed for two to three weeks in the dark at room temperature. Every week phosphate was measured. Sediment P release was calculated as

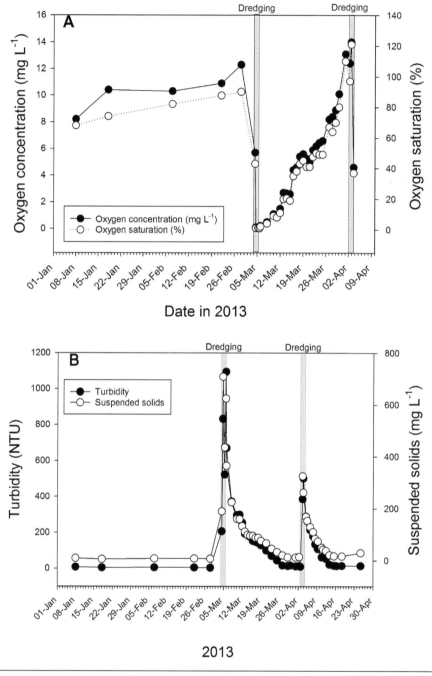

Figure 5.5   (A) Oxygen concentration (mg L^{-1}) and saturation (%) around two dredging events in pond Molenwiel (The Netherlands); (B) turbidity (NTU) and suspended solids concentration (mg L^{-1}).

the difference in phosphate concentration (mg P m^{-3}) multiplied by the water-column height in the cores (m) divided by the corresponding incubation time (d).

The pond did not immediately clear, as was expected. Instead, the water quality in the pond gradually improved. In 2016, sporadic submerged macrophytes (*Ceratophyllum demersum*) were observed,

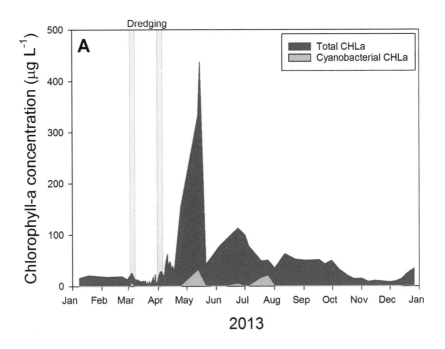

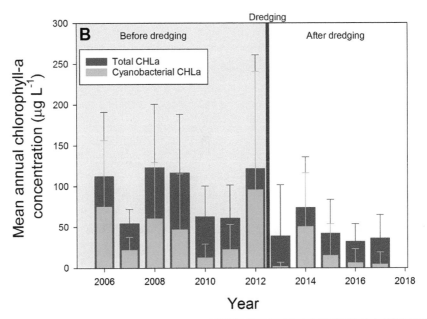

Figure 5.6 (A) Total- and cyanobacteria chlorophyll-*a* concentration (µg L⁻¹) in 2013 and (B) annual means in the period before (2006–2012) and after dredging (2013–2017) in pond Molenwiel, The Netherlands, associated with dredging events.

but these expanded in 2017 to about 10% coverage. Despite these promising developments and a clearly reduced cyanobacterial abundance in the last three years, the external nutrient sources, such as angling bait, and runoff were not controlled and the pond is expected to move back toward a cyanobacteria-dominated state. The dredging costs were around €120,000 (ca. $139,000 US) and total project costs,

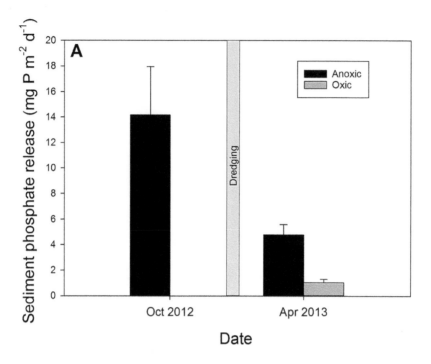

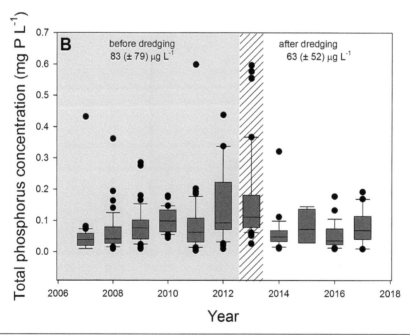

Figure 5.7   (A) Sediment phosphate release in pond Molenwiel, The Netherlands, under anoxic conditions before and after dredging including the release under oxic conditions after dredging (error bars indicate 1 SD) and (B) the water column total P concentrations in the period before dredging (2007–2012) and after dredging (2014–2017). The lower boundary of the boxes indicates the 25th percentile, lines within the boxes mark the medians, and the upper boundary of the box indicates the 75th percentile. Whiskers above and below the boxes indicate the 90th and 10th percentiles, respectively, while the dots above and below the boxes indicate outliers.

which included some softening of the banks and planting helophytes, came to around €200,000 (ca. $241,000 US). The cost of dredging was relatively high and around €140 ($162 US) per m³ sediment, which includes the transportation cost of trucking the collected sediment to a disposal site.

5.2.5   Sediment Removal to Improve Water Quality

In general, the most important reason for sediment removal is to improve navigation and access by physically deepening the water body. Sediment removal is viewed as regular maintenance, yet it is unclear how many projects are being planned and implemented. Sediment removal is also conducted to improve water quality and to restore lost recreational amenities, but information on effectiveness is sparse. Nonetheless, dredging to improve water quality dates back decades (see Table 5.1).

One of the first well-documented cases is Lake Trummen (Sweden: 100 ha), where in 1970 and 1971 a 1 m layer of sediment was removed (600,000 m³), together with 300,000 m³ lake water being pumped into settling ponds (Björk 1972). Water quality improved the years after dredging (Cronberg 1982). In Lake Trehörningen (Sweden: 65 ha) in 1975 and 1976, between 0.2 and 1 m thick sediment layer (320,000 m³) was removed by suction dredging (Ryding 1982). Total P (TP) in the water column was reduced by 50%, but algal dominance remained (Ryding 1982).

Table 5.1 Some literature examples of sediment removal to improve water quality (reduce eutrophication; further specified per case in text)

Site	Size (ha)	Dredge Type	Sediment Removed (m³)	Year	Reference
Lake Wuli	530	Cutter-suction	1,590,000	2002–2003	Liu et al. 2016
Lake Dongqian	1991	Suction	3,295,105	2009	Jing et al. 2015
Lake Javary	8	?	~120,000	2011	Veiga et al. 2013
Lake Hollingsworth	144	?	2,000,000	1997–2001	Poor 2010
Lake Yuehu	61	Suction	610,000	2006	Zhang et al. 2010
South Lake	96	?	807,278	2001–2004	Wang & Feng 2007
Vajgar fish pond	40	Suction	330,000	1991–1992	Pokorný & Hauser 2002
Staroměstský pond	15	Suction	120,000	?	Pokorný & Hauser 2002
Ulický fish pond	4.8	Suction	55,000	?	Pokorný & Hauser 2002
Stěpnický fish pond	10	Suction	45,000	?	Pokorný & Hauser 2002
Kamberk fish pond	20	Suction	170,000	?	Pokorný & Hauser 2002
City Park Lake	24	Hydraulic	100,000	1982–1983	Knaus & Malone 1984
University Lake	84	Hydraulic	360,000	1982–1983	Knaus & Malone 1984
Campus Lake	3.7	Dragline	24,000	1982–1983	Knaus & Malone 1984
College Lake	1.4	Dragline	4000	1982–1983	Knaus & Malone 1984
Lake Trehörningen	65	Suction	320,000	1975–1976	Ryding 1982
Lake Trummen	100	Suction	600,000	1970–1971	Björk 1972

A dredging experiment in the 3.3 ha Cockshoot Broad (UK 1982) removed about 70 cm sediment from about 2.5 ha of Broad and led to years of improved water quality (Moss et al. 1986; 1996). Dredging of four water bodies in the University Lakes System (Baton Rouge, Louisiana, USA) was undertaken from November 1981 to May 1983 removing between 4000 m^3 and 360,000 m^3 per lake of sediment (Knaus and Malone 1984). One of the lakes, City Park Lake (24 ha, 100,000 m^3 sediment removed), showed fewer algal blooms and fish kills the years after dredging, but TP returned to pre-restoration values 10 years later (Ruley and Rusch 2002). Dredging of Lake Geerplas between 1989 and 1991 (The Netherlands: 30 ha) lowered TP concentrations from around 0.4–1 mg L^{-1} to 0.1–0.2 mg L^{-1} (Van der Does et al. 1992). In parts of a polder system named "Wormer, Jisp and Nek" (The Netherlands: 5.7 ha) between November 1989 to February 1990 the top 35 cm of sediment was removed, which led to lower water column nutrients, less phytoplankton, and increased macrophytes compared to non-dredged sites (Hovenkamp-Obbema and Fiegen 1992). In their overview of lake restoration examples in Danish waters, Søndergaard et al. (2000b) listed one dredging case in Lake Brabrand where 500,000 m^3 sediment was removed over a seven-year period. Despite reduced internal P loading, the lake remained in a turbid state (Søndergaard et al. 2000b). Brouwer and Roelofs (2001) reviewed evidence from 15 lakes (between 0.5 and 100 ha) in The Netherlands in which dredging had been a successful restoration measure. Pokorný and Hauser (2002) reported on removal of 330,000 m^3 sediment from a 40 ha fish pond that reduced internal P loading and suppressed cyanobacteria for a few years. These same authors briefly mentioned four fish ponds (4.8–20 ha) where suction dredging was applied to improve water quality. Removal of 2 million m^3 of sediment from Lake Hollingsworth (USA: 144 ha) contributed to improved water quality for at least eight years (Poor 2010). More than 800,000 m^3 sediment was removed from South Lake (China: 96 ha) and with it more than 1880 metric tons of nitrogen (N) and 700 tons of P (Wang and Feng 2007). Dredging of Lake Yuehu (China: 61 ha) led to lower P, organic matter, and algal biomass, and promoted crustacean zooplankton (Zhang et al. 2010). In Lake Dongqian (China: 1991 ha) 33.1% of the sediment of this 1991 ha lake was removed to a depth of 50 cm, but sediment P fractionation showed a return to non-dredged conditions within three years (Jing et al. 2015). In Lake Wuli (China: 530 ha), the effect of dredging on sediment P release was temporary and lasted only about two years due to some years of continued external loading; but once external load was controlled, sediment TP remained < 500 mg P kg^{-1} in 2015 compared to > 2500 mg P kg^{-1} in 2002 prior to dredging (Liu et al. 2016).

In most cases where the positive effects of dredging on water quality were only temporary, ongoing external loading could be identified as the reason for limited longevity of the improved water quality. The majority of the examples in the literature deal with relatively large water bodies (see Table 5.1), but clearly dredging has also taken place in smaller water bodies. In The Netherlands, the costs of dredging activities over the period 2007 to 2027 under the jurisdiction of the National Water Authority (Rijkswaterstaat) are estimated at €332 million (ca. U.S. $370 million), yet with limited or unknown effects on the ecological quality of regional surface waters (Ligtvoet et al. 2008). The regional water authorities list dredging as standard maintenance on their websites; municipalities and lake owners implement dredging to improve water quality too, but in general, these actions are performed without full reporting and effective monitoring. In fact, in the majority of the small water bodies in The Netherlands, mostly urban ponds, surveillance and water quality monitoring are generally lacking (Waajen et al. 2014). For example, since 2006, a few dozen urban ponds have been part of a cyanobacterial bloom monitoring program and eight of them have experienced restoration activities in which dredging was the only, or an important, part of the restoration plan. In all eight ponds, dredging resulted in fewer or no cyanobacterial blooms during some years or up to at least eight years (see Figure 5.8). Further improvements are expected at such sites when combined with reduction in fish-stocking practices, following negotiations with fishing societies, to

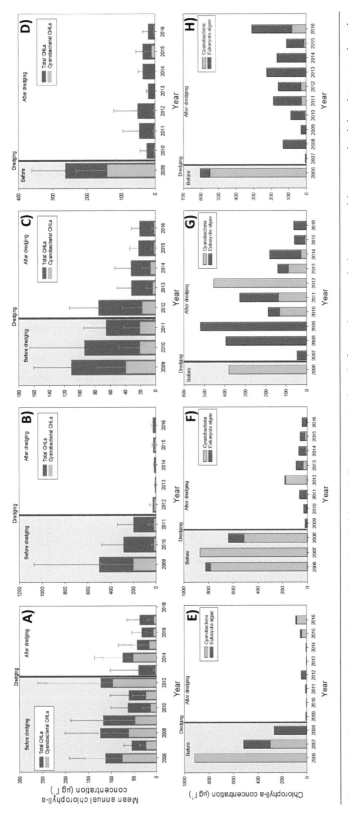

Figure 5.8 Mean annual cyanobacterial- and total chlorophyll-*a* concentrations in four urban ponds (upper row) in a pre-dredging (gray area) and post-dredging period (white area) as well as a summer snap-shot of cyanobacteria- and eukaryotic algal chlorophyll-*a* concentrations in four urban ponds (lower row). All ponds are in The Netherlands: (A) pond Molenwiel (Sint-Oedenrode), (B) pond Dongen (Dongen), (C) pond Stiffelio (Eindhoven), (D) pond Heesch (Heesch), (E) pond Anton van Duinkerkenpark (Bergen-op-Zoom), (F) pond Etten-Leur (Etten-Leur), (G) pond Looveijver (Someren), (H) pond Bennekom (Bennekom).

prevent over-stocking and feeding, which exacerbate poor water quality conditions. Nevertheless, those ponds are not isolated and are subjected to several diffuse inputs of nutrients from the catchment, which will make repeated intervention inevitable.

5.3 AERATION AND OXYGENATION

Iron (Fe) may control sediment P release under oxidizing conditions at the sediment water interface (see Chapter 1) (Smolders et al. 2006). Hypolimnetic aeration and oxygenation are designed to limit the onset of anoxia, and therefore promote the Fe-P trap in sediments, which commonly develops in the hypolimnion of stratified eutrophic lakes. Hypolimnetic aeration uses air to improve oxygen conditions, whereas oxygenation involves addition of pure oxygen. Hypolimnetic aeration was first implemented in Swiss Lac de Bret in 1947 (Mercier 1948; Mercier and Gay 1949). After 70 years of hypolimnetic oxygen enrichments, several reviews assessing effectiveness have been produced (e.g., Fast et al. 1976; Beutel and Horne 1999; Cooke et al. 2005; Singleton and Little 2006; Bormans et al. 2016). The hypolimnetic oxygen enrichers include three categories of devices: (1) airlift aerators, (2) *double bubble* contact systems or Speece cones, and (3) deep oxygen injection systems or bubble plume diffusors (Cooke et al. 2005; Singleton and Little 2006) (see Figure 5.9).

Airlift aerators are run as full-lift or partial-lift systems (see Figure 5.9). Full-lift designs inject compressed air in a cylinder in the deep hypolimnion, after which the bubbles transport the hypolimnetic water to the surface. The oxygen-enriched water is then forced back into the hypolimnion (Beutel and Horne 1999; Cooke et al. 2005; Singleton and Little 2006). Partial-lift systems operate similarly but the water moves only partially up the water column and is then directed back, while the bubbles are funnelled to the water surface (Beutel and Horne 1999; Cooke et al. 2005; Singleton and Little 2006). Double bubble contact systems or Speece cones are submerged, down-flow aerators where oxygen and hypolimnetic water are pumped down the cone. Injection systems consist of

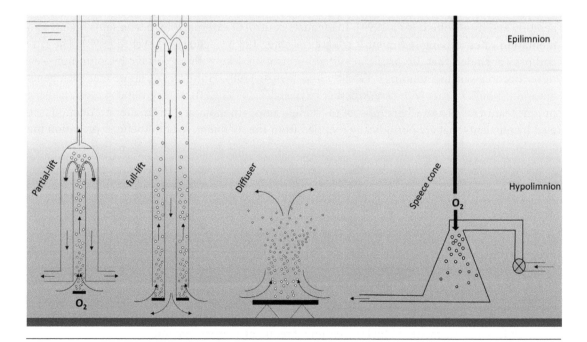

Figure 5.9 Different types of commonly used hypolimnetic aerators (After Cooke et al. 2005)

linear or circular diffusers with relatively low gas flow and small bubbles to prevent destratification (Beutel and Horne 1999; Cooke et al. 2005; Singleton and Little 2006). This is to prevent warming of the sediment, elimination of cold-water fish habitat, and impairment of water use in industry because it has become too warm (Beutel and Horne 1999).

Hypolimnetic oxygen enrichment is one of the most widespread lake restoration measures. Recently, Tammeorg et al. (2017) pointed out that in Finland alone, about 100 lakes are receiving aeration to control anoxia-related sediment P release. Despite many decades of using hypolimnetic oxygen enrichment to control internal P loading, its effects are mixed. Partial-lift and full-lift devices in Medical Lake (USA) were able to reduce hypolimnetic TP concentrations from around 400 µg L^{-1} to around 50 µg L^{-1} during aeration (Soltero et al. 1994). Multilayer aeration in Lake Shenipsit (USA) reduced mean TP by about 67% (Kortmann 1994). Total P in Lake Serraia (Italy) was reduced after hypolimnetic oxygen injection became operational (Toffolon et al. 2013). In contrast, in Lake Waccabuc, hypolimnetic aeration had no effect, which was ascribed to ongoing external P loads (Garrell et al. 1977). In five Minnesota Lakes, destratification aeration led to significantly higher TP concentrations than in six non-aerated control lakes (Beduhn 1994). In Newman Lake (USA) hypolimnetic oxygenation using a Speece cone was attributed as a major factor in lowering TP by reducing internal P cycling (Moore et al. 2012). However, the simultaneous operation of alum microfloc injection could also have had a strong impact on P (Moore et al. 2012). In the same lake, initially the phosphate and TP concentrations increased during aeration (Thomas et al. 1994). Based on their analysis of Finnish lakes, Tammeorg et al. (2017) concluded that hypolimnetic aeration may have limited use in lake restoration and that effectiveness depends on lake-specific characteristics, as is the case with all measures.

From the literature, several reasons emerge regarding why hypolimnetic aeration may not be effective in the control of internal P loading. Not all P may be originating from the anoxic hypolimnion and P from shallow parts of the lake may fuel eutrophication (Tammeorg et al. 2017). Iron availability may be insufficient to precipitate all phosphate rapidly (Lean et al. 1986). Underestimating sediment oxygen demand and overestimating oxygen supply could result in undersized aeration systems (Soltero et al. 1994). Inadequate control of external P load maintains high P sedimentation rates, driving sediment P release (Gächter 1987; Gächter and Wehrli 1998). The latter authors concluded that "an artificial system—eutrophic lake with an aerobic hypolimnion—will never become self-maintaining" (Gächter and Wehrli 1998). However, we argue that such generalization should be met with care; whether oxygenation will be effective or not depends on a lake's unique features. In cases where most of the annual autochthonous organic matter production is fueled by nutrients that are periodically recycled from the sediment, hypolimnetic oxygenation may be sufficient to break this cycle. We demonstrate this with the case study of Lake Ouderkerkerplas (The Netherlands).

Since 2010, the cold hypolimnetic water from Ouderkerkerplas (70 ha), a former sand excavation site with a maximum depth of 40 m, has been used to cool offices in the southern part of Amsterdam during summer months. As a result, the electrical power company NUON achieves a 75% CO_2 emission reduction, but discharge of the cool but nutrient-laden hypolimnetic water back to the epilimnion was not permitted. In response, an oxygen diffusing system was installed 1 m above the bottom to raise hypolimnion oxygen concentrations above 5 mg L^{-1} and to reduce P concentrations. Hypolimnetic and epilimnetic TP concentrations have dropped sharply and in seven years went from 6.2 to less than 1 µmol P L^{-1} (~190 to < 31 µg L^{-1}) (see Figure 5.10). As a consequence, cyanobacterial blooms no longer occur and the strongly reduced pelagic primary production results in much less organic matter to fuel the sediment reducing oxygen demand.

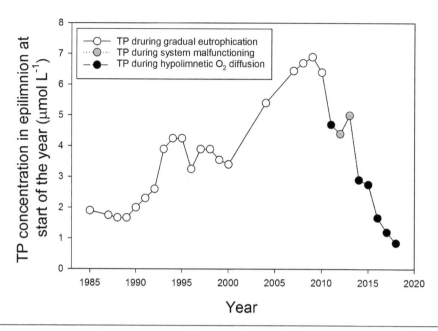

Figure 5.10   Course of the epilimnetic TP concentration in Lake Ouderkerkerplas (The Netherlands). Open symbols indicate the development of TP during gradual eutrophication. Black symbols are years in which hypolimnetic oxygen diffusors were operating, grey symbols indicate some malfunctioning in the system.

5.4   HYPOLIMNETIC WITHDRAWAL

Discharge of hypolimnetic water during late summer or early fall coinciding with the highest nutrient concentrations in the hypolimnion will remove those nutrients (P, NH_4) from the lake. To prevent negative impacts on downstream waters, this withdrawal water preferably undergoes treatment based on chemical precipitation and flocculation of nutrients and metals in a wastewater treatment plant (Nürnberg 2007). If hypolimnetic removal exceeds the nutrient input, it will effectively decrease P or N retention (Nürnberg 2007). In an extensive review, Nürnberg (2007) evaluated all accessible and known cases of hypolimnetic withdrawal up until that time. In most cases, water was withdrawn from the deepest point in the lake, targeting the highest P concentrations during stratification. The timing of withdrawal and the proportion of hypolimnion water being discharged were the main factors regulating effectiveness of the approach (Nürnberg 2007). A prerequisite is that sufficient water input compensates for the volume of withdrawal; Hupfer and Scharf (2002) indicated that the water retention time should be less than five years for this approach to be considered. If destratification and warming of hypolimnetic water can be avoided, and external P loads are controlled, hypolimnetic withdrawal can be viewed as a successful, relatively low-cost technique for internal loading control in stratifying lakes (Nürnberg 2007).

5.5   CHEMICAL ADDITION

The use of chemicals to counteract internal P loading dates back many decades. Materials used include two main categories: (1) chemicals to oxidize the upper sediment layer, and (2) chemicals that inactivate P directly. The basis for chemical inactivation of P lies in the chemical characteristics of phosphate, which can form salts with aluminum, calcium, iron, lanthanum, or other metals that have varying degrees of solubility, while such immobilization for nitrate or ammonium is not possible.

5.5.1 Sediment Oxidation

The oxidation of the upper sediment layer using oxygenated water has been addressed in Section 5.3. The underlying carbohydrate oxidation by bacteria drives oxygen consumption and redox in bed sediments, while providing the bacteria with energy (Wetzel 2001). Oxygen is the most preferred electron acceptor, but if it is depleted due to higher consumption than replenishment via photosynthesis and re-aeration, anoxia develops and bacteria begin to utilize alternative electron acceptors. The sequence of these alternative electron acceptors is based on the energy it will yield to the bacteria. Under anoxic conditions, the first preferred electron acceptor is NO_3^-, followed by Mn(IV), Fe(III), SO_4^{2-}, and finally CO_2 (Wetzel 2001). The sequence of preferred electron acceptors implies that as long as nitrate is available, denitrifying bacteria will drive carbohydrate oxidation before it becomes energetically favorable for dissimilatory manganese and iron reduction. The consequence of Fe(III) reduction is the release of phosphate, bound to ferric iron complexes, to the water (Lijklema 1977). For instance, in Lake Mathews (USA), nitrate available in the anoxic hypolimnion delayed sediment P release by about 45 days following the onset of anoxia (Beutel et al. 2008). In anoxic sediment cores, the delay in the onset of P release was amplified as nitrate input increased, while at a dose of 61 g N m^{-2}, P release was controlled completely and water column P concentrations decreased by 2.4 times (Foy 1986). Likewise, the onset of P release in Upper Mystic Lake (Boston, MA, USA) was apparently triggered by depletion of nitrate, rather than oxygen (Hemond and Lin 2010).

Based on the concept of preferred alternative electron acceptors, calcium nitrate, iron chloride, and slaked lime were injected into the sediment of Lake Lillesjön (Sweden) to reduce easily degradable organic matter and therefore sediment oxygen demand, as well as to improve the binding capacity for P (Ripl 1976). Ten years after intervention, sediment oxygen demand was still low and P release was strongly reduced (Ripl 1986). Calcium nitrate addition to Long Lake (USA) did not improve water quality, which was most probably due to ongoing external loading (Noonan 1986). In Lake Lyng (Denmark), the addition of granulated calcium nitrate was less effective than dissolved calcium nitrate; the phosphate, TP, and iron accumulation in the hypolimnion in treatment years was 35–82%, 45–77% and 31–61%, respectively, of that in non-treatment years, although hypolimnetic ammonium concentrations increased 166–265% (Søndergaard et al. 2000a).

In a laboratory experiment, calcium nitrate additions strongly reduced P in the water column (by 75%) and in the sediment pore water (by 89%) (Yamada et al. 2012). Although after 145 days almost all of the added N (98%) was removed from the system (Yamada et al. 2012), strongly elevated ammonium, nitrite, and nitrate concentrations exerted severe toxicity toward the waterflea *Ceriodaphnia silvestrii* and the midge larvae *Chironomus xanthus* (Janke et al. 2011; Yamada et al. 2012). In the laboratory experiment of Liu et al. (2017), calcium nitrate added to sediment collected from the Shajing River (China) strongly reduced sediment P release and improved the P binding capacity, but also led to strongly elevated nitrate concentrations in the pore water and over-lying water column.

Elevated ammonium, nitrite, and/or nitrate concentrations, as observed with calcium nitrate additions, may come with risks for aquatic animals (Camargo et al. 2005; Camargo and Alonso 2006). A way to reduce the risk of a potentially toxic N addition in sediments and hypolimnion is the use of slow nitrate release compounds (Wauer et al. 2005). The slow release compound Depox®, which consists of a colloid $(FeOOH)_n$ matrix and nitrate at a molar ratio Fe:N of 0.3 releases nitrate slowly over a period of weeks. It has been shown to effectively control sediment P release in enclosures, even a year after treatment (Wauer et al. 2005). Moreover, desulphurication and methanogenesis were also inhibited (Wauer et al. 2005).

5.5.2   Liquids to Bind P

The most common liquids used to counteract eutrophication are solutions of aluminum or iron salts, which are discussed in detail in Cooke et al. (2005) and Jiang and Graham (1998), who also elaborate on pre-polymerized inorganic coagulants. In addition to lake restoration, these solutions are used typically as a coagulant for potable water production and sewage treatment.

To reduce internal P load, the added liquid formulations should form an active barrier on the sediment. Aluminum hydroxide flocs are, however, easily resuspended (Egemose et al. 2010), which may reduce binding efficiency (Huser 2017). To overcome such uncontrolled distribution and to minimize side effects in the water column, sediment injection of polyaluminum chloride has been developed (Schüz et al. 2017). Injection precludes $Al(OH)_3$ flocs to strip P out of the water column. Because $Al(OH)_3$ begins to crystalize after forming it also starts to lose binding capacity (Berkowitz et al. 2006; de Vincente et al. 2008), which may hamper targeting potentially releasable P, such as organic forms. Likewise, Boers et al. (1982) mixed Fe(III) chloride with the surface sediments of the shallow Lake Groot Vogelenzang (The Netherlands) using a water-jet. Meanwhile, underwater injection systems, such as Proteus, are available to inject P binding solutions directly into the sediment (Wísniewski et al. 2010).

5.5.3   Solid Phase P-Binders

Over the past two decades, a profusion of solid-phase P adsorbents, in particular for application in aquatic systems or in constructed wetlands designed to intercept P from catchments (e.g., Fastner et al. 2016), has emerged utilizing a range of precursor materials and substrates (Gibbs et al. 2010; Mackay et al. 2014; Douglas et al. 2016; Noyma et al. 2016; Prashantha Kumar et al. 2018; Mucci et al. 2018). Driving the development of P adsorbents is the notion that if P can be made limiting to a sufficient extent, the effects of eutrophication can be minimized (e.g., Lürling and van Oosterhout 2013); P limitation can induce a reduction in phytoplankton biomass and/or a change to less problematic species or ideally, a return to macrophyte dominant clear water.

The application of P adsorbents to freshwater systems to manage the effects of eutrophication constitutes a challenging nexus spanning a range of considerations (e.g., Le Moal et al. 2019). These considerations include meeting water quality goals, public health requirements, perceptions and acceptance, and operational performance with the goal of restoration of ecosystem function, and in many cases, public amenity. This is particularly so as more efficient and environmentally acceptable methods are sought to combat the effects of eutrophication in aquatic systems.

More recently developed P adsorbents may range from modified clays or zeolites (Gibbs and Ozkundakci 2010; Fan et al. 2017, Osalo et al. 2013; Yin et al. 2011), natural and modified soils (Pan et al. 2011a, b; Noyma et al. 2016), mining and mineral processing by-products (Douglas et al. 2012a, b; Zheng et al. 2004; Wang et al. 2018), or related materials including Fe-oxides (Lalley et al. 2016). Critically, solid phase P adsorbents should also be considered as part of a strategy of complementary geoengineering techniques (Mackay et al. 2014). Such is the case in the combination of either Al salts or dredging with the application of Lanthanum-Modified Bentonite (LMB) as a P adsorbent (Lürling and Faassen 2012; Waajen et al. 2016).

As outlined earlier in this chapter, these solid phase P adsorbents rely primarily on the formation of sparingly soluble salts of aluminum, calcium, iron, lanthanum, or combinations in a variety of mineralogical forms (see Figure 5.11). On this basis, solid-phase P adsorbents will be discussed in terms of the element or elements and mineralogy involved in the binding of soluble P.

The composition and mineralogy of a potential P adsorbent are major determinants of its functionality, effectiveness, and potential viability. The goal is to be able to match the attributes of the P adsorbent to the nature of the system to ensure safe, effective P reduction over a variety of timescales, for instance for immediate P reduction prior to or during an algal bloom, or for sustained reduction over yearly to decadal timescales. These attributes have been previously elucidated by Douglas et al. (2016) and are summarized below.

In assessing solid-phase P adsorbents it is prudent to consider relevant attributes including origin, material composition, particle size and density, as well as its performance under a variety of pH and redox conditions that extend beyond those expected so as to avoid potential release of accumulated P. Changes in redox status following application are fundamentally important as many freshwater systems are characterized by large external loads of labile organic carbon and/or high rates of generation of organic carbon with microbially mediated organic degradation. This can often lead to sustained periods of water column anoxia and reducing conditions, in particular under periods of sustained water column stratification. Similarly, P adsorbents have to be resistant to changes in

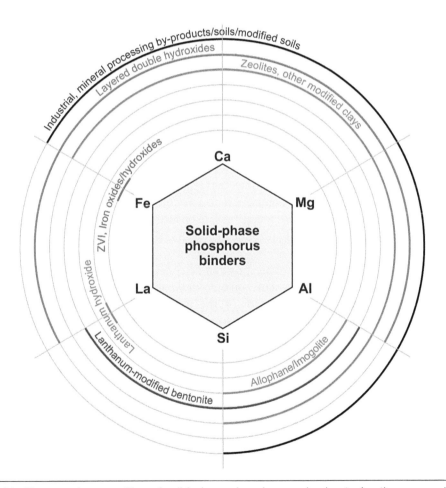

Figure 5.11   Types and composition of solid-phase phosphorus adsorbents. Lanthanum-modified bentonite (La-Al-Si) and other lanthanum containing compounds (La).

pH such as may occur during algal blooms, the influx of new waters during flood events, or during tidal salt wedge propagation in estuarine environments.

Particle morphology and, in particular, particle size distribution and density, and the intended outcome of the sediment capping, are also critical (e.g., Zhang et al. 2016). It may be advantageous that the P adsorbent is rapidly conveyed to the bottom sediments by virtue of particle size and/ or high density, and after settling and compaction, its cohesive strength; these properties render the adsorbent resistant to resuspension and transport. Conversely, these characteristics may also increase the rate of burial and potentially the effective loss of the P adsorbent from the sediment-water interface, which with locally dynamic pH and redox gradients and microbial activity, is often the most active zone of nutrient regeneration. The P adsorbent should also not be of a thickness, completeness of coverage, or resistance to penetration/bioturbation so as to smother aquatic biota or otherwise adversely affect biogeochemical processes other than the intended purpose of P removal (Steinman and Ogdahl 2012).

Of paramount importance is the ecotoxicity of the P adsorbent. Examples of materials with an inherent ecotoxicological burden may include those derived from mining, mineral processing, or industrial processes such as fly ash (e.g., Wang et al. 2016a) from which zeolite or other minerals are formed. For example, fly ash may often contain substantial enrichments of metals and metalloids (e.g., Stiernstrom et al. 2013; Belviso et al. 2015; Smolka-Daielowsak and Fiedor 2018) that may potentially become bioavailable over the lifetime of the material following application. Similarly, red mud derived from alumina production may contain a suite of potentially toxic elements and on occasion, may be enriched in radionuclides (Douglas et al. 2012a; Wendling et al. 2012).

In considering the application of a P adsorbent, a suitably scaled set of performance assessments are required, so as to identify any unforeseen outcomes prior to full implementation. These may range, in increasing order of scale: (1) complexity, time, and cost from laboratory testing of water and sediment samples; (2) monitoring and review of water quality data; (3) core reactors to mesocosm studies incorporating many of the major elements (water, sediment, biota, atmospheric exchange, mixing, stratification) of the aquatic system (e.g., Douglas et al. 2008; Pan et al. 2011a, b); and (4) scenario modeling of intervention outcomes (e.g., dose size and frequency) of P adsorbents. With increasing scale and complexity comes an increased understanding of the performance of, and confidence in, the P adsorbent, but also greater cost in ensuring necessary due diligence to achieve a specific ecological outcome for a particular aquatic system, albeit based on a shifting baseline (e.g., Duarte et al. 2009).

In broad terms, the impact of eutrophication can be considered on five levels: (1) the loss of, or changes in, ecosystem function; (2) a status quo, *do nothing*, approach; (3) the implementation of technologies to reduce external nutrient loads; (4) the implementation of existing P-adsorbent technologies; and (5) the development of novel interventions or management technologies. Quantification of the eutrophication costs are difficult, and may be highly system-specific, but some estimates have been made (e.g., Hamilton et al. 2013).

Costs may also accumulate as there may be delays in evaluation, through laboratory and field trials over a variety of scales, ecotoxicological testing, regulatory submission and approval (with possible imposition of conditions), and licensing and commercialization, which is often required to support optimization and upscaling of P adsorbent manufacture. These steps may have to be repeated in new jurisdictions, placing an additional burden on the applicant, notwithstanding further trials and validation, before there is regulatory and public acceptance, even before there is a commercial return on investment. A prime example is the novel LMB P adsorbent where considerable patience and substantial investment to underpin commercialization was required prior to eventual widespread international regulatory approval and application. LMB was developed in the

1990s by CSIRO (Australia), patented in 2002 (Douglas 2002), and commercialized as Phoslock® a few years later.

Concomitant public and social acceptance coupled with leadership and governance and effective strategy and policy settings (e.g., Gross and Hagy 2017) are pivotal in the successful application of a P adsorbent to an aquatic system. Critical in this process is to make available the scientific assessment in a transparent, user-friendly form. This is particularly important for P adsorbents such as fly ash, which may be considered waste materials or those not considered to be naturally-occurring, such as LMB. Ideally, the dissemination of information will commence in parallel with the laboratory/pilot trial stages such that key stakeholders will be kept fully informed of the progress to facilitate an environmental partnership, and garner acceptance prior to full-scale implementation of the P adsorbent material.

A multitude of studies are available that document the use of P adsorbents in a range of applications (e.g., Douglas et al. 2004; Egemose et al. 2010; Gibbs et al. 2010, 2011; Belyaeva and Haynes 2012; Pan et al. 2012; Spears et al. 2013a; Haynes 2015; Copetti et al. 2016; Mucci et al. 2018). In general, P adsorbents applied to aquatic systems, including as active constituents within constructed wetlands, can be divided into a range of categories based on composition. Six elements: Ca, Mg, Al, Si, La, and Fe, in a range of mineralogical and geochemically reactive forms, dominate the spectrum of materials used for the removal of P in aquatic systems. These are summarized in Figure 5.11 and are discussed in the following paragraphs.

5.5.3.1   Lanthanum-modified bentonite (La-Al-Si) and other lanthanum (La) containing compounds

Perhaps the most widely studied P adsorbent over the past decade is LMB (Douglas et al. 2008; Robb et al. 2003; Kuroki et al. 2014; Copetti et al. 2016; Dithmer et al. 2016; Spears et al. 2013b, 2016; Epe et al. 2017; Funes et al. 2018). Extensive laboratory and field trials have resulted not only in a considerably improved understanding of both its operational performance and limitations in application, but also an increasing degree of public acceptance as tangible benefits are realized following LMB application and other complementary nutrient management strategies. Major applications of LMB, principally to eutrophic lakes, have occurred worldwide (see Figure 5.12). As with all *in-situ*

Figure 5.12 Application of 300 tonnes of lanthanum-modified bentonite to Lake Goldap (130 ha) in Poland in 2017 (Image © PET).

treatment technologies using P absorbents, effective and sustained management of catchment P sources is critical to prolonged success, which if not addressed have the potential to limit in-lake nutrient management strategies, even in the short term (e.g., Cole and Spears 2018).

The development of LMB has also spawned a range of related technologies that may involve the application of LMB with other P absorbents such as Fe (Ding et al. 2018) or the development of a range of La, La oxide, La-hydroxide, or multi-element composite materials (e.g., Huang et al. 2014; Xie et al. 2014; Li et al. 2015; Yu and Chen 2015; Chen et al. 2016; Lai et al. 2016; Wang et al. 2016b; Qiu et al. 2017).

5.5.3.2   Industrial, Mineral Processing By-Products and Soils (Fe-Ca-Mg-Al-Si)

Soils have been extensively studied for the removal of P from natural waters *in-situ* in aquatic systems, in catchments, and in constructed wetlands (Arias et al. 2000; Degens et al. 2000; Xu et al. 2012; Kim et al. 2014; Dai and Pan 2014; Pan et al. 2002, 2013). Chitosan-modified soils or cationic starch-modified soils have been investigated, with the focus on the flocculation of cyanobacteria (and P contained therein) in lakes (Pan et al. 2006, 2011a, b; Noyma et al. 2016; Shi et al. 2015) although these modified soils may also simultaneously remove substantial soluble P (Dai and Pan 2014). Similarly, steel slags have been investigated as a material to intercept the release of P from marine sediments (Jo and Park 2017).

5.5.3.3   Layered Double Hydroxides (Mg-Fe-Al)

Layered double hydroxides, and in particular hydrotalcites, an Mg-Al form, occur naturally but may also be synthesized from industrial by-products or *de novo* (Douglas et al. 2016). The removal of P from waters by layered double hydroxides has been extensively studied (Miyata 1980; Misra and Perrotta 1992; Shin et al. 1996; Seida and Nakano 2000). A major obstacle to the use of hydrotalcites for P removal in natural and/or wastewaters is the selectivity for (bi)carbonate over P (Miyata 1980; Shin et al. 1996). To date, few studies have examined the application of hydrotalcites as an amendment in aquatic ecosystems; a possible barrier to application is the potential for P to be released, as observed in studies of layered double hydroxides as slow-release fertilizers in soils (e.g., Halajnia et al. 2013; Everaet et al. 2016; Moraes et al. 2018).

5.5.3.4   Allophane/Imogolite (Al-Si)

Similar to layered double hydroxides, the noted P adsorption of natural allophanes and imogolites present in soils and their analogues prepared in the laboratory present an intriguing prospect for their application to aquatic systems. As with layered double hydroxides, questions over the long-term stability of absorbed P during mineralogical transformation (e.g., Harsh et al. 2002), the effect of pH on P uptake, and the interference of competing anions places into question their potential for application in aquatic systems. One study, however, suggests the prospect of a substantial P uptake capacity from soil core trials (Gibbs et al. 2010), while other studies suggest that allophane may effectively reduce dissolved P from natural (Whitely 2017) or wastewaters (Yuan and Wu 2006). Wetlands containing soils with an allophanic component have also been successful in removing dissolved P (Degens et al. 2000; Qualls and Heyvaert 2017).

5.5.3.5   Zeolites, Other Modified Clays (Al-Si)

The application of both natural and modified zeolites (Gibbs and Ozcundacki 2010; Sun et al. 2011; Mucci et al. 2018), and in combination with other materials such as calcite (Lin et al. 2011), has been extensively studied. The application of both natural and modified zeolites due either to the incorporation of Al into the cage structure (Aqual-P) or combined with a La(OH)$_3$ precipitate (Ning et al. 2008) is, in part, an effort to utilize the natural NH$_4^+$ ion-exchange capacity of zeolite in combination with P-uptake. This allows it to be used as a dual N and P absorbent, and hence coupled reduction that confers a range of potential advantages in the management of eutrophication (Paerl 2009; Dodds and Smith 2016). One study indicated that the P-uptake capacity of Aqual-P was less than that of either allophane or LMB (Gibbs et al. 2011), while another indicated a similar result of Aqual-P relative to LMB, with both absorbing P across a range of environmentally relevant pHs (Mucci et al. 2018).

5.5.3.6   ZVI, Iron Oxides/Hydroxides (Fe)—but Also Magnetic Materials

The last decade has witnessed a resurgence in research surrounding the application of a variety of traditional iron-based oxy/hydroxide minerals (Fuchs et al. 2018; Funes et al. 2018), combination/hybrid materials (Zamparas et al. 2013; Kralchevska et al. 2016), designer nanomaterials zerovalent and magnetic iron-based materials (de Vincente et al. 2010; Almeelbi and Bezbaruah 2012; Sleiman et al. 2016; Funes et al. 2017; Fang et al. 2018) for the removal, and in the case of magnetic materials, recovery following adsorption of P from aquatic systems. Importantly, however, the majority of the zerovalent and magnetic iron-based technologies are currently only at the laboratory or microcosm stage of evaluation.

5.6   CONCLUSION

In this chapter, an overview was given of the most common categories of internal load management. Sediment dredging, excavation, or hypolimnetic water withdrawal are straightforward means of removing excessive nutrients stored in lake beds. Dredging and excavation come with considerable costs, while withdrawal of hypolimnetic water may require treatment before being discharged downstream. Inactivating P release from sediments by stimulation of the iron trap via aeration/oxygenation or by chemical precipitation of phosphate is generally viewed as a cheaper alternative than sediment removal. The recent development of solid phase P adsorbents is noteworthy, and is driven by the notion that if P can be made limiting to a sufficient extent, the effects of eutrophication can be strongly reduced. Despite being sometimes viewed as an *end-of-the-pipe* approach, internal P load management is essential for water quality improvement in lakes with either legacy or seasonal internal P load issues, in lakes where diffuse loading is transporting nutrients via ground water, and in lakes where catchment measures are insufficient to counteract eutrophication. Control of internal P loading requires comprehensive, site-specific diagnoses (i.e., LSA) to ensure the measures are appropriate, cost-effective, and have public support.

5.7   REFERENCES

Almeelbi, T and Bezbaruah, A. 2012. Aqueous phosphate removal using nanoscale zerovalent iron. J Nanopart Res. 14:900.

Arias, CA; Del Bubba, M; Brix, H. 2000. Phosphorus removal by sands for use as a media in subsurface flow constructed reed beds. Water Res. 35:1159–1168.

Beduhn, RJ. 1994. The effects of destratification aeration on five Minnesota lakes. Lake Reserv Manage. 9:105–110.

Belviso, C; Cavalcante, F; Di Gennaro, S; Palma, A; Ragone, P; and Fiore, S. 2015. Mobility of trace elements in fly ash and in zeolitised coal fly ash. Fuel. 144:369–379.

Belyaeva, ON and Haynes, RJ. 2012. Use of inorganic wastes as immobilizing agents for soluble P in green waste-based composts. Environ Sci Poll Res. 19:2138–2150.

Berkowitz, J; Anderson, MA; and Amrhein, C. 2006. Influence of aging on phosphorus sorption to alum floc in lake water. Water Res. 40:911–16.

Beutel, MW and Horne, AJ. 1999. A review of the effects of hypolimnetic oxygenation on lake and reservoir water quality. Lake Reserv Manage. 15:285–297.

Beutel, MW; Horne, AJ; Taylor, WD; Losee, RF; and Whitney, RD. 2008. Effects of oxygen and nitrate on nutrient release from profundal sediments of a large, oligo-mesotrophic reservoir, Lake Mathews, California. Lake Reserv Manage. 24:18–29.

Björk, S. 1972. Swedish Lake Restoration Program Gets Results. Ambio. 1:153–165.

Boers, P; Van der Does, J;Quaak, M; Van der Vlugt, J; and Walker, P. 1982. Fixation of phosphorus in lake sediments using iron(III)chloride: experiences, expectations. Hydrobiologia. 233:211–212.

Bormans, M; Maršálek, B; and Jančula, D. 2016. Controlling internal phosphorus loading in lakes by physical methods to reduce cyanobacterial blooms: a review. Aquat Ecol. 50:407–422.

Brouwer, E and Roelofs, JGM. 2001. Degraded softwater lakes: possibilities for restoration. Restor Ecol. 9:-155–166.

Camargo, JA; Alonso, A; and Salamanca, A. 2005. Nitrate toxicity to aquatic animals: a review with new data for freshwater invertebrates. Chemosphere. 58:1255–1267.

Camargo, JA and Alonso, A. 2006. Ecological and toxicological effects of inorganic nitrogen pollution in aquatic ecosystems: a global assessment. Environ Int. 32:831–849.

Carpenter, SR. 2005. Eutrophication of aquatic ecosystems: Bistability and soil phosphorus. Proc Natl Acad Sci. 102:10002–10005.

Carpenter, SR; Ludwig, D; and Brock, WA. 1999. Management of eutrophication for lakes subject to potentially irreversible change. Ecol Appl. 9:751–771.

Chen, M; Huo, C; Li, Y; and Wang, J. 2016. Selective adsorption and efficient removal of phosphate from aqueous medium with graphene-lanthanum composite. Sustain Chem Engin. 4:1296–1302.

Cole, S and Spears, BM 2018. Assessment of sediment phosphorus capping to control nutrient concentrations in English lakes—project summary. Bristol, UK: Environment Agency. (Project SC120064/R9, CEH Project no. C04646).

Cooke, GD; Welch, EB; Peterson, S; and Nichols, SA. 2005. Restoration and management of lakes and reservoirs. Boca Raton (FL): CRC Press.

Copetti, D; Finsterle, K; Marziali, L; Stefani, F; Tartari, G; Douglas, G; Reitzel, K; Spears, B; Winfield, I; Crosa, G; et al. 2016. Eutrophication management in surface waters using lanthanum-modified bentonite: a review. Water Res. 97:162–174.

Cronberg, G. 1982. Changes in the phytoplankton of Lake Trummen induced by restoration. Hydrobiologia. 86:185–193.

Dai, LC and Pan, G. 2014. The effects of red soil in removing phosphorus from water column and reducing phosphorus release from sediment in Lake Taihu. Wat Sci Technol. 5:1052–1058.

Degens, BP; Schipper, LA; Claydon, JJ; Russell, JM; and Yeates, GW. 2000. Irrigation of allophanic soil with dairy effluent for 22 years: responses of nutrient storage and soil biota. Aust J Soil Res. 38:25–35.

de Vincente, I; Huang, P; Andersen, F; and Jensen, H. 2008. Phosphate adsorption by fresh and aged aluminum hydroxide: consequences for Lake Restoration. Environ Sci Technol. 42:6650–6655.

de Vicente, I; Merino-Martos, A; Cruz-Pizarro, L; and de Vicente, J. 2010. On the use of magnetic nano and microparticles for lake restoration. J Hazard Mater. 181:375–381.

Ding, S; Sun, Q; Chen, X;Liu, Q; Wang, D; Lin, J; Zhang, C; and Tsang ,CW. 2018. Synergistic adsorption of phosphorus by iron in lanthanum modified bentonite (Phoslock®: New insight into sediment phosphorus immobilisation. Water Res. 134:32–43.

Dithmer, L; Nielsen, UG; Lürling, M; Spears, BM; Yasseri, S; Lundberg, D; Jensen, ND; and Reitzel, K. 2016. Responses in sediment phosphorus concentrations and composition across 10 lakes following applications of lanthanum modified bentonite. Water Res. 97:101–110.

Dodds, WK and Smith, VH. 2016. Nitrogen, phosphorus, and eutrophication in streams. Inland Waters. 6:155–164.

Douglas, GB. 2002. US Patent 6350383: Remediation Material and Remediation Process for Sediments.

Douglas, GB; Adeney, J; Johnston, K; Wendling, LA; and Coleman, S. 2012a. Investigation of major, trace element, nutrient and radionuclide mobility in a mining by-product NUA-amended soil. J Environ Qual. 41:1818–1834.

Douglas, GB; Hamilton, DP; Robb, MS; Pan, G; Spears, BM; and Lurling, M. 2016. Guiding principles for the development and application of solid-phase phosphorus adsorbents for freshwater ecosystems. Aquat Ecol. 50:385–405.

Douglas, GB; Robb, MS; Coad, DN; and Ford, PW. 2004. A review of solid phase adsorbents for the removal of phosphorus from natural and wastewaters. In: Valsami-Jones E., editor. Phosphorous in Environmental Technology: Principles and Applications. London: IWA Publishing. p. 291–320.

Douglas, GB; Robb, MS; and Ford, PW. 2008. Reassessment of the performance of mineral-based sediment capping materials to bind phosphorus: a comment on Akhurst et al. (2004). Mar Freshwater Res. 59:836–837.

Douglas, GB; Wendling, LA; and Coleman, S. 2012b. Productive use of steelmaking by-product in environmental applications I: mineralogy and major and trace element geochemistry. Miner Eng. 35:49–56.

Downing, JA. 2014. Limnology and oceanography: two estranged twins reuniting by global change. Inland Waters. 4:215–232.

Duarte, CM; Conley, DJ; Carstensen, J; and Sanchez-Camacho, S. 2009. Return to neverland: shifting baselines affect eutrophication restoration targets. Estuaries Coasts. 32:29–36.

Egemose, S; Reitzel, K; Andersen, FØ; and Flindt, MR. 2010. Chemical lake restoration products: sediment stability and phosphorus dynamics. Environ Sci Technol. 44:985–991.

Epe, TS; Finsterle, K; and Yasseri, S. 2017. Nine years of phosphorus management with lanthanum modified bentonite (Phoslock) in a eutrophic, shallow swimming lake in Germany. Lake Reserv Manage. 33:119–129.

Everaet, M; Warrinnier, R; Baken, S; Gustafsson, J-P; De Vos, D; and Smolders, E. 2016. Phosphate-exchanged Mg-Al layered double hydroxides: a new slow-release fertilizer. Sustain Chem Engin. 4:4280–4287.

Eviner, VT and Hawkes, CV. 2008. Embracing variability in the application of plant–soil interactions to the restoration of communities and ecosystems. Restor Ecol. 16:713–729.

EU (European Union). 2000. Directive 2000/60/EG of the European Parliament and of the Council establishing a framework for the Community action in the field of water policy of 23 October. PB L327 of 22 December 2000 CBS, 2014.

Fan, Y; Li, Y; Wu, D; Li, C; and Kong, H. 2017. Application of zeolite/hydrous zirconia composite as a novel sediment capping material to immobilize phosphorus. Water Res. 123:1–11.

Fang, L; Liu, R; Li, J; Xu, C; Huang, L-Z; and Wang, D. 2018. Magnetite/Lanthanum hydroxide for phosphate sequestration and recovery from lake and the attenuation effects of sediment particles. Water Res. 130:243–254.

Fast, AW; Lorenzen, MW; Asce, M; and Glenn, JH. 1976. Comparative study with costs of hypolimnetic aeration. J Environ Engin Div. EE6:1175–1187.

Fastner, J; Abella, S; Litt, A; Morabito, G; Vörös, L; Pálffy, K; Straile, D; Kümmerlin, R; Matthews, D; Phillips, MG; et al. 2016. Combating cyanobacterial proliferation by avoiding or treating inflows with high P load—experiences from eight case studies. Aquat Ecol. 50:367–383.

Foy, RH. 1986. Suppression of phosphorus release from lake sediments by the addition of nitrate. Water Res. 20:1345–1351.

Fuchs, E; Funes, A; Saar, K; Reitzel, K; and Jensen, HS. 2018. Evaluation of dried amorphous ferric hydroxide CFH-12® as agent for binding bioavailable phosphorus in lake sediments. Sci Total Environ. 628-9:990–996.

Funes, A; Arco, A; Álvarez-Manzaneda, I; de Vicente, J; and de Vicente, I. 2017. A microcosm experiment to determine the consequences of magnetic microparticles application on water quality and sediment phosphorus pools. Sci Total Environ. 579:245–253.

Funes, A; Martínez, FJ; Álvarez-Manzaneda, I; Conde-Porcuna, JM; de Vicente, J; Guerrero, F; de Vicente, I. 2018. Determining major factors controlling phosphorus removal by promising adsorbents used for lake restoration: a linear mixed model approach. Water Res. 141:377–386.

Gächter, R. 1987. Lake restoration. Why oxygenation and artificial mixing cannot substitute for a decrease in the external phosphorus loading. Schweiz Z Hydrol. 49:170–185.

Gächter, R and Wehrli, B. 1998. Ten years of artificial mixing and oxygenation: no effect on the internal phosphorus loading of two eutrophic lakes. Environ Sci Technol. 32:3659–3665.

Garrell, MH; Confer, JC; Kirschner, D; and Fast, AW. 1977. Effects of hypolimnetic aeration on nitrogen and phosphorus in a eutrophic lake. Water Resour Res. 13:343–347.

Gibbs, M and Özkundakci, D. 2010. Effects of a modified zeolite on P and N processes and fluxes across the lake sediment–water interface using core incubations. Hydrobiologia. 661:21–35.

Gibbs, MM; Hickey, CW; and Özkundakci, D. 2010. Sustainability assessment and comparison of efficacy of four P-inactivation agents for managing internal phosphorus loads in lakes: sediment incubations. Hydrobiologia. 658:253–275.

Gibbs, MM; Hickey, CW; and Özkundakci, D. 2011. Sustainability assessment and comparison of efficacy of four P-inactivation agents for managing internal phosphorus loads in lakes: sediment incubations. Hydrobiologia. 658:253–275.

Gross, G and Hagy, JD III. 2017. Attributes of successful actions to restore lakes and estuaries degraded by nutrient pollution. J Environ Manage. 187:122–136.

Halajnia, A; Oustan, S; Najafi, N; Khataee, RA; and Lakzian, A. 2013. Adsorption–desorption characteristics of nitrate, phosphate and sulfate on Mg–Al layered double hydroxide. Appl Clay Sci. 80-81:305–312.

Hamilton, DP; Wood, SA; Dietrich, DR; and Puddick, J. 2013. Costs of harmful blooms of freshwater cyanobacteria. In: Sharma, NK; Rai, AK; and Stal, LJ, editors. Cyanobacteria: an economic perspective. Chichester (UK): John Wiley & Sons. p. 245–256.

Hamilton, DP; Salmaso, N; and Paerl, HW. 2016. Mitigating harmful cyanobacterial blooms: strategies for control of nitrogen and phosphorus loads. Aquat Ecol. 50:351–366.

Harsh, J; Chorover, J; and Nizeyimana, E. 2002. Allophane and imogolite. In: Soil Mineralogy with environmental applications. Madison (WI): Soil Science Society of America. p. 291–322.

Haynes, RJ. 2015. Use of industrial wastes as media in constructed wetlands and filter beds—prospects for removal of phosphate and metals from wastewater streams. Crit Rev Environ Sci Technol. 45:1041–1103.

Heisler, J; Glibert, PM; Burkholder, JM; Anderson, DM; Cochlan, W; Dennison, C; Dortch, Q; Gobler, CJ; Heil, CA; Humphries, E; Lewitus, A; Magnien, R; Marshall, HG; Sellner, K; Stockwell, DA; Stoecker, DK; and Suddleson, M. 2008. Eutrophication and harmful algal blooms: a scientific consensus. Harmful Algae. 8:3–13.

Hemond, HF and Lin, K. 2010. Nitrate suppresses internal phosphorus loading in an eutrophic lake. Water Res. 44:3645–3650.

Hovenkamp-Obbema, IRM and Fiegen, W. 1992. The effects of dredging and fish stocking on the trophic status of shallow, peaty ditches. Hydrobiologia. 233:225–233.

Huang, W-Y; Li, D; Liu, Z-Q; TAo, Q; Zhu, Y; Yang, J; and Zhang, Y-M. 2014. Kinetics, isotherm, thermodynamic, and adsorption mechanism studies of La(OH)$_3$-modified exfoliated vermiculites as highly efficient phosphate adsorbents. Chem Engin J. 236:191–201.

Hupfer, M and Scharf, BW. 2002. Chapter VI-2.1: Seentherapie: Interne Massnahmen zur Verminderung der Phosphorkonzentration. In Steinberg C, Calmano W, Klapper H, Wilken R-D, editors. Handbuch Angewandte Limnologie. Landsberg/Lech (Germany): Verlag. p. 1–68.

Huser, BJ. 2017. Aluminum application to restore water quality in eutrophic lakes: maximizing binding efficiency between aluminum and phosphorus. Lake Reserv Manage. 33:143–151.

Huser, BJ; Futter, M; Lee, JT; and Pernie, M. 2016. In-lake measures for phosphorus control: The most feasible and cost-effective solution for long-term management of water quality in urban? Water Res. 97:142–152.

Ibelings, BW; Fastner, J; Bormans, M; and Visser, PM. 2016. Cyanobacterial blooms. Ecology, prevention, mitigation and control: Editorial to a CYANOCOST Special Issue. Aquat Ecol. 50:327–331.

Janke, H; Yamada, TM; Beraldo, DAS; Botta, CMR; Nascimento, MRL; and Mozeto, AA. 2011. Assessment of the acute toxicity of eutrophic sediments after the addition of calcium nitrate (Ibirité reservoir, Minas Gerais-SE Brazil): initial laboratory experiments. Braz J Biol. 71:903–914.

Jiang, JQ, Graham, NJD. 1998. Pre-polymerized inorganic coagulants and phosphorus removal by coagulation—a review. Water SA 24:237–244.

Jing, L; Liu, X; Bai, S; Wu, C; Ao, H; and Liu, J. 2015. Effects of sediment dredging on internal phosphorus: a comparative field study focused on iron and phosphorus forms in sediments. Ecol Engin. 82:267–271.

Jo, S-W and Park, S-J. 2017. Applicability assessment of steel slag as reactive capping material for blocking phosphorus release from marine sediment. J Korean Soc Agric Engin. 56:11–17.

Kim, D; Ryoo, KS; Hong, YP; and Choi, JH. 2014. Evaluation of loess capability for adsorption of total nitrogen (TN) and total phosphorus (TP) in aqueous solution. Bull Korean Chem Soc. 35:2471–2476.

Knaus, RM and Malone, RF. 1984. An historical overview of a successful lakes restoration project in Baton Rouge, Louisiana. Lake Reserv Manage. 1:412–415.

Knott, NA; Aulbury, JP; Brown, TH; and Johnston, EL. 2009. Contemporary ecological threats from historical pollution sources: impacts of large-scale resuspension of contaminated sediments on sessile invertebrate recruitment. J Appl Ecol. 46:770–781.

Kortmann, RW. 1994. Oligotrophication of Lake Shenipsit by layer aeration. Lake Reserv Manage. 9:94–97.

Kralchevska, RP; Prucek, R; Kolaří k, J; Tuček, J; Machala, L; Filip, J; Sharma, VK; and Zbořil, R. 2016. Remarkable efficiency of phosphate removal: ferrate(VI)-induced *in situ* sorption on core-shell nanoparticles. Water Res. 103:83–91.

Kuroki, V; Bosco, GE; Fadini, PS; Mozeto, AA; Cestari, AR; and Carvalho, WA. 2014. Use of a La(III)-modified bentonite for effective phosphate removal from aqueous media. J Hazard Mater. 274:124–131.

Lai, L; Xie, Q; Chi, L; Gu, W; and Wu, D. 2016. Adsorption of phosphate from water by easily separable Fe_3O_4@ SiO_2 core/shell magnetic nanoparticles functionalized with hydrous lanthanum oxide. J Colloid Interf Sci. 465:76–82.

Lalley, J; Han, C; Li, X; Dionysiou, D; and Nadagouda, M. 2016. Phosphate adsorption using modified iron oxide-based sorbents in lake water: kinetics, equilibrium and column tests. Chem Engin J. 284:1386–1396.

Lean, DRS; McQueen, DJ; and Story, VA. 1986. Phosphate transport during hypolimnetic aeration. Arch Hydrobiol. 108:269–280.

Le Moal, M; Gascuel-Odoux, C; Menesguen, A; Souchon, Y; Etrillard, C; Levain, A; Moatar, F; Pannard, A; Souchu, P; and Lefebvre, A, et al. 2019. Eutrophication: an old wine in a new bottle. Sci Total Environ. 651:1–11.

Li, G; Chen, D; ZhA o, W; and Zhang, X. 2015. Efficient adsorption of phosphate on La-modified tourmaline. J Environ Chem Engin. 3:515–522.

Ligtvoet, W; Beugelink, G; Brink, C; Franken, R; and Kragt, F. 2008. Kwaliteit voor Later. Ex ante evaluatie Kaderrichtlijn Water. PBL publicatienummer 50014001/200, ISBN: 978-90-6960-203-5.

Lijklema, L. 1977. The role of iron in the exchange of phosphorus between water and its implications for phosphorus management strategies. In Golterman, HL, editor. Interactions between sediments and fresh water. The Hague: Dr. W. Junk. p. 313–317.

Lin, J; Zhan, Y; and Zhu, Z. 2011. Evaluation of sediment capping with active barrier systems (ABS) using calcite/zeolite mixtures to simultaneously manage phosphorus and ammonium release. Sci Total Environ. 409:638–646.

Liu, C; Zhong, J; Wang, J; Zhang, L; and Fan, C. 2016. Fifteen-year study of environmental dredging effect on variation of nitrogen and phosphorus exchange across the sediment-water interface of an urban lake. Environ Poll. 219:639–648.

Liu, X; TA o, Y; Zhou, K; Zhanga, Q; Chen, G; and Zhang, X. 2017. Effect of water quality improvement on the remediation of river sediment due to the addition of calcium nitrate. Sci Total Environ. 575:887–894.

Lürling, M and Faassen, EJ. 2012. Controlling toxic cyanobacteria: effects of dredging and phosphorus-binding clay on cyanobacteria and microcystins. Water Res. 46:1447–1459.

Lürling, M; Mackay, E; Reitzel, K; and Spears, BM. 2016. Editorial–A critical perspective on geo-engineering for eutrophication management in lakes. Water Res. 97:1–10.

Lürling, M and van Oosterhout F. 2013. Controlling eutrophication by combined bloom precipitation and sediment phosphorus inactivation. Water Res. 47:6527–6537.

Mackay, EB; Maberly, SC; Pan, G; Reitzel, K; Bruere, A; Corker, N; Douglas, G; Egemose, S; Hamilton, D; Hatton-Ellis, T; Huser, B; Li, W; Meis, S; Moss, B; Lürling, M; Phillips, G; Yasseri, S; and Spears, BM. 2014. Geo-engineering in lakes: welcome attraction or fatal distraction? Inland Waters. 4:349–356.

Manap, N and Voulvoulis, N. 2015. Environmental management for dredging sediments—The requirement of developing nations. J Environ Manage. 147:338–348.

Mercier, P. 1948. Aération partielle sous-lacustre d'un lac eutrophe. Verh Internat Verein Limnol. 10:294–297.

Mercier, P and Gay, S. 1949. Station d'aération au lac de Bret. Schweiz Z Hydrol. 11:423–429.

Misra, C and Perrotta, AJ. 1992. Composition and properties of synthetic hydrotalcites. In Clays and Clay Minerals. 40:145–150.

Miyata, S. 1980. Physico-chemical properties of synthetic hydrotalcites in relation to composition. In Clays and Clay Minerals. 28:50–56.

Moore, BC; Cross, BK; Beutel, M; Dent, S; Preece, E; and Swanson, M. 2012. Newman Lake restoration: A case study Part III. Hypolimnetic oxygenation. Lake Reserv Manage. 28(4): 311–327.

Moraes, PI; Tavares, SR; Vaiss, VS; and LeitA o, AA. 2018. Investigation of sustainable phosphate release in agriculture: structural and thermodynamic study of stability, dehydration and anionic exchange of Mg-Al-HPO_4 layered double hydroxide by DFT calculations. Appl Clay Sci. 162:428–434.

Moss, B. 2010. Ecology of freshwaters. A view for the twenty-first century. 4th ed. Chichester (UK): John Wiley & Sons Ltd.

Moss, B; Balls, H; Irvine, K; and Stansfield J. 1986. Restoration of two lowland lakes by isolation from nutrient-rich water sources with and without removal of sediment. J Appl Ecol. 23:391–414.

Moss, B; Stansfield, J; Irvine, K; Perrow, M; and Phillips, G. 1996. Progressive restoration of a shallow lake: a 12-year experiment in isolation, sediment removal and biomanipulation. J Appl Ecol. 33:71–86.

Mucci, M; Maliaka, V; Noyma, NP; Marinho, MM; and Lurling, M. 2018. Assessment of possible solid-phase phosphate sorbents to mitigate eutrophication: influence of pH and anoxia. Sci Total Environ. 619–620:1431–1440.

Newcombe, CP and MacDonald, DD. 1991. Effects of suspended sediments on aquatic ecosystems. N Am J Fish Manage. 11:72–82.

Nimptsch, J; Woelf, S; Osorio, S; Valenzuela, J; Moreira, C; Ramos, V; Castelo-Branco, R; Nuno Leão, P; and Vasconcelos, V. 2016. First record of toxins associated with cyanobacterial blooms in oligotrophic North Patagonian lakes of Chile—a genomic approach. Int Rev Hydrobiol. 101:57–68.

Ning, P; Bart, HJ; Li, B; Lu, X; and Zhang, Y. 2008. Phosphate removal from wastewater by model-LaIII zeolite adsorbents. J. Env Sci. 20:670–674.

Noonan, TA. 1986. Water quality in long lake, Minnesota, following Riplox sediment treatment. Lake Reserv Manage. 2:131–137.

Noyma, NP; de Magalhães, L; Furtado, LL; Mucci, M; van Oosterhout, F; Huszar, VLM; Marinho, MM; and Lürling, M. 2016. Controlling cyanobacterial blooms through effective flocculation and sedimentation with combined use of flocculants and phosphorus adsorbing natural soil and modified clay. Water Res. 97:26–38.

Nürnberg, GK. 2007. Lake responses to long-term hypolimnetic withdrawal treatments. Lake Reserv Manage. 23:388–409.

OECD. 2018. Wastewater treatment (indicator). doi: 10.1787/ef27a39d-en (Accessed on 07 September 2018).

Oldenborg, KA and Steinman, AD. 2019. Impact of sediment dredging on sediment phosphorus flux in a restored riparian wetland. Sci Total Environ. 650:1969–1979.

O'Neil, JM; Davis, TW; Burford, MA; and Gobler, CJ. 2012. The rise of harmful cyanobacteria blooms: the potential roles of eutrophication and climate change. Harmful Algae. 14:313–334.

Osalo, TP; Merufinia, E; and Saatlo, ME. 2013. Phosphorus removal from aqueous solutions by bentonite: Effect of Al_2O_3 addition. J Civ Eng Urban. 3:317–322.

Osgood, RA. 2017. Inadequacy of best management practices for restoring eutrophic lakes in the United States: guidance for policy and practice. Inland Waters. 7:401–407.

Paerl, H. 2009. Controlling eutrophication along the freshwater-marine continuum: dual nutrient (N and P) reductions are essential. Estuaries Coasts. 32:593–601.

Paerl, HW. 2014. Mitigating harmful cyanobacterial blooms in a human- and climatically-impacted world. Life. 4:988–1012.

Paerl, HW; Gardner, WS; Havens, KE; Joyner, AR; McCarthy, MJ; Newell, SE; Qin, B; and Scott, JT. 2016. Mitigating cyanobacterial harmful algal blooms in aquatic ecosystems impacted by climate change and anthropogenic nutrients. Harmful Algae. 54:213–222.

Paerl, HW and Paul, V, 2012. Climate change: links to global expansion of harmful cyanobacteria. Water Res. 46:1349–1363.

Pan, G; Chen, J; and Anderson, DM. 2011a. Modified local sands for the mitigation of harmful algal blooms. Harmful Algae. 10:381–387.

Pan, G; Dai, L; Li, L; He, L; Li, H; Bi, L; and Gulati, RD. 2012. Reducing the recruitment of sedimented algae and nutrient release into the overlying water using modified soil/sand flocculation-capping in eutrophic lakes. Environ Sci Technol. 46:5077–5084.

Pan, G; Krom, MD and Herut, B. 2002. Adsorption-desorption of phosphate on airborne dust and riverborne particulates in East Mediterranean seawater. Environ Sci Technol. 36:3519–3524.

Pan, G; Krom, MD; Zhang, M; Zhang, X; Wang, L; Dai, L; Sheng, Y; and Mortimer, RJG. 2013. Impact of suspended inorganic particles on phosphorus cycling in the Yellow River (China). Environ Sci Technol. 47:9685–9692.

Pan, G; Yang, B; Wang, D; Chen, H; Tian, BH; Zhang, ML; Yuan, XZ; and Chen, JA. 2011b. In-lake algal bloom removal and submerged vegetation restoration using modified local soils. Ecol Eng. 37:302–308.

Pan, G; Zou, H; Chen, H; and Yuan, X. 2006. Removal of harmful cyanobacterial blooms in Taihu Lake using local soils III. Factors affecting the removal efficiency and an in situ field experiment using chitosan-modified local soils. Environ Pollut. 141:206–212.

Pandey, J and Yaduvanshi, MS. 2005. Sediment dredging as a restoration tool in shallow tropical lakes: cross analysis of three freshwater lakes of Udaipur, India. Environ Control Biol. 43:275–281.

Peterson, SA. 1982. Lake restoration by sediment removal. Water Resour Bull. 18:423–435.

Pokorný, J and Hauser, V. 2002. The restoration of fish ponds in agricultural landscapes. Ecol Eng. 18:555–574.

Poor, ND. 2010. Effect of lake management efforts on the trophic state of a subtropical shallow lake in Lakeland, Florida, USA. Water Air Soil Pollut. 207:333–347.

Prashantha Kumar, TKM; Mandlimath, TR; Sangeetha, P; Revathi, SK; and Ashok Kumar, S.K. 2018. Nanoscale materials as sorbents for nitrate and phosphate removal from water. Environ Chem Lett. 16:389–400.

Qiu, H; Liang, C; Yu, J; Zhang, Q; Song, M; and Chen, F. 2017. Preferable phosphate sequestration by La(III) (hydro)oxides modified wheat straw with excellent properties in regeneration. Chem Eng J. 315:345–354.

Qualls, RG and Heyvaert, AC. 2017. Accretion of nutrients and sediment by a constructed stormwater treatment wetland in the Lake Tahoe basin. J Am Water Resour Assoc. 53:1495–1512.

Ripl, W. 1976. Oxidation of polluted lake sediment with nitrate: a new lake restoration method. Ambio. 5:132–135.

Ripl, W. 1986. Internal phosphorus recycling mechanisms in shallow lakes. Lake Reserv Manage. 2:138–142.

Robb, MR; Greenop, B; Goss, Z; Douglas, GB; and Adeney, JA. 2003. Application of Phoslock™, an innovative phosphorus binding clay, to two Western Australian waterways—preliminary findings. Hydrol Proc. 494:237–243.

Ruley, JE and Rusch, KA. 2002. An assessment of long-term post-restoration water quality trends in a shallow, subtropical, urban hypereutrophic lake. Ecol Eng. 19:265–280.

Ryding, S-O. 1982. Lake Trehörningen restoration project. Changes in water quality after sediment dredging. Hydrobiologia. 92:549–558.

Ryding, S-O and Forsberg C. 1976. Six polluted lakes: A preliminary evaluation of the treatment and recovery processes. Ambio. 5:151–156.

Sato, T; Qadir, M; Yamamoto, S; Endo, T; and Zahoor, A. 2013. Global, regional, a country level need for data on wastewater generation, treatment, and use. Agr Water Manage. 130:1–13.

Schindler, DW. 2006. Recent advances in the understanding and management of eutrophication. Limnol. Oceanogr. 51:356–363.

Schultz, J; Rydin, E; and Huser, BJ. 2017. A newly developed injection method for aluminum treatment in eutrophic lakes: effects on water quality and phosphorus binding efficiency. Lake Reserv Manage. 33:152–162.

Seida, Y and Nakano, Y. 2000. Removal of humic substances by layered double hydroxide containing iron. Water Res. 34:1487–1494.

Sharpley, A; Jarvie, HP; Buda, A; May, L; Spears, B; and Kleinman, P. 2013. Phosphorus legacy: overcoming the effects of past management practices to mitigate future water quality impairment. J Environ Qual. 42:1308–1326.

Shi, WQ; Tan, WQ; Wang, LJ; and Pan, G. 2015. Removal of *Microcystis aeruginosa* using cationic starch modified soils. Water Res. 97:19–25.

Shin, HS; Kim, MJ; Nam, SY; and Moon, HC. 1996. Phosphorus removal by hydrotalcite compounds (HTLcs). Water Sci Tech. 34:161–168.

Singleton, V and Little, JC. 2006. Designing hypolimnetic aeration and oxygenation systems—a review. Environ Sci Technol. 40:7512–7520.

Sleiman, N; Deluchat, V; Wazne, M; Mallet, M; Courtin-Nomade, A; Kazpard, V; and Baudu, M. 2016. Phosphate removal from aqueous solution using ZVI/sand bed reactor: behaviour and mechanism. Water Res. 99:55–65.

Smolders, AJP; Lamers, LPM; Lucassen, ECHET; Van Der Velde, G; and Roelofs, JGM. 2006. Internal eutrophication: how it works and what to do about it—a review. Chem Ecol. 22:93–111.

Smolka-Danielowska, D and Fiedor, D. 2018. Potentially toxic elements in fly ash dependently of applied technology of hard coal combustion. Environ Sci Pollut Res. 25:25091–25097.

Soltero, RA; Sexton, LM; Ashley, KI; and McKee, KO. 1994. Partial and full lift hypolimnetic aeration of Medical Lake, way to improve water quality. Water Res. 28:2297–2308.

Søndergaard, M; Jensen, JP; and Jeppesen, E. 1999. Internal phosphorus loading in shallow Danish lakes. Hydrobiologia. 408/409:145–152.

Søndergaard, M; Jeppesen, E; and Jensen, JP. 2000a. Hypolimnetic nitrate treatment to reduce internal phosphorus loading in a stratified lake. Lake Reserv Manage. 16:195–204.

Søndergaard, M; Jeppesen, E; Jensen, JP; and Lauridsen, T. 2000b. Lake restoration in Denmark. Lakes Reserv: Res Manage. 5:151–159.

Spears, BM; Meis, S; Anderson, A; and Kellou, M. 2013a. Comparison of phosphorus (P) removal properties of materials proposed for the control of sediment p release in UK lakes. Sci Total Environ. 442:103–110.

Spears, BM; Lürling, M; Yasseri, S; Castro-Castellon, AT; Gibbs, M; Meis, S; McDonald, C; McIntosh, J; Sleep, D; and Van Oosterhout, F. 2013b. Lake responses following lanthanum-modified bentonite clay Phoslock® application: an analysis of water column lanthanum data from 16 case study lakes. Water Res. 47:5930–5942.

Spears, BM; Mackay, EB; Yasseri, S; Gunn, IDM; Waters, KE; Andrews, C; Cole, S; de Ville, M; Kelly, A; Meis, S; et al. 2016. Lake responses following lanthanum-modified bentonite (Phoslock®) application: a meta-analysis of water quality and aquatic macrophyte responses across 18 lakes. Water Res. 97:111–121.

Steinman, AD and Ogdahl, ME. 2012. Macroinvertebrate response and internal phosphorus loading in a Michigan Lake following alum treatment. J Environ Qual. 41:1540–1548.

Stiernstrom, S; Linde, M; Hemstrom, K; Wik, O; Ytreberg, E; Bengtsson, B-E; and Breitholtz, M. 2013. Improved understanding of key elements governing toxicity of energy ash eluates. Waste Manage. 33:842–849.

Stroom, JM and Kardinaal, WEA. 2016. How to combat cyanobacterial blooms: strategy toward preventive lake restoration and reactive control measures. Aquat Ecol. 50:541–576.

Sun, S; Wang, L; Huang, S; Tu, T; and Sun, H. 2011. The effect of capping with natural and modified zeolites on the release of phosphorus and organic contaminants from river sediment. Front Chem Sci Eng. 5:308–313.

Sychra, J and Adámek, Z. 2011. The impact of sediment removal on the aquatic macroinvertebrate assemblage in a fishpond littoral zone. J Limnol. 70:129–138.

Tammeorg, O; Möls, T; Niemistö, J; Holmroos, H; and Horppila, J. 2017. The actual role of oxygen deficit in the linkage of the water quality and benthic phosphorus release: Potential implications for lake restoration. Sci Total Environ. 599–600:732–738.

Thomas, JA; Funk, WH; Moore, BC; and Budd, WW. 1994. Short term changes in Newman Lake following hypolimnetic aeration with the Speece Cone. Lake Reserv Manage. 9:111–113.

Toffolon, M; Ragazzi, M; Righetti, M; Teodoru, CR; Tubino, M; Defrancesco, C; and Pozzi, S. 2013. Effects of artificial hypolimnetic oxygenation in a shallow lake. Part 1: Phenomenological description and management. J Environ Manage. 114:520–529.

Van der Does, J; Verstraelen, P; Boers, P; Van Roestel, J; Roijackers, R; and Moser, G. 1992. Lake restoration with and without dredging of phosphorus-enriched upper sediment layers. Hydrobiologia. 233:197–210.

Van Liere, L; Gulati, RD; Wortelboer, FG; and Lammens, EHHR. 1990. Phosphorus dynamics following restoration measures in the Loosdrecht Lakes (The Netherlands). Hydrobiologia. 191:87–95.

van Loosdrecht, MCM and Brdjanovic, D. 2014. Anticipating the next century of wastewater treatment. Science. 344:1452–1453.

Veiga, MM; Silva, DM; and Veiga, LBE. 2013. Managing water quality in a polluted lake of southeast Brazil. Int J Sus Dev Plann. 8:158–172.

Vlasbom, W. 2003. Dredging equipment and technology, Delft University of Technology. https://dredging.org/content/content.asp?menu=1000_53.

Waajen, G; van Oosterhout, F; Douglas, G; and Lürling, M. 2016. Management of eutrophication in Lake De Kuil (The Netherlands) using combined flocculant—lanthanum modified bentonite treatment. Water Res. 97:83–95.

Waajen, GWAM; Faassen, EJ; and Lürling, M. 2014. Eutrophic urban ponds suffer from cyanobacterial blooms: Dutch examples. Environ Sci Pollut Res. 21:9983–9994.

Wang, XY and Feng, J. 2007. Assessment of the effectiveness of environmental dredging in South Lake, China. Environ Manage. 40:314–322.

Wang, Z; Fan, Y; Qu, F; Wu, D; and Kong, H. 2016a. Synthesis of zeolite/hydrous lanthanum oxide composite from coal fly ash for efficient phosphate removal from lake water. Micropor Mesopor Mat. 222:226–234.

Wang, Z; Shen, D; Shen, F; and Li, T. 2016b. Phosphate adsorption of lanthanum loaded biochar. Chemosphere. 150:1–7.

Wang, G; Wang, Y; and Zhang, Y. 2018. Combination effect of sponge iron and calcium nitrate on severely eutrophic urban landscape water: an integrated study from laboratory to fields. Environ Sci Pollut Res. 25:8350–8363.

Wauer, G; Gonsiorczyk, T; Kretschmer, K; Casper, P; Koschel, R. 2005. Sediment treatment with a nitrate-storing compound to reduce phosphorus release. Water Res. 39:494–500.

Wendling, LA; Douglas, GB; Coleman, S; and Yuan, Z. 2012. Nutrient and dissolved organic carbon removal from water using mining and metallurgical by-products. Water Res. 46:2705–2717.

Wetzel, R. 2001. Limnology. Lake and River Ecosystems. 3rd ed. San Diego: Academic Press.

Whitely, S. 2017. Eutrophication in coastal New Zealand lakes and the mitigation potential of phosphorus immobilisation using clay based amendments. B Agric. Sci. (Hons) Dissertation, Lincoln University, 41pp.

Wiśniewski, R; Slusarczyk, J; Kaliszewski, T; Szulczewski, A; and Nowacki, P. 2010. "Proteus", a new device for application of coagulants directly to sediment during its controlled resuspension. Verh Internat Verein Limnol. 30:1421–1424.

Xie, J; Wang, Z; Lu, S; Wu, D; Zhang, Z; and Kong, H. 2014. Removal and recovery of phosphate from water by lanthanum hydroxide materials. Chem Eng J. 254:163–170.

Xu, D; Ding, S; Sun, Q; Zhong, J; Wu, W; and Jia, F. 2012. Evaluation of in situ capping with clean soils to control phosphate release from sediments. Sci Total Environ. 438:334–341.

Yamada, TM; Sueitt, APE; Beraldo, DAS; Botta, CMR; Fadini, PS; Nascimento, MRL; Faria, BM; and Mozeto, AA. 2012. Calcium nitrate addition to control the internal load of phosphorus from sediments of a tropical eutrophic reservoir: Microcosm experiments. Water Res. 46:6463–6475.

Yin, H; YeYun, Y; Zhang, Y; and Fan, C. 2011. Phosphate removal from wastewaters by a naturally occurring, calcium-rich sepiolite. J Hazard Mater. 198:362–369.

Yu, X; Grace, MR; Sun, G; and Zou, Y. 2018. Application of ferrihydrite and calcite as composite sediment capping materials in a eutrophic lake. J Soils Sediments. 18:1185–1193.

Yu, Y and Chen, JP. 2015. Key factors for optimum performance in phosphate removal from contaminated water by a Fe–Mg–La tri-metal composite sorbent. J Colloid Interf Sci. 445:303–311.

Yuan, G and Wu, L. 2006. Allophane nanoclay for the removal of phosphorus in water and wastewater. Sci Technol Adv Mat. 8:60–62.

Zamparas, M; Deligiannakis, Y; and Zacharias, I. 2013. Phosphate adsorption from natural waters and evaluation of sediment capping using modified clays. Desalin Water Treat. 51:2895–2902.

Zhang, C; Zhu, MY; Zeng, GM; Yu, ZG; Cui, F; Yang, ZZ; and Shen, LQ. 2016. Active capping technology: a new environmental remediation of contaminated sediment. Environ Sci Pollut Res. 23:4370–4386.

Zhang, S; Zhou, Q; Xu, D; Lin, J; Cheng, S; and Wu, Z. 2010. Effects of sediment dredging on water quality and zooplankton community structure in a shallow of eutrophic lake. J Environ Sci. 22:218–224.

Zheng, L; Li, X; and Liu, J. 2004. Adsorptive removal of phosphate from aqueous solutions using iron oxide tailings. Water Res. 38:1318–1326.

Section II

Case Studies from Around the World

OBSERVED AND MODELED INTERNAL PHOSPHORUS LOADS IN STRATIFIED AND POLYMICTIC BASINS OF A MESOTROPHIC LAKE IN CANADA

Gertrud Nürnberg[1]

Abstract

Observed internal phosphorus (P) load was determined from total P (TP) increases of the euphotic layer for 32 years of two sections of mesotrophic Lake Simcoe, Canada. Modeled internal load was determined from laboratory release rates (RRs) and two measures of the spatial and temporal extent of hypoxia. Hypoxia was determined: (1) from dissolved oxygen (DO) profiles (< 3.5 mg L^{-1} DO) as the hypoxic factor (HF); and (2) modeled as an active release area factor (AA). Both modeled internal load estimates were similar to each other (~95 mg m^{-2} summer^{-1}) but substantially lower than observed estimates (151 mg m^{-2} summer^{-1}) in the stratified bay. This discrepancy was likely caused by underestimated RRs, because the HF and the HF-based internal load estimates were both significantly correlated with the observed internal load estimates throughout 1980–2011. In the polymictic basin, the modeled estimates based on predicted AA (82 mg m^{-2} summer^{-1}) were similar (not significantly different, t-test) to observed estimates (86 mg m^{-2} summer^{-1}). Internal load based on HF (41 mg m^{-2} summer^{-1}) was smaller, presumably because HF underestimates the extent of sediment anoxia due to frequent exchanges with aerated water in the polymictic basin. But model performance differed with time periods and predicted internal load based on modeled AA estimates were similar to observed internal load in both basins in 2000–2011. Further, the smoothing of observed and modeled variables by using three-year averages drastically increased their correlations, demonstrating that these estimates are more meaningful for groups of years and may not apply well to individual years.

Key words: Internal phosphorus load, summer; release rate; hypoxic factor model; Lake Simcoe, Ontario

[1] Freshwater Research, 3421 Hwy 117, RR.1, Baysville, Ontario, P0B 1A0 Canada. E-mail: gkn@fwr.ca.

6.1   INTRODUCTION

This contribution presents several relatively simple evaluations of internal P loading depending on the availability of monitoring data and other lake information. The advantages of simple modeling methods are not just limited effort and cost, but more importantly, the verification of model results by comparison with observed estimates (see Chapter 3). However, observed estimates are subject to variability and uncertainty, which has to be accounted for in addition to modeling uncertainties (Brett et al. 2016). Exploration of variabilities in internal load estimates include dependencies on weather-related variables, on confounding factors including other internal sources such as sediment resuspension and macrophyte senescence, and on methodological idiosyncrasies. Uncertainty in internal load estimates is especially large in polymictic systems, where timing and sources of sediment P release are not readily visible.

This study investigates errors and uncertainties in the observed and predicted internal load estimates of deep and stratified Kempenfelt Bay and the large polymictic Main Basin of Lake Simcoe, Ontario, Canada. Detailed descriptions of the individual internal load estimates for 32 years and characteristics in all basins of Lake Simcoe are presented in Nürnberg et al. (2013a; 2013b).

6.2   METHODS

6.2.1   Lake Characteristics

Lake Simcoe is a large (722 km^2) mesotrophic lake (see Figure 6.1) located in a densely populated area of southern Ontario. Its high external P inputs have been managed since the 1990s in an effort to prevent the occurrence of end-of-summer hypolimnetic DO levels that were lethal to cold-water fish species (Young et al. 2011).

Of the three distinct lake sections, this study concentrates on the Main Basin comprising the largest surface area, and Kempenfelt Bay, a fjord-like basin. Even though Lake Simcoe is relatively deep at 41 m maximum depth, the morphometric index (mean depth, m, divided by the square root of area, km^2) of its sections is variable (see Table 6.1). The higher morphometric index for Kempenfelt Bay indicates a stronger stratification compared to the low value for the Main Basin that indicates frequent exchange between the bottom and surface layers.

Table 6.1   Stations, morphometry, and average TP of the studied sections of Lake Simcoe. The station names correspond with Figure 6.1. (Revised from Nürnberg et al. 2013a, with permission).

	Kempenfelt Bay	**Main Basin**	**Total***
Stations	K39,K42	E51,S15,K45	All
Catchment basin area (km^2)	133.9	2,042.2	2,898.6
Area (km^2)	35.7	641.5	722.4
Volume (10^6 m^3)	918.1	9,789.1	11,043.3
Maximum Depth (m)	41	38	41
Mean Depth (m)	25.7	15.3	15.3
Morphometric Index (m km^{-1}), $z/A_o^{0.5}$	4.30	0.60	n.m.
TP (May–Sep, µg L^{-1})**	13.5	13.5	13.6

*Total includes Cooke's Bay
**1980–2011 average of euphotic (composite) samples, no long-term trend
n.m., not meaningful
Source for morphometry: Ontario Ministry of Natural Resources

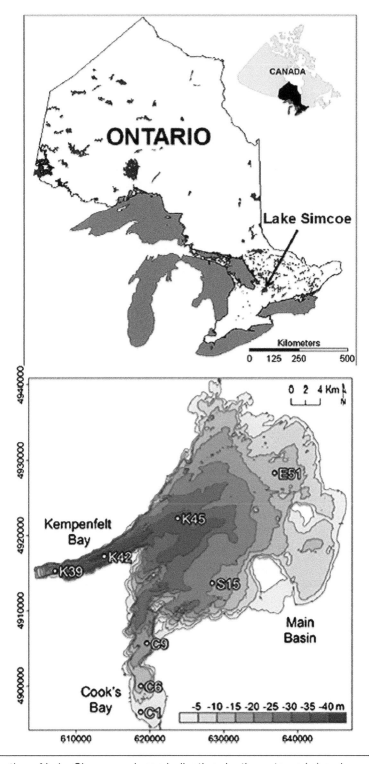

Figure 6.1 Location of Lake Simcoe and map indicating depth contours (m) and sample stations (circle) for the three sections (Source: Ontario Ministry of Natural Resources and Lake Simcoe Region Conservation Authority). (Redrawn from Nürnberg et al. 2013b, with permission.)

Lake Simcoe is well-studied (Palmer et al. 2013), and its water quality has been extensively monitored since 1980 by local and regional agencies. Methods pertinent to sampling and data analysis are described in Nürnberg et al. (2013a; 2013b).

6.2.2 Observed Internal Load from In Situ Phosphorus Increases

Variations in the TP mass in the water column over the summer and fall describe an internal P load (L_{int_1}), when external sources are at steady state or considered explicitly in the calculations. L_{int_1} was determined from Equation 6.1 (Nürnberg et al. 2013a).

$$L_{int\_1} = (TP\_t_2 \times V\_t_2 - TP\_t_1 \times V\_t_1) / (A_o) \qquad \text{[Eq. 6.1]}$$

where t_1 is initial date (in July) and t_2 is date at end of period (in October or early November). V_t is the corresponding volume and A_o is the corresponding surface area for each lake basin (see Table 6.1). TP_t is the corresponding concentration of two different measures of water column TP, as explained next.

In these calculations, euphotic zone TP concentrations (TP_{eu}) for each of the lake sections were estimated from volumetrically prorated composite data of eight stations for 32 years (1980–2011) (see Figure 6.1) (Nürnberg et al. 2013b). Composite samples consisted of water samples taken with a tube from the surface to a depth of 2.5 times Secchi disk transparency to a maximum sampling depth (e.g., 15 m in Kempenfelt Bay).

Water column TP averages (TP_{avg}) were calculated for Kempenfelt Bay and the Main Basin (see Equation 6.2), where discrete depth samples (usually taken at 1, 5, and 10 m above bottom from 2000 to 2011) were combined with the respective section TP_{eu}.

$$TP_{avg} = (TP_{eu} \times V_{eu} + TP_1 \times V_1 + TP_2 \times V_2 + TP_3 \times V_3) / V_0 \qquad \text{[Eq. 6.2]}$$

where TP_1, TP_2, and TP_3 are TP concentrations of the three discrete depth hypolimnetic samples and V_i is the volume of the corresponding water layer with V_{eu}, the volume of the euphotic depth (i.e., 0–25 m for K42 and 0–20 m for K45).

L_{int_1} is the only measure of internal loading that is solely based on observed input in this study; and it can be considered to be as close to an *observed* estimate as is possible. The two different estimates of lake TP concentration, TP_{eu} and TP_{avg}, were used to illuminate errors and uncertainty associated with L_{int_1}. The corresponding internal load estimates were called L_{int_1eu} and L_{int_1avg}.

6.2.3 Modeled Internal Load from Sediment Release Rates and Sediment Release Area

Predicted internal load was determined as the product of a release rate and a factor representing the annual spatial and temporal extent of the sediments involved in P release (see Equation 6.3) (Nürnberg 1987).

$$L_{int\_2} = RR \times Factor \qquad \text{[Eq. 6.3]}$$

where RR is the anoxic areal P release rate per anoxic sediment surface area ($mg\ m^{-2}\ d^{-1}$), and Factor is the number of days per season an area the size of the lake is sufficiently hypoxic (see Section

6.2.3.2) to release P. Factors were either obtained as calculations from DO profiles (hypoxic factor, HF) or modeled as active sediment AA. The corresponding internal load estimates were called L_{int_2HF} and L_{int_2AA}.

6.2.3.1 RR

Annually variable summer RRs (RR_i, Equation 6.4, n = 3, R = 0.96, after logarithmic transformation of RR values to base 10) were determined from temperature-RR relationships for Lake Simcoe (based on sediment incubation and release experiments) (Loh et al. 2013) and mean August to October water temperature ($T_{Aug-Oct}$, °C) of the hypolimnion (Nürnberg et al. 2013a).

$$\log (RR_i) = 0.314 + 0.0304 \times T_{Aug-Oct} \qquad \text{[Eq. 6.4]}$$

6.2.3.2 HF and AA

Observed HF for the prediction of internal load (see Equation 6.3) was determined throughout 1980–2011 with Equation 6.5 (in units of days summer^{-1}) (Nürnberg 2004).

$$HF = \sum_{i=1}^{n} \frac{t_i \times a_i}{A_o} \qquad \text{[Eq. 6.5]}$$

where t_i is the period of hypoxia (days), a_i is the corresponding lake area that is hypoxic (m^2), A_o is the surface area (m^2), and n is the number of periods with different depths for the specific hypoxic level considered. The threshold of hypoxia chosen for the prediction of internal load according to Equation 6.3 was 3.5 mg L^{-1} DO in Lake Simcoe, because of an observed homogenous and thick hypoxic layer (Nürnberg et al. 2013b). Such an unusually thick hypoxic layer (average annual maximum extent of 20 m above bottom in both lake sections) was explained by the dynamic nature of the Lake Simcoe water masses as expected by the small depth to lake area ratio and caused by observed internal seiches and high winds (Chowdhury et al. 2015; Cossu et al. 2017). In lakes with a more solid stratification, a lower threshold of 1–2 mg L^{-1} hypolimnetic DO concentration would more adequately reflect anoxic sediment-water interfaces (see Chapter 3) (Nürnberg 2009).

In Lake Simcoe, L_{int_2HF}, which is based on DO profiles, is applicable only to Kempenfelt Bay, because HF underestimates the active sediment release area in shallower systems exposed to aeration during mixing. Direct estimates of sediment-water interface anoxia were not available, which is typical in polymictic lakes. Thus, L_{int_2HF} delivers a low or, perhaps, minimum estimate for the Main Basin.

A more realistic estimate of the extent and duration of the sediment release area in the polymictic Main Basin, where even under mixed conditions a large sediment surface area can be anoxic during quiescent periods despite aerated overlying water, is the active sediment release area model (AA, days summer $^{-1}$) (see Equation 6.6) (Nürnberg 2019).

$$AA = -36.2 + 50.1 \log (TP_{summer}) + 0.762 \, z/A_o^{0.5} \qquad \text{[Eq. 6.6]}$$

where TP_{summer} is the May to September average TP_{eu}, and $z/A_o^{0.5}$ is a morphometric index, with z, mean depth (m), and A_o, surface area (km^2) of the respective lake basins. Nürnberg (2005) found that such a release area can be predicted from a model originally developed for the anoxic factor (using a DO threshold of 1–2 mg L^{-1}) for stratified lakes. L_{int_2AA} is based on AA and presents a modeled internal load estimate in both Lake Simcoe sections according to Equation 6.3.

6.2.4 Statistics

T-tests were performed to determine any significant differences between the different internal load estimates. Linear regression parameters were inspected to see whether the regression line is different from the 1:1 line of perfect prediction between observed and predicted values. A running three-year average produced smoothed time series. Significance levels of 0.05 or lower were reported. Statistical analyses were conducted with the SYSTAT statistical program, Version 13 for Windows.

6.3 RESULTS AND DISCUSSION

6.3.1 In Situ *Observed* Internal Load (L_{int_1})

L_{int_1eu} could be computed for the whole study period of 1980–2011 and presents the longest period of observed internal load. Areal internal loads in Kempenfelt Bay were 1.75 times that of the Main Basin (see Table 6.2), but because of the 18-fold greater area of the Main Basin (see Table 6.1), internal loading from the Main Basin contributes the largest portion to Lake Simcoe. These rates are similar to other mesotrophic lakes and much smaller than those of eutrophic lakes (Nürnberg 1987; Steinman and Ogdahl 2015; Tammeorg et al. 2017).

An increase in hypolimnetic P mass from mid-summer to fall can be attributed to sediment release if it exceeds net external inputs (i.e., the proportion of external load that has not been removed via sedimentation) during that period. In contrast, an epilimnetic P increase might be due to high tributary inputs, P release from macrophyte senescence, and sediment resuspension. According to Nürnberg et al. (2013a), July–October external inputs were less than mass increases in 10 out of 16 years in the Kempenfelt Bay euphotic zone and only 12% of external load could be expected to contribute to lake water TP because of settling processes.

In stratified regions, such as Kempenfelt Bay, increases in TP_{eu} are expected to underestimate internal load because hypolimnetic increases below the euphotic zone are not fully considered. The 2000–2011 Kempenfelt Bay's average of internal load computed from increases in TP_{eu} (L_{int_1eu}) is

Table 6.2 Mean and SE for different internal load estimates (mg m^{-2} yr^{-1}) for 1980–2011 and 2000–2011. L_{int_1eu} is an observed internal load based on in situ increases in the euphotic TP concentration, while observed L_{int_1avg} is based on volumetric water column average TP concentration (see Equation 6.1 and Equation 6.2). L_{int_2HF} is a modelled internal load based on DO profiles applicable to stratified Kempenfelt Bay that underestimates internal load in polymictic Main Basin. L_{int_2AA} is a modelled internal load based on Equation 6.3, where Factor is the active release area (AA, see Equation 6.6) that is applicable to stratified and polymictic lake sections.

	Observed				Predicted			
Basin	L_{int_1eu}		L_{int_1avg}		L_{int_2HF}		L_{int_2AA}	
1980–2011, n=32								
Kempenfelt Bay	151	(23.9)	n.a.		94	(10.3)	95	(3.1)
Main Basin	86	(17.7)	n.a.		41	(5.0)*	82	(3.4)
2000–2011, n=12								
Kempenfelt Bay	101	(20.1)	120	(24.1)	67	(15.6)	99	(3.8)
Main Basin	76	(20.5)	61	(16.8)	26	(6.3)*	84	(4.0)

n.a., not applicable

*Only in Main Basin is L_{int_2HF} significantly different from L_{int_1} (paired t-test, p < 0.05) slightly lower, 101 ± 20.1 mg m^{-2} yr^{-1} (mean $\pm$ standard error, SE), compared to that from TP$_{avg}$ (L$_{int\_1avg}$) of 120 ± 24.1 mg m^{-2} yr^{-1} (see Table 6.2). But standard errors are large so that the paired t-test does not indicate any significant difference ($p = 0.323$, $n = 12$). Also, the significant regression indicates a relationship close to the 1:1 line between the two estimates, because the slope is not significantly different from one and the constant not significantly different from zero (see Figure 6.2, top).

In the mixed Main Basin of Lake Simcoe, these estimates of internal loading are expected to be similar. While the 2000–2011 average L$_{int\_1eu}$ is higher at 76 ± 20.5 mg m^{-2} yr^{-1} compared to L$_{int\_1avg}$ of 61 ± 16.8 mg m^{-2} yr^{-1} (see Table 6.2), a paired t-test does not reveal a significant difference ($p = 0.465$, $n = 12$). The regression between the two estimates is only marginally significant ($p = 0.09$) (see Figure 6.2, bottom).

Estimates of L$_{int\_1}$ based on TP$_{avg}$ are 20% higher than when based on TP$_{eu}$ in Kempenfelt Bay, but 20% lower in the Main Basin. At least in stratified Kempenfelt Bay, these L$_{int\_1}$ estimates are correlated and fall close to the 1:1 line, but Main Basin estimates do not. Clearly, the observed estimates of L$_{int\_1}$ are not without error.

Nürnberg et al. (2013a) expressed the following reasons for the uncertainty of L$_{int\_1}$ estimates: (1) limited availability of hypolimnetic samples to document extreme values; (2) euphotic depth

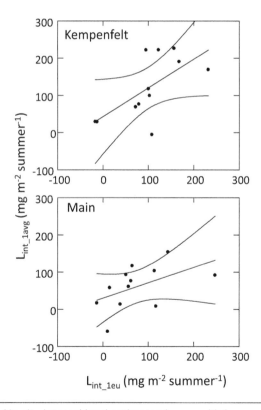

Figure 6.2 Comparison of in situ internal load estimates L$_{int\_1eu}$ with L$_{int\_1avg}$ (available for 2000–2011, $n = 12$) for stratified Kempenfelt Bay (top, $R^2 = 0.41$, $p < 0.02$, $n = 12$, or without outlier of 2006: $R^2 = 0.56$, $p = 0.01$, $n = 11$. In both regressions, the slope is not significantly, n.s., different from 1, and the constant is n.s. different from 0). In the polymictic Main Basin the regression is only marginally significant with the slope significantly different from 1 and the intercept n.s. different from 0 (bottom, $R^2 = 0.26$, $p = 0.09$). L$_{int\_1eu}$ is based on euphotic layer TP$_{eu}$ and L$_{int\_1avg}$ on water-column average TP$_{avg}$. Regression lines and 95% confidence bands are shown.

changes throughout the summer; (3) variable and unconsidered exchange with other lake sections and external loads; (4) possibility of increased TP_{eu} by resuspended sediments; and (5) loss of sediment-derived P via re-sedimentation and export during the calculation period. Whereas uncertainties 1–4 were considered non-directional and minor, uncertainty 5 decreased internal load estimates, so that the presented estimates can be considered conservative. These uncertainties are probably more pronounced in shallow Main Basin, where dreissenid mussels, macrophyte senescence (Gudimov et al. 2015), and wind action (Cossu et al. 2017) could also contribute to seasonal TP changes.

6.3.2   Predicted Internal Load Components (L_{int_2})

All L_{int_2} predictions include laboratory-determined RRs adjusted for temperature (see Equation 6.4). But there is no significant correlation between L_{int_1eu} and hypolimnetic temperature used in the modeling of RR. The RRs may be conservative or underestimated in Kempenfelt Bay (average: 4 mg m^{-2} d^{-1}, range: 3.0–4.6) for several reasons, discussed previously by Nürnberg et al. (2013a). These include: (1) low experimental rates because of possible rapid adsorption of P in the RR experiments; (2) low temperature adjustment because the mean water temperature may have occasionally been higher caused by seiches; and (3) earlier RRs before P abatement measures began in 1990 may have been higher compared to the experimental rates of 2011; low RRs would yield low L_{int_2} estimates (see Section 6.3.3).

There is some controversy about the importance of the mechanism of anoxic phosphate release from iron-oxy-hydroxides contributing to internal loading in lakes (Hupfer and Lewandowski 2008). If sediment P release is caused by the release of phosphate from iron-oxy-hydroxides under anoxic sediment conditions into the hypoxic overlying water, then stratified Kempenfelt Bay should have a better correlation of L_{int_1} with measured hypoxia than the polymictic Main Basin, where the water above the sediment is intermittently aerated even though sediment surfaces may be anoxic. To test this theory, I regressed L_{int_1eu} on HF (HF values are presented in Nürnberg et al. 2013a). These regressions are significant for Kempenfelt Bay and less so for the Main Basin (see Figure 6.3) (n = 32, Kempenfelt Bay: R^2 = 0.20, p < 0.01; Main Basin: R^2 = 0.14, p < 0.05). Regressions are much improved by using smoothed values for three-year running averages (Kempenfelt Bay: R^2 = 0.52, p < 0.0001; Main Basin: R^2 = 0.17, p < 0.02). These significant relationships support the use of HF

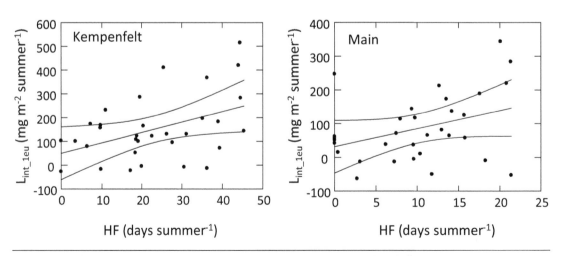

Figure 6.3   Observed internal load (L_{int_1eu}) versus HF in Kempenfelt Bay (R^2 = 0.20, p < 0.01) and the Main Basin of Lake Simcoe (R^2 = 0.17, p < 0.02) for 1980–2011 (n = 32). Regression lines and 95% confidence bands are shown.

in the prediction of internal load in stratified Kempenfelt Bay. But as expected, HF is less useful in the mixed Main Basin.

As described in Nürnberg et al. (2013b), long-term average HF (23.0 ± 2.43 days summer^{-1}) and AA (23.4 ± 0.63 d summer^{-1}) were similar in stratified Kempenfelt Bay and paired t-tests between annual values did not show any significant difference (n = 32, p = 0.872). But, as expected, HF (10.2 ± 1.21 d summer^{-1}) was significantly lower than AA (20.5 ± 78.5 d summer^{-1}) in the mixed Main Basin (paired t-test, n = 32, p < 0.0001), presumably because HF here underestimates the extent of sediment anoxia due to frequent exchanges with aerated water.

Annual values of AA and HF were not significantly correlated with each other in Kempenfelt Bay or in the Main Basin. This is understandable because the AA model (see Equation 6.6) has been developed on a cross-sectional data base of many lakes and predicts long-term averages, but does not necessarily apply to individual annual values. Also, the only variability captured in the AA model is due to annual TP differences that were small during the study period. Morphometric relationships do not change markedly in lakes, rendering a rather constant predicted AA with time in the investigated sections of Lake Simcoe (Nürnberg et al. 2013b). Because the Lake Simcoe AA model was verified by the Kempenfelt Bay long-term average HF, it is the best predictor of active sediment areas in the mixed Main Basin.

The steadiness of AA is carried over into the L_{int_2AA} estimates that also have little variability compared to L_{int_2HF} (see Figure 6.4). Obviously, the benefits of using L_{int_2HF} rather than L_{int_2AA} include observed variability in temporal patterns and trends. For example, Nürnberg et al. (2013a)

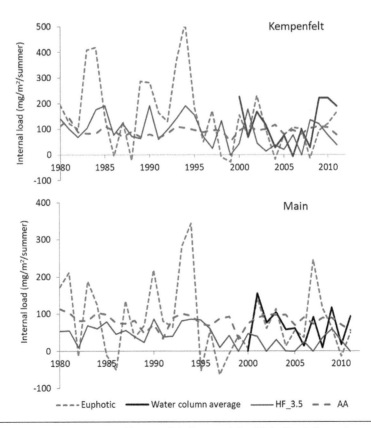

Figure 6.4 Internal load estimates for Kempenfelt Bay and the Main Basin based on euphotic (L_{int_1eu}) and water column average (L_{int_1avg}) TP concentration, and the product of RRs with HF (L_{int_2HF}) and with AA, active area (L_{int_2AA}). (Redrawn from Nürnberg et al. 2013a, with permission).

determined a significant annual decrease of 1.41 mg m^{-2} yr^{-1} (Mann-Kendall, p < 0.01) or 1.46 mg m^{-2} yr^{-1} (linear regression, p < 0.01, R^2 = 0.23, n = 32) for L_{int_2HF}, but not for L_{int_2AA} in the Lake Simcoe basins combined.

6.3.3   Comparison of Internal Load Estimates

While it is important to consider uncertainty in observed values (Brett et al. 2016) (see Chapter 3), observations are the only direct way for the verification of model estimates.

I evaluated the performance of the L_{int_2} model by comparison of model results with L_{int_1eu} values (see Table 6.2). Long-term average L_{int_2HF} was significantly lower than L_{int_1eu} in Kempenfelt Bay (t-test, p < 0.02) and in Main Basin (t-test, p < 0.02), but L_{int_2AA} was similar when compared to L_{int_1eu} in Main Basin. Both L_{int_2} estimates were similar to each other and lower than L_{int_1eu} in Kempenfelt Bay.

L_{int_2HF} for Kempenfelt Bay was significantly correlated with observed *in situ* L_{int_1eu} (n = 32, R = 0.43, p < 0.02) and smoothing of both variables by using 3-year averages drastically increases the correlation (n = 32, R = 0.71, p < 0.0001) (see Figure 6.5). This demonstrates that these estimates are more meaningful for conditions across groups of years and may not apply well to individual years. A similar procedure also increases the correlation coefficient for Main Basin and yields a significant relationship only after smoothing (n = 32, R = 0.39, p < 0.05). However, both the untransformed and the smoothed (see Figure 6.5) regressions in both lake sections have constants significantly different from zero and slopes that are significantly different from 1, indicating that observed and predicted values are not in close agreement.

The much lower predicted than observed internal load in Kempenfelt Bay (see Table 6.2) (L_{int_2HF} or L_{int_2AA} are 0.63 of L_{int_1}) is unexpected, because the method for calculating L_{int_1} includes the settled portion that would have occurred throughout the period of calculation. In contrast, L_{int_2} estimates do not include any settling and are considered gross estimates (Nürnberg 2009). The discrepancy is most likely caused by underestimated RR as discussed in Section 6.3.2, because earlier

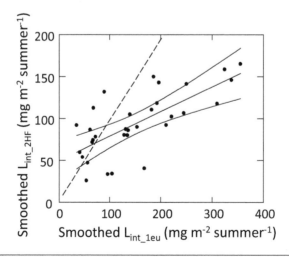

Figure 6.5   Comparison of Kempenfelt Bay observed internal load L_{int_1eu} with predicted L_{int_2HF} after smoothing of both variables by using three-year averages. The solid line represents regression line [y = 49.6 (9.3) + 0.29 (0.05) x] embraced by the 95% confidence band; the broken line represents perfect prediction (1:1 line).

RR may have been higher. Underestimation of the HF term of Equation 6.3 is less likely, because of the significant correlation of L_{int_1} with HF and also because of the similarity of HF with predicted AA. That the experimentally derived RR (in 2011) is more adequate for recent years is supported by more similar L_{int_2AA} estimates compared to L_{int_1} in both basins for a subset of more recent years (2000–2011) (see Table 6.2). Also, L_{int_2HF} in Kempenfelt Bay is not significantly different from L_{int_1} for that period.

Different from Kempenfelt Bay, L_{int_1} is similar to L_{int_2AA} in Main Basin, which is consistent with the applied RR. It is also possible that polymictic Main Basin internal loading is controlled by more than redox-related P release. Some of these processes may re-adsorb P, including through the resuspension of P adsorbing material in shallow areas (Cyr et al. 2009) of mesotrophic lakes (Tammeorg et al. 2017), or chemically controlled dissolution processes may increase desorption in more eutrophic, wind exposed lakes (Tammeorg et al. 2016, 2017).

Similarly, the L_{int_2} model (see Equation 6.3) was supported by L_{int_1} estimates based on the difference between maximum hypolimnetic P mass, close to the end of anoxic stratification, and before the onset of hypolimnetic anoxia, for 34 lake years in 8 stratified lakes with a range of trophic states (Nürnberg 1987).

6.3.4 Factors Influencing Magnitude and Frequency of Internal Loading in Lake Simcoe and Further Approaches to Estimate Internal Load

Nürnberg et al. (2013a) determined two periods (1980–1994 and 1995–2011) when changes (decreases) of L_{int_1} averages coincided with dreissenid invasion (1995), external load abatement (decreased external input by 53% after 1998), and decreased hypoxia (decreased HF by 56–66%) in separate and combined sections of Lake Simcoe. This is also evident from smaller internal load average estimates for the recent period of 2000–2011 versus the whole period of 1980–2011 (see Table 6.2). Obviously, these factors influence the magnitude of internal loading, as do other well-known factors such as temperature (Liu et al. 2017), stratification (Tammeorg et al. 2017), and trophic state (Nürnberg 2009) (see Chapters 1 and 23).

Even though two independent approaches, *in situ* observed L_{int_1} and predicted L_{int_2}, were used to estimate internal load in Lake Simcoe, other approaches can be used to support these results. Therefore, Nürnberg (2009) also proposed to determine an annual P mass balance that considers the difference between incoming and outgoing P corrected for settling, in order to indicate internal load (see Chapter 3). This approach yields net estimates (after a year of settling) that can be converted to gross estimates for comparison with the approaches used here, as was done in previous studies (Nürnberg et al. 2012; Nürnberg and LaZerte 2016). But complete annual mass balance data were not available for Lake Simcoe basins. While accessible external loads could be used for comparison with internal load estimates (Nürnberg et al. 2013b), there were no outflow loads or P export data available for the individual lake sections so that this approach could not be used.

Another approach is a more detailed sediment diagenetic modeling that involves extensive model input (see Chapter 3). Such a non-steady state reactive transport diagenetic model has been applied to Lake Simcoe (McCulloch et al. 2013; Gudimov et al. 2015). Internal load estimates for the Main Basin are much lower than presented here, most likely because model input was constrained by rates computed as Fick's diffusion (Dittrich et al. 2013), which is known to underestimate sediment RRs (see Chapter 3) and, possibly by underestimating the anoxic sediment surfaces by using oxygen conditions of the lake water.

6.4   CONCLUSIONS

Differences in thermal stratification cause different internal load model functioning. In the polymictic basin of mesotrophic Lake Simcoe, increases in euphotic zone TP (i.e., observed internal load) were best predicted by a model that considered P RRs and predicted active sediment area. In the stratified basin, observed internal load was underestimated, depending on the period of observations. Comparison of observed internal load estimates with modeled internal load estimates over 32 years for both sections demonstrated that the modeled estimates are more meaningful for groups of years and may not apply well to individual years.

6.5   ACKNOWLEDGMENTS

Former work used in this study was funded by Environment Canada's Lake Simcoe Clean-Up Fund. Data provision and support by staff of Lake Simcoe Region Conservation Authority and the Ontario Ministry of the Environment are gratefully acknowledged. Comments from Bruce LaZerte and the editors are greatly appreciated.

6.6   REFERENCES

Brett, MT; Ahopelto, SK; Brown, HK; Brynestad, BE; Butcher, TW; Coba, EE; Curtis, CA; Dara, JT; Doeden, KB; Evans, KR; et al. 2016. The modeled and observed response of Lake Spokane hypolimnetic dissolved oxygen concentrations to phosphorus inputs. Lake Reserv Manage. 32:243–255.

Chowdhury, MR; Wells, MG; and Cossu, R. 2015. Observations and environmental implications of variability in the vertical turbulent mixing in Lake Simcoe. J Great Lakes Res. 41:995–1009.

Cossu, R; Ridgway, MS; Li, JZ; Chowdhury, MR; and Wells, MG. 2017. Wash-zone dynamics of the thermocline in Lake Simcoe, Ontario. J Great Lakes Res. 43:689–699.

Cyr, H; McCabe, SK; and Nürnberg, GK. 2009. Phosphorus sorption experiments and the potential for internal phosphorus loading in littoral areas of a stratified lake. Water Res. 43:1654–1666.

Dittrich, M; Chesnyuk, A; Gudimov, A; McCulloch, J; Quazi, S; Young, J; Winter, J; Stainsby, E; and Arhonditsis, GB. 2013. Phosphorus retention in a mesotrophic lake under transient loading conditions: Insights from a sediment phosphorus binding form study. Water Res. 47:1433–1447.

Gudimov, A; Kim, D-K; Young, JD; Palmer, ME; Dittrich, M; Winter, JG; Stainsby, E; and Arhonditsis, GB. 2015. Examination of the role of dreissenids and macrophytes in the phosphorus dynamics of Lake Simcoe, Ontario, Canada. Ecol Inform. 26:36–53.

Hupfer, M and Lewandowski, J. 2008. Oxygen controls the phosphorus release from lake sediments—a long-lasting paradigm in limnology. Int Rev Hydrobiol. 93:415–432.

Liu, Q; Ding, S; Chen, X; Sun, Q; Chen, M; and Zhang, C. 2017. Effects of temperature on phosphorus mobilization in sediments in microcosm experiment and in the field. Appl Geochem. 88:158–166.

Loh, PS; Molot, LA; Nürnberg, GK; Watson, SB; and Ginn, B. 2013. Evaluating relationships between sediment chemistry and anoxic phosphorus and iron release across three different water bodies. Inland Waters 3:105–117.

McCulloch, J; Gudimov, A; Arhonditsis, G; Chesnyuk, A; and Dittrich, M. 2013. Dynamics of P-binding forms in sediments of a mesotrophic hard-water lake: Insights from non-steady state reactive-transport modeling, sensitivity and identifiability analysis. Chem Geol. 354:216–232.

Nürnberg, GK. 1987. A comparison of internal phosphorus loads in lakes with anoxic hypolimnia: laboratory incubations versus hypolimnetic phosphorus accumulation. Limnol Oceanogr. 32:1160–1164.

Nürnberg, GK. 2004. Quantified hypoxia and anoxia in lakes and reservoirs. Sci World J 4:42–54.

Nürnberg, GK. 2005. Quantification of internal phosphorus loading in polymictic lakes. Verh Intern Verein Limnol. 29:623–626.

Nürnberg, GK. 2009. Assessing internal phosphorus load—problems to be solved. Lake Reserv Manage. 25:419–432.

Nürnberg, GK. 2019. Quantification of anoxia and hypoxia in water bodies (2). In: Water Encyclopedia. John Wiley & Sons, Inc.

Nürnberg, GK and LaZerte BD. 2016. More than 20 years of estimated internal phosphorus loading in polymictic, eutrophic Lake Winnipeg, Manitoba. J Great Lakes Res. 42:18–27.

Nürnberg, GK; LaZerte, BD; Loh, PS; and Molot, LA. 2013a. Quantification of internal phosphorus load in large, partially polymictic and mesotrophic Lake Simcoe, Ontario. J Great Lakes Res. 39:271–279.

Nürnberg, GK; Molot, LA; O'Connor, E; Jarjanazi, H; Winter, JG; and Young, JD. 2013b. Evidence for internal phosphorus loading, hypoxia and effects on phytoplankton in partially polymictic Lake Simcoe, Ontario. J Great Lakes Res. 39:259–270.

Nürnberg, GK; Tarvainen, M; Ventelä, A-M; and Sarvala, J. 2012. Internal phosphorus load estimation during biomanipulation in a large polymictic and mesotrophic lake. Inland Waters 2:147–162.

Palmer, ME; Hiriart-Baer, VP; North, RL; and Rennie, MD. 2013. Toward a better understanding of Lake Simcoe through integrative and collaborative monitoring and research. Inland Waters 3:47–50.

Steinman, AD and Ogdahl, ME. 2015. TMDL reevaluation: reconciling internal phosphorus load reductions in a eutrophic lake. Lake Reserv Manage. 31:115–126.

Tammeorg, O; Horppila, J; Tammeorg, P; Haldna, M; and Niemistö, J. 2016. Internal phosphorus loading across a cascade of three eutrophic basins: A synthesis of short- and long-term studies. Sci Total Environ. 572:943–954.

Tammeorg, O; Möls, T; Niemistö, J; Holmroos, H; and Horppila, J. 2017. The actual role of oxygen deficit in the linkage of the water quality and benthic phosphorus release: Potential implications for lake restoration. Sci Total Environ. 599–600:732–738.

Young, JD; Winter, JG; and Molot, L. 2011. A re-evaluation of the empirical relationships connecting dissolved oxygen and phosphorus loading after dreissenid mussel invasion in Lake Simcoe. J Great Lakes Res. 37, Suppl. 3:7–14.

CHAPTER 7

INTERNAL PHOSPHORUS LOADS IN SUBTROPICAL SHALLOW LAKES: TWO FLORIDA LAKES AS CASE EXAMPLES

K. Ramesh Reddy[1], Todd Z. Osborne[1,2], Dean R. Dobberfuhl[3], and Laura K. Reynolds[1,4]

Abstract

Long-term phosphorus (P) loading to lakes has resulted in accumulation of P in sediments. Internal nutrient loading from sediments of shallow lakes has become a major concern in restoration programs. In this chapter, we review the results of the research conducted on two highly managed subtropical shallow lakes, Lake Apopka (located in Central Florida, USA) and Lake Okeechobee (located in South Florida, USA). We focus on: (a) the storage of P in sediments; (b) the influence of biogeochemical processes regulating P reactivity and mobility; and (c) the role of legacy P in soils of the drainage basin and lake sediments in regulating restoration activities in these shallow lakes. Both Lake Apopka and Lake Okeechobee are P impacted with high P concentrations in both sediments and overlying water. Substantial storage of P in soft sediments promotes resuspension with regularity and watershed P storages indicate significant legacy P sources that can make restoration challenging. Restoration practices include: fish biomass harvest; reestablishment of submerged aquatic vegetation; chemical treatments; and watershed management activities. Legacy P and current environmental conditions in both lakes suggest restoration efforts may take longer to effect change than anticipated.

Key Words: Internal Nutrient Load; Legacy Nutrients; Sediment Nutrient Cycling; Submerged Aquatic Vegetation; Water Quality

[1] Wetland Biogeochemistry Laboratory, Soil and Water Sciences Department, Institute of Food and Agricultural Sciences, University of Florida, Gainesville, Florida, USA. email: krr@ufl.edu

[2] Whitney Laboratory of Marine Sciences, University of Florida, Gainesville, Florida, USA.

[3] Saint Johns River Water Management District, Palatka, Florida, USA.

[4] Coastal Ecology Laboratory, Soil and Water Sciences Department, Institute of Food and Agricultural Sciences, University of Florida, Gainesville, Florida, USA.

7.1 INTRODUCTION

Eutrophication of lakes can be attributed to: (1) increased external inputs from allochthonous sources of nutrients from point and nonpoint sources; and/or (2) accelerated internal nutrient cycling associated with changes in environmental conditions of bed sediments and the overlying water column (Carpenter et al. 1998; Reddy et al. 1996, 2011; Steinman et al. 2004; Jeppesen et al. 2007). For many lakes, eutrophication is often linked to only allochthonous sources of nutrients; however, autochthonous nutrient sources can be equally important. This is especially so in eutrophic lakes where large reserves of organic and inorganic bound nutrients are stored in sediments. Eutrophication of lakes resulting from allochthonous nutrient loading has led to increased efforts to monitor and assess status and trends in water quality and ecological conditions of these systems. For example, in the United States, state and federal agencies have initiated a number of programs to develop management strategies to reduce anthropogenic nutrient loads from lake watersheds.

Lakes are one of the final recipients of nutrients discharged from adjacent uplands, wetlands, and streams. Since many lakes are P limited, movement of this nutrient is of particular concern to environmental managers. As P moves through uplands and wetlands, it undergoes various biogeochemical transformations, and much of it remains within these systems (Reddy and Delaune 2008). Nonpoint sources of P dominate eutrophication processes in many lakes. Thus, in many situations, alternative land use management practices in upland ecosystems have been implemented in an effort to reduce the overall load to receiving water bodies. Effective nutrient control strategies can be implemented effectively only if the storage, fate, and transport of nutrients in uplands, ditches, canals, wetlands, and streams of the watershed/drainage basin are well understood. In this chapter, we review the results of the research conducted on two highly managed subtropical shallow lakes, Lake Apopka (located in Central Florida) and Lake Okeechobee (located in South Florida). We focus on: (a) the storage of P in soils and sediments; (b) the influence of biogeochemical processes regulating P reactivity and mobility; and (c) the role of legacy P in soils of the drainage basin and lake sediments in regulating restoration activities in these shallow lakes.

7.2 LAKE APOPKA

Lake Apopka (125 km^2; mean water depth = 1.6 m and mean water residence time = 7.5 years) located in the Upper Ocklawaha River Basin in Central Florida (latitude 28° 37′ N; longitude 81° 38′ W) is the fourth largest lake in Florida and is a headwater lake for the Harris Chain of Lakes (Hoge et al. 2003; Coveney et al. 2005; see Figure 7.1). During the early 1900s, Lake Apopka was dominated by abundant submerged and emergent aquatic vegetation and clear water that promoted a national reputation for sportfishing. As a result of the demand for row crop production, approximately 80 km^2 of marsh on the north end of the lake was diked off from the lake and drained for agriculture. Intense agricultural activities in the drained area (8,000 ha) and associated nutrient loading from the farms have impacted the lake. Soils in the farming area are characterized as organic soils or Histosols, classified as Everglades muck, depressional Euic, hyperthermic Typic Haplohemists (Furman et al. 1975). Upon draining, these organic soils were subjected to soil subsidence, biological oxidation, and compaction, resulting in a new elevation of the farming area that was approximately 2–3 m below the surface of the lake. Because of oxidation of the drained muck soils, surface elevations in the farm areas are currently below lake level.

To maintain optimal soil moisture content of organic soils to grow crops, water was controlled by a system of field ditches, canals, and large pumps. During periods of excess rainfall, water was

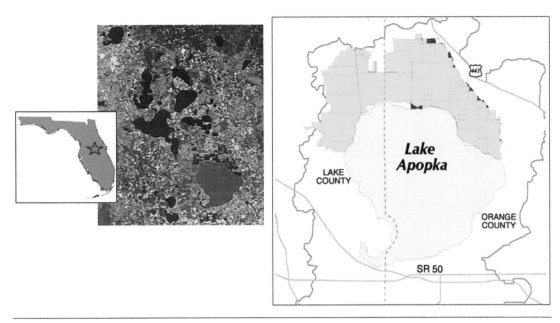

Figure 7.1   Map of Lake Apopka (*Source*: St. Johns River Water Management District).

pumped back into the lake. The installation of detention ponds on several farms allowed recycling of water on the farm and greatly reduced the necessity of pumping into Lake Apopka. Overall, the muck farms on the north shore contributed approximately 85% of the total P (TP) load to the lake. With the passage of legislation in 1996, farms were purchased through 1999 using state and federal funds. This was followed by conversion of these farmlands into wetland areas, with the goal of eliminating agricultural pumping and reducing P loads to the lake. Prior to 1996, land on the north shore of the lake was used mostly for agriculture; after purchase of these lands, this area was converted into re-flooded wetlands by 2013.

7.2.1   Long-Term Phosphorus Loads to the Lake

External nutrient loading included discharges from agricultural lands, wet and dry atmospheric deposition, basin runoff, Apopka Springs, the Winter Garden wastewater treatment plant, and groundwater. Agricultural activities on organic soils located around the north shore of the lake were the major source of nutrient loading from the catchment to the lake. During the dry season, lake water was used to irrigate crops grown on the north shore of the lake. Prior to the establishment of the restoration program, external loads accounted for approximately 62 metric tons of P per year, with agricultural drainage contributing approximately 86% of the total load, 5% from atmospheric deposition, and the remaining 4% from other unaccounted sources (Coveney 2000, 2016; see Figure 7.2). The St. Johns River Water Management District established a restoration P loading target for Lake Apopka of approximately 16 metric tons of P per year (Coveney et al. 2005). This loading target was derived through input-output modeling to meet a restoration goal for TP concentration in lake water of 0.055 mg L^{-1} (Coveney 2000). To date, restoration programs aided in substantial reduction in P loads to the lake and the P loading has met the total maximum daily load (TMDL) target (15.9 metric tons P per year) for 13 years out of 17 years during 2000–2017 (see Figure 7.2), although TP concentrations remain above the 0.055 mg L^{-1} restoration goal (see Section 7.2.2).

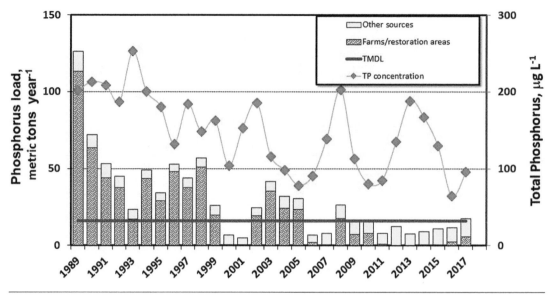

Figure 7.2   Long-term trends in phosphorus loads to Lake Apopka (Data from the St. Johns River Water Management Distirct).

7.2.2   Long-Term Surface Water Phosphorus Concentrations

Total P concentrations ranged from 200 μg L^{-1} during 1989–1994 and decreased to 100 μg L^{-1} by the year 2000. After that period, surface water TP levels showed peaks during 2002, 2008, and 2013 (see Figure 7.2). This was probably due to multiyear cycles of dry and wet conditions. Total P concentrations increased during each period of low lake stage and decreased as the lake stage reached a normal range (Coveney 2016). In essence, TP mass does not appreciably change but the increases and decreases in lake volume drive concentration changes. Over the long-term (1987–2015), surface water TP composition of Lake Apopka was 91% particulate P (PP), 5% dissolved organic P (DOP), and 4% dissolved inorganic P (soluble reactive P, SRP, also referred to as dissolved reactive P, DRP) (Coveney 2016).

7.2.3   Phosphorus Accumulation in Sediments

Total sediment depth varied significantly as reported in two spatial sediment core sampling studies conducted (1969 and 1987) in the lake (Schneider and Little 1969; Reddy and Graetz 1991). Depending on the consistency and physical structure, sediment profile was divided into several horizons. These included: unconsolidated floc (UCF), consolidated floc (CF), peat, sand, clay, and marl. Both UCF and CF were formed as result of allochthonous particulate organic matter and sedimentation of detrital material from phytoplankton and aquatic macrophytes. Peat, sand, clay, and marl represent native sediment types of the lakebed. In this paper, we review the importance of both UCF and CF layers of the sediments since P stored in these sediments can potentially be released into the overlying water column. Long-term nutrient loading resulted in hyper-eutrophic conditions in the lake with high densities of cyanobacteria, low water transparency, elimination of submerged vegetation, modified fish community, and accretion of nutrient-rich, floc sediments (Coveney et al. 2005). Benthic sediments of the lake are highly organic with low bulk densities, mainly due to deposition of particulate matter associated with cyanobacteria (see Table 7.1).

Table 7.1 Selected physico-chemical properties of surface sediments of Lake Apopka (Reddy and Graetz 1991). TP = total phosphorus; TN = total nitrogen; TC = total carbon; UCF = unconsolidated floc; CF = consolidated floc; Thickness = sediment layer thickness.

Parameter	Units	UCF	CF	Peat	Sand	Clay	Marl
Thickness	cm	32 ± 19	82 ± 35	20 ± 21	18 ± 14	9 ± 5	13 ± 11
Bulk density	g cm^{-3}	0.035 ± 0.014	0.086 ± 0.029	0.096 ± 0.051	0.582 ± 0.461	0.540 ± 0.241	0.299 ± 0.163
TP	mg kg^{-1}	970 ± 430	600 ± 370	380 ± 240	420 ± 380	290 ± 200	450 ± 350
TN	g kg^{-1}	23.7 ± 7.5	22.4 ± 4.3	26.1 ± 8.3	6.0 ± 5.0	4.2 ± 3.3	8.8 ± 5.9
TC	g kg^{-1}	302 ± 83	325 ± 80	409 ± 112	72 ± 61	39 ± 30	139 ± 71

The P content of the upper sediment layer averaged 1.0 mg P g^{-1} dry weight in 1987 (Reddy and Graetz 1991; Schelske 1997). In a recent study, Torres et al. (2014) reported sediment TP content of 1.3 mg P g^{-1} dry weight in 2005, based on sediment cores collected in the western zone of Lake Apopka. Based on the spatial study reported by Reddy and Graetz (1991), approximately 1,370 and 5,270 metric tons of P were stored in UCF and CF sediments, respectively. The UCF depth increased by 22 cm or an estimated sediment accretion rate of 1.15 cm year^{-1} during the period of 1968 and 1987, based on spatial sediment sampling conducted during these two years. Average TP concentration of UCF sediments sampled during 1987 was 970 mg kg^{-1} and bulk density was 0.035 g cm^{-3}; using these data, P accretion rates were estimated at 0.39 g P m^{-2} year^{-1} (Coveney 2000). Similar values of 0.35–0.39 g P m^{-2} year^{-1} were estimated by using ^{210}Pb dating techniques and physico-chemical and microfossil characteristics of sediments (Schelske 1997). During the same time period, estimated external P loading to the lake was 0.55 g P m^{-2} year^{-1} (Coveney 2000). The estimated P accretion rates accounted for approximately 70% of the external load, suggesting the importance of sedimentation as one of the storage mechanisms of P and other nutrients in sediments.

7.3   LAKE OKEECHOBEE

Lake Okeechobee (26° 58′ N, 80° 50′ W) is the largest lake (1732 km^2) in the southern United States (see Figure 7.3). This shallow (average depth 2.7 m), eutrophic lake is used for agricultural and residential water supply, flood control, commercial and sport fishing, and supports wildlife habitat for native species such as the American alligator and endangered species such as the snail kite. There is an extensive littoral zone on the western side of the lake that is adjacent to a sand bottom area in the shallow pelagic region of the lake. A large area of peat sediments exists on the south end of the lake with a limestone rock outcrop abutting just to the north. In the deeper pelagic region, there is an extensive area of flocculent mud sediments. The lake receives water and nutrients that originate in a 1.4 million hectare watershed, which is comprised of six sub-basins: Upper Kissimmee, Lower Kissimmee, Taylor Creek/Nubbin Slough, Lake Istokpoga, Indian Prairie, and Fish Eating Creek (see Figure 7.3). Approximately, 90% of the water and the nutrients are discharged into the lake from sub-basins north of the lake. Three sub-basins situated west, east, and south of the lake, in general, do not contribute water and nutrients to the lake (Welch et al. 2019). Land use in the watershed is primarily agriculture (51%) with natural areas (31%) and urban development (10%) accounting for the majority of the balance. Approximately 18% and 15% of the Lake Okeechobee watershed exists as wetlands and open water bodies, respectively, with these systems being the final recipients of discharges from adjacent terrestrial systems (FDEP 2014).

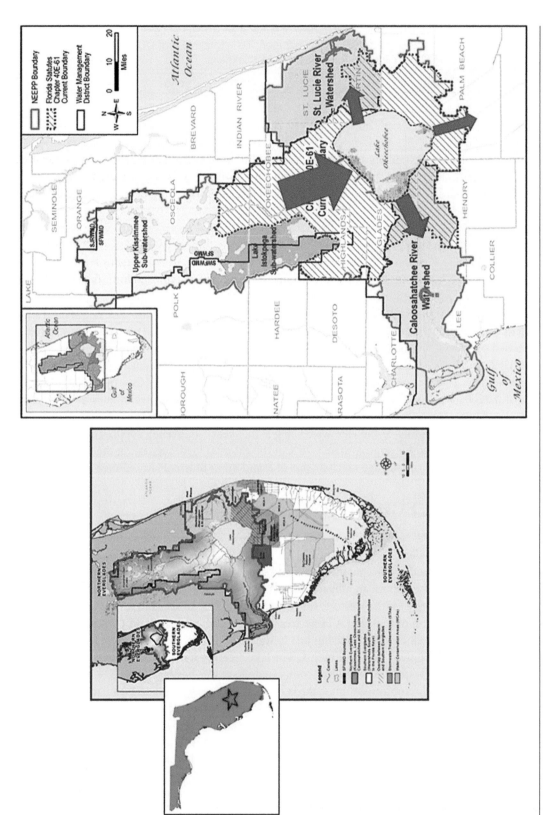

Figure 7.3   Map of Lake Okeechobee (*Source*: South Florida Water Managment District).

7.3.1   Long-Term Nutrient Loads to the Lake

As a part of nutrient load reduction to the lake, the Florida Department of Environmental Protection (FDEP) established a TMDL of 140 metric tons per year (FDEP 2014), which includes 35 metric tons of P per year from atmospheric deposition, resulting in an allowable P load from the watershed of 105 metric tons per year.

Approximately 11,000 metric tons of P per year are imported into the Lake Okeechobee watershed, with half being exported and the other half accumulating in the watershed and stored in soils (HDR 2010). Approximately 170,000 metric tons of P is stored in Lake Okeechobee watershed soils (Reddy et al. 2011; SWET 2008a, b). Phosphorus stored in soils is considered a legacy pool that can contribute to overall P load to the lake even after P imports are decreased. Legacy P can substantially extend the time required for a wetland or aquatic system to recover from an impaired state and/or revert to a stable condition (Reddy et al. 2011). Thus, legacy P in the Lake Okeechobee watershed could sustain current P loading rates of 500 metric tons per year for more than two centuries. Clearly, there is a need to fully recognize the potential contribution of legacy P as restoration programs are developed and implemented. Annual P loading rates reported for Lake Okeechobee are based on a water year (WY), which is defined as the period from May 1 to April 30. For example, May 1, 2017 to April 30, 2018 is defined as WY2018 (see Figure 7.4).

Historically, the major source of nutrients to the Greater Everglades Ecosystem watershed was from atmospheric deposition, with minimum secondary nutrient inputs through infrequent sheet flow across the northern Everglades wetlands from Lake Okeechobee. Presently, approximately two thirds of the P load from Lake Okeechobee is discharged to the east and west to the St. Lucie and Caloosahatchee estuaries, respectively. Agricultural and urban intensification in the Greater Everglades

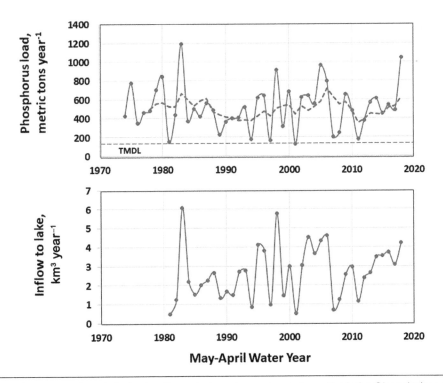

Figure 7.4   Long-term trends in hydraulic and phosphorus loads from the Lake Okeechobee watershed to the lake (Data form South Florida Water Management District).

Ecosystem watershed has led to excessive nutrient loads to natural systems contained within the region. For example, over the past three decades, TP loads to Lake Okeechobee were in excess of 500 mt per year, with the exception of a few dry years (Reddy et al. 2011). These loads are approximately 3.6 times the annualized TMDL of 140 mt per year. Total P loading rates into Lake Okeechobee varied over time because of a combination of climatic conditions, land-use changes, and changes in water management conditions. Jawitz and Mitchell (2011) reported that 80% of the P load to Lake Okeechobee is delivered in approximately 73 days in any particular year. External P loads are significantly correlated with the hydraulic loading rates to the lake: loads less than 200 metric tons per year occur when water discharge is approximately 1.233 km^3 per year (1 million acre-feet per year), whereas up to 600 metric tons year^{-1} occur when water discharge is approximately 3.7 km^3 year^{-1} (3 million acre-feet per year) to the lake.

7.3.2 Long-Term P Surface-Water P Concentrations

Annual average TP concentration in the water ranged from 50 to 100 µg L^{-1} during the WY1975–1995 (Havens and James 2005) (see Figure 7.5). The more recent five-year moving TP average (2012–2016) was 117 µg L^{-1}, which is approximately in the pre-hurricane range prior to 2004. Three hurricanes—Frances, Jeanne, and Wilma—that passed through during 2004–2005, resulted in elevated levels of TP concentrations up to 205 µg L^{-1} in the water column (see Figure 7.5). This was due to resuspension of surface sediments, which increased the turbidity and release of P into the water column from suspended particulate matter. Total P decreased to 93, 118, and 150 µg L^{-1} during WY2012, WY2016, and WY2017, respectively. During WY2018, the after effects of hurricane Irma (in October 2017) once again elevated in-lake TP concentrations to 203 µg L^{-1} (see Figure 7.5). Total P concentration in the lake is cumulatively influenced by various factors including external P loads and internal cycling involving sediments, water depth, phytoplankton, algae, and macrophytes. Earlier studies suggested the overriding importance of internal P cycling controlling year-to-year variations of in-lake TP concentrations of water column (James et al. 1997; Havens and James 2005).

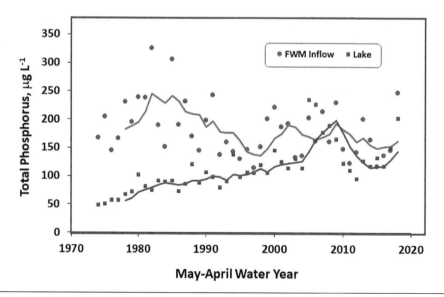

Figure 7.5 Long-term trends in inflow and in-lake total phosphorus concentrations in Lake Okeechobee (Data from South Florida Water Management District).

7.3.3 Phosphorus Accumulation in Sediments

One direct consequence of eutrophication in Lake Okeechobee has been the substantial accumulation of P-enriched sediment throughout the lake basin. In 1998, a comprehensive survey reported that the upper 10 cm of mud sediments within the lake contained an estimated 28,600 metric tons of P (Fisher et al. 2001). Within large, shallow lakes such as Lake Okeechobee, sediment resuspension driven by wind-induced water movement may play a significant role in nutrient cycling since particulate bound forms of P can be released into the overlying water column (Lowe et al. 1999; Søndergaard et al. 2003; Bachmann et al. 2005). Given that over 40% of Lake Okeechobee's benthic zone has been found to be covered with P-enriched mud sediments, resuspension of these sediments may be one of the most important mechanisms affecting P availability within this lake ecosystem (James et al. 1995a, b). In fact, past research has suggested that internal P loading has delayed ecological responses within the lake to reduced P loading from the surrounding watershed even after implementation of a variety of restoration activities, including agricultural best management practices (BMPs) (James et al. 1995b, 2005; Fisher et al. 2005).

Between September 2004 and October 2005, three hurricanes (Frances, Jeanne, and Wilma) passed immediately over the lake creating large seiches, waves, and currents. Post-storm observations of the lake indicated that the mud sediments were remixed to depths between 10 and 20 cm (Pollman and Engstrom 2005; James et al. 2005) and that these mud sediments were resuspended and transported around the lake. Given that these sediments were highly P-enriched and the known importance of sediment resuspension in P cycling within Lake Okeechobee, the distribution of these sediments is of direct concern for lake management and restoration efforts aimed at reducing its current eutrophic status. In a subsequent study conducted in 2006 to determine spatial distribution and associated changes in sediment TP, Cohen et al. (2007) reported no significant change in sediment TP values from the previous synoptic surveys conducted in 1988 and 1998. Sediment TP concentrations ranged from 220 to 1030 mg kg^{-1}, with higher values measured in mud zone sediments and lower values observed in sand and peat sediments (see Table 7.2). Further supporting the assertion that wind-driven resuspension is a major factor in Lake Okeechobee, Cohen et al. (2007) reported significant declines in mud depths across the center of the lake and increases in mud depth in fringing areas. Interaction between lake sediment and water is a major contributor to water quality conditions observed in various regions of the lake (James et al. 2008). Mapping the spatial distribution of sediment P and its changes over time continues to be important to assess the status and trends in sediment quality and provide some measures of restoration success.

Table 7.2 Selected physico-chemical properties of surface sediments of Lake Okeechobee (Fisher et al. 2001). TIP = total inorganic P; TP = total phosphorus; TN = total nitrogen; TC = total carbon.

Parameter	Units	Mud	Sand	Littoral	Peat
Bulk density	g cm^{-3}	0.27 ± 0.28	1.3 ± 0.3	0.33 ± 0.45	0.19 ± 0.09
TIP	mg kg^{-1}	777 ± 274	233 ± 361	138 ± 117	172 ± 138
TP	mg kg^{-1}	1034 ± 376	224 ± 271	472 ± 493	250 ± 209
TN	g kg^{-1}	9 ± 4	0.37 ± 0.51	18 ± 16	20 ± 6
TC	g kg^{-1}	143 ± 54	10.3 ± 9.9	222 ± 191	358 ± 95

7.4   INTERNAL PHOSPHORUS LOADS

Lakes often serve as final recipients of water and nutrients discharged from adjacent uplands and wetland ecosystems. Because many lakes are P-limited, loading of this nutrient causes major concern to environmental managers developing strategies to reduce loads from external sources. Alternative land management practices including BMPs are now in place or are being developed to reduce P loads. Once the external P loads from the watershed are curtailed through the implementation of BMPs and other improved practices, key questions that often are asked include: (1) will lakes respond to P load reduction? (2) if so, how long will it take for these systems to recover and reach their background condition or alternate stable conditions? and (3) are there any economically feasible management options to hasten the recovery process? Relationships between watershed input and output for various nutrients including P are moderately well understood. However, internal processes regulating reactivity and mobility of P in these lakes' sediments are not well understood.

In shallow lakes, P flux across the sediment-water interface occurs in two different modes, depending on meteorological and hydrodynamic conditions. During calm days when vertical turbulent mixing and bottom shear stress are insufficient to resuspend surface sediment, dissolved P moves via passive diffusion and advection. The processes affecting P exchange at the sediment water interface include: (1) diffusion and advection due to wind-driven currents; (2) diffusion and advection due to flow and bioturbation; (3) processes within the water-column (mineralization, sorption by particulate matter, and biotic uptake and release); (4) diagenetic processes (mineralization, sorption, and precipitation dissolution) in bottom sediments; and (5) redox conditions at the sediment-water interface. During windy periods, resuspension and deposition of sediments may be an important mode of P transfer to the water column (Reddy et al. 1996). Because sediment resuspension events are transitory, P flux by this process may occur at shorter-time scales but at more rapid rates compared to diffusive flux.

Sediment P is easily recycled into the lake's water column either through diffusion of dissolved forms or resuspension of particulate forms. However, the net transport of P is from water column into sediments. Although sediments act as a net sink for P, release of dissolved inorganic P into the overlying water can occur if the concentration of interstitial P exceeds that of the overlying water. This is especially true if surface sediments are anaerobic, reducing the ability of iron to bind to inorganic P, thus increasing the amount of interstitial P that can be released to the water column (Moore et al. 1998). Relative importance of P transfer due to diffusive flux and resuspension flux must be quantified to accurately estimate annual P flux from internal sources. This internal load can extend the time required for ecosystems to return to their original trophic condition. Though marked reductions have been achieved in P loads from external sources, in many aquatic systems, including shallow lakes, internal loading from the P stored in sediments has become a major concern. Lake sediments serve as sinks for particulate P and serve as sources for dissolved P (see Figure 7.6). In shallow lakes such as Lake Apopka and Lake Okeechobee, these concentration gradients are often disrupted by wind-driven sediment resuspension and bioturbation (Reddy et al. 1996; James et al. 1997; Fisher et al. 2005).

Internal loads of nutrients are usually measured using benthic chambers and pore water equilibrators under field conditions and using intact sediments cores incubated under controlled environmental conditions (see Chapter 2). Potential DRP fluxes measured in selected lakes and other aquatic systems in Florida are summarized in Table 7.3.

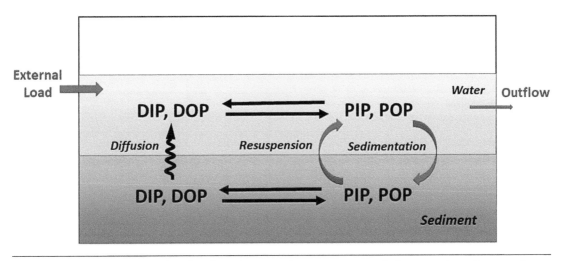

Figure 7.6   Schematic showing the phosphorus exchange processes between water column and sediments.

Table 7.3   Potential DRP fluxes between sediments and overlying water column of selected lakes and other aquatic systems in Florida, USA. Unless specified, overlying water column in incubated sediment cores was aerated during phosphorus flux measurements under controlled conditions. Peepers = Pore-water equilibrators.

Ecosystem	P Flux—mg P m^{-2} day^{-1}	Reference
Lake Apopka	2.27	Moore et al.1991
Lake Apopka—sediment pore water	1.69	Moore et al.1991
Lake Apopka—sediment pore water	0.97 ± 0.59	Reddy et al.1996
Lake Okeechobee—mud zone—1989–90	1.69 ± 0.69	Moore et al. 1998
Lake Okeechobee—mud zone—1999	1.01 ± 0.03	Fisher et al. 2005
Lake Okeechobee—mud zone—1999–Peepers	0.83 ± 0.11	Fisher et al. 2005
Lake Okeechobee—mud zone—2000	0.41	Reddy et al. 2007
Lake Okeechobee—sand zone—1989–90	0.02 ± 0.09	Moore et al. 1998
Lake Okeechobee—sand zone—1999	0.62 ± 0.29	Fisher et al. 2005
Lake Okeechobee—littoral zone—1989–90	0.79 ± 0.84	Moore et al. 1998
Lake Okeechobee—littoral zone—1999	0.37 ± 0.36	Fisher et al. 2005
Lake Okeechobee—peat zone—1989–90	2.22 ± 0.91	Moore et al. 1998
Lake Okeechobee—peat zone—1999	2.12 ± 1.16	Fisher et al. 2005
Lake Okeechobee—peat zone—2000–Peepers	0.38 ± 0.17	Fisher et al. 2005
Kissimmee River—Inflow—1989–90	0.34 ± 1.77	Moore et al. 1998
Kissimmee River—Inflow—1999	0.58 ± 0.05	Fisher et al. 2005
Taylor Creek—Inflow—1989–90	0.40 ± 2.51	Moore et al. 1998
Taylor Creek—Inflow—1999	0.39 ± 0.23	Fisher et al. 2005
Lake Cypress	0.07 ± 0.14	Martin 2004

Continued

Ecosystem	P Flux—mg P m^{-2} day^{-1}	Reference
Lake Hatchineha	0.30 ± 0.41	Martin 2004
Lake Istokpoga	0.30 ± 0.40	Martin 2004
Lake Kissimmee	0.07 ± 0.02	Martin 2004
Lake Tohopekaliga	0.42 ± 0.73	Martin 2004
Indian River Lagoon	0.16 to 1.54	Reddy et al. 2001
Indian River Lagoon—Peepers	0.04 to 1.60	Reddy et al. 2001
St. Johns River—oxic water column	−0.13 to 0.60	Malecki et al. 2004
St. Johns River—anoxic water column	2.35 to 11.7	Malecki et al. 2004

7.4.1   Lake Apopka

Surface sediments in Lake Apopka represent UCF (see Table 7.1) with an average depth of approximately 30 cm, with an underlying CF of approximately 80 cm (Reddy and Graetz 1991). The UCF (surface layer) sediment is actively involved in nutrient exchange with the overlying water column. Earlier studies showed that DRP profiles of pore water showed distinct gradients with approximately 8 cm of surface sediment characterized by a dissolved P-depleted zone, while the underlying CF sediments displayed steep gradients with depth (Reddy et al. 1996). These results suggested that DRP transport from the surface sediments was due to active sediment resuspension, while diffusive flux regulated the upward mobility of DRP in sediments below 8 cm depth. The UCF sediment layer is derived from phytoplanktonic material, while the underlying CF layer is derived from both phytoplankton and macrophytes (Waters et al. 2005). It has been suggested that the UCF sediments, especially at the sediment-water interface, may include viable meroplankton that may be active in depleting DRP from the pore water of surface sediment layers (Carrick et al. 1993). Spatial patterns in diffusive flux of DRP from sediment to the overlying water column ranged from 0.2 to 1.9 mg P m^{-2} day^{-1}, with an average value of 1 mg P m^{-2} day^{-1} (see Table 7.3), with minimal seasonal variations (Reddy et al. 1996). Dissolved reactive P release during sediment resuspension and settling (simulated using sediment cores under laboratory conditions) ranged from 0.2 to 3.3 mg P m^{-2} h^{-1}, an order of magnitude higher than diffusive flux of DRP (Reddy et al. 1996). Under field conditions, biologically active phytoplankton and meroplankton at the sediment water interface can rapidly assimilate any DRP released during resuspension and settle on sediment surface after the resuspension is stopped. Given the uncertainty of the role of DRP release during sediment resuspension as result of low DRP concentrations in the pore water of surface sediments, we concluded that in Lake Apopka, overall P flux from sediments is regulated by diffusion in response to concentration gradients across the sediment-water interface.

Reactive P (RP) in Lake Apopka sediments includes biologically and chemically active compounds (dissolved inorganic P, dissolved organic P, particulate inorganic P, and particulate organic P). In contrast, nonreactive P (NRP) includes P that is biologically not available and buried in sediments. Operationally defined chemical fractionation schemes have been routinely used to identify RP and NRP pools in sediments (Olila et al. 1995; Reddy et al. 2011). Key chemical extractants used in the identification of organic and inorganic pools in sediments are 0.1 to 0.5 M NaOH and 0.5 to 1 M HCl. Sediment P not extracted by acid and alkali is considered as residual P or operationally defined as NRP, whereas P that is extracted is considered as RP. For all practical purposes, NRP is very slowly available and is not considered in estimating internal loads in lakes. Acid-extractable P represents inorganic P-associated metals such as Ca, Mg, Fe, and Al, while alkali-extractable P represents organic P including monoester P and diester P. In Lake Apopka sediments, inorganic P is primarily dominated by Ca- and Mg-bound P, and organic P is dominated by phosphodiesters and phosphomonoesters (Torres et al. 2014). We used the estimated P storage in surface UCF sediments to determine the potential lag time to recovery resulting from the legacy P presently stored in benthic sediments (Reddy and Graetz 1991). Based on these observations, we estimated approximately 80% of the TP in sediments is in the RP pool and 20% is in the NRP pool (Torres et al. 2014). Within the RP pool, approximately 30% to 40% of the TP is in exchangeable P and microbial biomass P, which is most readily available. Based on these initial conditions, we performed two scenarios to estimate the role of legacy sediment P in regulating internal loading from sediments to the water column (see Figure 7.7). Using the estimated P flux rates from sediments and assuming 25% and 100% of the RP is available for release, we estimated lag times in recovery of the lake toward background status or reach some alternate stable state.

Based on diffusive P flux reported, we estimated approximately 44 metric tons of P per year is released from sediments to the overlying water column. This internal load will last for six years if 25% of the RP in sediments is assumed to be available for release from sediments to the overlying water column. However, the lag time extends to 25 years if 100% of the RP is available for release. As a first approximation, these estimates suggest that sediments will continue to support an internal load of 44 metric tons of P per year for at least six to 25 years. However, it should be noted that internal regeneration of dissolved P will continue to occur in sediments through several biogeochemical processes (Reddy and Delaune 2008). In a phytoplankton-dominated system such as Lake Apopka, both C and N fixation will continue to maintain primary productivity until P limitation occurs in

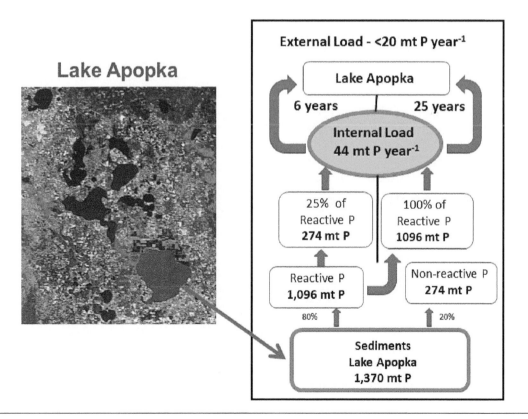

Figure 7.7   Role of legacy phosphorus in the Lake Apopka in determining the lag time for recovery.

the system. Over a period of time, phytoplankton in the lake will be limited by P availability resulting in deposition of high C/P and N/P ratio planktonic matter. Deposition of P-limited material on the sediment surface can create a boundary layer with low P that can potentially serve as a sink for P that is released from subsurface sediments and decrease the P flux into the water column.

Lake Apopka, once with clear water with abundant growth of macrophytes, shifted to turbid water dominated by cyanobacteria blooms and loss of submerged aquatic vegetation. Similar regime shifts have been noted in many shallow lakes in other parts of the world (Ibelings et al. 2007). This regime shift is accelerated by historic external nutrient loads including P in excess of 60 metric tons P per year from adjacent agricultural activities north of Lake Apopka. However, subsequent reduction in external loads to less than 20 metric tons per year did not result in substantial improvement in water quality and the lake is still turbid and dominated by phytoplankton. This trend was attributed to wind-driven sediment resuspension and release of P into the water column (Coveney et al. 2005). Currently, management efforts are underway to establish submerged aquatic vegetation in the littoral zones of the lake.

7.4.2   Lake Okeechobee

In Lake Okeechobee, mud, sand, and littoral sediments are dominated by mineral matter, while sediments south of the lake are dominated by peat deposits (Fisher et al. 2001). The mud sediments occupy 46% of the lake and are dominated by Ca- and Mg-bound P (65% of total P), followed by residual P (28% of total P), with low exchangeable (2% of total P) and Fe- and Al-bound P (5% of total P) (Olila et al. 1995). The peat and littoral sediments of Lake Okeechobee, however, have higher exchangeable P (9% to 10% of TP) and Fe- and Al-bound P (6% to 18% of total P). Pore water DRP concentrations in surface layers of all sediment types increased between 1988 and 1998 spatial sampling events (Fisher et al. 2001). High DRP concentrations in pore water are due to the decreased capacity of sediments to adsorb P as a result of saturation with P. However, an increase of DRP was not reflected in P flux from sediments to the overlying water column (Fisher et al. 2005). Two mechanisms that control exchangeable P have been suggested for Lake Okeechobee mud sediments under different redox conditions. Under oxic conditions, Fe appears to control the amount of exchangeable P, while under anoxic conditions, Ca and Mg seem to control exchangeable P (Moore and Reddy 1994; Olila et al. 1995). The data suggest that approximately 35% of sediment TP is accounted for in the NRP pool, whereas 65% of TP is in the RP pool and biologically available over a range of time scales.

Phosphorus flux measured during the years 1988–89 and 1999–2000 ranged from 0.14 to 1.9 mg P m^{-2} day^{-1} (Fisher et al. 2005; Moore et al. 1998; Reddy et al. 2007). These data sets suggest that over the time periods tested, internal P loading from sediments has not changed appreciably. Internal DRP fluxes are approximately equivalent to the estimated external loads of 500 metric tons per year (Moore et al. 1998; Fisher et al. 2005). Similarly, the significance of internal P loads in regulating eutrophication has been demonstrated in other shallow lakes (Graneli 1999; Sondergaard et al. 1999, 2003; see Chapters 8, 12, 13, 14, and 16). Therefore, internal P fluxes into the water column can offset any water column responses to external load reductions.

The mud zone in Lake Okeechobee covers an area greater than 80,000 hectares (46% of the lake surface area). During 1998 sediment sampling, approximately 21,390 metric tons of P were stored in the top 10 cm of the mud zone sediment. Based on these initial conditions, we performed two scenarios to estimate the role of mud zone sediment legacy P in regulating internal loads from sediments and resulting eutrophic conditions in the water column (see Figure 7.8). Using the estimated P flux rates from sediments and assuming 25% or 65% of the RP is available for release, we estimated lag times in recovery of the lake toward background status or to reach some alternate stable state.

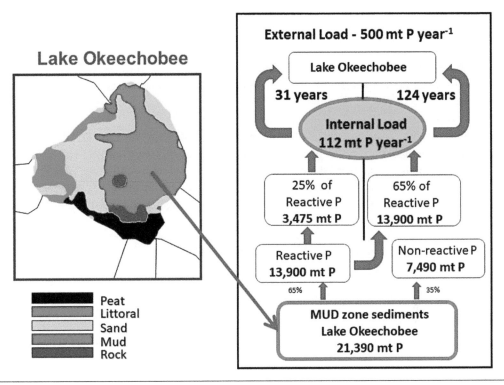

Figure 7.8   Role of legacy phosphorus in the Lake Okeechobee in determining the lag time for recovery.

Based on DRP fluxes reported, we estimated approximately 112 metric tons per year is released from sediments to the overlying water column (Reddy et al. 2011). At this P flux, a 31-year lag time results if 25% of the RP is assumed to be available for release from sediments to the overlying water column. However, the lag time extends to 124 years if 65% of the RP is available for release. Under these estimates, mud zone sediments will continue to support an internal load of 112 metric tons per year for 31 to 124 years.

Lake Okeechobee is shallow, turbid, eutrophic, and approximately 14 times larger than Lake Apopka, with distinct sediment types including mud, sand, peat, and littoral zone. Major long-term stresses on the lake include excessive P loads, extreme water level fluctuations, and rapid spread of exotic and nuisance plants in the littoral zone (Welch et al. 2019). In addition, the lake has been affected by extreme events such as hurricanes, which created extensive sediment resuspension resulting in degraded water quality, destroyed fish habitat including loss of submerged and emergent aquatic vegetation, and disrupted links in the food web (Welch et al. 2019). The Lake Okeechobee Protection Plan developed by interagency groups including the South Florida Water Management District (SFWMD), FDEP, and Florida Department of Agriculture and Consumer Services (FDACS) provides guidance in implementation of restoration goals (SFWMD et al. 2004).

7.5   GENERAL DISCUSSION

In terms of mass balances (measured based on input/ouput), the amount of P transferred to lakes may not be measurable in P budgets of the drainage basin. Algal biomass (measured in terms of chlorophyll *a*) is often used as an indicator of eutrophication because of its relationship with increased P loading to lakes. Since algae can obtain carbon (C) (and some can obtain nitrogen (N)) from the atmosphere, addition of P regulates the growth of algal biomass, resulting in alteration of stoichiometry of macro-elements such as C, N, and P and ultimately trophic conditions of the lake. Phosphorus and other nutrients stored in algal biomass are cycled within the water column during decomposition, and the P associated with recalcitrant dead algal biomass settles on the bottom and becomes an integral part of sediments. Although bottom sediments function as a major storage reservoir of P, they can at times function as a source of P to the water column. Both biotic and abiotic reactions regulate DRP concentration of the water column. Concentration of TP, and various P pools, are generally higher in recent sediments, and decrease with depth, suggesting the influence of P loading (Engstrom et al. 2006). Although accretion of sediment bound P suggests that P flux is downward (i.e., from water column to sediments), the DRP flux is upward (i.e., from sediments to water column) in response to concentration gradients established at the sediment-water interface (Moore et al. 1998).

Coupled biotic and abiotic processes are involved in regulating reactivity and mobility of DRP in sediments (Reddy and Delaune 2008; Spears et al. 2007). In Lake Okeechobee, some of this internal load is attributed to potential anoxia in surface sediments where P release occurs through the redox-mediated reduction of iron from ferric phosphate and release of P (Moore and Reddy 1994; Nikolai and Dzialowski 2014; Søndergaard et al. 2003). In Lake Okeechobee sediments, P solubility is governed by sorption and precipitation with iron oxides under aerobic conditions, while under anaerobic conditions, solubility is regulated by precipitation with calcium oxides (Moore and Reddy 1994). Lake Apopka sediments are high in organic matter and calcium carbonate, and have low levels of reactive iron. In these sediments, P solubility is regulated by pH and precipitation with calcium compounds (Olila and Reddy 1995, 1997). In addition, enzymatic hydrolysis and mineralization of organic P also governs release of DRP from sediments (Torres et al. 2014).

Effective management of water and legacy nutrient loads from watersheds have proven effective in improving water quality of receiving aquatic ecosystems, including shallow lakes (Janssen et al. 2014). Recovery responses were noted over a shorter time period in smaller lakes than in larger shallow lakes. Spatial extent of alternate stable states is governed by a combination of lake size, spatial heterogeneity, and internal connectivity (Janssen et al. 2014). In Lake Okeechobee, approximately 40% of the lake is governed by the mud zone, which at present, regulates the overall water quality through a combination of sediment resuspension of fine sediments and internal load. Suspended particles in Lake Okeechobee are rich in Mg-bearing minerals, including sepiolite and palygorskite (Harris et al. 2007). This small sized, low density material that is suspended from mud zone sediments is biologically inert and maintains turbid conditions in a shallow lake such as Lake Okeechobee (Harris et al. 2007).

7.6   RESTORATION ACTIVITIES

One of the key programs for removal of P from Lake Apopka has been harvesting rough fish (mainly gizzard shad). The St. Johns River Water Management District has demonstrated that harvesting rough fish is a cost-effective tool to remove P that benefits impacted lakes (Godwin et al. 2011; Fulton et al. 2015; Fulton 2018). Rough fish harvest benefits the lake in multiple ways. First, it is a permanent, direct removal of P that is contained in fish tissues. Second, reducing rough fish standing stock reduces benthic feeding and subsequent P excretion. Third, rough fish harvest reduces the number of fish causing bioturbation, which can resuspend sediments and negatively affect water clarity. Since 1993, over 12,000 metric tons of fish have been harvested, representing a direct removal of almost 100 metric tons of internal P.

Another restoration activity on Lake Apopka is a recirculating flow way that uses wetland cells to remove P and suspended sediments (Dunne et al. 2015). The flow way was constructed on former subsided farm fields. Lake water flows into four independent wetland cells by gravity and treated water is collected in a pump basin and pumped back out to the lake. Over a span of 13 years of operation, the flow way has removed 40,000 metric tons of suspended solids and 30 metric tons of P. Most of this material is likely derived from internal sediment resuspension and associated algal production.

Re-establishing or increasing the cover of submerged aquatic vegetation (SAV) has the potential to dramatically improve environmental conditions (Lefcheck et al. 2018). Water column concentrations of dissolved inorganic phosphorus are lower within large, established SAV beds than outside of these meadows (Gurbisz et al. 2017). This reduction is due primarily to plant assimilation into tissue and is facilitated by a reduction of water flow within these meadows increasing water residence times and contact between water masses and plant tissue. These meadows can also alter P cycles through modification of pH. During photosynthesis, plants use dissolved carbon dioxide and bicarbonate ions, increasing water column pH, which promotes $CaCO_3$ precipitation and co-precipitation of P moving it into the sediment where it is more stable than in the water column (Dierberg et al. 2002).

In addition to reducing water column P concentrations, plants increase several other ecosystem functions and services—reduction of flow can increase sedimentation and increase water clarity (Hansen and Reidenbach 2012) and the structure provided by plants can act as habitat for ecologically and economically important fish and invertebrate species (Barbier et al. 2011).

Unfortunately, the current environmental condition in both Lake Apopka and Lake Okeechobee is unfavorable to SAV establishment. Major threats to plant survival are algal growth associated with high nutrient loads and unconsolidated sediments resulting in low-light conditions. Finding windows of opportunity or investing in creative strategies for initial plant establishment are worthwhile, nonetheless, because of the potential positive feedbacks associated with SAV. The increased water clarity associated with SAV promotes additional SAV expansion (Orth et al. 2012). Further, increased density results in lower flow and more efficient nutrient removal through assimilation. These positive feedbacks, however, can take time to establish, often resulting in a lag period between planting and the return or enhancement of ecosystem function (McGlathery et al. 2012; Reynolds et al. 2016).

7.7   CONCLUSIONS

Long-term P applications to watersheds to support agricultural activities have resulted in substantial P accumulation in soils. Water and nutrient loading from watersheds into lakes has resulted in accumulation of P in sediments. In many drainage basins, reduction of external loads to lakes alone did not result in their recovery, primarily due to internal nutrient loads from sediments. This has become a major concern in restoration programs. Based on the available data, results suggest that Lake Apopka showed significant improvement after the external loads were reduced by two thirds. However, the lake still remains turbid due to dominance of phytoplankton and frequent resuspension of benthic sediments. Lake Okeechobee, almost 14 times larger than Lake Apopka, has not shown significant improvements in surface total P concentrations. In spite of several P reduction programs implemented in the watershed, this lake consistently receives substantial external loads as a result of legacy P in the watershed. This is compounded by internal loads from benthic sediments and the lake still remains turbid. In both lakes, establishment of SAV is critical to improve water quality and reduce turbidity. Current restoration programs should help to achieve this goal.

Management of these lakes should focus on reduction of external loads (especially for Lake Okeechobee), which will ultimately have a positive effect in reducing the internal load. The internal load (a consequence of past excessive external loads) can extend the time required for the lake to reach its restoration goals. The lag time for recovery should be considered in developing alternate adaptive management strategies for the lake. To determine the main factors regulating internal P loading, it is critical that we have a thorough understanding of the dynamics of physical, chemical, and biological processes at the sediment-water interface, at the whole lake scale.

7.8 REFERENCES

Bachmann, RE; Hoyer, MV; Vinzon, SB; and Canfield, DE. 2005. The origin of the fluid mud layer in Lake Apopka Florida. Limnol Oceanogr. 50:629–635.

Barbier, EB; Hacker, SD; Kennedy, C; Koch, EW; Stier, AC; and Sillman, BR. 2011. The value of estuarine and coastal ecosystem services. Ecol Monogr. 81:169–193. https://doi.org/10.1890/10-1510.1.

Carpenter, SR; Caraco, NF; Correl, DL; Howarth, RW; Sharpley, N; and Smith, VH. 1998. Nonpoint pollution of surface waters with phosphorus and nitrogen. Ecol Appl. 8:559–568.

Carrick, HJ; Aldridge, FJ; and Schelske, CL. 1993. Wind influences phytoplankton biomass and composition in a shallow, productive lake. Limnol Oceanogr. 38:1179–1192.

Cohen, MJ; Osborne, TZ; Vogel, WJ; Daoust, RJ; James, RT; Yan, Y. 2007. Lake Okeechobee sediment quality mapping project. Final Report ST060576-WO01. Submitted to South Florida Water Management District, West Palm Beach, Florida, USA. 30 pages.

Coveney, MF. 2000. Sedimentary phosphorus stores, accumulation rates, and sedimentation coefficients in Lake Apopka: Prediction of the allowable phosphorus loading rate. Technical Memorandum. St. Johns River Water Management District, Palatka, FL, U.S.A.

Coveney, MF. 2016. Water quality changes in Lake Apopka, Florida and the St. Johns River Water Management District's restoration program. Technical Memorandum 56. St. Johns River Water Management District, Palatka, Florida, USA. p. 25.

Coveney, MF; Lowe, EF; Battoe, LE; Marzolf, ER; and Conrow, R. 2005. Response of a eutrophic, shallow subtropical lake to reduced nutrient loading. Freshwat Biol. 50:1718–1730 (and Erratum Freshwater Biology 50: 2167).

Dierberg, FE; DeBusk, TA; Jackson, SD; Chimney, MJ; and Pietro, K. 2002. Submerged aquatic vegetation-based treatment wetlands for removing phosphorus from agricultural runoff: response to hydraulic and nutrient loading. Water Res. 36:1409–1422.

Dunne, EJ; Coveney, MF; Hoge, VR; Conrow, R; Naleway, R; Lowe, EF; and Battoe, L. 2015. Phosphorus removal performance of a large-scale constructed treatment wetland receiving eutrophic lake water. Ecol Eng. 79:132–142.

Engstrom, DR; Schottler, SP; Leavitt, PR; and Havens, KE. 2006. Revaluation of the cultural eutrophication of Lake Okeechobee using multiproxy sediment records. Ecol Appl. 16:1194–1206.

FDEP. 2014. Basin management action plan for the implementation of total maximum daily loads for total phosphorus by the Florida Department of Environmental Protection, Tallahassee, Florida, USA.

Fisher, MM; Reddy, KR; and James, RT. 2001. Long-term changes in the sediment chemistry of a large shallow subtropical lake. Lake Reserv Manage. 17: 217–232.

Fisher, MM; Reddy, KR; and James, RT. 2005. Internal nutrient loads from sediments in a shallow, subtropical lake. Lake Reserv Manage. 21:338–349.

Fulton, RS. 2018. Nutrient loading and water quality trends in the Upper Ocklawaha Basin lakes through 2016. Technical Memorandum 54-3. St. Johns River Water Management District, Palatka, Florida, USA. p. 145.

Fulton, RS; Godwin, WF; and Schaus, MH. 2015. Water quality changes following nutrient loading reduction and biomanipulation in a large shallow subtropical lake, Lake Griffin, Florida, USA. Hydrobiologia. 753: 243–263.

Furman, AL; White, HO; Cruz, OE; Russell, WE; and Thomas, BP. 1975. Soil Survey of Lake County Area, Florida. In: USDA/NRCS in cooperation with the University of Florida, Agricultural Experiment Station.

Godwin, WF; Coveney, MF; Lowe, EF; and Battoe, L. 2011. Improvements in water quality following bio-manipulation of gizzard shad (*Dorosoma cepedianum*) in Lake Denham, Florida. Lake Reserv Manage. 27:287–297.

Granéli, W. 1999. Internal phosphorus loading in Lake Ringsjön. Hydrobiologia. 404:19–26.

Gurbisz, C; Kemp, WM; Cornwell, JC; Sanford, LP; Owens, MS; and Hinkle, DC. 2017. Interactive effects of physical and biogeochemical feedback processes in a large submersed plant bed. Estuaries and Coasts. 40:1626–1641.

Hansen, JCR and Reidenbach, MA. 2012. Wave and tidally driven flows in eelgrass beds and their effect on sediment suspension. Mar Ecol Prog Ser. 448:271–287.

Harris, WG; Fisher, MM; Cao, X; Osborne, T; and Ellis, L. 2007. Magnesium-rich minerals in sediment and suspended particulates of South Florida. Water bodies: Implications for turbidity. J Environ Qual. 36:1670–1677.

Havens, KE and James, RT. 2005. The Phosphorus mass balance of Lake Okeechobee, Florida: Implications for eutrophication management. Lake Reserv Manage. 21:139–148. DOI: 10.1080/07438140509354423.

HDR Team. 2010. Task 2—Nutrient budget analysis for the Lake Okeechobee Watershed. Task 4—Final report submitted May 2010 to SFWMD, West Palm Beach, Florida, USA.

Hoge, VR; Conrow, R; Stites, DL; Coveney, MF; Marzolf, ER; Lowe, EF; and Battoe, LE. 2003. SWIM Plan for Lake Apopka, Florida. St. Johns River Water Management District, Palatka, Florida, USA. p. 61.

Ibelings, BW; Portielje, R; Eddy, HR; Lammens, R; Noordhuis, R; van den Berg, MS; Joosse, W; and Meijer, ML. 2007. Resilience of alternative stable states during the recovery of shallow lakes from eutrophication: Lake Veluwe as a Case Study. Ecosystems. 10:4–16. DOI: 10.1007/s10021-006-9009-4.

James, RT; Bierman, VJ; Erickson, MJ; and Hinz, SC. 2005. The Lake Okeechobee Water Quality Model (LOWQM) enhancements, calibration, validation and analysis. Lake Reserv Manage. 21:231–260.

James, RT; Chimney, MJ; Sharfstein, B; Engstrom, DR; Schottler, SP; East, T; and Jin, K-R. 2008. Hurricane effects on a shallow lake ecosystem, Lake Okeechobee, Florida (USA). Fund Appl Limnol. 172(4):273–287.

James, RT; Jones, BL; and Smith, VH. 1995a. Historical trends in the Lake Okeechobee ecosystem II. nutrient budgets. Arch Hydrobiol Suppl. 107:25–47.

James, RT; Martin, J; Wool, T; and Wang, PF. 1997. A sediment resuspension and water quality model of Lake Okeechobee. J Am Water Res Assoc. 33:661–678.

James, RT and Pollman CD. 2011. Sediment and nutrient management solutions to improve the water quality of Lake Okeechobee. Lake Reserv Manage. 27:28–40. DOI: 10.1080/07438141.2010.536618.

James, RT; Smith, VH; and Jones, BL. 1995b. Historical trends in the Lake Okeechobee ecosystem III. water quality. Arch Hydrobiol Suppl. 107:49–69.

Janssen, ABG; Teurlincx, S; An, S; Janse, JH; Paerl, HW; and Mooij, WM. 2014. Alternative stable states in large shallow lakes? J Great Lakes Res. 40:813–826.

Jawitz, JW and Mitchell, J. 2011. Temporal inequality in catchment discharge and solute export. Water Resour Res. 47: W00J14. DOI:10.1029/2010WR010197.

Jeppesen, E; Søndergaard, M; Meerhoff, M; Lauridsen, TL; and Jensen, JP. 2007. Shallow lake restoration by nutrient loading reduction—some recent findings and challenges ahead. Hydrobiologia. 584:239–252.

Lefcheck, JS; Orth, RJ; Dennison, WC; Wilcox, DJ; Murphy, RR; Keisman, J; Gurbisz, C; Hannamh, M; Landry, JB; Moore, KA; Patrick, CJ; Testak, J; Welle, DE; and Batiuk, RA. 2018. Long-term nutrient reductions lead to the unprecedented recovery of a temperate coastal region. Proc Nat Acad Sci. 115(14):3658–3662. www.pnas.org/cgi/doi/10.1073/pnas.1715798115.

Lowe, EF; Battoe, LE; Coveney, MF; and Stites, DL. 1999. Setting water quality goals for restoration of Lake Apopka: inferring past conditions. Lake Reserv Manage. 15:103–120.

Malecki, LM; White, JR; and Reddy, KR. 2004. Nitrogen and phosphorus flux rates from sediment in the lower St. Johns River estuary. J Environ Qual. 33:1545–1555.

Martin, C. 2004. Phosphorus flux from sediments in the Kissimmee River chain of lakes. MS. Thesis. University of Florida, Gainesville, Florida. USA.

McGlathery, KJ; Reynolds, LK; Cole, LW; Orth, RJ; Marion, SR; and Schwarzchild, A. 2012. Recovery trajectories during state change from bare sediment to eelgrass dominance. Mar Ecol Prog Ser. 448:209–221.

Moore, PA and Reddy, KR. 1994. The role of Eh and pH on phosphorus geochemistry in sediments of Lake Okeechobee, Florida. J Environ Qual. 23:955–964.

Moore, PA; Reddy, KR; and Fisher, MM. 1998. Phosphorus flux between sediment and overlying water in Lake Okeechobee, Florida: spatial and temporal variations. J Environ Qual. 27:1428–1439.

Moore, PA; Reddy, KR; and Graetz, DA. 1991. Phosphorus geochemistry in the sediment-water column of a hypereutrophic lake. J Environ Qual. 20:869–875.

Nikolai, SJ and Dzialowski, AR. 2014. Effects of internal phosphorus loading on nutrient limitation in a eutrophic reservoir. Limnologica. 49: 33–41.

Olila, OG and Reddy, KR. 1995. Influence of pH on phosphorus sorption in oxidized lake sediments. Soil Sci Soc Am J. 59:946–959.

Olila, OG and Reddy, KR. 1997. Influence of Redox potential on phosphorus-uptake by sediments in two subtropical eutrophic lakes. Hydrobiologia. 345:45–57.

Olila, OG; Reddy, KR; and Harris, WG. 1995. Forms and distribution of inorganic phosphorus in sediments of two shallow eutrophic lakes in Florida. Hydrobiologia. 129:45–65.

Orth, RJ; Moore, KA; Marion, SR; Wilcox, DJ; and Parrish, DB. 2012. Seed addition facilitates eelgrass recovery in a coastal bay system. Mar Ecol Prog Ser. 448: 177–195.

Pollman, CD and Engstrom, DR. 2005. Assessment of sediment mixing In Lake Okeechobee, FL by radioisotopic methods. In partial fulfillment of contract number PO P501114, submitted to: South Florida Water Management District, West Palm Beach, FL. p. 20.

Reddy, KR and Delaune, RD. 2008. *Biogeochemistry of Wetlands: Science and Applications.* CRC Press.

Reddy, KR; Fisher, MM; and Ivanoff, D. 1996. Resuspension and diffusive flux of nitrogen and phosphorus in a hypereutrophic lake. J Environ Qual. 25:363–371.

Reddy, KR; Fisher, MM; Pant, H; Inglett, PW; and White, JR. 2001. Nutrient exchange between sediment and overlying water column. *In* Indian River lagoon hydrodynamics and water pore water concentrations of nutrients and their diffusive fluxes at quality model: Nutrient storage and transformations in sediments. St. Johns River Water Manage. Dist., Palatka, Florida, USA.

Reddy, KR; Fisher, MM; Wang, Y; White, JR; and James, RT. 2007. Potential effects of sediment dredging on internal phosphorus loading in a shallow, subtropical lake. Lake Reserv Manage. 23:27–38.

Reddy, KR and Graetz, DA. 1991. Internal nutrient budget for Lake Apopka. Special Publication SJ 91-SP-6, plus Addendum. St Johns River Water Management District, Palatka, FL.

Reddy, KR; Newman, S; Osborne, TZ; White, JR; and Fitz, HC. 2011. Phosphorus cycling in the Everglades ecosystem: legacy phosphorus implications for management and restoration. Crit Rev Environ Sci Technol. 41:149–186.

Reynolds, LK; Waycott, M; McGlathery, KJ; and Orth, RJ. 2016. Ecosystem services returned through seagrass restoration. Restor Ecol. 24:583–588.

Schelske, CL. 1997. Sediment and phosphorus deposition in Lake Apopka. Special Publication SJ97-SP21. St Johns River Water Management District, Palatka, FL.

Schneider, RF and Little, JA. 1969. Characterization of bottom sediments and selected nitrogen and phosphorus sources in Lake Apopka, Florida. United States Department of Interior, Technical Programs, Athens, GA. p. 35.

SFWMD, FDEP, and FDACS. 2004. Lake Okeechobee Protection Program, Lake Okeechobee Protection Plan. South Florida Water Management District, West Palm Beach, FL; Florida Department of Environmental Protection, Tallahassee, FL; and Florida Department of Agriculture and Consumer Services, Tallahassee, FL.

Soil Water Engineering and Technology (SWET), Inc. 2008a. Legacy phosphorus abatement plan for project entitled "Technical Assistance in Review and Analysis of Existing Data for Evaluation of Legacy Phosphorus in the Lake Okeechobee Watershed." West Palm Beach, Fla.: South Florida Water Management District.

Soil Water Engineering and Technology (SWET), Inc. 2008b. Task 2: Evaluation of Existing Information. For Project Entitled: Technical Assistance in Review and Analysis of Existing Data for Evaluation of Legacy Phosphorus in the Lake Okeechobee Watershed. Final Report to SFWMD, West Palm Beach, FL.

Søndergaard, M; Jensen, JP; and Jeppesen, E. 1999. Internal phosphorus loading in shallow Danish lakes. Hydrobiologia. 408/409:145–152.

Søndergaard, M; Jensen, JP; and Jeppesen, E. 2003. Role of sediment and internal loading of phosphorus in shallow lakes. Hydrobiologia. 506:135–145.

Spears, BM; Carvalho, L; Perkins, R; Kirika, A; and Paterson, DM. 2007. Sediment phosphorus cycling in a large shallow lake: spatio-temporal variation in phosphorus pools and release. Hydrobiologia. 584:37–48. DOI 10.1007/s10750-007-0610-0

Steinman, AD; Rediske, R; and Reddy, KR. 2004. The importance of internal phosphorus loading in Spring Lake, Michigan. J Environ Qual. 33:2040–2048.

Torres, IC; Turner, BL; and Reddy, KR. 2014. The chemical nature of phosphorus in subtropical lake sediments. Aquat Geochem. 20: 437–457.

Waters, MN; Schelske, CL; Kenney, WF; and Chapman, AD. 2005. The use of sedimentary algal pigments to infer historic algal communities in Lake Apopka, Florida. J Paleolimnol. 33:53–71.

Welch, Z; Zhang, J; and Jones, P. 2019. Chapter 8B: Lake Okeechobee watershed annual report. South Florida Environmental Report (SFER). https://www.sfwmd.gov/science-data/scientific-publications-sfer. Vol. I. pp. 1–53. South Florida Water Management District, West Palm Beach, FL, USA.

ALUM TREATMENT DID NOT IMPROVE WATER QUALITY IN HYPEREUTROPHIC GRAND LAKE ST. MARY'S, OHIO

Geraldine Nogaro[1], Amy J. Burgin[2], Astrea Taylor[3], and Chad R. Hammerschmidt[3]

Abstract

Grand Lake Saint Mary's (GLSM) is a large (52 km²) and shallow hypereutrophic lake in western Ohio, USA. Many years of phosphorus (P) loading from the watershed have induced eutrophication in GLSM and associated blooms of harmful cyanobacteria and fish kills. The lake's water quality decline has caused an economic loss to the region since it is also a state park used for camping, boating, hunting, and picnicking. In 2011, the Ohio (USA) Environmental Protection Agency decided to treat the central area of the lake with aluminum sulfate (alum) and sodium aluminate as a buffer (21 mg Al L⁻¹, 44 g Al m⁻²) to inactivate excess P and potentially reduce toxic algae blooms. The objectives of the current study were to: (1) estimate the P mass balance in GLSM to determine the relative importance of internal and external loadings in 2011, when an alum treatment was applied to the lake; and (2) investigate the effects of the alum treatment on metal and nutrient (P) cycling and invertebrate communities in GLSM. Our results suggest that alum is not the best solution to prevent harmful algal blooms and restore water quality in GLSM because external P sources were substantially higher than benthic loadings in 2011 (i.e., 70 ± 7% of external P sources, subdivided into 65 ± 5% from watershed runoff, 4 ± 2% from point-source discharges, and 1 ± 0.2% from direct atmospheric deposition versus 30 ± 14% from benthic release). Addition of alum did not significantly reduce P and chlorophyll *a* concentrations measured immediately before and after

continued

---

[1] Corresponding author (G. Nogaro). Present address: EDF Research and Development, National Hydraulic and Environment Laboratory (LNHE), 6 quai Watier, 78401 Chatou, FRANCE. Contact: geraldine.nogaro@gmail.com

[2] Department of Ecology and Evolutionary Biology, The University of Kansas, 1200 Sunnyside Avenue, Lawrence, KS 66045, USA, burginam@gmail.com

[3] Department of Earth and Environmental Sciences, Wright State University, 3640 Colonel Glenn Highway, Dayton, OH 45435, USA

the one-month alum application. However, alum additions significantly increased pH, filtered aluminum, and particulate manganese in surface water but did not affect the abundance of biota in sediment. GLSM underwent a large addition of alum resulting in no improvement of water quality. Best management practices need to be implemented in the watershed for controlling P export to the lake.

Key Words: Alum, sediment, eutrophication, internal loading, phosphorus.

8.1   INTRODUCTION

GLSM is a shallow hypereutrophic lake in northwestern Ohio, USA (see Table 8.1). GLSM is Ohio's largest inland lake with a surface area of 52 km² and an average depth of 1.5 m. GLSM has a watershed area of 241 km², including eight sub-catchments that are composed of poorly drained silt and clay soil (Geib 1910; Priest 1979) and largely dominated by agriculture and livestock operations. GLSM was constructed in the 1840s as a reservoir for the Miami and Erie Canal. The lake drains into two primary outlets: (1) Beaver Creek River, which drains the west side of the lake to the Wabash River; and (2) the Miami and Erie Canal, from the east side of the lake, which flows into the St. Mary's River (see Figure 8.1).

GLSM is used as a recreational area for boating, fishing, swimming, and water skiing. The state park is also known for year-round activities such as camping, boating, hunting, and picnicking. Agricultural activities in the watershed have resulted in high P loadings to the lake and excess algal growth (Hoorman et al. 2008). Many years of high P loadings to the lake have induced frequent harmful algal blooms of toxic cyanobacteria in the summer, negatively impacting recreational activities, businesses, and property values around the lake.

Key factors suspected to impact water quality of GLSM are heavy agricultural land use and to a much lesser degree, potentially poorly maintained septic systems in the watershed. The land use within the GLSM watershed is mostly cropland (85% row crops and 9% pasture/hay) and associated

Table 8.1   Summary of GLSM characteristics

Parameters	Unit	GLSM
Average depth	m	1.5
Lake area	km²	52
Watershed area	km²	241
Altitude	m	265
Coordinates	DMS	40°31′46″N; 84°28′24″W
Trophic state		hypereutrophic
Length	km	13
Width	km	4.5
Estimated volume	m³	8.3×10^7
Estimated water retention time	days	429 in 2009 480 in 2010 180 in 2011

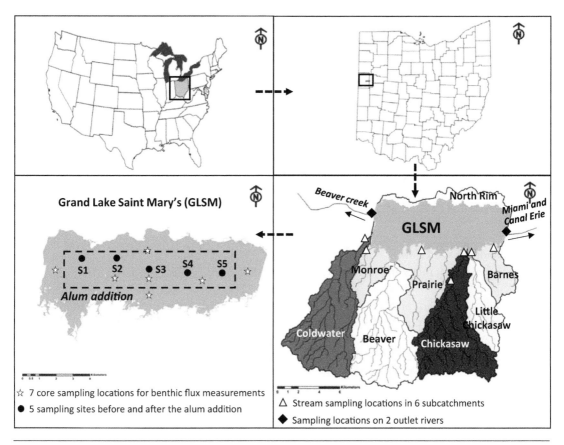

Figure 8.1   Upper left panel: map of USA showing state of Ohio in gray; Upper right panel: State of Ohio with outlined county where GLSM is located; Lower right panel: Blow-up of GLSM and watershed, showing sub-catchments, outflows, and the watershed sampling locations. White triangles identify stream sampling locations in six sub-catchments that drain into GLSM. Black diamonds indicate sampling locations on two outlet rivers that drain GLSM (Beaver Creek River and Miami and Erie Canal). Arrows shows that the two rivers are sending water out of the lake; Lower left panel: GLSM with alum application location (within dashed box) and the five sampling sites (i.e., sites S1 to S5 from the west side to the east side of the lake shown by the black dots). White stars identify the seven sediment locations for benthic flux cores in May 2011, before alum treatment.

livestock operations, whereas forest and residential areas cover only about 3% and 2%, respectively (Hoorman et al. 2008). Agricultural crops include mostly corn, soybeans, hay pasture, and wheat. There are also 286 livestock operations with an average animal density of 241 units per km^2 (OEPA 2007; Hoorman et al. 2008).

In an attempt to improve the lake's water quality, aluminum sulfate (alum) was added to the central part of the lake (19.6 km^2—about 40% of lake area) in June 2011 to scavenge P from the water column and limit internal P loadings from sediment, with the aim of reducing the potential for harmful algae blooms. The alum dose was applied from June 2 to 29, 2011 and corresponded to 21 mg Al L^{-1} (44 g m^{-2}) from alum and sodium aluminate (used as a buffer) combined (Tetra Tech 2012).

The objectives of this study were to: (1) estimate the mass balance of P in GLSM to determine the relative importance of internal and external P loadings in 2011, when an alum treatment was applied to the lake; and (2) investigate the short-term effects of the alum treatment on metal and P cycling and invertebrate communities in GLSM. The P mass balance was estimated using P fluxes from point-source discharge into inlet streams, inputs from watershed runoff, benthic remobilization, direct atmospheric deposition, outflow of water from the lake through two outlets, and sedimentation. The effects of the alum treatment were investigated by sampling surface water, sediment, and pore water at five sites in GLSM immediately before and after the alum addition.

Because the objective of the alum treatment was to help control internal P loadings from sediment, based on results from prior studies (Cooke et al. 2005), we expected that internal P loadings would be much greater than the external P loadings. Based on our previous study (Nogaro et al. 2013), we also hypothesized that alum, which is intended to target P, would reduce P and chlorophyll *a* concentrations in the water column. Moreover, we hypothesized that alum addition would have unintended consequences, including potentially increasing pH and metal (Al and Mn) concentrations, and decreasing invertebrate community abundance in sediment.

8.2   METHODS

8.2.1   Study Site

Six wastewater treatment plants (WWTPs) discharge effluent into streams that drain to GLSM, although none discharge directly to the lake. GLSM has a large fetch (14.5 km), is well-mixed, and does not thermally stratify throughout the ice-free season, typically from about April to November (cf. Filbrun et al. 2013). GLSM ultimately drains north to Lake Erie and south to the Ohio River. GLSM has been classified as a eutrophic lake for decades (USEPA 1975). Nitrogen (N):P molar ratios in inlet streams and lake water in late summer (mean = 4N:1P; USEPA 1975; Hoorman et al. 2008) are disproportionately low in GLSM compared to typical stoichiometric requirements (16N:1P as per Redfield et al. 1963), and have promoted the growth of cyanobacteria, some of which can fix N_2 (Berman-Frank et al. 2003; Cooke et al. 2005).

8.2.2   Estimation of P Mass Balance

The P mass balance for GLSM was estimated from six main fluxes: (1) point-source discharge into inlet streams; (2) inputs from watershed runoff; (3) benthic flux; (4) direct atmospheric deposition; (5) outflow of water from the lake through two outlets; and (6) sedimentation (closing term). These fluxes were estimated from measurements during multiple sampling events. Potential inputs of P from septic systems and groundwater to GLSM are likely to be insignificant because the clay soils surrounding and underlying the lake are poorly drained (Geib 1910; Priest 1979). To our knowledge, there are no data on the potential source from bird and wildlife *deposits* that could also affect GLSM's P fluxes.

8.2.2.2   Point-Source Discharges

Multiple municipal and private entities, particularly WWTPs, discharge P into GLSM streams. Each discharger is required to monitor water volume and periodically measure total P (TP) in the discharge. Annual inputs from point sources were estimated as the product of entity-specific water discharge volume and mean TP concentration in 2011. Four of the 10 dischargers were not required to measure P (accounting for 2.5% of total water flow). To account for likely P inputs from the entities that did not measure P, fluxes were estimated as the product of measured water discharge and average TP concentration in wastewater from similar dischargers into GLSM inlet streams. Error was estimated for each entity as the product of total discharge volume and standard deviation of mean TP concentration.

8.2.2.3   Watershed Inputs

Watershed inputs of P were examined by sampling streams draining into GLSM (Coldwater, Monroe, Prairie, Chickasaw, Little Chickasaw, and Barnes Creeks); water was not collected from Beaver Creek because it did not have a suitable sampling location (described later). Water was sampled from each stream in May (n = 3), June, July, and September (n = 1 each), with sampling periods representative of relatively low, medium, and high water (i.e., storm event) flows. Stream water was sampled at locations about 100 m upstream from the mouth of the lake (see Figure 8.1). In addition to the mouth, stream water was also sampled upstream on the Chickasaw Creek at a U.S. Geological Survey (USGS) gauging station (402913084285400, data available at http://waterdata.usgs.gov). Water was collected from the middle of each stream with a clean pole sampler and rigorously cleaned bottles (Hammerschmidt et al. 2011). Samples were transported to Wright State University on ice and were either analyzed on the day of collection or preserved for later analysis of TP, soluble reactive P (SRP), and total soluble P (TSP; < 0.2 μm). Particulate P was calculated as the difference between total and TSP, and represents the fraction of P sorbed to particles greater than 0.2 μm in diameter.

Areal watershed discharge from each of the sub-catchments was estimated from the mean instantaneous discharge measured in 2011 at the gauge in Chickasaw Creek, which was 1.075 m^3 sec^{-1}, with an upstream watershed of 42.48 km^2. The resultant areal flux of water from the watershed upstream of the gauge was 7.98 $\times 10^8$ L of water per km^2 yr^{-1} for 2011, or a runoff velocity of 0.80 m yr^{-1}, which is comparable to velocities (0.76–0.86 m yr^{-1}) measured in two streams located 100 km south of GLSM (Naik and Hammerschmidt 2011).

Because the other sub-catchments have similar geological characteristics, land uses, and rainfall as the Chickasaw sub-catchment (Hoorman et al. 2008), the water yield from the Chickasaw sub-catchment was used to predict the annual delivery of water from the entire watershed to GLSM in 2011, which was estimated to be 1.92×10^8 m^3, and similar to estimates made by others (OEPA 2007). Fluxes of P from each sub-catchment were estimated as the product of annual water discharge (estimated for each sub-catchment) and the discharge-weighted TP concentration near the mouth of each stream. The discharge-weighted P concentration was determined from linear regression analyses of TP measured in each stream versus instantaneous water discharge measured simultaneously at the Chickasaw Creek gauge. Uncertainty of the flux estimates for each sub-catchment was estimated similarly as the product of annual water discharge and the standard error of slope of the regression analysis for each stream. The North Rim, which does not have a channelized drainage system, and Beaver Creek sub-catchments values were estimated from average areal flux of TP from the six sub-catchments with measured P concentrations. Annual point-source inputs (all of which were upstream from stream sampling locations) were subtracted from each relative watershed total.

8.2.2.4   Benthic Flux

Fluxes of P from lake sediments to the overlying water were quantified with benthic flux chambers (Hammerschmidt and Fitzgerald 2008). These consisted of undisturbed cores of lake sediment and overlying water that were incubated for up to 7.5 days in a dark water bath at 21°C, which is near the average benthic water temperature in GLSM, from March 30 to October 31 (19.7 ± 6.1°C; USGS central buoy B-1, data available at http://waterdata.usgs.gov) with constant stirring of overlying water. Sediment cores and lake water were sampled from seven locations in GLSM on May 5, 2011 before the alum treatment of the lake (see Figure 8.1). Lake water for use in the chambers was sampled near the sediment-water interface at each site (Taylor 2012; see Figure 8.1).

Cores were equipped with caps that had open vents to the atmosphere and magnetic stir bars (Hammerschmidt and Fitzgerald 2008) that mixed overlying water at 25 rotations per minute, mimicking the natural movement of lake water and preventing a static benthic boundary layer. Overlying water was sampled from each flux chamber after predefined incubation periods to quantify the fluxes of SRP and TSP. After removal of water during each sampling period, stored lake water from the same site (kept in a dark refrigerator) was added gently to the flux cores to replace the removed volume (Hammerschmidt and Fitzgerald 2008). The oxic benthic flux of P was estimated as the product of lake area and mean (±SE) areal P flux determined from the benthic cores sampled in May.

To account for P release under hypoxic conditions during the summer, P fluxes were estimated as the product of different internal loading rates for hypereutrophic lakes (i.e., 10, 20, and 30 mg m^{-2} d^{-1}; Nürnberg and LaZerte 2004) and the numbers of days of low oxygen levels (< 1 mg L^{-1} and < 2 mg L^{-1}) from April to October 2011 in GLSM (i.e., respectively 3.5% and 8% of the time, USGS central buoy B-1, http://waterdata.usgs.gov). These different rates allowed us to calculate a range of potential internal loading rates under hypoxic conditions and have a better understanding of the possible uncertainties in estimating internal loading in GLSM.

Estimated internal load at GLSM in 2011 was calculated by scaling up the mean P fluxes during oxic and hypoxic conditions for the entire lake area and multiplying by the percentage of days under oxic and hypoxic conditions. Uncertainty of the internal loading flux was estimated from the standard deviation of the full range of P fluxes under oxic and hypoxic conditions.

8.2.2.5   Atmospheric Deposition

Rainwater was sampled during three precipitation events with trace-metal clean equipment between March and May 2012 in Fairborn, OH, which is about 100 km south of GLSM, but atmospheric fluxes of P are presumed comparable based on regional similarities in deposition of P (data from the U.S. National Atmospheric Deposition Program at http://nadp.sws.uiuc.edu). Rainwater was analyzed for TP and SRP. Atmospheric fluxes directly to GLSM were estimated as the product of the mean measured concentration in rainwater and average wet deposition depth (1.3 ± 0.13 m) measured at seven locations near GLSM in 2011 (data available from the U.S. National Weather Service at http://www.weather.gov).

8.2.2.6   Lake Outflow

Two primary outflows export P from GLSM: (1) Beaver Creek River, which drains the west side of the lake to the Wabash River; and (2) the Miami and Erie Canal, from the east side of the lake, which flows into the St. Mary's River (see Figure 8.1). Annual export of P through the two outflows was estimated as the product of water volume outflow and mean TP concentration measured in water near the outflows, from May 2011 to May 2012 (n = 6 for Beaver Creek River, n = 3 for the Miami and Erie Canal). Water outflow in 2011 was estimated to be 2.1×10^8 m^3, based on the difference between water inputs to the lake (watershed runoff + direct precipitation = 2.6×10^8 m^3) and estimated loss by evaporation (0.86 m yr^{-1} or 4.8×10^7 m^3; Farnsworth et al. 1982). Uncertainty of the export flux was estimated from the standard deviation of measured levels of TP in the outflow.

8.2.2.7   Sedimentation

Benthic sedimentation is a sink for P and represents the closing term for most P budgets in lakes (Cooke et al. 2005). The burial flux of P in GLSM was estimated by difference from all other measured fluxes.

8.2.3   Effects of Alum Addition on GLSM

8.2.3.1   *Sampling of Water, Sediment, and Organisms*

Surface water and sediment cores were sampled at five representative sites in GLSM (see Figure 8.1) during the day before and after alum treatment to measure biogeochemical characteristics. At each site, we measured physicochemical parameters, dissolved ions, and both filtered and particulate metals in three replicate water samples and sediment cores. We also collected three replicate Ponar grabs (400 cm^2) before (i.e., on May 5, 2011) and after (i.e., on July 14, 2011) alum addition (from June 2 to 29, 2011) to estimate invertebrate abundance and main taxa (i.e., chironomids and oligochaetes).

Samples of surface and pore water were analyzed for metals, nutrients, and gases. Standard colorimetric techniques were used to quantify SRP (Wetzel and Likens 1991). Dissolved and particulate P (after digestion with concentrated HNO_3) were determined by inductively coupled plasma mass spectrometry (ICPMS; Maher et al. 2003). Chlorophyll *a* was measured following standard methods (APHA et al. 1995). *In vivo* phycocyanin fluorescence (i.e., cyanobacteria pigment activity) was measured using a field YSI 6600 probe (YSI Inc., Yellow Springs, OH). Dissolved organic carbon (DOC) was measured by infrared combustion analysis (Sharp et al. 1995). Filtered and particulate metals were measured by ICPMS (USEPA 1994). Dissolved oxygen (DO) concentrations were measured with a membrane inlet mass spectrometer (MIMS) (Kana et al. 1994). Blanks, standards, and duplicate samples (10% of samples) were analyzed to determine the accuracy of each analysis as a part of the quality assurance protocol.

8.2.4   Statistical Methods

Concentrations of TP, chlorophyll *a*, phycocyanin, DOC, and DO, and pH values measured in surface water between sampling periods were tested with two-way nested analyses of variance (i.e., nested ANOVAs) with site and alum treatment (nested within site) as the main effects. Concentrations of filtered and particulate Al, and filtered and particulate Mn measured in surface water also were tested with two-way nested ANOVAs with site and alum treatment (nested within site) as the main effects. Comparisons of total dissolved P, filtered Al, and filtered Mn concentrations measured in pore water were tested with three-way nested ANOVAs using site, alum treatment (nested within site), and depth of the sediment core as the main effects. Chironomid and oligochaete abundance also were tested with two-way nested ANOVAs with site and alum treatment (nested within site) as main effects. If significant differences were detected, then Tukey post-hoc tests were performed to determine which treatments differed. When necessary, data were log-transformed before statistical analysis to meet the assumptions of homoscedasticity and normality. All statistical analyses were performed with R software (R Development Core Team 2016).

8.3   RESULTS

8.3.1   Estimation of P Mass Balance

Total loadings of P to GLSM in 2011 were estimated to be 92,200 kg ± 19,400 kg, with 65% ± 5% of loadings attributed to watershed runoff, 30% ± 14% to benthic flux (internal loading), 4% ± 2% to point-source discharges, and 1% ± 0.2% to direct atmospheric deposition (see Figure 8.2). We estimated that 62% ± 7% of P in the water column was exported to river outflow and the remaining 38% ± 10% was lost to benthic deposition, assuming steady state. The specifics of how we calculated this mass balance are described later.

Point-source discharges were a relatively minor source of P to GLSM, contributing a combined total of 3,400 kg ± 1,500 kg of P in 2011 (see Table 8.2). Inputs from dischargers not required to measure P were estimated to be about 100 kg in 2011, based on average P concentrations for similar entities and relative discharge volume.

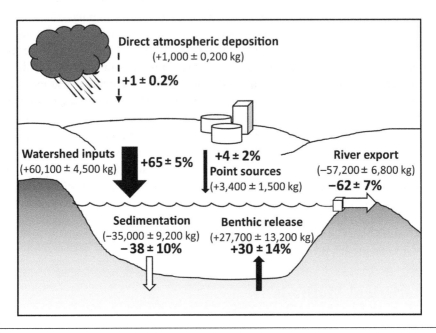

Figure 8.2   Phosphorus sources in kg of P and mass balance for GLSM in 2011. Black arrows indicate P inputs; white arrows indicate P losses.

Table 8.2   Point-source discharge inputs of P to streams that drain into GLSM in 2011

Discharging Entity[†]	Mean Total P in Discharge (mg L⁻¹)	Water Volume Discharged (m³ yr⁻¹)	Total P Flux (kg yr⁻¹)
Montezuma Club WWTP	1.37 ± 0.78	1.26×10^6	1,730 ± 980
St. Henry WWTP	1.30 ± 0.23	7.73×10^5	1,010 ± 174
Mike's Sanitation	4.23 ± 2.38	8.68×10^4	367 ± 207
Northwood WWTP	1.10 ± 0.93	7.05×10^4	77 ± 65
Chickasaw WWTP[‡]	1.50	5.42×10^4	81
Celina Landfill	0.19 ± 0.08	6.84×10^2	2.1 ± 0.1
Philothea WWTP[§]	NA	3.61×10^4	NA (69 ± 47)
Chapel Hill WWTP[§]	NA	1.64×10^4	NA (31 ± 21)
Marion School[§]	NA	4.80×10^3	NA (9 ± 6)
Elk Club[§]	NA	3.18×10^2	NA (0.6 ± 0.4)
		Total =	3,400 ± 1,500

[†]WWTP = Wastewater treatment plant
[‡]No statistical P data (n = 1).
[§]Entity not required to report P concentration. Parenthetical value represents the product of entity's annual water discharge and average total P concentration for similar industries.

Sub-catchment-based P loadings to GLSM from the watershed were estimated to be 60,100 kg ± 4,500 kg in 2011 (see Table 8.3). Estimates of P flux were based on discharge-weighted TP concentrations for each stream at the mean instantaneous discharge rate at the Chickasaw gauge during 2011 (1.075 m^3 sec^{-1}), which, in addition to the median flow rate (0.17 m^3 sec^{-1}), was within the range of flows during sampling periods. Stream waters sampled during the period of lowest measured discharge (0.13 m^3 sec^{-1}) had a mean concentration of 221 kg ± 76 µg TP L^{-1}. While loadings of P to GLSM varied largely as a function of sub-catchment area, areal watershed yields of P were similar among each of the sub-catchments (see Table 8.3). Similarities of areal P yields among the sub-catchments might be expected as a result of either similarities in land uses or due to assumptions about water runoff fluxes and precipitation among sub-catchments. The mean annual yield of P from the watershed was 250 mg m^{-2} ± 20 mg m^{-2} in 2011. Greater than average precipitation in 2011 (1.3 ± 0.13 m versus annual average of 0.91 ± 0.06 m from 1961 to 1990; OEPA 2007) may have increased the watershed P loadings to GLSM, whereas lower watershed yields and fluxes might be expected during drier years. Atmospheric deposition was an insignificant source of P to GLSM compared to watershed and benthic fluxes. Total P in rainwater averaged 14.2 µg L^{-1} ± 2.8 µg L^{-1} ($n = 3$) with about 30% of the TP as SRP in rainwater. The mean concentration of TP in rainwater was within the range of levels determined in precipitation at other rural locations in North America (7–20 µg L^{-1}; Rose 1993, Shaw et al. 1989). The product of precipitation depth in 2011 (1.3 m ± 0.13 m), mean concentration of TP in rainwater, and area of GLSM equated to a source of 1,000 kg ± 200 kg directly to the lake surface in 2011 (see Figure 8.2).

Outflow was the largest measured loss of P from GLSM. The mean measured concentration of TP in water draining from GLSM was 260 µg L^{-1} ± 31 µg L^{-1}. Concentrations at the outflows were similar to those of lake surface water from June to July 2011 (266 µg L^{-1} ± 29 µg L^{-1}, $n = 43$, $p = 0.4$). Calculated water outflow from GLSM was estimated to be 2.14 × 10^8 m^3, which exported 57,200 kg ± 6,800 kg of P from the lake in 2011 (see Figure 8.2).

Benthic fluxes were the second greatest source of P to GLSM in 2011. Concentrations of TSP in overlying water increased over time in each benthic flux chamber. Nearly all (mean = 98%) of the P mobilized from the lake sediment to overlying water was as SRP (data not shown). Under oxic conditions, benthic fluxes of P averaged 0.33 mg m^{-2} d^{-1} ± 0.11 mg m^{-2} d^{-1} from the seven locations in GLSM that were sampled in May before alum treatment. Scaling of mean benthic P flux measured

Table 8.3 Watershed fluxes and loadings of P to GLSM in 2011

Sub-Catchment	Sub-Catchment area (km^2)	Water Flux (m^3 yr^{-1})	P Loadings to GLSM (kg yr^{-1})	Areal P Flux (mg m^{-2} yr^{-1})
Coldwater Creek	50.0	4.0 × 10^7	14100 ± 1000	280
Monroe Creek	12.8	1.0 × 10^7	3500 ± 200	270
Beaver Creek[†]	52.8	4.2 × 10^7	12100 ± 1000	230
Prairie Creek	31.1	2.5 × 10^7	9400 ± 600	300
Little Chickasaw Creek	13.0	1.0 × 10^7	2700 ± 300	210
Chickasaw Creek	53.0	4.2 × 10^7	11000 ± 900	210
Barnes Creek	14.1	1.1 × 10^7	3600 ± 200	250
North Rim[†]	14.1	1.1 × 10^7	3700 ± 300	260
Total	241	1.92 × 10^8	60,100 ± 4,500	–

[†]Total P was not determined in these sub-catchments: mean discharge-weighted concentration from each of the other sampled streams was multiplied by the water flux specific to this sub-catchment to estimate P loadings.

in May to the entire year and lake area equated to an annual input of 6,300 kg ± 2,100 kg under oxic conditions. The estimated benthic flux range of P under hypoxic conditions was estimated between 6,600 kg and 45,550 kg (mean ± SD = 21,800 ± 14,000; n = 6), with the variability resulting from differences in the product of different internal loading rates for hypereutrophic lakes (Nürnberg and LaZerte 2004) and the numbers of days of low oxygen levels from April to October 2011 in GLSM (see methods). Accordingly, the internal loading of P to GLSM in 2011 under oxic and hypoxic conditions was estimated to be 27,700 kg ± 13,200 kg with a flux of 1.5 mg m^{-2} d^{-1} ± 0.7 mg m^{-2} d^{-1} (see Figure 8.2).

Sedimentation of P was the closing term in our steady-state mass balance. By difference between all sources and other losses of P from GLSM, sedimentation was estimated to be 35,000 kg yr^{-1} ± 9,200 kg yr^{-1}. The uncertainty of the closing term was estimated as the geometric mean of uncertainties associated with all other fluxes into and out of the lake.

8.3.2   Effects of Alum Addition on GLSM's Water Quality

Unexpectedly, TP concentration increased between pre- and post-alum additions at all sites in surface water in GLSM (two-way nested ANOVA, alum [nested in site] effect, P < 0.05) (see Figure 8.3A). Chlorophyll *a* concentrations in the surface water were not affected by the alum addition (two-way nested ANOVA, alum [nested in site] effect, P = 0.318) (see Figure 8.3B), although a decrease may have occurred at Site 5 after the alum addition in comparison with the pre-alum addition. Phycocyanin fluorescence (i.e., cyanobacteria pigment) increased in surface water at Sites 1, 2, and 5 after the alum addition (two-way nested ANOVA, alum [nested in site] effect, P < 0.01) (see Figure 8.3C) while no significant differences were detected at Sites 3 and 4. DOC in surface water also increased at all sites between immediately before and after the alum addition (two-way nested ANOVA, alum [nested in site] effect, P < 0.001) (see Figure 8.3D). Among all sites, DO concentrations in the surface water were not different before and after alum addition (two-way nested ANOVA, alum [nested in site] effect, P = 0.991) (see Figure 8.3E). In addition to not reducing P and chlorophyll *a* concentrations, the alum addition had other effects on water quality. The pH of surface water increased significantly at all sites after alum addition, up to greater than pH 9 (two-way nested ANOVA, alum [nested in site] effect, P < 0.001) (see Figure 8.3F).

Concentrations of filtered Al in surface water increased greatly (note the log scale) at all sites after alum addition compared to the pre-alum addition (two-way nested ANOVAs, alum [nested in site] effect, P < 0.001) (see Figure 8.4A), whereas particulate Al in surface water decreased at all sites after alum compared to the pre-alum addition (two-way nested ANOVAs, alum [nested in site] effect, P < 0.001) (see Figure 8.4B). Filtered Mn in surface water was not affected by the alum addition (two-way nested ANOVAs, alum [nested in site] effect, P = 0.776) (see Figure 8.4C) while particulate Mn in the surface water increased with alum addition at all sites, with Mn largely dominated by its particulate form (two-way nested ANOVAs, alum [nested in site] effect, P < 0.001) (see Figure 8.4D).

Total dissolved P in pore water was low (i.e., 5–20 μg L^{-1}), even before alum addition, and decreased after alum addition, except at Site 4, compared to the pre-alum site (three-way nested ANOVA, alum [nested in site] × depth effect, P < 0.05) (see Figure 8.5A). Mean values of filtered Al in pore water declined at all sites and depths after alum addition (three-way nested ANOVA, alum [nested in site] effect, P < 0.001) (see Figure 8.5B), although the declines at Site 4 were not

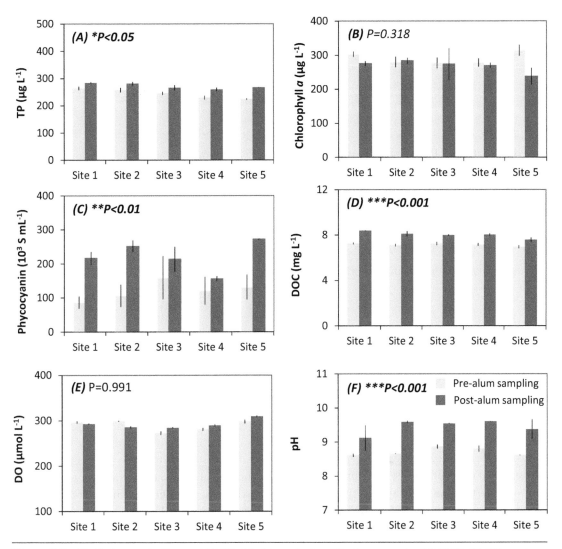

Figure 8.3   (A–F) Concentrations of (A) TP, (B) chlorophyll a, (C) phycocyanin, (D) DOC, (E) DO, and (F) pH values in the surface water at five sampling sites of GLSM before and after (i.e., 27 days) the alum addition in June 2011 (mean ± 1 SE, *n* = 3). For each panel (A, B, etc.), P value shows the alum [nested in site] effect of two-way nested ANOVA at ***P < 0.001, **P < 0.01, *P < 0.05 and P > 0.05 (non-significant).

statistically significant. Filtered Mn in pore water decreased at all sites after alum addition except, again, at Site 4 where Mn in pore water increased after alum addition (three-way nested ANOVA, alum [nested in site] effect, depth effect, P < 0.001) (see Figure 8.5C). Chironomid and oligochaete abundances measured in the sediment increased at all sites after the alum addition (sampling on July 14) compared to before the alum addition (May 5) (two-way nested ANOVA, alum [nested in site] effect, P < 0.001) (see Figure 8.6).

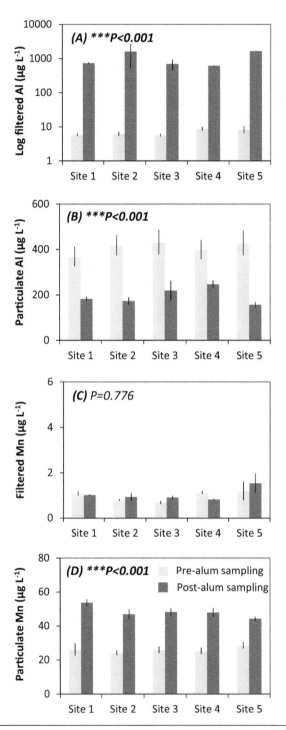

Figure 8.4   (A–D) (A) Concentrations of filtered aluminum (Al), (B) particulate aluminum (Al), (C) filtered manganese (Mn), and (D) particulate manganese (Mn) measured in the surface water at five sampling sites of GLSM before and after (i.e., 27 days) the alum addition in June 2011 (mean ± 1 SE, $n = 3$). Please note the filtered Al concentration (panel A) is on a logarithmic scale. For each panel (A, B, etc.), P value shows the alum [nested in site] effect of two-way nested ANOVA at ***P < 0.001 and P > 0.05 (non-significant).

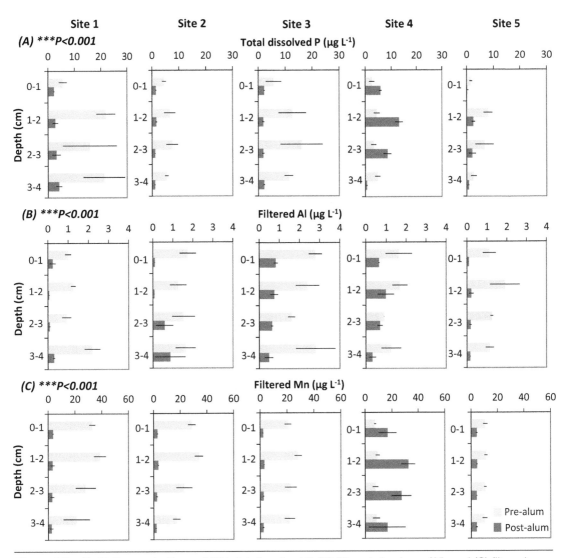

Figure 8.5   (A–C) Concentrations of (A) total dissolved P, (B) filtered aluminum (Al), and (C) filtered manganese (Mn) in the pore water at five sampling sites of GLSM before and after (i.e., 27 days) the alum addition in June 2011 (mean ± 1 SE, $n = 3$). For each panel series (A, B, etc.), P value shows the alum [nested in site] effect of three-way nested ANOVA at ***P < 0.001.

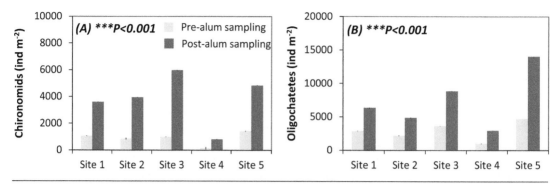

Figure 8.6   (A, B) Chironomid and oligochaete abundance from the invertebrate survey performed at five sampling sites of GLSM before (i.e., 5 May 2011) and after (i.e., 14 July 2011) the alum addition (mean ± 1 SE, n=3). For both panels, P value shows the alum [nested in site] effect of two-way nested ANOVA at ***P < 0.001.

8.4   DISCUSSION

8.4.1   Estimation of P Mass Balance

We expected that the magnitude of internal P loading to GLSM would be significantly greater than external P loadings because the main objective of alum treatment was to reduce excess P in the surface water from internal P release at the water-sediment interface (Cooke et al. 2005).

8.4.1.1   Factors Influencing Magnitude and Frequency of Internal Loading

In 2011, estimated benthic fluxes of P averaged 1.5 ± 0.7 mg m^{-2} d^{-1} in GLSM, which is within the range of those estimated for other hypereutrophic lakes but on the low end of anoxic conditions for eutrophic lakes (Nürnberg and LaZerte 2004). Benthic P flux estimates for eutrophic lakes are variable (0.1–30 mg m^{-2} d^{-1}) and can depend on benthic oxygen concentrations and methods for estimating P release rates (Cooke et al. 1977, 1993; Steinman et al. 2004; Chowdhury and Al Bakri 2006).

Temperature and DO concentrations are primary factors affecting the kinetics of biogeochemical reactions, such as P remineralization (Boström et al. 1982; Steinman et al. 2009; Filbrun et al. 2013; see Chapter 1). Incubation temperature for the flux chambers (21°C) was similar to the average benthic water temperature in GLSM from March 30 to October 31, 2011 (19.7 ± 6.1°C). Moreover, DO concentrations in the flux chamber water was similar to oxic conditions observed at the sediment-water interface in GLSM (> 6 mg L^{-1}); *in situ* measurements of DO in GLSM at the sediment-water interface were greater than 1 and 2 mg L^{-1} for 96.5% and 92% of 2011, respectively (USGS central buoy B-1, data available at http://waterdata.usgs.gov). Anoxia can exacerbate fluxes of P from lake sediments (Phillips et al. 1994; Nürnberg 1985; Auer et al. 1993; Steinman et al. 2004; Cooke et al. 2005) by causing P to be liberated from redox-sensitive minerals, namely $FePO_4$ (Boström et al. 1988). However, colder temperatures during winter would be expected to slow rates of organic matter decomposition and P remobilization from sediments (Cooke et al. 2005; Steinman et al. 2009). Additionally, DO concentrations at the sediment-water interface are unknown under the ice in GLSM. Phosphorus release under hypoxic conditions during the summer was taken into account in our P flux calculations as the product of internal loading rate under hypoxic conditions (i.e., 10, 20, and 30 mg m^{-2} d^{-1}; Nürnberg and LaZerte 2004) and the number of days of low oxygen levels (< 1 and < 2 mg L^{-1}) from April to October 2011 (i.e., 13 days, USGS central buoy B-1, http://waterdata.usgs.gov).

8.4.1.2   Factors Influencing the Relative Importance of Internal Versus External Loadings

The mean areal yield of P from the watershed was 250 mg m^{-2} ± 20 mg m^{-2} in 2011. This yield is twice as much as that estimated for the watershed in 1975 (100 mg m^{-2}; USEPA 1975), but is less than that estimated for the watersheds of other hypertrophic lakes in the region (550–890 mg m^{-2} yr^{-1}; Wilson and Walker 1989). Above average precipitation in 2011 (1.3 m versus annual average of 0.91 m) may have increased the watershed P loadings to GLSM, whereas lower watershed yields and fluxes might be expected during drier years.

Inputs of P from point-source discharges and direct atmospheric deposition are in good agreement with previous estimates for GLSM (OEPA 2007; GLWWA 2009). Benthic remobilization can be a significant source of P in some eutrophic lakes (Cooke et al. 1993, 2005), but runoff from the GLSM watershed delivered twice as much P to the lake as was released from the benthic sediments. During years with average precipitation, runoff of P from the watershed might be less than that in 2011, and benthic remobilization might have greater relative significance. However, our estimates suggest that reductions of P loadings from watershed runoff would have a greater impact on GLSM water quality than alum treatment efforts to attenuate benthic remobilization. Outflow of water from GLSM removed 62% ± 7% of the P imported to the lake in 2011, with the remaining 38% ± 10% buried to the sediments. The large amount of P transferred from GLSM to stream and river networks may have implications for nutrient management and water quality impairments downstream of the lake.

Our P budget for GLSM, in 2011, suggests that external P loadings are more significant than internal sources compared with estimates from prior studies (see Table 8.4). Depending on the analysis, estimated external inputs of P to GLSM in 2011 varied considerably. Estimated P loads included 43,000 kg yr^{-1} (Tetra Tech 2012), 132,000 kg yr^{-1} (Filbrun et al. 2013), and 60,100 kg yr^{-1} ± 4,500 kg yr^{-1} (this study). The other studies did not estimate uncertainty so we cannot discern whether there are real differences in our estimates. The estimated P budget by Tetra Tech (2012) was calculated from May 2010 to May 2011. 2010 was a particularly dry year compared to 2011, which could explain the lower estimation of external P loads compared to our results. Filbrun et al. (2013) used P loads measured from a single inlet stream in Chickasaw Creek to estimate P loadings from the entire watershed of GLSM and did not estimate internal P loading. In contrast, we estimated P loads from measured concentrations in six different streams of GLSM's watershed but only from April to October 2011. Nevertheless, even if our external P loads were underestimated, our conclusion remains that in 2011 the TP external loads were the major contributor of P inputs (at least 70% ± 7%) compared to internal loadings (30% ± 14%). The fraction of external watershed runoff was 65% ± 5%, whereas point-source discharges and direct atmospheric deposition were 4% ± 2% and 1% ± 0.2%, respectively. The different results of these three studies demonstrate the uncertainty in estimating P budgets and the need for comprehensive sampling and consistent methodology.

Ideally, a lake mass balance for P should include measurements of total inputs, outputs, and changes in lake storage (net P sedimentation or net P internal loading). Additionally, we suggest that the lake P budget should be estimated over at least three years to account for annual stream flow variability. Nonpoint P sources should be estimated from streams located in each of the different sub-catchments as well as stream discharges under multiple low- and high-flow conditions. Internal loading should be experimentally measured under oxic and hypoxic conditions to account for potentially greater P release under low DO levels in the summer.

Table 8.4   Estimates of P-mass balance for GLSM. *Dash indicates that the data was not estimated in the study.*

P—Mass Balance	Year	External P Loads (kg yr^{-1})	P River Export (kg yr^{-1})	Annual Internal P Loads (kg yr^{-1})	P Sedimentation Rate (kg yr^{-1})
This study	2011	60,100 ± 4,500	57,200 ± 6,800	27,700 ± 13,200	35,000 ± 9,200
Filbrun et al. (2013)	2009	38,400	—	—	—
	2010	36,300	—	—	—
	2011	131,600			
Tetra Tech (2012)	May 2010– May 2011	42,691	25,780	14,552 (estimation)	31,463
OEPA (2007)	1996	116,000	—	—	—
	1997	83,000	—	—	—
	1998	90,000	—	—	—
	1999	65,000	—	—	—
	2000	66,000	—	—	—
	2001	84,000	—	—	—
	2002	87,000	—	—	—
	2003	130,000	—	—	—
	2004	87,000	—	—	—
	2005	148,000	—	—	—
	2006	96,000	—	—	—
USEPA (1975)	1973	21,670	15,020	—	6,650

Dash indicates that the data was not estimated in the study.

8.4.1.3   Potential Management Strategies to Address Internal Loading in the Context of Future Environmental Change

Even though our estimates of P inputs to GLSM may differ from those of Filbrun et al. (2013), whose uncertainty of loading was not determined, the results of our study suggest similar management actions. We, and Filbrun et al. (2013), recommend reducing external P loadings from the watershed before further adding alum to the lake. That is, if external additions of P to GLSM remain uncontrolled, there can be little expectation that the addition of alum will improve water quality. Alum additions are a short-term solution targeting internal P loadings. Harmful algae blooms and water quality problems are expected to remain for several years especially in a context of future climate change, as cyanobacteria respond positively to warmer temperatures (Paerl and Otten 2013). Best management practices of nutrient runoff control need to be implemented in the watershed of GLSM to reduce the main contributor of P loadings (i.e., external sources) to the lake.

8.4.2   Effects of Alum Addition on GLSM

8.4.2.1   Effects of Alum on P Cycling and Algae Concentrations

Based on our previous study (Nogaro et al. 2013), we hypothesized that alum would affect chemical and biological parameters such as reducing P and chlorophyll *a* concentrations in the water column.

However, the results of this study suggest that alum addition did not affect chlorophyll a concentrations at most of the sampling sites; a decrease in chlorophyll a was statistically detected only at Site 5 after the alum addition in comparison with the pre-alum sampling date. The increase of phycocyanin fluorescence (i.e., cyanobacteria pigment activity) at Sites 1, 2, and 3 and the slight increase of TP concentration after the alum addition compared to pre-alum sampling suggest that the alum addition did not remove enough P from the water column and did not limit the growth of cyanobacteria during the spring/summer 2011. Although DO in surface water was not affected by the alum addition, increases of phycocyanin, DOC, and TP concentrations suggest that cyanobacteria increased their biomass throughout the month of June 2011, after the alum application. Our hypothesis that alum would decrease P and algae concentrations in the surface water was not supported by our results. However, and expectedly, total TSP in pore water decreased with alum addition, except at Site 4, suggesting that alum might be more effective at reducing excess P in sediment pore water than in surface water in GLSM (see Figure 8.5A).

The limited effect of the alum treatment on reducing P and chlorophyll a in surface water of GLSM is an unusual finding because other studies have found opposite results (Kennedy and Cooke 1982; Cooke et al. 2005). One explanation for this lack of improved water quality following alum application in GLSM may be that the area of the lake treated was too small; i.e., only 40% of the lake (central) was treated (see Figure 8.1). The final alum concentration in the central part of the lake was 21 mg Al L^{-1} (nominally, from both from alum and sodium aluminate [Tetra Tech 2012]), which was in the range of concentrations used in other eutrophic lakes (5–30 mg Al L^{-1}, Cooke et al. 2005). However, if this concentration is scaled to the entire lake area, it corresponds to the low end of the range typically used (i.e., 8.4 mg Al L^{-1}), which may not have been enough to effectively treat the lake. Also, TP concentrations remained high, and almost unchanged between pre- and post-treatment, probably due to the high external P loadings, which might explain the growth of cyanobacteria during the summer and associated high chlorophyll a concentrations. Moreover, the buffering of surface water was excessive during the alum addition, which may have reduced the efficacy of alum to inactivate excess P. At the pH of the alum treatments (most > 9), Al is estimated to be speciated mostly as $Al(OH)_4^-$, which will not complex PO_4^{2-}.

8.4.2.2   *Unintended Consequences of Alum*

The large increase in pH caused by the alum addition at the five sampling sites (i.e., pH : 9) (see Figure 8.3F) probably reduced the efficacy of aluminum to bond with P and induced unintended effects on metal concentrations in surface and pore water. The pH increase was associated with much greater concentrations of filtered Al in surface water after the alum addition (i.e., 614–1650 µg Al L^{-1}) compared to those before alum addition (i.e., 5.8–8.9 µg L^{-1}) (see Figure 8.4A). Particulate Al in surface and pore water was also affected by the alum addition; there were lower particulate Al concentrations in surface water and greater filtered concentrations in pore water (see Figures 8.4B and 8.5B, respectively), suggesting a substantial release of Al that was loosely bound to colloids and sediment particles. This is consistent with Al speciation, which is mainly controlled by pH, with soluble Al hydroxides $(Al(OH)_4^-)$ dominating at alkaline pH (> 9), insoluble Al hydroxide floc $(Al(OH)_{3(s)})$ at pH between 6 and 8, and free toxic Al^{3+} at low pH (Kennedy and Cooke 1982). Our results also agree with our previous study, which focused on a pilot study of alum addition in GLSM's near-shore bays; this work showed a large increase of pH and filtered Al in surface water and Al in pore water after alum addition in comparison to control bays that did not receive alum (Nogaro et al. 2013). Other studies have shown that when alum additions are associated with increased pH, filtered Al may increase, which may reduce the sorption capacity of alum for P and thereby its effectiveness for improving water quality (Berkowitz et al. 2005; Reitzel et al. 2013).

Our results suggest that addition of alum also affected metal cycling besides Al, such as Mn. For example, addition of alum was associated with an increase of particulate Mn in surface water (see Figure 8.4D) and a decrease of filtered Mn in pore water (see Figure 8.5C, except at Site 4). The redox state and speciation of Mn is controlled by Eh and pH (Jablońska-Czapla 2015), and the conditions of GLSM would favor the formation and precipitation of manganese oxide-hydroxide (MnO-OH; Jablońska-Czapla 2015). In our study, the increase of pH associated with the alum addition to GLSM may have induced precipitation of manganese in the form of low-solubility oxides and hydroxides.

Lastly, our hypothesis that alum addition would decrease invertebrate community abundance in the sediment was not supported by observations. Our invertebrate survey in 2011 immediately prior to (May 5) and after (July 14) addition of alum to GLSM showed that the abundance of the main taxa (i.e., chironomid larvae and oligochaete worms) in GLSM sediments increased during this time period (see Figure 8.6). This result may be due to a seasonal growth effect; that is, more invertebrates may be present in July due to population dynamics rather than any effect of the alum treatment. Thus, without a control lake, which was not available, we cannot conclude whether there was any potential effect of alum treatment on invertebrate abundance. Previous studies on benthic invertebrate responses to alum addition have shown short-term reduction of invertebrate abundance after alum addition followed by a recovery in the following years (Smeltzer et al. 1999; Steinman and Ogdahl 2008, 2012). The lack of alum effect on benthic invertebrates may be due to the low alum quantity and incorrect pH buffering in GLSM, which probably resulted in a thin and poorly distributed alum floc. The effect of such thinner floc may have been dampened since the main mechanism affecting benthic invertebrates is the smothering by the alum floc (Steinman and Ogdahl 2008, 2012). Several years of invertebrate survey may be needed to understand the ecological impacts of alum addition in eutrophic freshwater ecosystems.

8.4.2.3   Why Did Alum Treatment Not Improve Water Quality GLSM?

Our present and previous studies (Nogaro et al. 2013) have shown that the alum addition had very limited efficacy in either removing excess P from surface water or improving water quality in GLSM. These conclusions differ from the unrealized optimism reported by contractors of the state (i.e., Tetra Tech Inc., Seattle, WA). In spite of the clear increase of P and phycocyanin at all sites in the lake before and after the alum treatment in June 2011, the contractor claimed a 26% reduction in whole lake P and a 37% reduction of internal loading in 2011 due to alum, compared to 2010 (Tetra Tech 2012). The same contractors also stated that the limited effectiveness of the GLSM alum treatments in removing P from the water column in 2010 and 2011 was due to very high algal abundance at the start of the treatments in September 2010 and June 2011 (Tetra Tech 2012). However, several factors could explain the difference in the lake water P concentrations and internal loadings between 2010 and 2011 other than the alum treatment, including nutrient contents and ratios, temperature, residence time, light, turbulence, and vertical mixing (Paerl and Otten 2013). Environmental conditions were different between 2010 and 2011; 2010 was much drier and warmer than 2011, with 2011 having an estimated three-fold greater external P loading than in 2010 (Filbrun et al. 2013). Thus, the warmer temperatures and shorter residence time (180 versus 480 days in 2010 versus 2011, respectively) may have led to higher internal P loading during the summer in 2010 compared to 2011, which could explain the difference in P concentrations. As a consequence, we suggest that the alum treatment was not the best solution to restore water quality in hypereutrophic GLSM.

In April 2012, another partial lake alum treatment (with alum and sodium aluminate) was applied to GLSM in the same 40% central area as in June 2011 (23.6 mg Al L^{-1}, 49.6 g m^{-2}) (Tetra Tech 2013). Again, the limited effectiveness of the alum treatment for improving water quality in that area was attributed to interference effects from early high algal biomass due to: (1) an unusually mild winter with no ice cover; (2) spring warm weather; and (3) a lower dose of alum recommended to treat the lake (i.e., 2011 and 2012 doses together represented only 70% of the recommended treatment for the mid-lake area) (Tetra Tech 2013). The three alum additions (i.e., 2010 in three near shore bays and 2011 and 2012 in central in-lake) did not result in the predicted 50% reduction in TP concentrations in lake water. Given the limited success, the State of Ohio did not apply another alum treatment in following years.

Since the alum applications, GLSM has experienced harmful algae blooms every year during the summer, presumably due to high nutrient inputs from crop and livestock agriculture activities in its watershed. A recent study focused on the long-term effects of management priorities (winter manure application ban, filter strips, buffers, cover crops, manure storage structures, etc.) on water quality in the GLSM watershed (Jacquemin et al. 2018). Since 2011, significant reductions in total suspended solids, particulate P, SRP, nitrate N, and total Kjeldahl nitrogen concentrations from daily Chickasaw Creek have been measured at all flows (Jacquemin et al. 2018), suggesting that changes in land conservation and other best management practices can improve water quality in GLSM, and other similar lakes.

8.5   CONCLUSIONS

In 2011, GLSM had TP loadings of 92,200 kg, with 65% ± 5% attributed to watershed runoff (external loading), 30% ± 14% to benthic release (internal loading), 4% ± 2% to point-source discharges, and 1% ± 0.2% to direct atmospheric deposition. The P fluxes going out of the lake were estimated to be 62% ± 7% to river export and 38% ± 10% to sedimentation. Results from this steady-state P budget suggest that an alum addition was unlikely to prevent harmful algal blooms and restore water quality in GLSM because external P sources are significantly higher than internal loadings (i.e., 70% ± 7% versus 30% ± 14%, respectively in 2011).

Our results also showed that the addition of alum did not significantly reduce P concentration and cyanobacteria biomass in the lake in 2011, but did significantly increase pH, filtered Al, and particulate Mn in the surface water. In-lake additions of alum occurred in June 2011 and April 2012, but the lake remains hypereutrophic and regularly experiences blooms of toxic algae during the summer. We suggest that application of alum, at least as it was applied, is not an appropriate best management solution for P management in GLSM at present. Additional studies would be needed to determine if continuous, internal load control using small alum doses (see Chapter 5) may be an effective management strategy. However, given the characteristics of the watershed and GLSM, it may be that external P sources from the watershed must be better controlled, and become a smaller source than internal loading, before alum application makes sense.

8.6   ACKNOWLEDGMENTS

We thank Dr. Robert Hiskey, Dan Marsh, James Detraz, Matthew Konkler, Katlin L. Bowman, Sarah Harvey, and Deepthi Nalluri for help with either sampling or analysis. Kurt Thompson assisted with GIS data processing and mapping. Greg Koltun (USGS) provided information from *in situ* monitors, and Michelle Sharp (OEPA) provided point-source information. We also thank Dr. Alan Steinman for his valuable comments to the manuscript and his constructive suggestions. This work was supported by the Ohio Water Resources Research and Wright State University Research Initiation Programs as well as a student research grant from the Wright State University Graduate Student Assembly.

8.7 REFERENCES

[APHA] American Public Health Association, American Water Works Association (AWWA), Water Environment Federation (WEF). 1995. Standard methods for the examination of water and wastewater. 19th ed. Washington, D.C.

Auer, MT; Johnson, NA; Penn, MR; and Effler, SW. 1993. Measurement and verification of rates of sediment phosphorus release for a hypereutrophic urban lake. Hydrobiologia. 253(1–3):301–309.

Berkowitz, J; Anderson, MA; and Graham, RC. 2005. Laboratory investigation of aluminum solubility and solid-phase properties following alum treatment of lake waters. Water Res. 39:3918–3928.

Berman-Frank, I; Lundgren, P; and Falkowski, P. 2003. Nitrogen fixation and photosynthetic oxygen evolution in cyanobacteria. Res Microbiol. 154(3):157–164.

Boström, B; Jansson, M; and Forsberg, C. 1982. Phosphorus release from lake sediments. Arch Hydrobiol Beih Ergebn Limnol. 18:5–59.

Boström, B; Andersen, JM; Fleischer, S; and Jansson, M. 1988. Exchange of phosphorus across the sediment-water interface. Hydrobiologia. 170:229–244.

Chowdhury, M and Al Bakri, A. 2006. Diffusive nutrient flux at the sediment-water interface in Suma Park Reservoir, Australia. Hydrolog Sci J. 51(1):144–156.

Cooke, GD; McComas, MR; Waller, DW; and Kennedy, RH. 1977. The occurrence of internal phosphorus loading in two small eutrophic glacial lakes in northeastern Ohio. Hydrobiologia. 56(2):129–135.

Cooke, GD; Welch, EB; Martin, AB; Fulmer, DG; Hyde, JB; and Schrieve, GD. 1993. Effectiveness of Al, Ca, and Fe salts for control of internal phosphorus loading in shallow and deep lakes. Hydrobiologia. 253(1–3):323–335.

Cooke, GD; Welch, EB; Peterson, SA; and Nichols, SA. 2005. *Restoration and Management of Lakes and Reservoirs*. Boca Raton (FL): Taylor & Francis.

Farnsworth, RK; Thompson, ES; and Peck, EL. 1982. Evaporation atlas for the contiguous 48 United States. Washington DC: US Department of Commerce and National Oceanic and Atmospheric Administration Technical Report National Weather Service 33.

Filbrun, JE; Conroy, JD; and Culver, DA. 2013. Understanding seasonal phosphorus dynamics to guide effective management of shallow, hypereutrophic Grand Lake St. Marys, Ohio. Lake Reserv Manage. 29(3):165–178.

Geib, WJ. 1910. Soil survey of Auglaize County, Ohio. Washington DC: US Department of Agriculture.

[GLWWA] Grand Lake/Wabash Watershed Alliance. 2009. Grand Lake St. Marys/Wabash River Watershed Action Plan. Celina (OH).

Hammerschmidt, CR and Fitzgerald, WF. 2008. Sediment-water exchange of methylmercury determined from shipboard benthic flux chambers. Mar Chem. 109(1–2):86–97.

Hammerschmidt, CR; Bowman, KL; Tabatchnick, MD; and Lamborg, CH. 2011. Storage bottle material and cleaning for determination of total mercury in seawater. Limnol Oceanogr-Meth. 9:426–431.

Hoorman, J; Hone, T; Sudman Jr., T; Dirksen, T; Iles, J; and Islan, KR. 2008. Agricultural impacts on lake and stream water quality in Grand Lake St. Marys, western Ohio. Water Air Soil Poll. 193(1–4):309–322.

Jabłońska-Czapla, M. 2015. Manganese and its speciation in environmental samples using hyphenated techniques: A review. J Elem. 20(4):1061–1075.

Jacquemin, SJ; Johnson, LT; Dirksen, TA; and McGlinch, G. 2018. Changes in water quality of Grand Lake St. Marys Watershed following implementation of a distressed watershed rules package. J Environ Qual. 47:113–120.

Kana, TK; Darkangelo, C; Hunt, MD; Oldham, JB; Bennett, GE; and Cornwell, JC. 1994. Membrane inlet mass spectrometer for rapid high-precision determination of N_2, O_2, and Ar in environmental water samples. Anal Chem. 66:4166–4170.

Kennedy, RH and Cooke, DG. 1982. Control of lake phosphorus with aluminum sulfate: dose determination and application techniques. Water Resour Bull. 18:389–395.

Maher, W; Krikowa, F; Kirby, J; Townsend, AT; and Snitch, P. 2003. Measurement of trace elements in marine environmental samples using solution ICPMS: current and future applications. Aust J Chem. 56:103–116.

Naik, AP; and Hammerschmidt, CR. 2011. Mercury and trace metal partitioning and fluxes in suburban southwest Ohio watersheds. Water Res. 45(16):5151–5160.

Nogaro, G; Burgin, AJ; Schoepfer, VA; Konkler, MJ; Bowman, KL; and Hammerschmidt, CR. 2013. Aluminum sulfate (alum) application interactions with coupled metal and nutrient cycling in a hypereutrophic lake ecosystem. Environ Pollut. 176:267–274.

Nürnberg, GK. 1985. Availability of phosphorus upwelling from iron-rich anoxic hypolimnia. Arch Hydrobiol. 104(4):459–476.

Nürnberg, GK and LaZerte, BD. 2004. Modeling the effect of development on internal phosphorus load in nutrient-poor lakes. Water Resour Res 40:W01105, doi:10.1029/2003WR002410.

[OEPA] Ohio Environmental Protection Agency. 2007. Total maximum daily loads for the Beaver Creek and Grand Lake St. Marys watershed. Columbus (OH): Division of Surface Water.

Paerl, HW and Otten, TG. 2013. Harmful cyanobacterial blooms: causes, consequences, and controls. Microb Ecol. 65(4):995–1010.

Phillips, G; Jackson, R; Bennett, C; and Chilvers, A. 1994. The importance of sediment phosphorus release in the restoration of very shallow lakes (The Norfolk Broads, England) and implications for biomanipulation. Hydrobiologia. 275/276:445–456.

Priest, TC. 1979. Soil survey of Mercer County, Ohio. Washington DC: U.S. Department of Agriculture Soil Conservation Service.

R Development Core Team. 2016. R: a language and environment for statistical computing. R Foundation for Statistical Computing, Vienna, Austria. http://www.R-project.org.

Redfield, AC; Ketchum, BH; and Richards, FA. 1963. The influence of organisms on the composition of sea-water. In: Hill MN, editor. The Sea. Vol. 2. New York (NY): Interscience. pp. 26–77.

Reitzel, K; Jensen, HS; and Egemose, S. 2013. pH dependent dissolution of sediment aluminum in six Danish lakes treated with aluminum. Water Res. 47(3):1409–1420.

Rose, WJ. 1993. Water and phosphorus budgets and trophic state, Balsam Lake, northwestern Wisconsin, 1987–1989. Madison (WI): U.S. Geological Survey Water Resources Investigations Report 91–4125.

Sharp, JH; Benner, R; Bennett, L; Carlson, CA; Fitzwater, SE; Peltzer, ET; and Tupas, LM. 1995. Analyses of dissolved organic carbon in seawater: the JGOFS EqPac methods comparison. Mar Chem. 48:91–108.

Shaw, RD; Trimbee, AM; Minty, A; Fricker, H; and Prepas, EE. 1989. Atmospheric deposition of phosphorus and nitrogen in Central Alberta with emphasis on Narrow Lake. Water Air Soil Poll. 43(1–2):119–134.

Smeltzer, E; Kirn, RA; and Fiske, S. 1999. Long-term water quality and biological effects of alum treatment of Lake Morey, Vermont. Lake Reservoir Manage. 15(3):173–184.

Steinman, AD; Chu, X; and Ogdahl, M. 2009. Spatial and temporal variability of internal and external phosphorus loads in Mona Lake, Michigan. Aquat Ecol. 43:1–18.

Steinman, AD and Ogdahl, M. 2008. Ecological effects after an alum treatment in Spring Lake, Michigan. J. Environ Qual. 37:22–29.

Steinman, AD and Ogdahl, ME. 2012. Macroinvertebrate response and internal phosphorus loading in a Michigan lake after alum treatment. J Environ Qual. 41:1540–1548.

Steinman, A; Rediske, R; and Reddy, KR. 2004. The reduction of internal phosphorus loading using alum in Spring Lake, Michigan. J Environ Qual. 33(6):2040–2048.

Taylor, A. 2012. Phosphorus mass balance for hypereutrophic Grand Lake St Marys, Ohio. Master of Science [dissertation]. [Dayton (OH)]: Wright State University.

Tetra Tech. 2012. Final draft: Effectiveness of a 2011 alum treatment, future phosphorus load reductions and associated water quality monitoring for Grand Lake St. Marys. Columbus (OH): prepared for OEPA and USEPA.

Tetra Tech. 2013. Preliminary Assessment of Effectiveness of the 2012 Alum application Grand Lake St. Marys. Prepared for US EPA Region 5, Chicago IL.

[USEPA] United States Environmental Protection Agency. 1975. National eutrophication survey: report on Grand Lake St. Marys, Auglaize and Mercury counties, Ohio, EPA Region V. Working Paper No. 411. Corvalis (OR): Pacific Northwest Environmental Research Laboratory.

USEPA. 1994. Method 200.8, Revision 5.4, U.S. Environmental Protection Agency, Cincinnati, OH.

Wetzel, RG and Likens GE. 1991. Limnological Analyses, 2nd ed. Springer-Verlag, New York.

Wilson, CB and Walker Jr., WW. 1989. Development of lake assessment methods based upon the aquatic eco-region concept. Lake Reserv Manage. 5(2):11–22.

CHAPTER 9

INTERNAL POOLS AND FLUXES OF PHOSPHORUS IN DIMICTIC LAKE ARENDSEE, NORTHEASTERN GERMANY

Michael Hupfer[1], Andreas Kleeberg[2], and Jörg Lewandowski[3]

Abstract

The dimictic Lake Arendsee in northeastern Germany has suffered from severe eutrophication for many years. Low transparency, extensive development of cyanobacteria, and increasing oxygen deficiency in the hypolimnion characterize the water quality. This case study uses the results of long-term monitoring and detailed sediment investigations to quantify the phosphorus (P) pools and fluxes in the lake. Hypolimnetic P accumulation during summer stratification, with an average of 7.75 ± 1.52 mg P m^{-2} d^{-1} (2008–2017), is synchronized with epilimnion P loss due to sedimentation. The annual hypolimnetic P accumulation exceeds the mobile P pool in the sediment; thus, only the permanent P supply and rapid diagenesis can explain the observed hypolimnetic P accumulation. Of the total P released, 36.5% occurs during sedimentation in the water body. Despite the intense hypolimnetic P accumulation during summer stratification, the sediments cannot be considered as an *internal source*. Both the P mass balance including internal fluxes, and the repetitive measurement of newly deposited material above the artificial capping layer (time marker), have clearly shown that the sediment is a P sink. The present study demonstrates that high rates of hypolimnetic P accumulation are not always correlated with a large total or potentially mobile P pool in the sediment. In the case of Lake Arendsee, a long water residence time coupled with large amounts of P in the water body, which have accumulated over decades, indicate that short-term control without

continued

---

[1] Department of Chemical Analytics and Biogeochemistry, Leibniz Institute of Freshwater Ecology and Inland Fisheries; Berlin, D. E-mail for correspondence: hupfer@igb-berlin.de

[2] Department Geology, Soil, Waste, State Laboratory Berlin-Brandenburg; Berlin, D

[3] Department of Ecohydrology, Leibniz Institute of Freshwater Ecology and Inland Fisheries and Humboldt-University of Berlin, Geography Department; Berlin, D

internal load control is impossible. Therefore, a chemical P precipitation is expected to be effective for short-term improvement, particularly if external P load reduction is delayed or insufficient.

Key words: stratified lake, phosphorus mass balance, sediments, potentially mobile phosphorus, phosphorus retention, lake restoration

9.1   INTRODUCTION

For many years the selection of appropriate measures to reduce the severe eutrophication of Lake Arendsee (area 5.14 km², max. depth 49 m, mean depth 29 m) has been a significant challenge for water managers and scientists (Klapper 1976; Klapper 1992; Hupfer et al. 2016). The lake, located in northeastern Germany (52°53′21″ N, 11°28′27″ E), was formed by a collapsed salt doline, which resulted in an elliptical shape and steep slopes close to the shoreline. The littoral zone is very small (see Figure 9.1). The area below a depth of 30 m occupies approximately 60% of the total lake area. Primary limnological results were already published in 1896 (Halbfass 1896). Long-term biological, physical, and chemical parameter data have been available since 1976. In 2009, an automatic monitoring station was installed to continuously record weather and water data in high temporal resolution (Engelhardt and Kirillin 2014). The above-ground catchment area (29.5 km²) is dominated by agriculture (52.1%) and forestry (30.6%). The town of Arendsee is situated directly on the southwest shore of the lake (see Figure 9.1).

The theoretical water residence was calculated to be 56 years (Meinikmann et al. 2015). The hard water lake is mainly fed by groundwater (Meinikmann et al. 2013). Currently, four ditches, which drain adjacent agricultural areas, flow into the lake, and a small artificial drainage channel transports the water out of the lake. The sum of the different, separately determined external P sources was 1560 kg yr^{-1} (0.303 g m^{-2} yr^{-1}). More than 50% of this entire P load enters the lake via groundwater, which is enriched in P and passes below Arendsee (Meinikmann et al. 2015).

The lake has been strongly eutrophied since at least the middle of the last century (Scharf 1998). Water quality in summer is often affected by low transparency with Secchi depths of less than 1 m and extensive development of phytoplankton dominated by cyanobacteria such as *Planktothrix rubescens* (DC. ex GOMONT), diazotrophic *Anabaena flos-aquae* (BORY DE ST.-VICENT), and *Aphanizomenon flos-aquae* (L.). The assessment based on the phytoplankton community has revealed a *poor ecological status*; the requirements of the EU Water Framework Directive (WFD) for a *good ecological status* therefore are not being achieved at present. In addition, dissolved oxygen (O$_2$) in the hypolimnion at the end of summer stratification has continuously decreased over the last four decades. The volume-weighted O$_2$ concentration between 20 m and 48 m dropped from 4.45 ± 0.57 mg O$_2$ L^{-1} (1976 to 1985) to 2.42 ± 0.81 mg O$_2$ L^{-1} (2008 to 2017). Simultaneously, the upper boundary of the layer with a concentration of < 2 mg O$_2$ L^{-1} shifted upward from 42.4 ± 2.1 m to 33.9 ± 2.5 m, increasing the hypolimnetic volume from 7% to 27%. During summer stratification, a pronounced metalimnetic O$_2$ minimum forms every year with a thickness of a few meters (Kreling et al. 2017). Due to global warming, the temperature at the water surface has increased by an average of 0.52°C per decade since 1976, and the duration of the summer stratification has been prolonged by 8.2 days per decade.

Between 1975 and 2010, the concentration of total P (TP) steadily increased (see Figure 9.2). Since 2010, the TP concentration has stabilized at a high level and corresponds to a volume-weighted mean TP concentration of 186 ± 6 µg P L^{-1} (2008–2017). The construction of a domestic wastewater

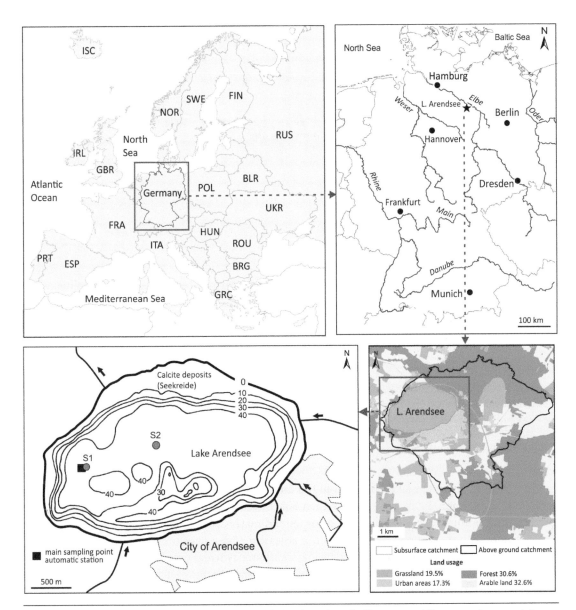

Figure 9.1 Upper pictures: regional setting of Lake Arendsee in Europe and Germany. Lower pictures: catchment area of Lake Arendsee and their usage structure. The bathymetric map of Lake Arendsee shows in the west of the lake, the main sampling point and location of the platform used for weather and water monitoring. S1 and S2 are the positions of the sediment sampling. The black arrows indicate the flow direction of the small water inlets and outlets.

treatment plant in the mid-1960s did not result in any obvious changes in water quality. Based on observed considerable hypolimnetic P accumulation during summer stratification, two internal restoration measures were implemented. Both were not successful. In the first attempt, a hypolimnetic water withdrawal was operated for 15 years (1976–1990) to increase the P output, where P-rich deep water was selectively discharged instead of P-poor surface water (Klapper 1976). However, due to the long water residence time of Lake Arendsee and the significant infiltration of lake water into the aquifer, the discharge volume is very low. Thus, the additional P output by hypolimnetic water withdrawal was too low to significantly reduce the TP content of the lake (see Figure 9.2).

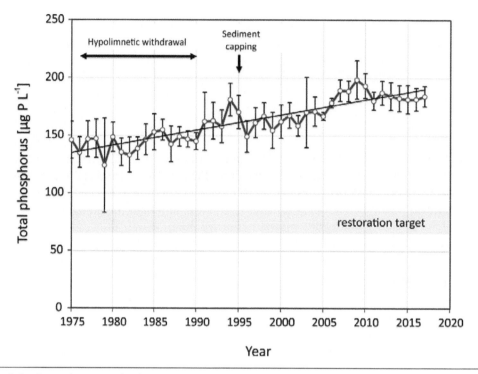

Figure 9.2 Annual means of volume-weighted total phosphorus (TP) concentrations (±SD) in Lake Arendsee over time since 1975. Neither the hypolimnetic withdrawal nor the sediment capping with calcareous mud from the littoral zone were able to stop the rising trend of TP. The current TP concentration is three times higher than the TP concentration required for achieving the lake's *good ecological status* (see Hupfer et al. 2016).

In the second attempt, a sediment capping to reduce the internal P load from the sediment was performed in Autumn 1995 (see Figure 9.2). A calcareous mud (Seekreide) from the littoral zone was extracted by suction dredgers, and the suspended material was distributed on the lake surface by floating pipelines. This technique led to the coverage of deep sediments with a 1–10 cm thick calcareous layer (Rönicke et al. 1998; Stüben et al. 1998). However, the insertion of the artificial barrier, intended to interrupt the internal P cycle, ultimately had no effect on the magnitude of P release from the sediment (Hupfer et al. 2000) or on the TP concentration in the lake (see Figure 9.2). To avoid future management failures, a detailed P balance was performed, and the internal P pools and vertical P fluxes were carefully analyzed. We focused on the following two questions: do the sediments act seasonally or annually as a P source? and (2) how long can the existing mobile P pool in sediment sustain a high P release from the sediment? In this chapter, we synthesize various studies on Lake Arendsee's P balance based on methods presented in Chapter 2, in the context of future management options utilizing both external and internal measures.

9.2 METHODS

9.2.1 Mass Balance Calculations

The lake internal P pool was calculated using TP concentrations in water samples taken at depths of 0, 5, 10, 15, 20, 25*, 30, 35*, 40, 45, and 48 m (*since 2015) at the deepest location of the lake (see Hupfer and Lewandowski 2005). For the 0–15 m and 15–48 m layers, the P pools were calculated from the TP concentration and the volume of the respective layer (see Chapter 2, Section 2.2.1). Data on the P mass in the hypolimnion refers to water layers below 15 m, as they are excluded from circulation during the entire summer stratification. The P release was determined based on the increase of the hypolimnetic P mass from the beginning of the stratification period (first routine sampling date in April) to the maximum hypolimnetic P mass at the end of summer. To calculate the sedimentation P supply toward the hypolimnion, a separate mass balance was calculated for the epilimnion. Due to the long water residence time, the epilimnetic P loss mainly results from sedimentation processes (Hupfer and Lewandowski 2005). Furthermore, while the mass flow by P sedimentation is often acknowledged, the P net input from epilimnetic sediments and from diffusive or convective P transport from the hypolimnion to the epilimnion can be neglected. At depths below 15 m (3.7 km^2), fluxes from the epilimnion (including the metalimnion) to the hypolimnion and increases of P masses in the hypolimnion were utilized to calculate release and sedimentation per time unit.

9.2.2. Gradients of Reactive Soluble P at the Sediment-Water Interface

For small-scale vertical determination of dissolved P at the sediment-water interface, dialysis samplers (peepers) were exposed monthly at a depth of 49 m between April and November 1996. The calculated P release rates (see Chapter 2, Section 2.2.2) were compared with the hypolimnetic P accumulation rates during the same period.

9.2.3. Potentially Mobile P Pool in the Sediment

The sediment mobile P pool was determined by a combination of a sequential P extraction method providing defined P fractions and the TP gradient method as described in Chapter 2, Section 2.3 (see also Hupfer et al. 2016). Two sediment cores were taken from the deepest point of the lake (S1) and 800 m east of this position, where the lake is approximately 40 m deep (S2) (see Figure 9.1). The uppermost 5 cm of the sediment was sliced into 0.5 cm and 1 cm layers. The single sediment layers were fractionated according to Psenner et al. (1984) (modified by Hupfer et al. 1995). The mobile P pool was calculated as the sum of P forms potentially contributing to P release, such as loosely adsorbed P (NH$_4$Cl-P), redox-sensitive P (BD-P), and organic-bound P (NaOH, nonreactive P: NRP) (Rydin 2000; Reitzel et al. 2005). The same cores and sediment layers were used to determine the mobile P based on the vertical TP gradient in the sediment. The pool was calculated for the upper layers as the difference between TP in each single layer and background TP in layers deeper than the endpoint (a vertical constant TP content) of early diagenesis. Both the differences in the TP content (gradient method) and the content of mobile P (fractionation method) were multiplied by the dry mass of the respective layer and summed to yield the mobile P mass per unit area.

9.2.4. Determination of P Retention Using Sediment Cores

The P retention rate was determined 15 times between April 1997 and September 2017, using sediment cores taken at four to seven randomly selected sites on each sampling date (see also Hupfer et al. 2016). All deposited material above the calcareous mud (sediment capping in 1995, mentioned previously) was separated. The dry mass and TP content of the material were then determined. The deposited P mass includes the mobile P pool. Theoretically, the increase of the P mass above the calcareous mud layer over time should follow a linear accretion, and the mobile P pool can be calculated as the y-axis intercept (Hupfer et al. 2016).

9.3   RESULTS

9.3.1. Hypolimnetic P Accumulation

The total P masses in the epilimnion and hypolimnion show an opposite annual pattern (see Figure 9.3). The P mass in the epilimnion decreases during summer stratification mainly due to sedimentation, whereas the P mass in the hypolimnion regularly increases after a short delay. The hypolimnetic P mass increased on average by 5.5 ± 0.8 t during summer stratification within a reference time of 179 ± 16 days (2008–2017). The hypolimnetic P increases are associated almost exclusively with soluble reactive P (SRP). At the end of the summer stratification, SRP accounts for $87.7 \pm 8.8\%$ of the total hypolimnetic P. In relation to the hypolimnetic sediment area, a P release rate of 7.75 ± 1.52 mg P m^{-2} d^{-1} (2008–2017) was calculated.

The epilimnetic P concentration decreased on average by 63% from 194 ± 9 µg P L^{-1} to an annual minimum of 72 ± 6 µg P L^{-1} from 2008 to 2017. The SRP concentration reached values below 10 µg P L^{-1} over longer periods because most of the dissolved P was taken up by phytoplankton and transformed into particulate P in biomass. To calculate the P supply toward the hypolimnion, a separate summer P mass balance of the epilimnion was calculated (see Table 9.1). This was done under the simplifying assumption that both the P import and the water exchange have no seasonal dynamic. Only the epilimnion receives external P input (Meinikmann et al. 2015). Thus, the P imported during stratification for the corresponding period is on average 0.77 t, until the epilimnion P minimum is reached (see Table 9.1), resulting in an annual P load of 1.56 t per year. Phosphorus is exported

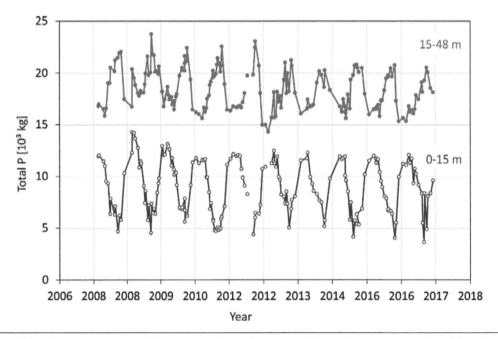

Figure 9.3   Seasonal course of P masses in the water layers 0–15 m (epilimnion) and 15–48 m (hypolimnion) in Lake Arendsee. The increase of total P in the hypolimnion and its decrease in the epilimnion were used to calculate the vertical P fluxes (compare with Table 9.1).

Table 9.1 Characteristics and results of the lake internal P balance for the epilimnion and the hypolimnion of Lake Arendsee during its summer stratification

	P in the lake (annual average)	P the epilimnion at the beginning of the summer stratification	Maximum decrease of P in the epilimnion during summer stratification	P loss by sedimentation out of the epilimnion*	P entering the hypolimnion from the overlying epilimnion	Hypolimnetic P accumulation**	Reference period for P_{hypo} (accu)	P release rate**
	P_{lake} x 10^3 kg	P_{epi} (start) x 10^3 kg	P_{epi} (loss) x 10^3 kg	P_{sed} (lake) x 10^3 kg	P_{sed} (hypo) x 10^3 kg	P_{hypo} (accu) x 10^3 kg	d	PRR mg m^{-2} d^{-1}
2008	27.7	11.5	6.8	7.4	5.3	5.6	190	7.97
2009	29.2	12.7	8.2	8.7	6.3	5.7	149	10.34
2010	28.4	12.6	8.0	8.5	6.1	5.0	186	7.27
2011	26.5	11.6	6.8	7.3	5.3	5.9	189	8.44
2012	27.5	12.1	7.7	8.3	6.0	6.3	185	9.20
2013	27.1	12.5	7.4	8.0	5.8	5.5	177	8.40
2014	26.8	11.8	7.2	7.7	5.5	3.7	189	5.29
2015	26.7	11.7	7.5	7.9	5.7	4.6	154	8.07
2016	26.7	11.3	7.6	8.2	6.0	4.3	196	5.93
2017	27.0	11.3	7.7	8.2	6.0	4.3	176	6.60
Average	**27.3**	**11.9**	**7.5**	**8.0**	**5.8**	**5.1**	**179**	**7.75**
SD	**0.9**	**0.5**	**0.5**	**0.5**	**0.3**	**0.8**	**16**	**1.52**

*Based on the P mass balance in the epilimnion (0–15 m) under consideration of the P input (total P import of 1.56 tons and proportionally related to the time until minimum cP$_{Epi}$ is reached) and P output (calculated by the mean cP$_{Epi}$ and the water discharge during the period until the minimum cP$_{Epi}$ is reached) related to the entire lake area.

**P released from both the lake sediment and from particulate matter in the water column imported from the epilimnion.

from the epilimnion as tributary outflow and surface water infiltration into the aquifer. The total water volume of the surface and subsurface output was multiplied by the mean P concentration in the epilimnion during stratification to calculate the P mass leaving the lake.

The results in Table 9.1 show that the hypolimnetic P accumulation and the P transport from the epilimnion into the hypolimnion are in general agreement with each other. On average, the vertical downward P mass transport is 0.7 t larger than the accumulation in the hypolimnion. This means that even during summer stratification, the sediment is not acting as an additional net internal P source. The slight fluctuations of the volume-weighted TP mass of the lake did not follow any seasonality and therefore cannot be correlated to the P release from the sediment.

The hypolimnetic P accumulation estimates include both the P release from the sediment—sediment release rate (SRR)—and P mobilized in the hypolimnion during the sedimentation of particles through the water column—water release rate (WRR) (see Chapter 2, Section 2.2). A model originally developed for oxygen by Livingstone and Imboden (1996) was adapted to distinguish between SRR and WRR. The approach assumes that the vertical exchange between the horizontally oriented water layers is negligible. In Lake Arendsee, there is a strong linearity between the P mobilization rate (per water volume) and the sediment surface-to-volume ratio for individual water layers between 15 m and 40 m (see Figure 9.4). This suggests that the sediment of the different hypolimnetic water layers is involved in hypolimnetic P accumulation with similar release rates. In the deepest layer between 40–48 m, the P release rate was lower than expected based on the sediment-to-volume ratio in almost all years considered between 2008 and 2017, resulting in a plateau of the curve (see Figure 9.4).

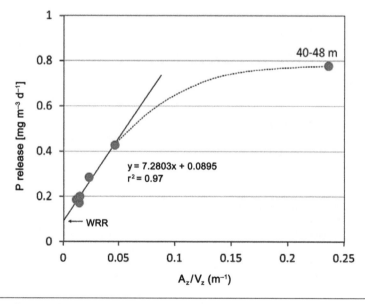

Figure 9.4 Phosphorus release rates during the summer stratification in different hypolimnetic water layers (15–20, 20–25, 25–30, 30–35, 35–40, 40–48 m) versus the corresponding area (A_z)-to-volume (V_z) ratio in Lake Arendsee in 2017. The intersection with the y-axis is the water release rate (WRR)—that is, the release in the water body occurring during sedimentation of particulate matter (see Chapter 2, Section 2.2). The dotted curve shows that the sediment in the deepest area of the lake (40–48 m) contributes at a smaller proportion to the P release. The distribution of the points implies that the P release rate of the deepest sediment (40–48 m) is lower compared to other areas.

The linear relationship in the sediment layers up to 40 m water depth was used to calculate the WRR because the impact of the sediment on P release is theoretically zero at the y intercept. The average WRR of Lake Arendsee is 0.121 ± 0.054 mg m^{-3} d^{-1} (2008–2017). The P release during sedimentation (= WRR) accounts for $36.5\% \pm 12.9\%$ of the hypolimnetic P accumulation during summer stagnation. The lower contribution of P release from the sediment below 40 m compared to the hypolimnetic sediment between 15 m and 40 m was at first surprising because the seasonal development of P vertical profiles at the main sampling point (z_{max}) and the observed high P concentrations close to the sediment during the summer stagnation suggest that the P release starts in deep, anaerobic water layers. This distinctly lower P release rate at deeper depths (below 40 m) compared to that at depths above 40 m could result from a mineralization rate that is lower under anoxic than oxic conditions. Another possible explanation is that mineralization already occurs in the water column above 40 m, so that little easily degradable organic P reaches depths of 40–48 m.

Long-term monitoring reveals that the TP concentrations in the hypolimnion and epilimnion show different slopes of increase with time (see Figure 9.5). In the upper water body (0–15 m), TP concentration has increased on average by 24% since 1978, whereas in deeper water (15–48 m), it increased by 45%. The different trends (see Figure 9.5) lead to a decrease of the stratification factor (ß) (see Chapter 2, Section 2.2.1), thus reducing the relative export of P from the system (relative to the total P mass in the lake). It is possible that an increased external P load will affect the hypolimnion more than the epilimnion because increased sedimentation can compensate for higher P loading in the latter. Therefore, an increased external P load and the corresponding downward transport processes of P might explain the changed P distribution with higher P found in the hypolimnion.

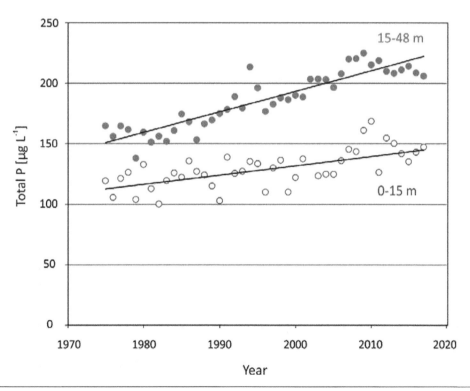

Figure 9.5 Long-term development of annual mean total P concentration in the upper (0–15 m) and the deeper (15–48 m) water body of Lake Arendsee (1975–2017).

However, climate change effects may also be important here. From 1975 to 2017, the stratification period was extended by about 35 days as a result of warming. Due to prolonged stratification, more P will be transported by sedimentation from the epilimnion to the hypolimnion.

In conclusion, the mass balance calculations show that P release is a fast process that occurs either during or immediately after sedimentation; i.e., P is already resupplied to a large extent in the pelagic zone. Thus, the P release is highly dependent on the current P supply by sedimention.

9.3.2. Quantification of P Release Rates by *In Situ* Measurements of Pore Water Profiles

The transition zone between sediment and water is characterized by steep gradients of dissolved substances and redox conditions (see Figure 9.6). In Lake Arendsee, the concentration of SRP in the sediment pore water often reaches values of more than 3 mg L^{-1}. The concentration of ammonium (NH_4^+) at the sediment-water interface shows a similar course, with maximum concentrations in the pore water around 15 mg L^{-1}. These high NH_4^+ concentrations are caused by an intensive decomposition process of organic material, such as ammonification. Usually, in June, nitrate (NO_3^-) is widely consumed near the bottom. The strong decrease of the sulphate (SO_4^{2-}) concentration at the immediate sediment-water interface indicates an intense dissimilatory sulphate reduction. Sulphate reduction is probably heavily involved in organic matter degradation. Due to the redox conditions in the sediment, reduced manganese (Mn^{2+}) accumulates in the sediment's pore water. The concentration of dissolved reduced iron (Fe^{2+}) is low in the uppermost sediment layers, and since ferrous iron is fixed in iron sulphides (FeS_x), it behaves differently compared to Mn^{2+}.

Based on the SRP gradients at the sediment-water interface, a mean P release of 5.21 ± 2.12 mg P $m^{-2} d^{-1}$ (n = 7) was calculated between May and October. During the same period, the P release rate as determined by hypolimnetic P accumulation was 7.97 mg $m^{-2} d^{-1}$. Relating this P release (WRR

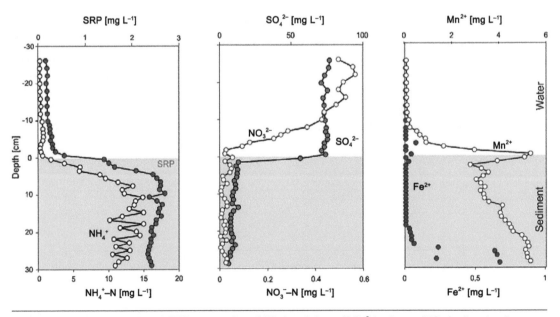

Figure 9.6   Vertical profile of SRP, ammonium (NH_4^+), sulphate (SO_4^{2-}), nitrate (NO_3^-), dissolved reduced iron (Fe^{2+}), and dissolved reduced manganese (Mn^{2+}) at the sediment-water interface of Lake Arendsee at 49 m water depth in June 1996.

and SRR) to the hypolimnetic volume results in 0.389 mg m^{-3} d^{-1}. Using the approach of Livingston and Imboden (1996) (see Figure 9.4), the rate of P already released in the water body (WRR) was found to be 0.098 mg m^{-3} d^{-1} in summer 1996. Consequently, 25.3% of the TP release is attributed to water release whereas the remaining 74.7% originates from the sediment. The latter value, as determined from the mass balance, was only 5.95 mg m^{-2} d^{-1}. The good agreement with the P release rates calculated via the SRP gradients (5.21 mg m^{-2} d^{-1}) is surprising, as such rates calculated from the SRP gradients often underestimate the actual P release (Urban et al. 1997). According to the approach of Livingston and Imboden (1996) (see Chapter 2, Figure 2.4), the P release from sediments below 40 m is lower than in other water depths. However, the inclusion of the highly-resolved SRP gradients at the sediment-water interfaces shows that the P release rate below 40 m is likely not below the hypolimnion average P release rate. Therefore, it is assumed that transport processes between the deepest layer and the layers above lead to a disproportional rate in depths of 40–48 m in relation to area to volume ratios (see Figure 9.4).

In conclusion, P release rates based on *in situ* determination of SRP gradients across the sediment-water interface can be meaningful and can supplement mass balance calculations, if gathered over longer periods.

9.3.3.   Quantification of Mobile P in Sediments

The TP in the sediment shows a strong decrease within the uppermost centimeters to a value between 1.0 and 1.1 mg P g^{-1} dw, which can be considered as the endpoint of early diagenesis. The mobile P pool, calculated as the difference between TP above this point and TP at each layer endpoint, summed up to 0.91 g m^{-2} at 40 m and 0.75 g m^{-2} at 49 m (see Figure 9.7). Additionally, the mobile P pool was calculated from the vertical distribution of P fractions down to 5 cm using the same cores as previously mentioned. The decrease in TP with increasing sediment age and depth, is mainly attributed to the decrease in reductive soluble P (BD-P) and organically bound P (NaOH-NRP). The other fractions change little during diagenesis.

This confirms that the mineralization of organic matter and P release from reduced iron compounds are the main P mobilization processes in operation. The top 0.5 cm is about six months old, and P diagenesis has already released 50% of the settled P (Hupfer and Lewandowski 2005). Diagenetic P release, therefore, follows an exponential curve. The mass of potentially mobile P based on fractionation method in the 0–5 cm layers was 3.48 g m^{-2} (48 m) and 3.23 g m^{-2} (40 m). This translates into P mass approximately three to four times higher than those resulting from the gradient method. It is assumed that the P fractionation method often overestimates the amount of P that is actually mobilized. It is well known that *mobile P* forms, determined by chemical extraction, are also found in deep(er) sediment layers, where P diagenesis is largely completed. Relatively strong extraction reagents used to simulate mobilization in a much shorter time frame than *in situ* contributes to this finding. Part of the organic P, determined as NaOH-NRP, is resistant to microbial degradation (Rydin 2000; Reitzel et al. 2007). Some studies have also shown that the redox sensitive P fraction (BD-P) is immobile, although strict anaerobic conditions prevail in the deeper sediment layers (Kleeberg and Grüneberg 2005). In Lake Arendsee, repeated investigations of the vertical TP distribution above the artificial calcareous (Seekreide) layer were carried out, which made it possible to observe the P mobilization in real-time (Hupfer et al. 2016). These investigations clearly indicate that the TP profile method provides more reliable values of P release for Lake Arendsee than the fractionation method. The values detected independent of these methods at depths of 48 m and 40 m are very similar; these values are reliable for at least approximately 60% of the hypolimnetic area. Additionally, seasonal variability of the mobile P pool in Lake Arendsee is

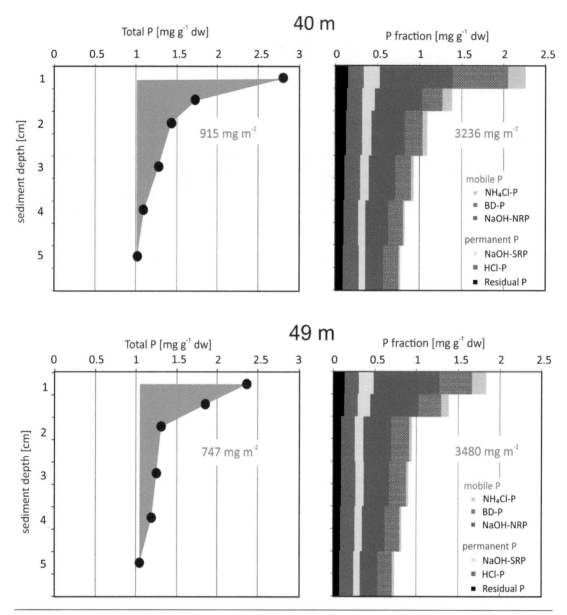

Figure 9.7   Mobile P pool in sediment cores taken in two different water depths of Lake Arendsee (40 m and 49 m) and determined by two different methods (TP profile and P fractionation).

relatively low (Hupfer et al. 2016). A comparison of the mobile P pool (calculated from TP profiles) and the mean P release per m² (see Table 9.1) shows that the mobile P pool would theoretically be exhausted within 106 days if sedimentation did not provide a supply of P. The time would actually be slightly longer because it is likely that only 64.5% of the hypolimnetic P accumulation is due to sediment P release. The calculation of the mobile P pool for the entire lake depends on the reference area, that is, the area for which the TP profiles at depths of 48 m and 40 m are considered representative. If they are considered representative for the entire hypolimnetic area, the extrapolated value for the mobile P pool is 3.06 t.

In conclusion, the L. Arendsee example shows that a rather simple TP profile method (integrated over a longer period of time) can more reliably provide the potential mobile P pool than the more complicated P extraction method.

9.3.4. Phosphorus Burial in the Sediment

To determine the permanent P burial in the sediment, we used the Seekreide layer, which originated from the failed sediment capping in 1995 and is a clearly visible time marker. Repeated measurements in the years following the in-lake measure show a linear increase in the dry mass of the newly deposited material (see Figure 9.8). On this basis, the annual sedimentation is 311 g dry mass m^{-2} yr^{-1}. Before the restoration attempt (between 1986 and 1995), a slightly higher sedimentation rate of 423 g dry mass m^{-2} yr^{-1} was determined at 48 m by means of a ^{137}Cs dated core, with the ^{137}Cs maximum caused by the reactor accident in Chernobyl (Hupfer and Lewandowski 2005). Nevertheless, the current rate since 1995 is very reliable since it is based on repeated measurements at several sampling sites. There are no systematic depth-dependent differences in the sedimentation rate in the areas with water depths between 40 m and 48 m.

In total, 8.18 g P m^{-2} were deposited between 1995 and 2017 (see Figure 9.8). Like the sediment dry mass, P mass deposition increased linearly over time. This P mass deposition contains not only the permanent P pool, but also the mobile P pool. The linear relationship between TP quantity and the elapsed time can be used to determine the mobile P pool: the y intercept, 0.79 g m^{-2}. This value tightly corresponds to the independently determined mobile P pool that is based on the TP profiles (0.82 g P m^{-2} d^{-1}) (see Figure 9.7).

According to the correlation observed in Figure 9.8, 0.34 g P m^{-2} yr^{-1} is retained in the sediment. This value agrees with the earlier determined burial rates (Hupfer and Lewandowski 2005) and seems to be a very constant rate over long periods of time. Depending on the sediment area to which this value is related, P retention between 0.7 t yr^{-1} (areas with a water depth below 40 m) or 1.24 t yr^{-1} (areas with a water depth below 15 m) is calculated for the entire lake. The role of shallow sediments (areas with a water depth shallower than 15 m) in contributing to P retention is unknown.

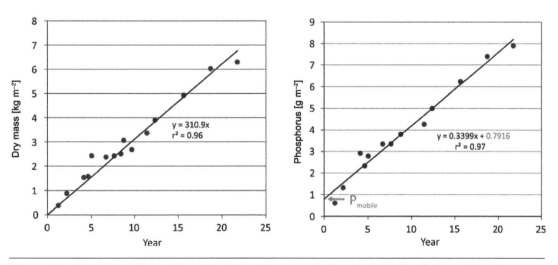

Figure 9.8   Long-term observation of sediment dry mass (left) and total P (right) above the calcareous mud (Seekreide) layer originating from a restoration attempt in 1995 (= year 0).

However, since such sediments are considerably sandy in most locations of the lake, their P retention capacity is likely quite small.

In conclusion, artificially introduced layers such as Seekreide in the present example can supplement other approaches for dating in following sediment accretion, which is the accumulation and retention of various sediment components over extended periods of time.

9.4   DISCUSSION

It is assumed that the phosphorus of Lake Arendsee is currently in a steady state, since there are only small inter-annual variations in P pools and P fluxes (see Figures 9.2, 9.3, 9.5, and 9.8). Figure 9.9 summarizes Lake Arendsee's P pools and P fluxes. The lake water contains 27.35 t P (see Table 9.1), which corresponds to a mean TP concentration of more than 180 µg P L^{-1} (see Figure 9.2). The temporarily bound P (mobile P) in the sediment is only 3.06 t.

The P concentration in the lake water shows a distinct seasonal dynamic that leads to strong losses of P from the epilimnion during stratification and synchronously to P accumulation in the hypolimnion (see Figure 9.3). Thus, there is an intensive internal P cycle leading to high P release rates in hypolimnetic water. The P enrichment in the hypolimnion during the summer stratification exceeds the amount of the mobile P pool in the sediment; thus, the P release can only be explained by the permanent supply of P from the epilimnion. Despite the high accumulation of P in hypolimnetic water during summer stratification, the sediment of Lake Arendsee cannot be considered an *internal source* because it is actually a net P sink (see Figures 9.7, 9.8, and 9.9). This insight is based on the P balance in the water body, which shows that during summer stratification the vertical transport by P sedimentation is higher than the P release from the sediment (see Table 9.1). The difference of an average of 0.7 t P agrees generally with the annual P retention of 1.24 t (> 15 m) as determined from sediment cores dated by the artificially introduced Seekreide layer.

Few reliable data are available for the internal P fluxes during the mixing period in winter. The P sedimentation measured by sediment traps at a depth of 40 m between November 1995 and April

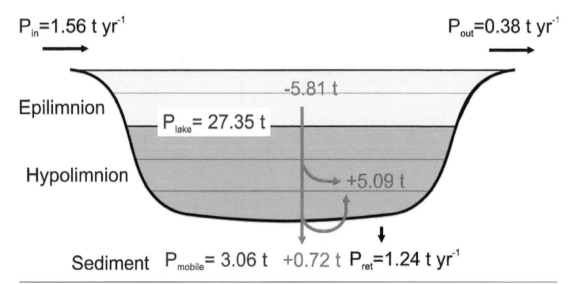

Figure 9.9   The P balance of Lake Arendsee with the P pools (P_{lake}, P_{mobil}), the annual fluxes (P_{in}, P_{out}, P_{ret}), and the summer fluxes (in red). P_{lake}: mass of P in the lake; P_{mobil}: mobile P pool in the sediment; P_{in}: P input; P_{out}: Poutput, P_{ret}: P retention.

1996 accounted for 36% of the annual total P sedimentation (Hupfer unpubl. data). Resuspension of deposited sediment probably plays a significant role during winter circulation. But this diagenetically altered material is likely to result in little or no P release during winter circulation. The balance data estimates that an annual average of 80–85% of the settled P is released during or after sedimentation. Long-term observations in the sediment have shown that P retention is constant (see Figure 9.8) and depends on the anaerobic P binding potential of the deeper sediment layers (see Figure 9.7). Phosphorus output must be considered as a further sink term in the P balance (see Figure 9.9). The P output results from the mean P concentration in the upper water body; the residence time of the water (independent of whether the water leaves the lake via groundwater or surface water) must be considered, as well. For Lake Arendsee, an annual P output of 0.38 t per year was calculated (2008–2017). Assuming current steady-state conditions prevail, the P input into Lake Arendsee must offset the sum of the losses both from P retention and P output. The sum of both loss terms is between 1.1 and 1.6 t per year, depending on the reference area for P retention in the sediment. This value tightly corresponds with the external load of 1.56 t per year as calculated by measuring and adding up the single P sources (Meinikmann et al. 2015). The present study shows that the external P load value can be reliably calculated if output and retention are known based on internal P balances and long-term trends of TP_{lake} (see Table 9.1).

These mass balance data can be used as input data for simple models to predict lake development or recovery when applying different management measures (Schauser et al. 2003; Hupfer et al. 2016). In the case of Lake Arendsee, a reduction of the external P load below the necessary critical loading estimates theoretically required to meet water quality targets, would lead only to a very slow recovery of the lake. Corresponding scenario analyses, using the one-box Model, have shown that an immediate trophic state improvement of lakes with such long water residence times is achievable only through in-lake P inactivation (Hupfer et al. 2016). In Lake Arendsee, such chemical inactivation would mainly aim at binding the P in the water body and depositing it permanently in the sediment. To calculate the dose, the mobile P pool in the sediment must be included and added to the water body P pool. A simple or repeated precipitation with an overdose can partly compensate a continual external load, providing time to implement the complex external measures required for a sustainable restoration. Internal measures are sustainable in the long-term only if the P input is reduced so that future input is below Lake Arendsee's critical threshold value. If no measures are carried out, oxygen conditions will continue to deteriorate, especially with current climate change; the *good ecological status*, according to the Water Framework Directive, will not be achieved due to the trophic situation and the phytoplankton composition.

9.5 CONCLUSIONS

The precise P balance analysis in stratified Lake Arendsee provides new scientific and practical insights that contradict traditional perspectives of sediment's role as an internal P source. A high P release rate observed in the hypolimnion can occur even if the mobile P pool in the sediment is small. The hypolimnetic P accumulation results mainly from the redistribution of P during stratification. This case study shows that P release can be a fast process and is mainly controlled by the mineralization of settling detritus; the temporary P pool in the sediment is thus small. In this case study, if the P supply from the epilimnion were to be cut off, the P pool would be depleted within less than four months. Longer stratification periods and prolonged duration of anoxic conditions due to climate effects intensify the internal redistribution of P. However, longer and more stable stratification reduces the availability of P in the euphotic zone, thereby prolonging conditions of P limitation. Thus, ongoing climate change could reduce P availability for phototrophic organisms while also potentially counteracting eutrophication. This study has shown that the internal P balance makes it possible to reliably determine the external P load regardless of the input sources (cf. Meinikmann et al. 2015).

The paradigm, *reduction of external P load sources is an obligatory prerequisite for in-lake measures*, cannot be maintained for lakes with long water residences times. Under certain circumstances, chemical P precipitation is also effective if P load reduction is delayed or insufficient. This allows the desired ecological status to be achieved much earlier, compared to external load reduction, alone. For Lake Arendsee, an internal measure such as chemical P precipitation is an appropriate option because P replenishment is very slow due to the long water residence time. Chemical P precipitation will allow for extra time to assess efficient external measures. However, external restoration measures are particularly difficult at Lake Arendsee because most of the P is discharged into the lake via groundwater.

Particularly for applied questions, the P pools and P fluxes of a lake must be considered in the proper context in order to avoid selecting inappropriate management measures. For example, it is not adequate to speak of an internal P load (resulting from the sediment) at Lake Arendsee because on an annual balance, the sediment is acting as a P sink. Due to the long water residence time in Lake Arendsee and the large amounts of P in the water body that have accumulated over decades, short-term responses in water quality following external control reductions, alone, are impossible. A gradual external P load reduction, even if the effect is delayed, will increase the necessity of in-lake P inactivation as an effective intervention.

9.6   ACKNOWLEDGMENTS

We are grateful to our colleagues Sylvia Jordan, Thomas Rossoll, Matthias Rothe, and Catherin Neumann for field work, and Christiane Herzog (IGB) for her conscientious laboratory work. Parts of this study were funded by the State Agency for Flood Protection and Water Management Saxony-Anhalt (LHW) and by the German Research Foundation (DFG, HU 740/5-1/2). We thank the Department of Lake Research of the Helmholtz Centre for Environmental Research (UFZ) and the LHW for providing monitoring data. The manuscript was influenced by many inspiring discussions with Werner Eckert (Kinneret Limnological Laboratory, Israel), Björn Grüneberg (BTU Cottbus, D), Kasper Reitzel, Henning Jensen (both: University of Southern Denmark, Odense, DK), Christof Engelhardt and Karin Meinikmann (both: IGB).

9.7   REFERENCES

Engelhardt, C and Kirillin, G. 2014. Criteria for the onset and breakup of summer lake stratification based on routine temperature measurements. Fundam Appl Limnol. 184(3):183–194.

Halbfass, W. 1896. Der Arendsee in der Altmark. Petermanns Mitteilungen aus J Perthes Geogr Anstalt. 42:173–187.

Hupfer, M; Gächter, R; and Giovanoli, R. 1995. Transformation of phosphorus species in settling seston and during early sediment diagenesis. Aquat Sci. 57(4):305–324.

Hupfer, M and Lewandowski, J. 2005. Retention and early diagenetic transformation of phosphorus in Lake Arendsee (Germany)—consequences for management strategies. Arch Hydrobiol. 164(2):143–167.

Hupfer, M; Pöthig, R; Brüggemann, R; and Geller, W. 2000. Mechanical resuspension of autochthonous calcite (Seekreide) failed to control internal phosphorus cycle in a eutrophic lake. Water Res. 34:859–867.

Hupfer, M; Reitzel, K; Kleeberg, A; and Lewandowski, J. 2016. Long-term efficiency of lake restoration by chemical phosphorus precipitation: Scenario analysis with a phosphorus balance model. Water Res. 97:153–161.

Klapper, H. 1976. Oligotrophierung eines tiefen, geschichteten Sees in einem Erholungsgebiet durch Ableitung des Tiefenwassers. Limnologica. 10:587–593.

Klapper, H. 1992. Calcite covering of sediment as a possible way of curbing blue-green algae. In: Sutcliffe, DW and Jones, JG, (Eds.). Eutrophication: research and application to water supply. Ambleside, UK, Freshwater Biological Association, pp. 107–111.

Kleeberg, A and Grüneberg, B. 2005. Phosphorus mobility in sediments of acid mining lakes, Lusatia, Germany. Ecol Engineering. 24(1–2):89–100.

Kreling, J; Bravidor, J; Engelhardt, C; Hupfer, M; Koschorreck, M; and Lorke, A. 2017. The importance of physical transport and oxygen consumption for the development of a metalimnetic oxygen minimum in a lake. Limnol Oceanogr. 62:348–363.

Livingstone, DM and Imboden, DM. 1996. The prediction of hypolimnetic profiles: a plea for a deductive approach. Can J Fish Aquat Sci. 53(4):924–932.

Meinikmann, K; Hupfer, M; and Lewandowski, J. 2015. Phosphorus in groundwater discharge—a potential source of lake eutrophication. J Hydrol. 524:214–226.

Meinikmann, K; Lewandowski, J; and Nützmann, G. 2013. Lacustrine groundwater discharge: combined determination of volumes and spatial pattern. J Hydrol. 502:202–211.

Psenner, R; Pucsko, R; and Sager, M. 1984. Die Fraktionierung organischer und anorganischer Phosphorverbindungen von Sedimenten—Versuch einer Definition ökologisch wichtiger Fraktionen. Arch Hydrobiol Suppl. 70:111–155.

Reitzel, K; Ahlgren, J; DeBrabandere, H; Waldeback, M; Gogoll, A; Tranvik, L; and Rydin, E. 2007. Degradation rates of organic phosphorus in lake sediment. Biogeochemistry 82(1):15–28.

Reitzel, K; Hansen, J; Andersen, FØ; Hansen, KS; and Jensen, HS. 2005. Lake restoration by dosing aluminum relative to mobile phosphorus in the sediment. Environ Sci Technol. 39(11):4134–4140.

Rönicke, H; Beyer, M; and Elsner, M. 1998. Seekreideaufspülung am Arendsee—ein neues Restaurierungsverfahren für überdüngte Hartwasserseen. GAIA. 7(2):117–126.

Rydin, E. 2000. Potentially mobile phosphorus in Lake Erken sediment. Wat. Res. 34(7):2037–2042.

Scharf, BW. 1998. Eutrophication history of Lake Arendsee (Germany). Palaeogeogr Palaeoclimatol Palaeoecol. 140:85–96.

Schauser, I; Lewandowski, J; and Hupfer, M. 2003. Decision support for the selection of an appropriate in-lake measure to influence the phosphorus retention in sediments. Water Res. 37(4): 801–812.

Stüben, D; Walpersdorf, E; Voss, K; Rönicke, H; Schimmele, M; Baborowski, M; Luther, G; and Elsner, W. 1998. Application of lake marl at Lake Arendsee, NE-Germany: first results of a geochemical monitoring during the restoration experiment. Sci Total Environ. 218:33–44.

Urban, NR; Dinkel, C; and Wehrli, B. 1997. Solute transfer across the sediment surface of a eutrophic lake: I. Porewater profiles from dialysis samplers. Aquat Sci. 59:1–2.

CHAPTER 10

STUDIES OF LEGACY INTERNAL PHOSPHORUS LOAD IN LAKE PEIPSI (ESTONIA/RUSSIA)

Olga Tammeorg[1,2], Jukka Horppila[1], Tõnu Möls[2], Marina Haldna[2], Reet Laugaste[2], and Juha Niemistö[1]

Abstract

Lake Peipsi is a polymictic eutrophic lake on the border between Estonia and Russia. Currently, the external nutrient loading is considerably lower than during the 1980s and 1990s. While lake water total nitrogen (TN) concentration has decreased, the total phosphorus (TP) concentration and the contribution of cyanobacteria to the phytoplankton has increased since the 1980s. Thus, it was hypothesized that internal phosphorus (P) loading was responsible for these latter changes. In the current study, we summarize the key findings of internal P loading studies in the lake. Specifically, these studies focused on assessing the magnitude of the internal P load, spatial and temporal variations in sediment resuspension and diffusive fluxes, and implications for lake water quality. In Lake Peipsi, the net estimates of the internal P load constituted only 2% of the gross sedimentation-based estimates that were driven by resuspension of TP. Sediment resuspension seems to affect water quality, both directly and via changes in diffusive fluxes. Such events are particularly important in August when there is a large pool of mobile P in sediments. Such combinations of P resuspension and diffusion amplify algal blooms from August through September, leading to increased amounts of fresh organic material available for resuspension in late September. The internal P load was similar to or in excess of the external P load, confirming that sediments are an important source of P to the lake. For the water quality in summer, internal P load was shown to have potentially higher importance than

continued

[1] Ecosystems and Environment Research Programme, University of Helsinki, P.O. Box 65, FI-00014, Helsinki, Finland. olga.tammeorg@helsinki.fi, horppila@helsinki.fi, juha.niemisto@helsinki.fi

[2] Centre for Limnology, Estonian University of Life Sciences, 61117 Rannu, Tartumaa, Estonia.

the external P load. Elevated net internal P loading from 1985 through 1989 and from 2001 through 2005 was due to elevated wave action (increasing sediment resuspension) and was the most likely reason for the observed deterioration of the water quality during these periods in Lake Peipsi.

Key Words: Lake Peipsi, gross sedimentation, resuspension of phosphorus, diffusive fluxes, redox potential

10.1 INTRODUCTION

10.1.1 Key Water Body Features

Lake Peipsi—in surface area (3,555 km²) the fourth largest lake and the largest transboundary lake in Europe—is located on the border between Estonia and Russia (see Figure 10.1). These countries share the major portion (Estonia 34% and Russia 58%) of the catchment area of the lake (47,800 km², including lake area). The lake has a mean depth of 7.1 m and a maximum depth of 15.3 m (located in the middle basin, Lake Lämmijärv). The southern basins of Lake Peipsi, Lake Lämmi- järv, and Lake Pihkva are generally shallower than the northernmost Lake Peipsi *sensu stricto* (*s.s.*). The total volume of water in Lake Peipsi is 25 km³ and the mean water residence time is about two years. The morphometric characteristics of the lake make it sensitive to natural water-level fluctua- tions. The average annual fluctuation amplitude of the water level is 1.15 m, while the maximum is 3.04 m (Jaani 2001). The latter entails a difference of 850 km² in surface area and 11.15 km³ in lake volume. Due to the large area and shallowness, thermal stratification in the lake is usually episodic and unstable. Although the three basins are usually oxygen-rich during the ice-free period, anoxic conditions may occur near the bottom layers on hot and calm summer days and during the ice-cover period. Ordinarily, Lake Peipsi is covered with ice from December to April.

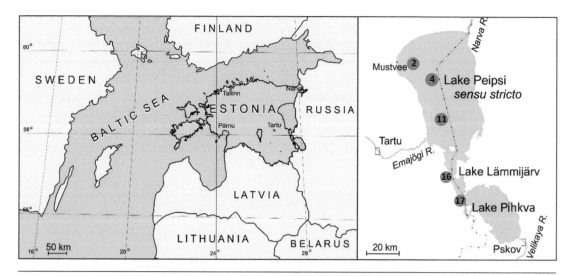

Figure 10.1 Location of the three basins of Lake Peipsi and sediment sampling sites.

Large amounts of fresh water are one of the most valuable characteristics of the lake. The shores of the lake have a high recreational value attracting many tourists from abroad (the number of tourists in Estonia only was ~300,000 according to 2016 census; Statistics Estonia 2017). Additionally, the lake is of great economic significance in terms of water supply for energy production, navigation, and fisheries. The Estonian waters of Lake Peipsi yield about 90% of the freshwater fish catch of the country (Statistics Estonia 2017). Providing a large variety of ecosystem services, Lake Peipsi and its surroundings have been a center of human activities for many years. The Rivers Velikaya (catchment area of 25,200 km^2) and Emajõgi (9,740 km^2) account for the bulk of the external nutrient loading (about 74% of TP and TN loading) into the lake (Loigu et al. 2008). Arable land constitutes about 50% of the Velikaya River catchment and about 35% of the Emajõgi River catchment (Loigu et al. 2008). In total, agricultural areas cover around 40% of the lake catchment. Forests and semi-natural areas are concentrated mainly in the northern and eastern parts of the lake's catchment. Southern basins of Lake Peipsi, receiving wastewaters from the largest city of the catchment Pskov and discharge from the Velikaya River, act as settling ponds for the northernmost Lake Peipsi *s.s.* (Nõges et al. 2003; Rumyantsev et al. 2006). The lake formed about 13,000 radiocarbon years ago as a large ice-dammed water basin (Raukas 2008) is one of the most productive lakes of the Baltic regions (Rumyantsev et al. 2006). Moreover, a spatial gradient occurs in the trophic state across the cascade of the lake basins (see Table 10.1), which also have differing morphology, hydrology, and biotic compositions.

Most of the lake water has been classified as mesotrophic in recent years (Starast et al. 2001). However, both long-term observations (expansion of reeds, massive blooms of cyanobacteria, shifts in species composition and dynamics of phytoplankton, drastic decrease of zooplankton abundance, fish kills; Kangur et al. 2012) and paleolimnological studies (Leeben et al. 2008, 2013; Kisand et al. 2017) revealed progressive eutrophication since the 1960s. The use of inorganic fertilizers peaked

Table 10.1 Main morphometric and water-quality characteristics of the three areas of Lake Peipsi (Lake Peipsi *sensu stricto*, Lake Lämmijärv, and Lake Pihkva). Water-quality variables are given in geometric means and 90% tolerance limits are given in brackets. These estimates correspond to the open water periods (Julian days 100–310 within each year) between 2006 and 2010 (Tammeorg et al. 2013)

Characteristic	Lake Peipsi *s.s.*	Lake Lämmijärv	Lake Pihkva
Surface area, km^2	2611	236	708
Surface area in Estonia/Russia, %	55/45	50/50	1/99
Maximum depth, m	12.9	15.3	5.3
Mean depth, m	8.3	2.5	3.8
Water volume, km^3	21.79	0.6	2.68
TP, µg P l^{-1}	38 (17–82)	67 (32–140)	116 (53–251)
TN, µg N l^{-1}	703 (417–1188)	896 (573–1401)	1143 (829–1577)
Chlorophyll *a* concentration, µg l^{-1}	18 (6–54)	33 (14–81)	63 (26–150)
Secchi depth, m	1.8 (1.0–3.2)	0.95 (0.6–1.5)	0.7 (0.4–1.0)
OECD (1982) classification	eutrophic	eutrophic/hypertrophic	hypertrophic

(300 kg ha^{-1}) in the late 1980s (Iital et al. 2005). However, drastic changes associated with the collapse of Soviet-type agriculture occurred in the catchment of Lake Peipsi in the 1990s. These changes included, primarily, considerable reductions in the use of fertilizers (2001 levels were only 11% of the late 1980s levels) and livestock production (about a two-fold decrease), as well as lower water consumption by people and industries, and improved wastewater treatment (Iital et al. 2005). The high external loading of nutrients persisted in the 1990s (Nõges et al. 2003; Nõges et al. 2007; Loigu et al. 2008). Nowadays, nutrient loading is considerably lower (by about 23% for TN and 34% for TP) than in the 1980s and 1990s. While lake water TN concentration decreased, the TP concentration has increased about 1.6-fold in all basins of Lake Peipsi in comparison with the 1980s (Tammeorg et al. 2016). Moreover, the share of cyanobacteria increased from 20% to 60% in Lake Peipsi *s.s.*, and from 30% to 90% in the southern parts of the lake in summer months (Haberman et al. 2010). Assuming that current P loading is close to the critical level as defined by Vollenweider (1975), which is 0.213 g m^{-2} yr^{-1} as calculated for Lake Peipsi by Nõges et al. (2003), the role of internal legacy P load in water-quality regulation warranted investigation.

In the present study, we summarize the results of the studies on recycling of P in Lake Peipsi that were pioneered in 2011 and presented in Tammeorg et al. (2013, 2015, 2016). Our specific objectives were:

1. To quantify internal P loads using different methods (mass balance, dated sediment cores);
2. To ascertain temporal and spatial variations in internal P loading and key factors behind these variations (e.g., redox potential measurements, diffusion, and sediment resuspension);
3. To analyze the implications of internal P loading for lake water quality.

10.2   METHODS

10.2.1   Gross Sedimentation, Resuspension Rates of Total and Potentially Bioavailable P

To measure gross sedimentation, sediment traps were placed 2 m above the bottom at sampling Stations 2, 4, and 11 (at depths of 8, 10, and 10 m, respectively) in Lake Peipsi *s.s.* and at Stations 16 and 17 (at depths of 14 and 7 m, respectively) in Lake Lämmijärv (see Figure 10.1) from May 24 to October 10 in 2011. The sediment traps of 4.4 cm in diameter and 44 cm in height were designed to be suitable for lake conditions (Bloesch and Burns 1980). Four replicate cylindrical sediment traps were emptied at 14-day intervals (Bloesch and Burns 1980). The material entrapped was dried at 60°C for approximately three days in the laboratory to obtain the dry weight.

Concurrently, depth-integrated water samples were collected using a 2 L Van Dorn sampler. The suspended solids (SS) samples were filtered through a Whatman GF/C filter (pore size 1.2 μm) and analyzed for dry weight and loss-on-ignition. An ammonium molybdate spectrophotometric method (EVS-ES 1189) was used to determine the concentrations of TP and soluble reactive P (SRP). The SRP samples were initially filtered through a Whatman mixed cellulose ester filter (pore size 0.45 μm). The samples for chlorophyll *a* (Chl *a*) were filtered through a Whatman GF/C filter (pore size 1.2 μm) and analyzed spectrophotometrically after extraction with ethanol.

The gross sedimentation of P for each site was calculated as the product of the mean gross sedimentation rate (as dry matter) with the mean TP concentration of the entrapped material, and the resuspension of P was calculated as the product of the mean TP concentration of the surface sediments and the resuspension rate (as dry matter). The TP concentrations of the entrapped material and surface sediments (the topmost 0–1 cm layer) were determined by an inductively coupled plasma-mass spectrometer after wet combustion with nitric acid and hydrogen peroxide. The re-suspension rate (as dry matter) was calculated according to the *label approach* of Gasith (1975). This approach is based on the organic fraction of the entrapped material, surface sediment, and suspended solids.

To estimate the contribution of resuspension to P release, the percentage of potentially bioavailable P in the surface sediments was multiplied with the rates of P resuspension. The value for potentially bioavailable P in the surface sediments (as a sum of the NH_4Cl-P and NaOH-P fractions) of Lake Peipsi *s.s.* (10%) was obtained from Punning and Kapanen (2009). However, the value was not available for Lake Lämmijärv; thus, the contribution of resuspension on P release was estimated only for Peipsi *s.s.*

10.2.2 Redox Potential of the Surface Sediments and Diffusive Fluxes of Phosphorus

To follow the mobilization of P within the surface sediments, sediment cores were collected using an HTH gravity corer (Renberg and Hansson 2008) for redox potential measurements and SRP determination in the sediment pore water.

For redox measurements, two to three sediment cores were collected at Station 4 in Lake Peipsi *s.s.* in: August 2013; March, August, and October 2014; and August and October 2015. At the same time (except 2013), the sediments were sampled in Lake Lämmijärv at Station 17. For Lake Pihkva Station 52, the sediment samples were obtained only in March and August 2014 due to changes in our access to Russian territorial waters. These stations were considered representative of a large part of the particular basin based on water quality characteristics (according to the long-term data series archived by the Center for Limnology, Estonian University of Life Sciences). The sediments were subsampled into plastic tubes (inner diameter = 3.5 cm, height = 14 cm); ⅔ of the tube volume was filled with sediments and the remaining ⅓ of tube volume filled with lake water. The tubes were sealed with caps immediately after the samples were taken and transferred to the lab in the thermo-isolated box, where they were kept at 4°C. The redox potential of the sediments was measured directly in the tubes with the redox sensor (Unisense RD100 microsensor, reference electrode Ag/AgCl) down to 3 cm below the sediment surface in 1 mm steps within 24 h after sampling. The measurements were repeated three times for each tube.

Sediment cores were concurrently sampled (with the exception of Lake Pihkva, where extreme weather conditions prevented the pore-water extraction in August) in 2014 and 2015 to measure the SRP concentrations in the pore water. Rhizon Soil Moisture Samplers (Rhizosphere Research Products, Wageningen, The Netherlands) were used to remove the pore water from the sediment at depths of 1, 2, and 3 cm. For the same purposes, the topmost 0–1 cm of the sediments was collected during the period of May 24 through October 10, 2011 at all sediment trap locations. The pore water from those sediments was separated by centrifugation and filtered through a Whatman mixed cellulose ester filter (pore size 0.45 μm). SRP concentrations were determined using an ammonium molybdate spectrophotometric method (EVS-ES 1189).

The diffusive SRP flux (J, mg P m^{-2} day^{-1}) was calculated according to Fick's first law of diffusion (Berner 1980):

$$J = \phi \times D_s \times dc/dz \qquad \text{[Eq. 10.1]}$$

where ϕ is the sediment porosity, D_s (cm^2 s^{-1}) the diffusion coefficient of phosphate, and *dc/dz* is the concentration gradient between the sediment pore water and the overlying water column. In calculations of the concentration gradient, we used the SRP concentrations of the pore water in the 1 cm surface sediments ($dz = 0.5$ cm). Data on the SRP concentrations in the lake water (sampled 0.5 m above the lake bottom) and instantaneous values of the other environmental variables potentially affecting P recycling at the sediment-water interface (i.e., water temperature, pH, dissolved oxygen) on the sampling dates were provided by the Estonian Environment Agency. A porosity value obtained from the sediment cores (uppermost 3 cm) of all three basins of Lake Peipsi was 95%. The diffusion coefficient for SRP at 25°C ($D_{25°C}$) in sediment-water systems is 6.12×10^{-6} cm^2 s^{-1} (Li and Gregory 1974). The temperature dependence of the $D_{25°C}$ was taken into account according to the Stokes-Einstein relation (Lewandowski and Hupfer 2005):

$$D_s = D_{25°C} \times v_{25°C} \times T / v_T \times T_{25°C} \qquad \text{[Eq. 10.2]}$$

where T = the temperature during sampling in Kelvins, $T_{25°C}$ = the temperature at 25°C in Kelvins (298.15 K), $v_{25°C}$ = the dynamic viscosity of water at 25°C (0.8903 g m^{-1} s^{-1}), and v_T = the dynamic viscosity of water at temperature T (g m^{-1} s^{-1}).

10.2.3   Internal Legacy P Loading Estimates Using Two Different Approaches

The internal load (IL) of TP in two basins of Lake Peipsi was calculated using the following mass balance model (Lappalainen and Matinvesi 1990):

$$IL = TP_{out} + GS + dm/dt - TP_{in} \text{ (mg P m}^{-2} \text{ day}^{-1}) \qquad \text{[Eq. 10.3]}$$

where IL = the internal load, TP_{in} = the external load, TP_{out} = the outflow, GS = the gross sedimentation, and dm/dt = the storage change of TP in the water column.

The values for TP_{in} and TP_{out} (presented as a mean for 2001–2005) for Lake Peipsi were calculated using daily water discharge data and monthly measured concentrations (Loigu et al. 2008). The monitored area covered 91% of the whole lake catchment area. Whole basin TP_{in} and TP_{out} were extrapolated for a specific basin in proportion to the percentage of its catchment area: thus, the fluxes of Peipsi *s.s.*, Lämmijärv, and Pihkva accounted for 28, 7, and 65%, respectively, of the total Lake Peipsi fluxes. To calculate the IL in Lake Peipsi *s.s.* and Lake Lämmijärv, we used the means of GS measured in 2011 and water-column TP concentrations at the beginning and end of the trap exposure period of the same year.

Another approach that enabled us to quantify the net internal P loading rate, IP_{tot}, in all three basins of Lake Peipsi was based on combining mass balance calculations and radiometrically dated sediment cores, and on the assumption that the difference of the retentions can be used to estimate the magnitude of internal loading (modified from Nürnberg 1984):

$$IP_{tot} = TP_{in} \times (R_{sed} - R_{mb} / R_{sed}), \qquad \text{[Eq. 10.4]}$$

where R_{mb} = the retention calculated using mass the balance approach and R_{sed} = the retention of P in the sediment that was estimated using sediment cores. These calculations covered two periods of steady state conditions (suggested by relatively low variation of the water-column TP concentrations within these periods): 1985–1989 and 2001–2005, which are referred to hereafter as Periods I and II, respectively. Thus, R_{mb}, (in mg m^{-2} yr^{-1}) in a lake was calculated according to Hupfer and Lewandowski (2008):

$$R_{mb} = TP_{in} - TP_{out} \qquad \text{[Eq. 10.5]}$$

R_{sed} (mg m^{-2} yr^{-1}) was calculated as the product of the concentration of TP in the sediment layer with the sedimentation rate. The sediment cores were collected using an HTH gravity corer (Renberg and Hansson 2008) in August 2014 from Station 4 of Lake Peipsi *s.s.*, Station 17 of Lake Lämmijärv, and Station 52 of Lake Pihkva located in the accumulation areas of the basins (see Figure 10.1). Each of the cores was sectioned into 0.5 cm slices up to a depth of 20 cm to cover the years for which data on external loads were available. All sediment samples (40 samples per lake) were freeze-dried and ground. The TP concentrations from those subsamples were further measured using inductively coupled plasma optical emission spectrometry (ICP-OES, Thermo Scientific iCAP 6000) after wet digestion with sulphuric acid and hydrogen peroxide. Moreover, subsamples from each core were analyzed for ^{210}Pb, ^{226}Ra, and ^{137}Cs by direct gamma (γ) assay using Ortec HPGe GWL series well-type coaxial low-background intrinsic germanium detectors (Appleby et al. 1986). ^{210}Pb was determined via its γ emissions at 46.5 keV, and ^{226}Ra by the 295 keV, and 352 keV γ-rays emitted by its daughter radionuclide ^{214}Pb following three-week storage in sealed containers to allow for radioactive equilibration. ^{137}Cs was measured by its emissions at 662 keV. The absolute efficiencies of the detectors were determined using calibrated sources and sediment samples of known activity. Corrections were made for the self-absorption effect of low-energy γ-rays within the sample (Appleby et al. 1992). As a result, sedimentation rates were modeled by dating sediment cores of Lake Lämmijärv and Lake Pihkva with ^{210}Pb and ^{137}Cs. Dating results for Lake Peipsi *s.s.* (published by Kapanen (2012)) were based on the same method and thus were used for the extrapolation of the sedimentation rates for the sediment core in the present study.

Defining R_{mb} through the difference between inflowing and outflowing TP and defining R_{sed} through the product of the sedimentation rate with the TP content of the sediments (Dillon and Evans 1993) enables the quantification of net internal P release in both stratifying lakes, for which such a method was originally introduced, and in non-stratifying lakes. This is because the method with such terms does not differentiate the reasons for the release of P. The unit of the net annual upward rate of TP (mg m^{-2} yr^{-1}) was converted to the flux value per day, assuming that about 80% of the annual load occurs in summer in north temperate lakes (Steinman et al. 2009).

10.2.4 Drivers behind the Benthic P Fluxes

Data on daily values for the environmental variables: water level (L); wind speed (mean and maximum, V and Vmax); and water temperature (T) were measured at the station located on the shore of Lake Peipsi *s.s.* (obtained from Estonian Environment Agency). These data were used in the analysis of the temporal variations in rates of sediment resuspension in 2011. Moreover, these environmental variables were compared between Period I and Period II, as potential factors influencing the variation in IP_{tot}. Assuming that similar changes between those two study periods in wind speed and water level (affecting lake depth) occurred in all basins, wave action was quantified as follows (Hamilton and Mitchell 1988):

$$\text{Wave action} = H^2/Z, \qquad \text{[Eq. 10.6]}$$

where H is the wave height and Z is the water depth.

The wave height was calculated as one-half the wavelength. The theoretical wavelengths were calculated with equations presented by Carper and Bachmann (1984) that consider the effective fetch

(Beach Erosion Board 1972) and wind velocity. The procedure was described in detail by Horppila et al. (2017).

10.2.5 Statistical Analyses

The effect of different weather variables (L, T, V, and Vmax) on sediment resuspension, and water quality variables (i.e., TP, SRP, SS, Chl *a*) was assessed using data from 2011. Initially, we eliminated the influence of sampling site and time on the potentially dependent variables by replacing the values of these variables with residuals (defined here as the difference between observed values and those predicted from the sampling site and time). This can result also in the partial elimination of the influence of weather factors due to their potential correlation with the sampling site and time. The analysis of the residuals has less uncertainty than the analysis of the raw data variables because it separates the effects of factors that are possibly associated with, but not necessarily depending causally on the weather, from the overall correlation between weather and water variables.

Given that the weather factors display great temporal variability but the response is likely to depend on the average weather conditions over some time, we averaged the weather variables over the time intervals from 1 to 8 days, considering the time intervals measured 0 to 10 days earlier than the sampling. As a result, (4 weather factors) × (8 averaging intervals) × (11 backward time shifts) = 352 means of weather factors were associated with each sampling time. The influence of a weather factor was identified by two indices: the number of days for which the parameter was averaged; and the delay of the effect of the weather factor. For example, V_{7_2} denotes the mean effect of the wind speed measured from two to eight days before sampling. Using predictors constructed from the weather factors, the residuals of each water quality variable were separately subjected to stepwise regression analysis. In the stepwise selection procedure, we used the critical significance level of $p = 0.001$ for either including a weather factor term or excluding it from the model after the inclusion of the next significant term.

In the current study, the relationships between different variables (e.g., diffusive fluxes and temperature in 2011, 2014, 2015; concentrations of TP and SS) were examined with the Pearson correlation analysis, whereby the $\log_2$ values were used to make data distribution close to normal. Between-basin and between-period differences, whenever referred to in the text, were tested with an analysis of variance. Means were compared with a post-hoc Tukey test.

In regressions of the TP and SRP concentrations, biomass of cyanobacteria with IP_{tot} and TP_{in}, data for Periods I and II were used (archives of the Center for Limnology of the Estonian University of Life Sciences). For Lake Pihkva, the only sources of the water quality data since the 2000s were Estonian-Russian joint expeditions in August. Thus, comparable data for all three basins of Lake Peipsi were available only for August.

10.3 RESULTS

10.3.1 Estimates of Total Sedimentary Fluxes of P

The mean gross P sedimentation rates during May 24 to October 11, 2011 were 34.3 mg P m^{-2} day^{-1} in Lake Peipsi *s.s.* and 105.5 mg P m^{-2} day^{-1} in Lake Lämmijärv. For the same period, the storage change of P (dm dt^{-1}) in the water column was 1.09 mg P m^{-2} day^{-1} in Lake Peipsi *s.s.* and 0.43 mg P m^{-2} day^{-1} in Lake Lämmijärv. These growing season measurements resulted in very high internal loading values of TP of 35.3 mg P m^{-2} day^{-1} in Lake Peipsi *s.s.* and 105.6 mg P m^{-2} day^{-1} in Lake Lämmijärv (see Table 10.2) at the present levels of external TP loading (0.22 and 0.64 mg P m^{-2} day^{-1} for Lake Peipsi s.s. and Lake Lämmijärv, respectively).

Calculated as the difference between the P input and the P output of the lake, the annual retention (R_{mb}) in Period II was 33 mg P m^{-2} yr^{-1}, 102 mg P m^{-2} yr^{-1}, and 307 mg P m^{-2} yr^{-1} in Lake Peipsi *s.s.*, Lake Lämmijärv, and Lake Pihkva, respectively. Similarly, the sediment core approach resulted in considerably higher P retention (R_{sed}) in the southern basins (630 mg P m^{-2} yr^{-1} and 1060 mg P m^{-2} yr^{-1}) than in Lake Peipsi *s.s.* (450 mg P m^{-2} yr^{-1}). Moreover, R_{sed} was on average five times higher than R_{mb}. Based on the difference in retention, net internal P loading estimates for the contemporary Period II were 74 mg P m^{-2} yr^{-1}, 207 mg P m^{-2} yr^{-1}, and 506 mg m^{-2} yr^{-1} in Lake Peipsi *s.s*, Lämmijärv, and Lake Pihkva, respectively. When converted to the units of mg m^{-2} day^{-1} the respective P fluxes were 0.66 mg m^{-2} day^{-1}, 1.84 mg m^{-2} day^{-1}, and 4.50 mg m^{-2} day^{-1}, and constituted 1%–2% of the GS-based internal TP loads. These internal P loads were higher than in Period I. The difference between Periods I and II was more pronounced in the southern basins (see Table 10.2).

Table 10.2 Estimates for the internal load of P given by two different methods (M1: based on gross sedimentation measurements; M2: based on difference in retention), resuspension of TP and bioavailable P, and diffusion of P. The values for the external load and outflow of TP used in the mass balance calculations are also indicated. Method 2 yields net internal load of P: IP_{tot} in mg P m^{-2} yr^{-1}. Values were converted to the unit of mg P m^{-2} day^{-1} assuming that about 80% of the annual internal loading occurs in summer (Steinman et al. 2009). Since 2014, diffusive flux calculations are based on the rhizon-filtered samples.

Study Period		L. Peipsi *s.s* mg P m^{-2} day^{-1}	L. Lämmijärv mg P m^{-2} day^{-1}	L. Pihkva mg P m^{-2} day^{-1}
2001–2005	External load	0.22	0.64	1.95
2001–2005	Outflow	0.11	0.34	1.11
1985–1989	External load	0.25	0.74	2.19
1985–1989	Outflow	0.08	0.23	0.70
2011*	Internal load, M1	35.28	105.63	
2001–2005	Internal load, M2	0.66	1.84	4.50
1985–1989	Internal load, M2	0.64	1.49	0.97
2011, growing season	Resuspension of TP	21.10	71.30	
2011, growing season	Resuspension of bioavailable P	2.1		
2011, growing season	Diffusive fluxes of P	0.23	0.21	
2014, March	Diffusive fluxes of P	0.13	0.35	0.14
2014, August	Diffusive fluxes of P	12.38	9.16	
2014, October	Diffusive fluxes of P	−0.01	−0.01	
2015, August	Diffusive fluxes of P	7.34	5.13	
2015, October	Diffusive fluxes of P	2.55	0.11	

*In these calculations, the data for the external loads and outflow of TP were from the period 2001–2005

10.3.2   Redox-Associated Sediment Phosphorus Mobilization and Transport by Molecular Diffusion and Resuspension

Irrespective of time and site of sampling, redox potential was approximately 400 mV at the sediment-water interface (see Figure 10.2). The redox potential generally decreased with sediment depth. However, the depth of the value critical for P release, that is, lower than 200 mV, varied depending on season, basin, and year. In March 2014, such values were not reached within the top 3 cm of the sediment core in Lake Lämmijärv and Lake Pihkva. However, the redox potential remained < 200 mV below a depth of 2.2 cm in deeper Lake Peipsi *s.s.* In August 2014, the redox potential decreased rapidly in all three parts of Lake Peipsi, down to a depth of 1 cm, where the values were 205 mV (±39 mV), 216 mV (±32 mV), and 180 mV (±2 mV) in Lakes Peipsi *s.s.*, Lämmijärv, and Pihkva, respectively. The redox potential in the deeper layers remained below 200 mV in all parts of Lake Peipsi. Redox conditions were lower in the surficial sediments in August 2013 when compared with August 2015. There was a nine-fold difference in redox potential between the depth of 3 cm and the sediment surface (with the values close to 200 mV at a depth of 0.5 cm) in Lake Peipsi *s.s.* in August 2013. Similar vertical changes in redox potential were observed in October 2014 in Lake Lämmijärv. Also in Lake Peipsi *s.s.*, October was the season of the greatest vertical drop in redox potential in 2014. In that basin, redox potential declined from 365 mV at the sediment surface down to 100 mV at a depth of 3 cm, reaching values close to 200 mV at a depth of 2 cm. In 2015, the critical value of 200 mV was reached in considerably deeper (approximately 3 cm) sediment layers in both Lake Peipsi *s.s.* and Lake Lämmijärv.

A decrease in redox potential from the surface to the deeper sediment layers in 2014 and 2015 generally occurred with an increase in SRP concentrations in the sediment pore water ($p = 0.001$) (see Figure 10.3). SRP concentrations were similar in the uppermost sediment layers of the different basins and were highest (939 µg L^{-1} ± 525 µg L^{-1}) in August ($p < 0.001$). In August, the SRP concentration of the pore water at the surface was, on average, 18 times higher than in March and about eight times higher than in October. The SRP concentration at the 3 cm depth was two-fold greater in August, six-fold greater in March, and three-fold greater in October compared to concentrations at the sediment surface. Despite the more reduced conditions of the sediments in 2014, higher SRP concentrations were observed in October 2015. Moreover, SRP concentrations remained close to zero within the uppermost 2-cm layer in Lake Peipsi *s.s.* Also, diffusive fluxes (see Table 10.2) showed large seasonal variations ($p < 0.001$), though no significant differences were found between 2014 and 2015. The diffusive fluxes were highest in August (mean values at the studied sites in 2014 and 2015: 8.5 mg P m^{-2} day^{-1} ± 4.9 mg P m^{-2} day^{-1}), with somewhat higher rates in 2014 than in 2015. The mean diffusive SRP flux of the three studied basins in March was only 2% of that in August. Unlike in October 2015, diffusive fluxes were negative in October 2014.

In 2011, the pore water SRP concentrations in the 1-cm layer of the surface sediments were much lower compared to 2014 and 2015. This resulted in considerably lower diffusive fluxes (see Table 10.2 and Figure 10.4B). However, similar spatio-temporal trends were observed in data of 2011 and 2014–2015. There were no significant changes in the diffusive fluxes of SRP (also the pore water SRP concentrations) between Lake Peipsi *s.s.* and Lake Lämmijärv. The highest diffusive fluxes (also the pore water SRP concentrations) occurred in summer. In 2011, when the measurements covered the whole growing season, the diffusive flux (also SRP concentration in the sediment pore water) was significantly higher until late July when compared with the later phase of the growing season (see Figure 10.4B).

Resuspension of P constituted the bulk of gross sedimentation in 2011 (62% in Lake Peipsi *s.s.* and 68% in Lake Lämmijärv). The resuspension rate of P varied between 3.9 mg P m^{-2} day^{-1} and

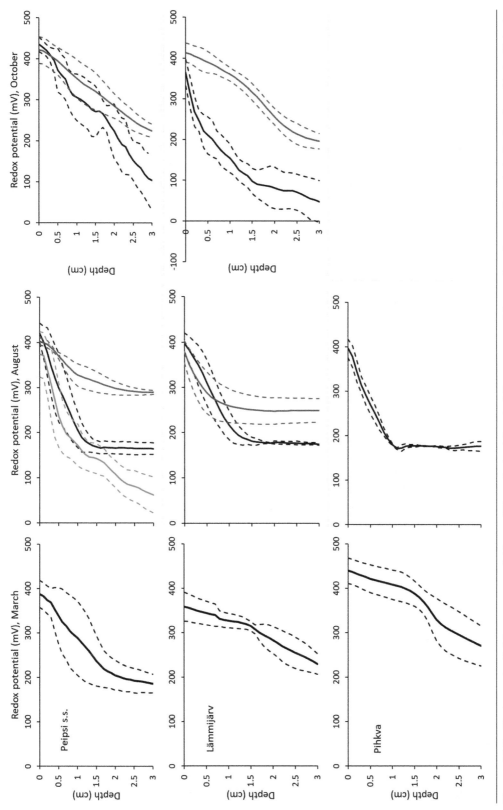

Figure 10.2   Redox potential (mean as a solid line ± SD as dashed lines) of the uppermost 3 cm of surface sediments in the basins of Lake Peipsi in August 2013 (green); in March, August, and October 2014 (black); and in August and October 2015 (blue) (modified from Tammeorg et al., 2016).

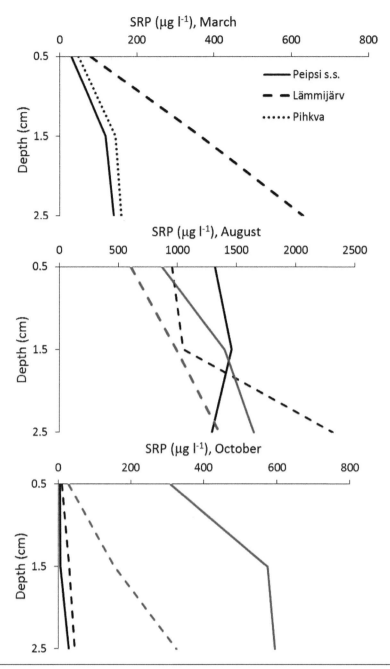

Figure 10.3 Mean SRP concentrations in the sediment pore water (the top-most 3 cm) in different basins of Lake Peipsi in March, August, and October 2014 (black), and in August and October 2015 (blue). Note different scales on the top x-axes (Modified from Tammeorg et al. 2016).

75.5 mg P m^{-2} day^{-1} in Lake Peipsi *s.s.* and was highest in September (see Figure 10.4A). The seasonal variations were less pronounced in Lake Lämmijärv where values varied between 47.8 mg P m^{-2} day^{-1} and 89.9 mg P m^{-2} day^{-1}, being on average 71.3 mg P m^{-2} day^{-1}. The resuspension rates of mobile P in Lake Peipsi *s.s* varied between 0.13 mg P m^{-2} day^{-1} and 7.55 mg P m^{-2} day^{-1}. They were

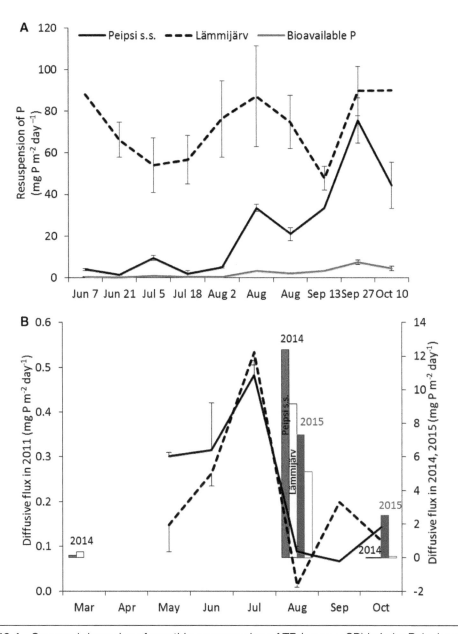

Figure 10.4   Seasonal dynamics of monthly resuspension of TP (mean ± SD) in Lake Peipsi s.s. and Lake Lämmijärv, and resuspension of bioavailable P in Lake Peipsi s.s. in 2011. (A) Diffusive fluxes in 2011 in Lake Peipsi s.s. (solid line) and in Lake Lämmijärv (dashed line) (Modified from Tammeorg et al. 2015). (B) Diffusive fluxes that were calculated less frequently for 2014 and 2015 in two basins of Lake Peipsi are shown with bars.

considerably lower and of similar magnitude to diffusive fluxes of P during May–July (mean ± SD, 0.44 mg P m⁻² day⁻¹ ± 0.36 mg P m⁻² day⁻¹) than in August–October (3.55 mg P m⁻² day⁻¹ ± 2.38 mg P m⁻² day⁻¹) (see Figure 10.4). On the other hand, the resuspension of potentially bioavailable P in the later part of the growing season in 2011 was of similar magnitude to the mean diffusive flux of August and October in 2014 and 2015 (5.57 mg P m⁻² day⁻¹).

10.3.3   Potential Contribution of Internal P Loading to Lake Water Quality

Seasonal variations in the SS concentration in 2011 agreed with those of resuspension of P in both basins of Lake Peipsi. Moreover, significant correlations were found between the concentrations of SS and TP ($r = 0.86$, $p < 0.001$), SS and Chl a ($r = 0.87$, $p < 0.001$), and TP and Chl a ($r = 0.85$, $p < 0.001$) in Lake Peipsi $s.s.$ In Lake Lämmijärv, SS concentration correlated well with the concentration of TP ($r = 0.81$, $p < 0.001$), but neither of them correlated with Chl a. SRP concentrations showed no significant temporal and between-basin changes in 2011.

In August of Period I, the mean water TP concentrations were 36 µg L^{-1}, 68 µg L^{-1}, and 82 µg L^{-1} in Lake Peipsi $s.s.$, Lake Lämmijärv, and Lake Pihkva, respectively. SRP concentration constituted 18% of TP in Lake Peipsi $s.s.$, and about 14% and 19% in Lake Lämmijärv, and Lake Pihkva, respectively. Biomass of cyanobacteria, as an August mean in Period I, was 2.2, 7.3, and 10.7 g m^{-3}. Between Periods I and II, TP concentration increased 1.3, 1.8, 2.1-fold, SRP concentration 1.4, 1.2, 2.5-fold, and cyanobacteria 2.0, 2.1, 1.9-fold in Lake Peipsi $s.s.$, Lämmijärv and Lake Pihkva, respectively. The between-basin and between-period differences were statistically significant. Moreover, water-column TP and SRP concentrations and biomass of cyanobacteria in August, correlated significantly with the IP$_{tot}$ (see Figure 10.5) in Lake Peipsi ($r = 0.927$, $p = 0.008$; $r = 0.853$, $p = 0.031$; $r = 0.851$, $p = 0.032$, respectively), while the corresponding correlations with the TP$_{in}$ were somewhat weaker, and not statistically significant (see Figure 10.5).

10.3.4   Environmental Conditions as Potential Drivers behind Variations in Internal P Loading

The pore water SRP concentration of the uppermost 1-cm sediment layer and diffusive P fluxes correlated negatively with DO ($r = -0.592$, $p < 0.0001$; $r = -0.623$, $p < 0.0001$) and positively with water temperature ($r = 0.515$, $p < 0.0001$; $r = 0.584$, $p < 0.0001$).

The highest rates of sediment resuspension were measured in the autumn, which was the season of exceptional weather events. These included the occurrence of the highest frequency of wind speeds exceeding 10 m s^{-1}, the maximum wind speed in 2011, and decreased water level (150 cm measured at 28 m a.s.l.). The regression analysis showed a significant effect of the weather factors prevailing approximately two weeks before water and sediment sampling on the studied variables (see Table 10.3). Wind speed was the best predictor of both sediment resuspension and SRP in the corresponding regression model. Water level was the best predictor of the TP, SS, and Chl a concentrations, and wind speed was selected as the next factor that improved the prediction of these water-quality variables (see increase of the corresponding model R^2 by partial R^2).

In the more recent Period II that had higher levels of internal P loading and deteriorating water quality, summer water temperature (19.4 ± 2.7°C) was higher compared to the earlier Period I (18.4 ± 2.4°C). Moreover, wind speed increased and water level decreased in Period II (Period I versus Period II: mean daily wind speed—2.1 versus 2.5 m s^{-1}; mean daily maximum wind speed—3.0 versus 4.1 m s^{-1}; mean water level—421.2 ± 33.6 cm versus 383.4 ± 35.3 cm). The wave action at the sediment surface of each basin of Lake Peipsi was approximately three times higher during Period II (Lake Peipsi $s.s.$: 5.5, Lake Lämmijärv: 13.0, and Lake Pihkva: 10.5) than during Period I.

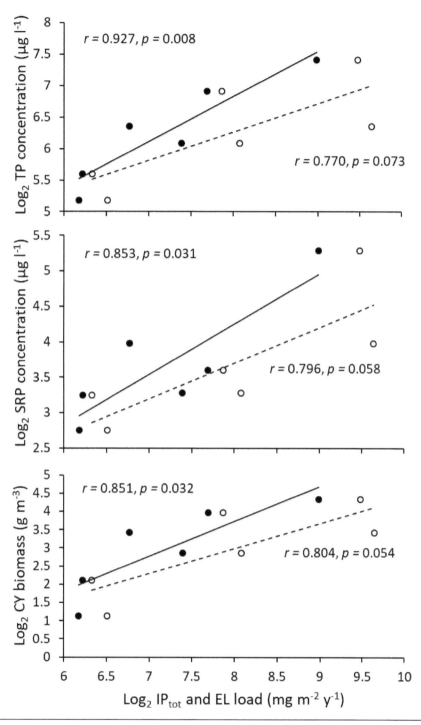

Figure 10.5 Correlations between the water-column TP or SRP concentrations or biomass of cyanobacteria in August and the internal P load (IP_{tot} filled circles; indicated by solid line) and the external P load (EL (TP_{in}) empty circles; indicated by dashed line) for Lake Peipsi (modified from Tammeorg et al. 2016). Values for Periods I and II for the three basins of Lake Peipsi are plotted.

Table 10.3   The most influential weather-factor terms selected by stepwise regression analysis and included in the regression model as predictors of the weather-dependent variables. L = water level, V = mean daily wind speed, and Vmax = maximal daily wind speed. The number of days over which a weather factor was averaged and the number of days prior to sampling date are shown as numerical subscripts. For example, V_{2_10} denotes the mean wind speed for two days (days 10 to 11), recorded 10 days before the measurement of the water variable. The partial R^2 of a factor term is the increment of the regression model R^2 when the corresponding term is added to the model term set (from Tammeorg et al. 2013).

Dependent Variable	Factor Term	Partial R^2	Model R^2	P
Resuspension	V_{1_5}	0.59	0.59	< 0.0001
	V_{2_10}	0.06	0.65	< 0.0001
	$Vmax_{1\_7}$	0.03	0.68	0.0004
Chlorophyll a (Chl a)	L_{1_1}	0.38	0.38	< 0.0001
	V_{3_2}	0.12	0.50	< 0.0001
	V_{2_0}	0.14	0.64	< 0.0001
	$Vmax_{2\_3}$	0.04	0.68	< 0.0001
Soluble reactive phosphorus (SRP)	$Vmax_{6\_6}$	0.37	0.37	< 0.0001
	V_{1_10}	0.07	0.44	< 0.0001
	$Vmax_{1\_3}$	0.07	0.51	< 0.0001
Suspended solids (SS)	L_{1_8}	0.56	0.56	< 0.0001
	V_{2_2}	0.10	0.66	< 0.0001
	V_{4_6}	0.05	0.70	< 0.0001
Total phosphorus (TP)	L_{1_0}	0.60	0.60	< 0.0001
	V_{2_2}	0.08	0.68	< 0.0001
	V_{1_8}	0.07	0.74	< 0.0001

10.4   DISCUSSION

Long-term mean estimates of the net internal P loading (IP_{tot}) were driven mainly by the values of the retention calculated from sediment cores (R_{sed}). One source of uncertainty in quantifying R_{sed} could be the effect of wind-induced waves on the vertical distribution of the sediments, affecting the accuracy of sediment chronology. Nevertheless, the sedimentation rates for Lake Pihkva in the present study are comparable with those that were documented by Kangur et al. (2007). Indeed, applying the sedimentation rates of accumulation area to the whole lake could result in the overestimation of R_{sed}. However, we consider this a minor factor in our study, given that the gross sedimentation measurements based on the use of the sediment trap method showed only a 1.1-fold difference in fluxes at sites close to the shore versus those in the central areas of Lake Peipsi *s.s.* (Tammeorg et al. 2013). Another source of uncertainty could be associated with the upward migration of P in the sediment, a common phenomenon (Carignan and Flett 1981; Carey and Rydin 2011). Nevertheless, these processes most likely affected the sediments dated to 2000s and 1980s similarly. This was indicated by the linear increase toward sediment surface in the TP concentrations in both Lake Peipsi *s.s.* and Lake Lämmijärv (Tammeorg et al. 2016). In Lake Pihkva, TP concentration remained constant in the sediments dated to cover the period from 1988 to 2007. These aspects justify a differentiation between Periods I and II in quantifying R_{sed} and IP_{tot}.

R_{sed} was considerably higher than P retention calculated by the mass balance approach (R_{mb}), which agrees with observations of other studies (Dillon and Evans 1993; Hupfer and Lewandowski 2005; Engstrom et al. 2009). The reported differences have generally been much lower for oligotrophic lakes (Dillon and Evans 1993) than for lakes with greater trophic status, such as those in our study. This supports the common knowledge that the dynamics of sediment P are closely linked to a lake's trophic state (Marsden 1989; Hupfer and Lewandowski 2008; Dittrich et al. 2013; Orihel et al. 2017). In our study, long-term averages of R_{mb} and R_{sed} for Periods I and II increased with increased trophic gradient across the cascade of basins, suggesting the reliability of the estimates. Moreover, higher R_{sed} for Period II was in accordance with increased trophy in Lake Peipsi compared with the late 1980s. Finally, our estimates of IP_{tot} (mean for the whole lake: 189 mg P m^{-2} yr^{-1}) were comparable with the estimates for the polymictic, eutrophic Lake Winnipeg (mean for the whole lake: 259 mg P m^{-2} yr^{-1}, Nürnberg and LaZerte 2016; 205 mg P m^{-2} yr^{-1}, Matisoff et al. 2017) and higher than the estimates for the polymictic, mesotrophic Lake Simcoe (mean for the whole lake: 62 mg P m^{-2} yr^{-1}) and for Lake Pyhäjärvi (61 mg P m^{-2} yr^{-1}, Nürnberg et al. 2012).

The mass balance calculations of the internal TP load in 2011 in Lake Peipsi were governed by the gross sedimentation rate of P. Due to high temporal variability, as commonly observed for gross sedimentation rates (Bloesch 1982; Horppila and Niemistö 2008), the rates measured during the growing season cannot automatically be applied over the whole year. Thus, the values of gross sedimentation of P cannot be compared straightforwardly with the long-term average estimates of P retention (R_{mb}, R_{sed}). Nevertheless, similar to the trends observed in the long-term average estimates of P retention, higher rates of gross sedimentation were observed in the more nutrient-rich southern basin of Lake Peipsi than in Lake Peipsi *s.s.*

By accounting for the bulk of the gross sedimentation, resuspension of P constituted 60% of the IL determined by the mass balance calculation in Lake Peipsi *s.s.* and 67% in Lake Lämmijärv. Thus, the resuspension of P was quantitatively the predominant internal process that affected cycling of particulate P in both basins of Lake Peipsi. Such conclusions have also been drawn for Lake Pyhäjärvi (Ekholm et al. 1997) and the Enonselkä basin of Lake Vesijärvi (Niemistö et al. 2012), where a similar approach was used to quantify the flux of TP from sediments. Finally, an important contribution of resuspension-induced P fluxes on lake water quality was reflected in the synchrony of the seasonal dynamics of resuspension of P with water-column concentrations of SS, TP, and Chl *a* in Lake Peipsi *s.s.* and partially in Lake Lämmijärv.

Given the governing role of sediment resuspension in Lake Peipsi, it is no surprise that gross sedimentation-based estimates of IL were about 160-fold of external P load (EL). The net estimates of IL were from 52% to 84% of the EL, indicating the high importance of the IL as a source of P, as has been observed in polymictic, eutrophic Lake Winnipeg and Lake Simcoe (Nürnberg et al. 2013; Nürnberg and LaZerte 2016). Moreover, the potentially higher importance of internal P loading compared to external P loading for lake water quality was demonstrated by stronger correlations of water quality variables in summer with IP_{tot} than with EL. In addition to providing particulate P, internal P loading seems to supply significant quantities of potentially bioavailable P to the water column during summer, contributing to cyanobacteria growth. A high dynamic ratio (the square root of the lake surface area in square kilometres divided by its mean depth in meters; Håkanson 1982) of 8.4 suggests strong interactions between the lake bottom and the surface water in Lake Peipsi, which enables continuous mixing. In such systems, with high turbidity and high water temperature, cyanobacteria often dominate over other phytoplankton groups (Bormans et al. 2016). Although EL directly contributes to the increase in lake water P, internal loads can have greater influence on algal biomass dynamics than external loads (Istvànovics et al. 2004; Steinman et al. 2009; Nürnberg 2009;

Nürnberg and LaZerte 2016) due to their timing and being in more bioavailable form (Nürnberg et al. 2013; Bormans et al. 2016). Therefore, our current findings suggest that internal P loading could have caused the deterioration of the water quality in Lake Peipsi in the 2000s. Such delayed responses of lake ecosystems after a reduction in external P load (Sas 1990; Jeppesen et al. 2005) have been observed worldwide.

The net estimate of the IL (for Period II) constituted only 2% of the gross sedimentation-based estimates (in 2011), suggesting the governing importance of particulate P in internal P fluxes, as found by Søndergaard et al. (2003). The measurements in August 2014 and 2015 showed that occasionally, the high diffusive fluxes of P can notably exceed the annual long-term mean and constitute a considerable part of the gross sedimentation-based estimates of internal P load (28% in Lake Peipsi *s.s.* and 7% in Lake Lämmijärv). The diffusive fluxes in 2011 in Lake Peipsi were similar to the close-to-zero benthic fluxes reported in other lakes worldwide (e.g., Nürnberg 1988; Orihel et al. 2017), but very low compared with fluxes of 2014 and 2015 in Lake Peipsi. Differences in the methods applied for sampling of the pore water SRP could cause notable differences in the diffusive fluxes, as sampling using rhizons enabled us to avoid oxygen contamination, resulting in higher SRP values (Seeberg-Elverfeldt et al. 2005), and a steeper concentration gradient between pore water and lake water. Nevertheless, even when the diffusive flux calculations were based on the pore water samples that were separated by centrifuging, the fluxes were at a level similar to that of EL in Lake Peipsi *s.s.* Moreover, particulate P may eventually become available for algae under conditions of high water pH, which favours the desorption of P from resuspended particles via ligand exchange reactions (Andersen 1975; Koski-Vähälä and Hartikainen 2001). On average, the resuspension rates of potentially bioavailable P were about nine-fold higher than the diffusion-induced P flux in 2011 in Lake Peipsi *s.s.* Similar trends were reported for other large, shallow lakes such as Lake Okeechobee (Havens et al. 2007) and Lake Apopka (Reddy et al. 1996). Interestingly, our results indicated also opposite trends for Lake Peipsi *s.s.*, when the mean diffusive fluxes of August and October in 2014–2015 were about 2.6 times higher than resuspension-induced mean flux determined for the growing season.

10.4.1   Mechanisms behind Internal Loading

An analysis of the seasonal changes in the redox potential, sediment pore water SRP concentration, and diffusive and resuspension flux measurements shed light into the complexity of the P mobilization and transport mechanisms driving sedimentary P release across the three basins of Lake Peipsi. The pronounced seasonality in the sediment pore water SRP concentrations and diffusive fluxes was largely due to temperature changes. The P retention capacity of the sediments is high in winter, while the oxidized surface layer is diminished in summer—when temperature, biological activity, and sedimentation of material increase (Søndergaard et al. 2003, 2013). Additionally, among the studied seasons, the redox barrier was the highest in March, suggesting the least favorable conditions for P release. The highest sediment pore water SRP concentrations and diffusive fluxes were measured in summer, when the decrease in the redox potential below the critical value of 200 mV occurred very close to the sediment surface. The importance of temperature was also apparent in decreasing redox conditions in August between 2013 and 2015, which coincided with the decrease in summer temperature. Moreover, lower summer temperatures in 2015 likely resulted in higher redox conditions in October of 2015, compared to conditions of 2014.

The measured redox profiles indicate a higher potential for P release rates in autumn 2014 compared to 2015, which seems to contradict trends in the calculated diffusive fluxes. However, disruption of the sediment surface to a depth of 2 cm was indicated by the sediment pore water profile measurements in October 2014 in Lake Peipsi *s.s.*, suggesting a resuspension event, which may influence the flux. Below the 2-cm-depth region, the SRP sharply increased indicating that P was being supplied into the upper sediment layers by an upward diffusive flux. This finding is supported by the concurrent redox measurements in Lake Peipsi *s.s.* that showed well-oxidized conditions in the upper 2-cm layer, but reduced conditions below the 2-cm depth. Similar processes could occur in Lake Lämmijärv, in which low negative diffusive fluxes were also observed. Analogous changes in the SRP concentration of the sediments, even with a deeper disruption zone, were described by Reddy et al. (1996) in large shallow Lake Apopka (Florida, USA; see Chapter 7).

Sediment disruption by resuspension results in much steeper concentration gradients favoring the release of P through diffusion. Moreover, the increase in the frequency and intensity of resuspension has been shown to increase benthic and near-bottom oxygen consumption, thereby stimulating organic material mineralization (Wainright and Hopkinson 1997) and creating favorable conditions for diffusion. Hence, resuspension events have particularly pronounced implications in summer/early fall when the pool of mobile P is particularly large. Interestingly, our measurements in 2011 showed an abrupt decrease in the diffusive fluxes in August, which coincided with the increase in the resuspension of P in Lake Peipsi *s.s.* This could result in an important source of P for the formation of algal blooms. Enhanced diffusion and high pH-mediated P release from resuspended sediment particles could then sustain the algal blooms through August and September, leading to high sedimentation rates of fresh organic material and subsequent resuspension. The highest rates of sediment resuspension were indeed observed in late September (Tammeorg et al. 2013). Diffusive fluxes calculated for August in 2014 and 2015 exceeded maximum resuspension fluxes of potentially bioavailable P, which implies that the cores were sampled after the recent resuspension event. In Lake Lämmijärv, sediment focusing could have obscured the seasonal changes in resuspension (Tammeorg et al. 2013). Nevertheless, the P recycling was likely to be governed by the processes similar to those in Lake Peipsi *s.s.*, as the rates of resuspension increased similarly in autumn.

Despite an important role of temperature in P cycling, the temporal dynamics of resuspension-induced P load were driven by wind-induced wave action, as indicated by the stepwise regression analysis. Hence, an increase in wave action during Period II, implying more severe resuspension events, can explain the increase in IP_{tot} in that period. Similarly, an increase in the IL over the long-term was attributed to the increase in wave action in shallow Lake Tuusulanjärvi (Horppila et al. 2017). A higher increase in wind activity between the two periods in southern basins of Lake Peipsi supports the more pronounced increase in IP_{tot} found in those basins in comparison with Lake Peipsi *s.s.* Southern basins of Lake Peipsi are more sensitive to water-level fluctuations due to their shallow depths. Regardless of the small increase in wind speed between Periods I and II (1 m s^{-1}), the water-level change of 0.4 m is proportionally higher in the shallower southern basins, resulting in a greater increase in wave action. Similarly, numerous studies have underlined the strong effects of water-level fluctuations for the P cycling of large and shallow lakes (Nõges and Nõges 1999; Pettersson et al. 2010).

There was a lack of significant spatial differences in the diffusive fluxes, possibly due to occurrence of oxic conditions in the surface sediments (Fisher et al. 2005), in all the basins of Lake Peipsi. However, notably higher resuspension rates of P were observed in Lake Lämmijärv than in Lake Peipsi *s.s.*, underlying the role of lake morphometry and lake trophy. Additionally, there was a clearly larger pool of mobile P in the anoxic zone of the more productive Lake Lämmijärv, when compared to other basins, in 2014. An increase in sediment enrichment with mobile P toward the more southern systems agrees with the relative increase in the IP_{tot} between Periods I and II. Thus, the large pool of mobile P in sediments can be an important reason for the delay in the water quality improvement after reduction in external P loading.

10.5 CONCLUSIONS

Depending on the estimation method, internal P load was similar to or exceeded the external P load, indicating that sediments are an important source of P to Lake Peipsi. For the water quality in summer, IL was shown to have potentially higher importance than the EL. In Lake Peipsi, the net estimates of the IL constituted only 2% of the gross sedimentation-based estimates.

Both seasonal measurements of P mobility in sediments and mass balance calculations indicated sediment resuspension as a key driver behind internal P loading. The mass balance calculations were dominated by gross sedimentation of P, and sediment resuspension constituted about 70% of the gross sedimentation. Moreover, seasonal measurements indicated an abrupt change in diffusive fluxes coinciding with an increase in sediment resuspension. Particularly pronounced implications of such events occurred in August, the period during which sediment mobile P was highest. The diffusive fluxes in August 2014 and 2015 exceeded the annual long-term mean and constituted a considerable part of the gross sedimentation-based estimates of IL. Enhanced diffusion and high pH-mediated P release from resuspended sediment particles could sustain the algal blooms in August–September, leading to increased amounts of fresh organic material that subsequently were available to resuspend. Such events were likely to result in the highest rates of sediment resuspension observed in late September.

Sediment resuspension was mainly driven by wave action, particularly pronounced at the time of decreased water level. In light of projected climate change that is likely to amplify symptoms of eutrophication via impacts on the ILs (e.g., increases in water temperature and wave action), continued declines in ELs are a prerequisite for the improvement of the lake water quality.

10.6 REFERENCES

Andersen, JM. 1975. Influence of pH on release of phosphorus from lake sediments. Arch Hydrobiol. 76:411–419.

Appleby, PG; Nolan, PJ; Gifford, DW; Godfrey, MJ; Oldfield, FJ; Anderson, NJ; and Battarbee, RW. 1986. [210]Pb dating by low background gamma counting. Hydrobiologia. 143(1):21–27.

Appleby, PG; Richardson, N; and Nolan, PJ. 1992. Self-absorption corrections for well-type germanium detectors. Nucl Instrum Meth B. 71(2):228–233.

Beach Erosion Board. 1972. Waves in inland reservoirs. Tech Memor. 132:1–129.

Berner, RA. 1980. Early diagenesis: A theoretical approach (No. 1). Princeton (NJ): Princeton University Press.

Bloesch, J. 1982. Inshore-offshore sedimentation differences resulting from resuspension in the eastern basin of Lake Erie. Can J Fish Aquat Sci. 39:748–759.

Bloesch, J and Burns, NM. 1980. A critical review of sedimentation trap technique. Schweiz Z Hydrol. 42 (1):15–55.

Bormans, M; Maršá lek, B; and Jančula, D. 2016. Controlling internal phosphorus loading in lakes by physical methods to reduce cyanobacterial blooms: a review. Aquat Ecol. 50(3):407–422.

Carey, CC and Rydin, E. 2011. Lake trophic status can be determined by the depth distribution of sediment phosphorus. Limnol Oceanogr. 56(6):2051–2063.

Carignan, R and Flett, RJ. 1981. Postdepositional mobility of phosphorus in lake sediments. Limnol Oceanogr. 26(2):361–366.

Carper, GL and Bachmann, RW. 1984. Wind resuspension of sediments in a prairie lake. Can J Fish Aquat Sci. 41(12):1763–1767.

Dillon, PJ and Evans, HE. 1993. A comparison of phosphorus retention in lakes determined from mass balance and sediment core calculations. Water Res. 27(4):659–668.

Dittrich, M; Chesnyuk, A; Gudimov, A; McCulloch, J; Quazi, S; Young, J; Winter, J; Stainsby, E; and Arhonditsis, G. 2013. Phosphorus retention in a mesotrophic lake under transient loading conditions: Insights from a sediment phosphorus binding form study. Water Res. 47(3):1433–1447.

Ekholm, P; Malve, O; and Kirkkala, T. 1997. Internal and external loading as regulators of nutrient concentrations in the agriculturally loaded Lake Pyhäjärvi (southwest Finland). Hydrobiologia. 345:3–14.

Engstrom, DR; Almendinger, JE; and Wolin, JA. 2009. Historical changes in sediment and phosphorus loading to the upper Mississippi River: mass-balance reconstructions from the sediments of Lake Pepin. J Paleolimn. 41(4):563–588.

EVS-EN 1189: Determination of phosphorus—Ammonium molybdate spectrometric method.

Fisher, MM; Reddy, KR; and James, RT. 2005. Internal nutrient loads from sediments in a shallow, subtropical lake. Lake Reserv Manage. 21(3):338–349.

Gasith, A. 1975. Tripton Sedimentation in eutrophic lakes—simple correction for the resuspended matter. Verh Internat Verein Limnol. 19(1):116–122.

Haberman, J; Haldna, M; Laugaste, R; and Blank, K. 2010. Recent changes in large and shallow Lake Peipsi (Estonia/Russia): causes and consequences. Pol J Ecol. 58:645–662.

Håkanson, L. 1982. Lake bottom dynamics and morphometry: the dynamic ratio. Water Resour Res. 18(5):1444–1450.

Hamilton, DP; Mitchell, SF. 1988. Effects of wind on nitrogen phosphorus and chlorophyll in a shallow New Zealand lake. Verh Int Ver Limnol. 23(1):624–628.

Havens, KE; Jin, KR; Iricanin, N; and James, RT. 2007. Phosphorus dynamics at multiple time scales in the pelagic zone of a large shallow lake in Florida USA. Hydrobiologia. 581(1):25–42.

Horppila, J; Holmroos, H; Niemistö, J; Massa, I; Nygrén, N; Schönach, P; Tapio, P; and Tammeorg, O. 2017. Variations of internal phosphorus loading and water quality in a hypertrophic lake during 40 years of different management efforts. Ecol Eng. 103:264–274.

Horppila, J and Niemistö, J. 2008. Horizontal and vertical variations in sedimentation and resuspension rates in a stratifying lake–effects of internal seiches. Sedimentology. 55:1135–1144.

Hupfer, M and Lewandowski, J. 2005. Retention and early diagenetic transformation of phosphorus in Lake Arendsee (Germany)– consequences for management strategies. Arch Hydrobiol. 164(2):143–167.

Hupfer, M and Lewandowski, J. 2008. Oxygen controls the phosphorus release from lake sediments—a long-lasting paradigm in limnology. Int Rev Hydrobiol. 93(4–5):415–432.

Iital, A; Stålnacke, P; Deelstra, J; Loigu, E; and Pihlak, M. 2005. Effects of large-scale changes in emissions on nutrient concentrations in Estonian rivers in the Lake Peipsi drainage basin. J Hydrol. 304(1–4):261–273.

Istvanovics, V; Osztoics, A; and Honti, M. 2004. Dynamics and ecological significance of daily internal load of phosphorus in shallow Lake Balaton, Hungary. Freshw Biol. 49(3):232–252.

Jaani, A. 2001. Hydrological regime and water balance. In: Nõges, T, (Ed.) Lake Peipsi: Meteorology hydrology hydrochemistry. Tartu: Sulemees Publishers. pp. 38–72.

Jeppesen, E; Søndergaard, M; Jensen, JP; Havens, KE; Anneville, O; Carvalho, L; Coveney, MF; Deneke, R; Dokulil, MT; Foy, B; et al. 2005. Lake responses to reduced nutrient loading—an analysis of contemporary long-term data from 35 case studies. Freshw Biol. 50(10):1747–1771.

Kangur, K; Kangur, A; and Raukas, A. 2012. Peipsi Lake in Estonia/Russia. In: Bengtsson, L; Herschy, RW; and Fairbridge, RW, (Eds.). Encyclopedia of lakes and reservoirs. Springer Netherlands. pp. 596–607.

Kangur, M; Kangur, K; Laugaste, R; Punning, JM; and Möls, T. 2007. Combining limnological and palaeolimnological approaches in assessing degradation of Lake Pskov. Hydrobiologia. 584(1): 121–132.

Kapanen, G. 2012. Pool of mobile and immobile phosphorus in sediments of the large shallow Lake Peipsi over the last 100 years. Environ Monitor Assess. 184(11):6749–6763.

Kisand, A; Kirsi, AL; Ehapalu, K; Alliksaar, T; Heinsalu, A; Tõnno, I; Leeben, A; and Nõges, P. 2017. Development of large shallow Lake Peipsi (North-Eastern Europe) over the Holocene based on the stratigraphy of phosphorus fractions. J Paleolimnol. 58(1):43–56.

Koski-Vähälä, J and Hartikainen, H. 2001. Assessment of the risk of phosphorus loading due to resuspended sediment. J Environ Qual. 30:960–966.

Lappalainen, K-M and Matinvesi, J. 1990. Järven fysikaalis-kemialliset prosessit ja aine-taseet. In: Ilmavirta, V. (Ed.). Järvien kunnostuksen ja hoidon perusteet. Yliopistopaino. Helsinki. pp. 54–84 (in Finnish).

Leeben, A; Tõnno, I; Freiberg, R; Lepane, V; Bonningues, N; Makarõtševa, N; Heinsalu, A; and Alliksaar, T. 2008. History of anthropogenically mediated eutrophication of Lake Peipsi as revealed by the stratigraphy of fossil pigments and molecular size fractions of pore-water dissolved organic matter. Hydrobiologia. 599(1):49–58.

Leeben, A; Freiberg, R; Tõnno, I; Kõiv, T; Alliksaar, T; and Heinsalu, A. 2013. A comparison of the palaeo-limnology of Peipsi and Võrtsjärv: connected shallow lakes in north-eastern Europe for the twentieth century especially in relation to eutrophication progression and water-level fluctuations. Hydrobiologia. 710(1):227–240.

Lewandowski, J and Hupfer, M. 2005. Effect of macrozoobenthos on two-dimensional small-scale heterogeneity of pore water phosphorus concentrations in lake sediments: a laboratory study. Limnol Oceanogr. 50(4):1106–1118.

Li, YH and Gregory, S. 1974. Diffusion of ions in sea water and in deep-sea sediments. Geochim Cosmochim Ac. 38:703–714.

Loigu, E; Leisk,Ü; Iital, A; and Pachel, K. 2008. In: Haberman, J; Timm, T; and Raukas, A, (Eds.). Pollution load and water quality of the Lake Peipsi basin. Peipsi. Tartu: Eesti Loodusfoto. pp. 179–199.

Marsden, MW. 1989. Lake restoration by reducing external phosphorus loading: the influence of sediment phosphorus release. Freshw Biol. 21(2):139–162.

Matisoff, G; Watson, SB; Guo, J; Duewiger, A; and Steely, R. 2017. Sediment and nutrient distribution and resuspension in Lake Winnipeg. Sci Tot Environ. 575:173–186.

Niemistö, J; Tamminen, P; Ekholm, P; and Horppila, J. 2012. Sediment resuspension: rescue or downfall of a thermally stratified eutrophic lake? Hydrobiologia. 686:267–276.

Nõges, T; Järvet, A; Kisand, A; Laugaste, R; Loigu, E; Skakalski, B; and Nõges, P. 2007. Reaction of large and shallow lakes Peipsi and Võrtsjärv to the changes of nutrient loading. Hydrobiologia. 599:253–264.

Nõges, P; Leisk,Ü; Loigu, E; Reihan, A; Skakalski, B; and Nõges, T. 2003. Nutrient budget of Lake Peipsi in 1998. Proc Acad Sci Estonia. Ecol Biol. 4:407–422.

Nõges, T and Nõges, P. 1999. The effect of extreme water level decrease on hydrochemistry and phytoplankton in a shallow eutrophic lake. In: Walz, N and Nixdorf B, (Eds.). Shallow Lakes' 98 Netherlands: Springer. pp. 277–283.

Nürnberg, GK. 1984. The prediction of internal phosphorus load in lakes with anoxic hypolimnia. Limnol Oceanogr. 29(1):111–124.

Nürnberg, GK. 1988. Prediction of phosphorus release rates from total and reductant-soluble phosphorus in anoxic lake sediments. Can J Fish Aquat Sci. 45(3):453–462.

Nürnberg, GK. 2009. Assessing internal phosphorus load–problems to be solved. Lake Reserv Manage. 25(4):419–432.

Nürnberg, GK and LaZerte, BD. 2016. More than 20 years of estimated internal phosphorus loading in polymictic eutrophic Lake Winnipeg Manitoba. J Great Lakes Res. 42(1):18–27.

Nürnberg, GK; LaZerte, BD; Loh, PS; and Molot, LA. 2013. Quantification of internal phosphorus load in large, partially polymictic and mesotrophic Lake Simcoe, Ontario. J Great Lakes Res. 39(2): 271–279.

Nürnberg, GK; Tarvainen, M; Ventelä, AM; and Sarvala, J. 2012. Internal phosphorus load estimation during biomanipulation in a large polymictic and mesotrophic lake. Inland Waters. 2(3):147–162.

Orihel, DM; Baulch, HM; Casson, NJ; North, RL; Parsons, CT; Seckar, DC; and Venkiteswaran, JJ. 2017. Internal phosphorus loading in Canadian fresh waters: a critical review and data analysis. Can J Fish Aquat Sci. 74:2005–2029.

Pettersson, K; George, G; Nõges, P; Nõges, T; and Blenckner, T. 2010. The impact of the changing climate on the supply and re-cycling of phosphorus. In: George, DG (Ed.). The impact of climate change on European lakes. Netherlands: Springer. pp. 121–137.

Punning, J and Kapanen, G. 2009. Phosphorus flux in Lake Peipsi sensu stricto. Eastern Europe Estonian J Ecol. 58(1):3–17.

Raukas, A. 2008. The composition and formation of the Lake Peipsi bottom sediments In: Haberman, J; Timm, T; and Raukas, A (Eds.). Peipsi. Tartu: Eesti Loodusfoto. pp. 93–99.

Reddy, KR; Fisher, MM; and Ivanoff, D. 1996. Resuspension and diffusive flux of nitrogen and phosphorus in a hypereutrophic lake. J Environ Qual. 25(2):363–371.

Renberg, I and Hansson, H. 2008. The HTH sediment corer. J Paleolimnol. 40(2):655–659.

Rumyantsev, VA; Kondraťev, SA; Basova, SL; Shmakova, MV; Zhuravkova, ON; and Savitskaya, NV. 2006. Chudsko-Pskovskii Lake Complex: monitoring and modeling phosphorus regime. Water Resour. 33: 661–669.

Sas, H. 1990. Lake restoration by reduction of nutrient loading: expectations experiences extrapolations. Verh Internat Verein Limnol. 24:247–251.

Seeberg-Elverfeldt, J; Schlüter, M; Feseker, T; and Kölling, M. 2005. Rhizon sampling of porewaters near the sediment-water interface of aquatic systems. Limnol Oceanogr-Meth. 3(8):361–371.

Søndergaard, M; Bjerring, R; and Jeppesen, E. 2013. Persistent internal phosphorus loading during summer in shallow eutrophic lakes. Hydrobiologia. 710(1):145–152.

Søndergaard, M; Jensen, JP; and Jeppesen, E. 2003. Role of sediment and internal loading of phosphorus in shallow lakes. Hydrobiologia. 506(1–3):135–145.

Starast, H; Milius, A; Möls, T; and Lindpere, A. 2001. In: Nõges, T (Ed.). Meteorology Hydrology Hydrochemistry. Tartu: Sulemees Publishers. pp. 97–131.

Statistics, Estonia. 2017. https://www.stat.ee/statistics.

Steinman, A; Chu, X; and Ogdahl, M. 2009. Spatial and temporal variability of internal and external phosphorus loads in Mona Lake Michigan. Aquat Ecol. 43(1):1–18.

Tammeorg, O; Horppila, J; Laugaste, R; Haldna, M; and Niemistö, J. 2015. Importance of diffusion and resuspension for phosphorus cycling during the growing season in large shallow Lake Peipsi. Hydrobiologia. 760(1):133–144.

Tammeorg, O; Horppila, J; Tammeorg, P; Haldna, M; and Niemistö, J. 2016. Internal phosphorus loading across a cascade of three eutrophic basins: A synthesis of short-and long-term studies. Sci Tot Environ. 572:943–954.

Tammeorg, O; Niemistö, J; Möls, T; Laugaste, R; Panksep, K; and Kangur, K. 2013. Wind-induced sediment resuspension as a potential factor sustaining eutrophication in large and shallow Lake Peipsi. Aquat Sci. 75(4):559–570.

Vollenweider, RA. 1975. Input-output models. Schweiz Z Hydrol. 37(1):53–84.

Wainright, SC and Hopkinson, CS. 1997. Effects of sediment resuspension on organic matter processing in coastal environments: a simulation model. J Marine Syst. 11(3):353–368.

CHAPTER **11**

PHOSPHORUS DYNAMICS AND ITS RELATIONSHIP WITH CYANOBACTERIAL BLOOMS IN LAKE TAIHU, CHINA

Liqiang Xie[1], Xiaomei Su[1,2], and Hai Xu[1]

Abstract

Excessive nutrient loading from both external and internal sources has promoted the growth of cyanobacteria in Lake Taihu and led to the frequent occurrence of large cyanobacterial blooms. We analyzed the long-term dynamics of total phosphorus (TP) and soluble reactive phosphorus (SRP), as well as the variations in the cyanobacterial cell density and chlorophyll a (Chl a) concentrations, in two stations of Lake Taihu, Meiliang Bay and Lake Center. In addition, we explored the relationship between phosphorus (P) dynamics and cyanobacterial blooms. The results showed that there were different seasonal patterns in the dynamics of P for the two sampling sites: TP concentrations were high in the summer at Meiliang Bay, but they were high in the winter at Lake Center. Autumn and summer were favorable seasons for the accumulation of cyanobacteria, and the cyanobacterial cell density in Meiliang Bay was much higher than that in Lake Center. The P content in the lake sediment has increased rapidly in recent years, suggesting that a reduction in internal P loading is both necessary and urgent in order to control cyanobacterial blooms in Lake Taihu.

Keywords: Phosphorus, cyanobacteria, eutrophication, algal bloom, Lake Taihu

11.1 INTRODUCTION

In the past few years, anthropogenic nutrient loading from both external and internal sources has resulted in the decline of water quality in Lake Taihu, China, increasingly promoting the development of cyanobacteria. With this excessive nutrient enrichment, cyanobacteria tend to grow rapidly and

[1] State Key Laboratory of Lake Science and Environment, Nanjing Institute of Geography and Limnology, Chinese Academy of Sciences, Nanjing 210008, China. Corresponding author: Liqiang Xie. E-mail: lqxie@niglas.ac.cn

[2] Jiangsu Provincial Key Laboratory of Environmental Engineering, Jiangsu Provincial Academy of Environmental Sciences, Nanjing 210036, China

blooms expand in intensity and magnitude. It has been widely shown that phosphorus (P) availability, which is often related to the trophic state of a water body, can be a limiting factor in the growth of freshwater algae (Schindler 1974; Sterner 2008; Schelske 2009). The Taihu Basin is one of the most industrialized and urbanized regions in China, containing only 0.4% of China's land area, while accounting for 11% of China's gross domestic product (Qin et al. 2007). Industrial development was initiated in the Taihu Basin in the 1960s, when TP loads were < 1,000 tons year^{-1} (t yr^{-1}) (Lai and Yu 2007). By 1988, TP loads had reached 2,000 t yr^{-1} (Huang et al. 2001). The TP concentration in Lake Taihu's water column actually declined from 1997 to 2001; this may be explained in part by the reduction of external P inputs from the catchment due to regulations imposed by the local government in 1995 (Chen et al. 2003).

The improvement in P concentration lasted only until 2001, after which wastewater discharge increased again (Qian and He 2009). Consequently, severe cyanobacterial blooms, followed by massive cyanotoxin production, have become an important ecological and human health issue, especially in the northern parts of Lake Taihu. These frequent blooms have led to serious environmental, economic, and societal problems; one example involves the highly publicized drinking water crisis in Wuxi in 2007 (Guo 2007; Qin et al. 2010). This has added to the urgency for researchers and government agencies to seek mitigation strategies and solutions in order to guarantee the safety of drinking water resources in Lake Taihu. Phosphorus is the primary limiting nutrient in winter and spring in Lake Taihu, while nitrogen (N) and P co-limit algal growth in summer and autumn (Xu et al. 2010). The control of P loading is thus a critical step in the restoration of the lake ecosystem. External nutrient loading has been the main water quality problem in Lake Taihu in recent years. However, once external P input is reduced, sediments that have acted as a P storage reservoir for decades are likely to become an important source. The P pool was estimated at 5,168 tons for the entire Lake Taihu and was closely related to pollution input and algal blooms (Zhu et al. 2013). The concentrations of total phosphorus ranged from 320 mg kg^{-1} to 2,481 mg kg^{-1} and the comprehensive pollution index indicated that the P contamination in sediment of Taihu was extremely high (Yu et al. 2017). The main purposes of this chapter are: (1) to describe the long-term changes of phosphorus concentration in the water column of Lake Taihu, outlining anthropogenic influences on the water quality; and (2) to discuss the mechanism underlying these changes, with an emphasis on the possible role of cyanobacterial blooms in shaping the patterns of the long-term dynamics of P in the lake (water and sediment).

11.2   MATERIALS AND METHODS

11.2.1   Study Site

The third largest freshwater lake in China—Lake Taihu—is located in the lower Yangtze River Delta in the east of China and has a water surface area of 2,338 km^2, a maximum width of 56 km, and a catchment area of 36,500 km^2. As a typical large, shallow lake, its average depth is 1.9 m and maximum depth is 2.6 m. It is dominated by a subtropical monsoon climate (Qin et al. 2007). Taihu basin is characterized by a complex set of river networks, with 117 rivers and tributaries draining into the lake. The annual freshwater input to this lake is about 8.8 × 10^9 m^3, and the water retention time of the lake is approximately 284 days (Qin et al. 2007). From 2005 to 2008, the area of agriculture and forestry decreased 6.68% and 1.44%, respectively, while the developed area increased 24.21%. These land use changes are due to the rapid development of the economy and societal growth in the catchment shortly after the reform and opening policy of the 1980s in China.

Two sampling sites were selected for this study (see Figure 11.1). Meiliang Bay is situated in the northern part of Taihu, and it is one of the most eutrophic regions of the lake. The surface area of the

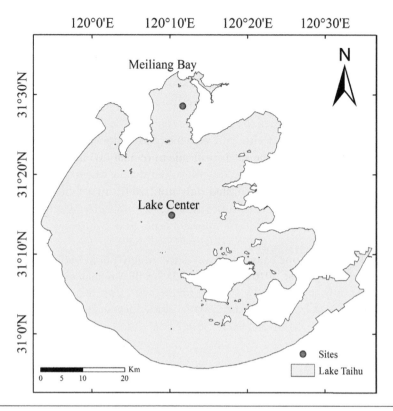

Figure 11.1   Map of Lake Taihu and two sampling sites.

bay is 132 km² and its mean depth is 2.0 m. Lake Center is located in the central part of the entire lake and it has lower nutrient concentrations due to advection. In addition, Lake Center is an open area that is favourable for dispersion of pollutants with the help of strong winds and water flow.

11.2.2   Phosphorus Analysis

Sediment cores were taken and then air dried, ground, and passed through a 200-mesh sieve for analyses. The P forms were analyzed using the sequential extraction method to fractionate P in sediments (Ruttenberg 1992) and modified by Li et al. (1998).

11.2.3   Data Source and Analysis

The data for environmental variables in this study were provided by Taihu Laboratory for Lake Ecosystem Research, which is located on the shore of Meiliang Bay and is responsible for the monthly and seasonal routine monitoring of Lake Taihu. Water sampling and environmental measurements were performed monthly from 2005 to 2012. Monthly surface water samples (upper 0.5 m) were collected from each station, stored in brown bottles, and immediately brought back to the laboratory in a portable refrigerator.

Chemical parameters included TP and SRP. SRP was determined using the molybdenum blue method (APHA et al. 1995). TP was analyzed using a combined persulfate digestion, followed by spectrophotometric analysis, as for SRP. Biological parameters included phytoplankton biomass, species composition, and Chl *a*. Chl *a* concentrations were determined spectrophotometrically after extraction in 90% hot ethanol (Pápista et al. 2002).

11.2.4 Phytoplankton Analysis

A 1-L water sample (integrated from throughout the water column) from each station was collected for phytoplankton analysis. Phytoplankton samples were preserved with Lugol's iodine solution (2% final concentration) and settled for 48 hours. Cell density was determined using a Sedgwick-Rafter counting chamber under a microscope at magnifications of 320–400 times. Phytoplankton species were identified according to Hu and Wei (2006). Biovolumes were calculated from cell numbers and cell size measurements. Conversion to biomass was based on 1 mm^3 of volume being equivalent to 1 mg of fresh weight biomass.

11.2.5 Statistical Analysis

Monthly nutrient and Chl *a* data were summarized as mean values for spring (March, April, and May), summer (June, July, and August), autumn (September, October, and November), and winter (December, January, and February) from 2005 to 2012. Spearman correlation analysis was performed to evaluate the relationship between cyanobacteria and environmental parameters using SPSS 20.0 statistical software.

11.3 RESULTS

11.3.1 The Dynamics of P Concentration from 2005 to 2012

There were distinct temporal dynamics in the seasonal variations of TP and SRP concentrations from 2005 to 2012 at the two sampling sites (see Figure 11.2). In Meiliang Bay, higher TP concentrations were observed in autumn and summer compared with spring and winter, whereas in Lake Center, higher values were usually observed in winter and spring (see Figure 11.2). In addition, the average concentrations of TP and SRP in Meiliang Bay (0.132 mg L^{-1} and 0.013 mg L^{-1}, respectively) were higher than those in Lake Center (0.099 mg L^{-1} and 0.010 mg L^{-1}, respectively) based on the data between 2005 and 2012. Since 2007, TP and SRP concentrations were highest in the winter at the Lake Center site, with the exceptions of SRP in 2010 and TP in 2012, when summer concentrations exceeded winter concentrations (see Figure 11.2). In contrast, seasonal TP and SRP concentrations were relatively low in winter in Meiliang Bay (see Figure 11.2).

11.3.2 The Variation in Cyanobacterial Cell Density from 2005 to 2012

In general, there were distinct temporal dynamics in the seasonal variations of cyanobacterial cell density and Chl *a* concentrations from 2005 to 2012 in the two sampling areas (see Figure 11.3). In Lake Center, cyanobacterial cell densities remained low from 2005 to 2010 and increased modestly in 2011 and 2012, especially in summer and autumn (see Figure 11.3). A similar pattern, but with greater densities, was observed in Meiliang Bay except for an initial maximum in summer of 2006 (see Figure 11.3). The average cyanobacterial cell density was 1.05×10^8 cells L^{-1} in Meiliang Bay, which was almost four times greater than that in Lake Center (2.61×10^7 cells L^{-1}). This generally

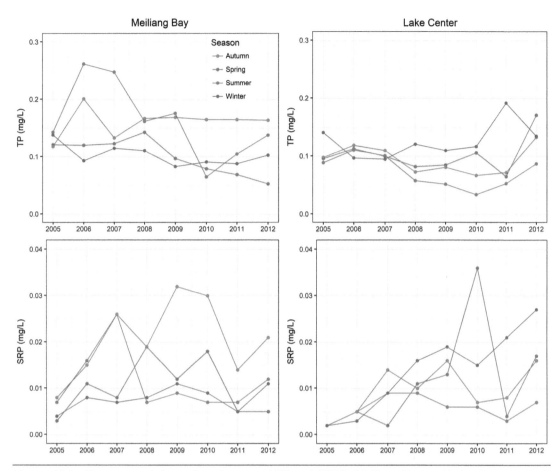

Figure 11.2   The temporal dynamics of seasonal TP and SRP in Meiliang Bay and Lake Center based on the data from 2005 to 2012 in Lake Taihu. Note the different y-axis scales for TP and SRP.

paralleled the difference in Chl *a* concentrations, with average values of 28.40 μg L^{-1} in Meiliang Bay and 13.95 μg L^{-1} in Lake Center (see Figure 11.3). Seasonally, the highest cyanobacterial cell densities and Chl *a* concentrations were observed in summer and autumn, while lower values were recorded in winter and spring, both at Meiliang Bay and at Lake Center (see Figure 11.3).

11.3.3   The Relationship between P Dynamics and Cyanobacterial Blooms

The relationship between seasonal average P concentrations (TP and SRP) and Chl *a* concentration and cyanobacterial cell density in the lake water column was explored by Spearman correlation analyses. Results showed that TP and Chl *a* were significantly and positively (both p < 0.05; r = 0.444 and 0.676, respectively) correlated with cyanobacterial cell density in Meiliang Bay, while SRP was not significantly (p > 0.05) correlated with cyanobacteria (see Table 11.1). In Lake Center, only Chl *a* was positively and significantly (p < 0.05; r = 0.389) correlated with cyanobacteria (see Table 11.1).

Interestingly, when data from Meiliang Bay and Lake Center were combined, cyanobacteria showed positive and significant relationships (p < 0.05) with TP, SRP, and Chl *a* (see Table 11.1).

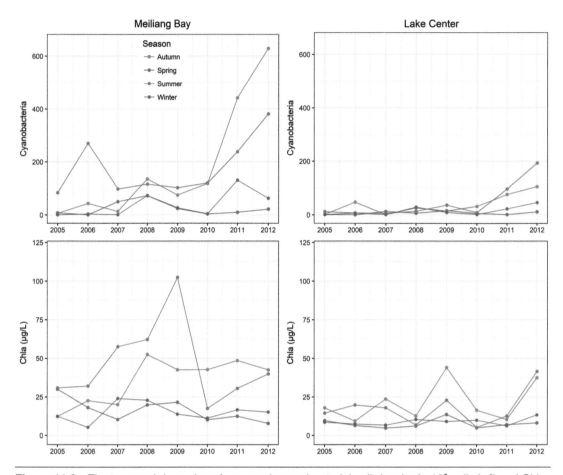

Figure 11.3   The temporal dynamics of seasonal cyanobacterial cell density (× 10⁶ cells L⁻¹) and Chl *a* concentrations in Meiliang Bay and Lake Center based on the data from 2005 to 2012 in Lake Taihu.

Table 11.1   Spearman correlation coefficients (r) for correlations between environmental factors and cyanobacterial cell density. ** and * indicate that correlation is significant at the 0.01 and 0.05 level (2-tailed), respectively.

Variable	Cyanobacteria		
	Meiliang Bay	**Lake Center**	**Combined Sites**
TP	0.444*	−0.141	0.275*
SRP	0.306	0.232	0.297*
Chl *a*	0.676**	0.389*	0.615**

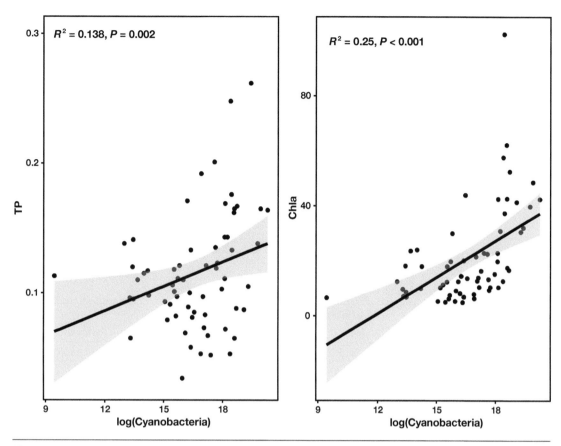

Figure 11.4 The relationships between cyanobacterial cell density and TP (left) and with Chl *a* concentration (right) in Lake Taihu from 2005 to 2012. Note log scale for the *x* axis. The gray shading indicates 95% confidence interval.

To further explore the relationship between environmental variables and cyanobacterial blooms, linear regression analysis was performed between cyanobacterial cell density and both TP and Chl *a* concentrations (see Figure 11.4). Both TP and Chl *a* were significantly and positively ($p < 0.01$) correlated with cyanobacterial cell density after combining all the data from Meiliang Bay and Lake Center, although there was considerable scatter in the plots.

11.3.4   The Phosphorus Variation in the Sediments

Numerous investigations on the nutrient content in Lake Taihu sediment have been made since the 1960s. During the five decades since 1960, there have been sustained increases in the mean content of TP in the sediment, although a plateau may have been reached recently (see Table 11.2). The mean sediment TP increased by 76.7% during the 55 years between 1960 and 2014. The yearly release of P was 21,000 metric tons, which is about 2 to 6 times the annual external load (Qin et al. 2006). In Meiliang Bay, the TP release rates increased from 2006 to 2016 (see Table 11.3). These results suggest that anthropogenically-derived pollution has affected the sediments in local areas of Lake Taihu.

Table 11.2 Total P content in the sediments (dry weight) from 1980 (Fan et al. 2000a; Jin et al. 2006; present study) to 2014 (sampling stations including Meiliang Bay, Gonghu Bay, Wuli Bay, Zhushan Bay, and Eastern Taihu Bay).

Year	TP Contents (mg/kg)	
	Range	Mean
1960	—	440
1980	370–670	520
1980–1991	400–1070	560
1995–1996	390–2370	580
1997–1999	280–2800	600
2002–2003	420–3408	1909
2013–2014	548–2894	1889

Table 11.3 The P release rates of Lake Taihu

Site	Wind Speed (m s^{-1})	Total P Release Rate (mg m^{-2} d^{-1})	Reference
Meiliang Bay	> 6	185	Qin et al. 2006
Meiliang Bay	7	1100	Wang et al. 2014
Meiliang Bay	2–16	460–5290	Wang et al. 2015
Lake Taihu	5	348	Huang et al. 2016

11.4 DISCUSSION

With a history of increasing eutrophication and global warming, Lake Taihu is now being disturbed by massive cyanobacteria blooms and their toxic metabolites during most times of the year (Su et al. 2015). Some studies have shown that the TP concentration is related to cyanobacterial blooms. An increased biomass of gas-vacuolate cyanobacteria has been attributed to nutrient-replete conditions, provided either in the water or from the sediments via internal P loading. In Lake Kasumigaura (Japan), TP and PO_4-P concentrations in the lake water column increased during a period of cyanobacterial blooms (Otsuki et al. 1984). In the freshwater tidal portion of the Potomac Estuary (USA), the recurrence of *Microcystis* blooms was related to the enhanced release of P from the sediments (Seitzinger 1991). Also, in Lake Donghu (China), declines of TP and PO_4-P concentrations in the water after the mid-1980s were coincident with the disappearance of cyanobacterial blooms (Xie and Xie 2002). Possible explanations for the increase in P concentrations in lake water during cyanobacterial blooms may be: (1) cyanobacteria cells increased the particulate P concentration in the lake water; and/or (2) the cyanobacterial bloom induced the release of dissolved P from the sediments (i.e., *Microcystis* blooms induced massive release of P from Fe-P complexes within the sediment, perhaps mediated by high pH caused by intense algal photosynthesis and/or depressed concentrations of nitrate nitrogen (Xie et al. 2003)).

One of the most important factors affecting P concentrations in lake water is P release from the sediments. It is widely known that internal P loading is a major source of P in hyper-eutrophic lakes (Xie et al. 2003). The release of P from lake sediments occurs either by mobilization of P from resuspended sediment particles, or after mobilization, to the dissolved pool in the sediment and a subsequent upward transport of the dissolved species (Boström et al. 1988; see Chapters 1 and 4).

Sediment resuspension is the most important factor affecting sediment phosphorus release in Lake Taihu (Qin et al. 2006; see Chapter 10). In disturbances caused by wind-wave action, the dissolved P in the sediment pore water is mixed into the overlying water column and is bioavailable, assuming no chemical inactivation occurs. Based on wave flume experiments, strong waves in Lake Taihu may resuspend the upper 10 cm of the sediment (Qin et al. 2004). Therefore, under strong wind conditions, we hypothesized that the upper 10 cm of sediment could be suspended into the water column, introducing both dissolved and particulate P. Zhu (2008) found that in the southwest central area, where wind-wave action is the most intense in Lake Taihu, there was no significant difference in the P content in pore water within the top 12 cm layer, suggesting there may be active exchange between the top 5–15 cm sediment layer and the overlying water. These results were similar to those obtained from the analyses of the distribution of P in pore water in some Lake Taihu areas in 1998 (Fan et al. 2000). According to Ruban et al. (2001), the potentially releasable P (NaOH+OP) in the sediment from Meiliang Bay accounted for 84% of sediment TP, and inorganic phosphorus (IP) was the main P fraction in the sediments of Lake Taihu. Jin et al. (2006) reported that for the sediment from the heavily eutrophic part of Taihu Lake, IP consisted mainly of NaOH-P (relative contribution of NaOH-P was 72%, 41%, 24%, and 12% of TP in Meiliang Lake, Wuli Lake, Gonghu Lake, and East Taihu Lake, respectively). In these areas, the P concentration increased rapidly with increasing pH; for example, P concentration increased from 1.38 mg kg^{-1} at pH = 7.14 to 201.09 mg kg^{-1} at pH = 11.99. This suggests that during the intense cyanobacterial bloom season, the higher pH, usually caused by the uptake of CO_2 during photosynthesis, may lead to more P release from the sediment (presumably from the NaOH-P fraction; Jin et al. 2006) into the lake water column.

Because large, shallow lakes are heavily influenced by wind-induced wave disturbance, it is difficult to estimate internal P loading directly (see Chapter 2). Kelderman et al. (2005) used mass balance budgets to estimate P accumulation in Taihu during 1998–2000 and found that 65% of the TP input reached the bed sediment, from where it has the potential to be released into the lake water. Qin et al. (2006) estimated internal P release rates in Lake Taihu to range from 0.22 to 5.67 mg m^2 d^{-1} under static conditions. Huang et al. (2016) used a dynamic P model that integrated hydrodynamic, wind-wave, and sediment transport to estimate very high TP release rates of 348 mg m^{-2} d^{-1} under a 5 m s^{-1} southeast wind and 364 mg m^{-2} d^{-1} under a 5 m s^{-1} northwest wind due largely to resuspension of P-rich sediment; overall, they estimated an average P release rate of 5.56 mg m^{-2} d^{-1}. The results of Qin's study were in reasonable agreement with the model results. Zhu et al. (2013) estimated the total algal available P pool at 5,168 tons, representing 30% of TP content in surface sediment. If the entire P pool was released into the overlying water, P concentrations in lake water would be increased by 0.844 mg L^{-1}, to almost seven times the current concentration. These studies indicated that the internal P loading of Taihu is, therefore, very important and must be addressed from a nutrient management perspective.

11.5 CONCLUSIONS

The P dynamics in the water of Lake Taihu are being driven both by external nutrient loading and internal cycling within the lake ecosystem. The P content in the lake sediment increased rapidly in recent years, suggesting that a reduction in internal P loading is both necessary and urgent in order to control cyanobacterial blooms in Lake Taihu.

11.6 ACKNOWLEDGMENTS

This research was jointly supported by National Natural Science Foundation of China (Grant No. 41877486), the "Major Science and Technology Program for Water Pollution Control and Treatment" of China (Grant No. 2018ZX 07208-008), Science and Technology Service Network Initiative (Grant No. KFJ-STS-ZDTP-038-3, and Open Research Fund of Jiangsu Province Key Laboratory of Environmental Engineering (Grant No. ZX2018007). We thank the Taihu Laboratory for Lake Ecosystem Research, and the Chinese Academy of Sciences for providing physicochemical and phytoplankton data.

11.7 REFERENCES

APHA, AWWA, WEF, 1995. Standard Methods for the Examination of Water and Wastewater, nineteenth ed. American Public Health Association, Washington DC.

Boström, B; Andersen, JM; Fleischer, S; and Jansson, M. 1988. Exchange of phosphorus across the sediment-water interface. Hydrobiologia. 170:229–244.

Chen, YW; Qin, BQ; Teubner, K; and Dokulil, MT. 2003. Long-term dynamics of phytoplankton assemblages: *Microcystis* domination in Lake Taihu, a large shallow lake in China. J Plankton Res. 25:445–453.

Fan, CX; Yang, LY; and Zhang, L. 2000. The vertical distributions of nitrogen and phosphorus in the sediment and interstitial water in Taihu Lake and their interrelations. J Lake Sci. 12:359–366 (In Chinese with English abstract).

Guo, L. 2007. Doing battle with the green monster of Lake Taihu. Science. 317:1166.

Hu, HJ and Wei, YX. 2006. The freshwater algae of China. Systematics, Taxonomy and Ecology. Science Press, Beijing (in Chinese).

Huang, L; Fang, HW; He, GJ; Jiang, HL; and Wang, CH. 2016. Effects of internal loading on phosphorus distribution in the Taihu Lake driven by wind waves and lake currents. Environ Pollut. 219:760–773.

Huang, Y; Fernando, RC; Mynett, AE; Fan, C; Pu, P; Jiang, J; and Dai, Q. 2001. The Water Environment and Pollution Control of Lake Taihu. Sciences Press (in Chinese), Beijing.

Jin, X; Wang, SR; Pang, Y; and Wu, FC. 2006. Phosphorus fractions and the effect of pH on the phosphorus release of the sediments from different trophic areas in Taihu Lake, China. Environ Pollut. 139:266–295.

Kelderman, P; Zhu, W; and Maessen, M. 2005. Water and mass budgets for estimating phosphorus sediment water exchange in Lake Taihu (China P.R.). Hydrobiologia. 544:167–175.

Lai, GL and Yu, G. 2007. A modeling-based assessment study on nutrients transport in Taihu basin during 1960s. J Grad School Chinese Acad Sci. 24:754–764 (in Chinese).

Li, Y; Wu, DN; and Xue, YX 19.98. A development sequential extraction method for different forms of phosphorus in the sediments and its environmental geochemical significance. Mar Environ Sci. 17:15–20 (In Chinese with English abstract).

Otsuki, A; Iwakuma, T; Kawai, T; and Aizaki, M. 1984. The trends in eutrophication of Lake Kasumigaura. Res Rep Natl Inst Environ Stud. 51:1–10 (in Japanese with English abstract).

Pápista,É; Ács,É; and Böddi, B. 2002. Chlorophyll-a determination with ethanol—a critical test. Hydrobiologia. 485:191–198.

Qian, YC and He, P. 2009. An analysis of the changes in the water quality in Taihu basin during 1998–2006. J Jiangxi Agr Univ. 31:370–374 (in Chinese).

Qin, BQ; Hu, WP; Chen, WM; et al. 2004. Process and Mechanism of Environmental Changes of the Taihu Lake. Science Press, Beijing (in Chinese).

Qin, BQ; Xu, PZ; Wu, QL; Luo, LC; and Zhang, YL. 2007. Environmental issues of Lake Taihu, China. Hydrobiologia 581:3–14.

Qin, BQ; Zhu, GW; Gao, G; Zhang, YL; Li, W; Paerl, HW; and Carmichael, WW. 2010. A drinking water crisis in Lake Taihu, China: linkage to climatic variability and lake management. Environ Management. 45(1):105–112.

Qin, BQ; Zhu, GW; Zhang, L; Luo, LC; Gao, G; and Gu, BH. 2006. Estimation of internal nutrient release in large shallow Lake Taihu, China. Science in China: Series D Earth Sciences. Supp. I 49:38–50.

Ruban, V; Lopez-Sanchez, JF; Pardo, P; Rauret, G; Muntau, H; and Quevauviller, Ph. 2001. Harmonized protocol and certified reference material for the determination of extractable contents of phosphorus in freshwater sediments-A synthesis of recent works. Fresen J Anal Chem. 370:224–228.

Ruttenberg, KC. 1992. Development of a sequential extraction method for different forms of phosphorus in marine sediments. Limnol Oceanogr. 37:1460–1482.

Schelske, CL. 2009. Eutrophication: focus on phosphorus. Science 324:722.

Schindler, DW. 1974. Eutrophication and recovery in experimental lakes: implications for lake management. Science 184:897–899.

Seitzinger, SY. 1991. The effect of pH on the release of phosphorus from Potomac Estuary sediments: implication for blue-green algal blooms. Estuar Coast Shelf Sci. 33:409–18.

Sterner, RW. 2008. On the phosphorus limitation paradigm for lakes. Int Rev Hydrobiol. 93:433–445.

Su, XM; Xue, QJ; Steinman, AD; Zhao, YY; and Xie, LQ. 2015. Spatiotemporal dynamics of microcystin variants and relationships with environmental parameters in Lake Taihu, China. Toxins 7:3224–3244.

Wang, JJ; Pang, Y; Li, YP; Huang, YW; Jia, JJ; Zhang, P; and Kou, XP. 2014. The regularity of wind-induced sediment resuspension in Meiliang Bay of Lake Taihu. Water Sci Technol. 70(1):167–174.

Wang, JJ; Pang, Y; Li, YP; Huang, YW; and Luo, J. 2015. Experimental study of wind-induced sediment suspension and nutrient release in Meiliang Bay of Lake Taihu, China. Environ Sci Pollut Res. 22:10471–10479.

Xie, LQ and Xie, P. 2002. Long-term (1956–1999) changes of phosphorus in a shallow, subtropical Chinese lake with emphasis on the role of inner ecological process. Wat Res. 36:343–349.

Xie, LQ; Xie, P; and Tang, HJ. 2003. Enhancement of dissolved phosphorus release from sediment to lake water by Microcystis blooms, an enclosure experiment in a hyper-eutrophic, subtropical Chinese lake. Environ Pollut. 122:391–399.

Xu, H; Paerl, HW; Qin, BQ; Zhu, GW; and Gao, G. 2010. Nitrogen and phosphorus inputs control phytoplankton growth in eutrophic Lake Taihu, China. Limnol Oceanogr. 55:420–432.

Yu, JJ; Yin, HB; Gao, YN; and Tang, WY. 2017. Characteristics of nutrient and heavy metals pollution in sediments of Taihu watershed. China Environ Sci. 37(6):2287–2294.

Zhu, GW. 2008. Sediment-Water Exchange and Its Significance. Qin, B.Q. (Ed.), Lake Taihu, China. Chapter 4: 151–196. Springer Science Business Media.

Zhu, MY; Zhu, GW; Li, W; Zhang, YL; Zhao, LL; and Gu, Z. 2013. Estimation of the algal-available phosphorus pool in sediments of a large, shallow eutrophic lake (Taihu, China) using profiled SMT fractional analysis. Environ Pollut. 173:216–223.

CHAPTER 12

LOCH LEVEN, UK: LONG-TERM (1985 TO 2016) PHOSPHORUS DYNAMICS IN A SHALLOW LAKE AND ITS IMPLICATIONS FOR WATER-QUALITY MANAGEMENT

Bryan M. Spears[1,2], Stephen C. Ives[1], and Linda May[1]

Abstract

Loch Leven is a shallow eutrophic lake in the United Kingdom with a history of eutrophication problems. Here, we document the response of internal phosphorus (P) cycling within the lake to a reduction in catchment P load from 5.25 mg total P (TP) m^{-2} d^{-1} (1985) to between 1.44 mg TP m^{-2} d^{-1} and 2.39 mg TP m^{-2} d^{-1} (1995, 2005, 2015). Since 1989, internal loading has resulted in elevated summer TP concentrations, although the magnitude of the summer peak load has varied between about 3.5 and 0.3 mg P m^{-2} d^{-1}. This variation can be explained, at least partly, by fluctuations in spring wind speed, temperature, and summer precipitation, mediated through a series of complex ecological and biogeochemical interactions. We present an empirical model that predicts the effects of future climate change on internal loading and offer recommendations for the development of short-term forecasting approaches linked to large-scale teleconnection indices, such as the North Atlantic Oscillation (NAO).

Keywords: Sediment, eutrophication, internal loading, recovery, climate change, teleconnection, land-use change, catchment management.

12.1 INTRODUCTION

Loch Leven (see Figure 12.1) is a large (13.3 km^2 surface area), shallow (mean depth 3.9 m), eutrophic freshwater lake in east-central Scotland (latitude 56°10′N, longitude 3°30′W). The lake has high conservation value (Site of Special Scientific Interest, Ramsar, Special Area of Conservation, Natura

[1] Centre for Ecology & Hydrology Edinburgh, Penicuik, Midlothian, EH26 0QB, UK.

[2] Corresponding author: Bryan Spears Email spear@ceh.ac.uk.

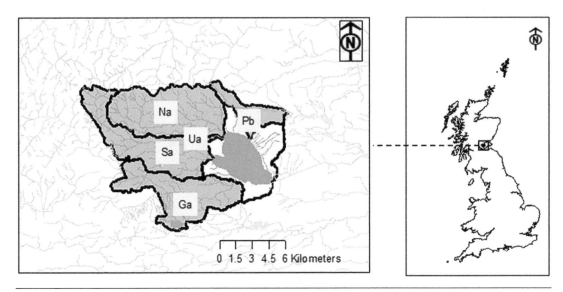

Figure 12.1 Map of Loch Leven showing its location in Scotland and the surface water catchment delineated into sub-catchments included in this study; these are drained by the Ury Burn (Ua); South Queich (Sa); Pow Burn (Pb); North Queich (Na); and the Gairney Water (Ga). The outflow is to the southeast of the lake.

2000) as an important breeding ground for overwintering waterfowl (Carss et al. 2012) and as good habitat for aquatic plants (Dudley et al. 2012) and fish (Winfield et al. 2012). Loch Leven has a long and well-documented history of management to improve the fishery and reduce eutrophication problems (May and Spears 2012). In previous decades, management efforts have focused, mainly, on the reduction of P loading from the catchment to improve water quality (May et al. 2012). In addition, the Loch Leven Long-Term Monitoring Programme has continued, since 1968, to produce one of the most comprehensive limnological data sets in the world for studying ecosystem scale responses to environmental change.

By the 1970s and 1980s, decades of elevated P inputs to Loch Leven from its catchment had resulted in high water column total P (TP) concentrations, increased phytoplankton biomass, and decreased water clarity (Bailey-Watts and Kirika 1999; Carvalho et al. 2012). As a result, macrophyte diversity and extent had declined (Dudley et al. 2012; May and Carvalho 2010). In an attempt to improve water quality, measures to reduce external P inputs to the loch by 60% were implemented between the 1970s and 1990s (Bailey-Watts and Kirika 1999; May et al. 2012; Spears and May 2015). Specifically, these included the control and diversion of P inputs from a woollen mill resulting in an estimated reduction of 6.3 metric tons TP yr^{-1} (1.3 mg TP m^{-2} lake surface area d^{-1}) by 1987, and the introduction of P stripping facilities at four wastewater treatment works achieving an estimated further reduction of 3.3 metric tons yr^{-1} (0.7 mg TP m^{-2} d^{-1}) between 1993 and 1997 (May et al. 2012). The lake responded to these measures very slowly, but by 2008 it had shown a marked improvement in water quality, including a reduction in TP concentrations, a lower phytoplankton biomass, and an increase in the diversity and extent of macrophytes (Spears et al. 2012; Spears and May 2015). This suggests a recovery period of at least 20 years at this site (1989–2008) (Sharpley et al. 2013).

It is generally accepted that this prolonged recovery has been the result of re-equilibration processes (*internal loading*), whereby sediment P that has accumulated over periods of high external loading is released into the overlying water column during the period of recovery in shallow lakes (Sas 1989). However, little is known of the factors that have regulated the magnitude and frequency of internal loading in Loch Leven since then, while the catchment load has been low and relatively stable. In shallow lakes, it has been hypothesized that internal loading may be disrupted (capped) relatively quickly through the establishment of natural buffering systems at the sediment-water interface (e.g., macrophytes and benthic algae) (Scheffer 2001; Mehner et al. 2008; Spears et al. 2008). However, during the period of recovery, Loch Leven has also been affected by climate change, as indicated by the warming of its surface waters (0.7°C per decade) at a rate that is higher than that of the associated air temperature (O'Reilly et al. 2015). It is unclear to what extent climate and land use stressors interacted to regulate internal loading in Loch Leven during the recovery period, and in turn, whether variations in local weather conditions were regulated by regional scale atmospheric phenomena, a process known as *teleconnection*. Indicators of relevant large-scale weather phenomena include the North Atlantic Oscillation (NAO), the Arctic Oscillation (AO), and the East Atlantic Index (EAI), all of which are expected to relate to fluctuations in weather in the UK (Comas-Bru and McDermott 2014; Rust et al. 2018). In general, positive phases of the NAO are associated with warmer, wetter, and stormier weather conditions than negative phases. Strongly negative phases of the AO index are characterized by a higher likelihood of cold and wintry conditions in the UK compared to strongly positive phases. Finally, positive phases of the EAI are characterized by a greater likelihood of drier winters in the UK compared to negative phases. The NAO is considered to be the primary teleconnection phenomenon affecting weather patterns in the UK, although interactions between the NAO, AO, EAI, and other teleconnection processes are expected to occur.

If significant relationships between these indicators, local weather conditions, and the intensity of internal loading in Loch Leven are found, then these would need to be taken into account in future management plans at this site, and may be relevant to other lake management programmes where P mitigation, alone, is the current objective. The objectives of this chapter are: (1) to review catchment loading and P-retention data from surveys conducted in 1985, 1995, 2005, and 2015/2016 to assess responses in loading to land-use change; (2) to examine water column TP fluctuations, associated with internal loading processes, allowing the estimation of within-year (2004–2005) and longer-term (1989–2016) variations in the magnitude of internal P loading during recovery; and (3) to investigate the drivers of variation in internal P loading against long-term data on weather and water quality during the recovery period.

12.2 METHODS

12.2.1 Estimating Catchment P Load and P Retention

Data from catchment P loading surveys conducted in 1985, 1995, 2005, and 2015/16 were extracted from the Loch Leven Long-Term Monitoring database for inflow streams draining five sub-catchments and for the lake outflow (L). The inflow streams were the Ury Burn (Ua), the South Queich (Sa), the Pow Burn (Pb), the North Queich (Na), and the Gairney Water (Ga) (see Figure 12.1). Methods for collection of water samples, P analyses, and hydrological determination are described in detail by May et al. (2012) and May et al. (2017). In general, water samples were collected at eight-day intervals in all years.

For each year, estimates of the daily, monthly, and annual inputs of TP, filterable reactive P (FRP) and filterable unreactive P (FURP) from the monitored streams were calculated using the linear interpolation methods of Stevens and Smith (1978) and Ferguson (1986), as outlined by Defew (2008), Spears and May (2015), and May et al. (2017). These data were used to produce annual estimates of P retention for each year, with monthly resolution, by subtracting the mass of P entering the lake through these inflows from the mass of P leaving via the outflow, with estimates of point sources included.

The 1985, 1995, and 2005 surveys commenced in January and ceased in December. In 2015/2016, the survey commenced in July 2015 and ceased in June 2016. For consistency of presentation, the data are presented here as January to June 2016 and July to December 2015. As such, they are not representative of a calendar year, in contrast to the earlier surveys.

Collectively, the inflows that were sampled drained an estimated 84% of the surface water catchment of the lake. However, during the years included in this analysis, the volume of water entering the lake through these inflows accounted for about 93% of the volume of water leaving the lake via its outflow. Rain falling directly onto the surface of the lake and groundwater sources were not included in this study because it was assumed that they represented only a minor source of water to the lake (Bailey-Watts and Kirika 1999).

12.2.2 Within-Year Variation in Sediment-Water P Flux Estimates

Diffusive FRP flux estimates across the sediment-water interface were estimated roughly monthly between April 2004 and April 2005. Intact sediment cores were collected from six sites along a depth transect (2 m, 2.5 m, 3.5 m, 5.5 m, 10 m, and 22 m overlying water depth). Cores were collected from a boat using a Jenkin Surface Sediment sampler and returned to the laboratory for processing the same day. Bottom water samples were removed from about 1 cm above the sediment surface using a syringe and filtered using a Whatman GF/C filter prior to FRP analysis. The upper 3 cm of sediment was extruded and homogenized for centrifugation prior to filtration, as previously mentioned, for FRP analysis. Diffusive flux was estimated using Fick's Law and corrected for water temperature on each sample date using the Stokes-Einstein equations as described by Sinke et al. (1990), with tortuosity calculated following Fisher and Reddy (2001), and diffusive flux coefficients taken from Lerman (1979). The flux calculations follow the same general approach as that described in more detail by Tammeorg et al. (see Chapter 10). General Linear Modeling, Minitab version 17, was used to assess the independent and interaction effects of season and depth on soluble reactive P (SRP) flux, following log_{10} transformation of the flux estimates.

We acknowledge that there are limitations in all methods for estimating sediment P flux with respect to drawing inferences on whole-lake functioning (see Chapter 2). With respect to the limitations of diffusive flux estimates using concentration gradients, we have shown previously that the biological community living at the sediment-water interface may retard the apparent diffusive flux as calculated here (Spears et al. 2008). We present these results in the context of flux estimates made using laboratory-controlled intact sediment core incubation assays for Loch Leven during the same sampling period (Spears et al. 2007, 2012). For comparison, ranges of FRP flux from assays conducted using cores collected from sediment at water depths of 3.5 m on August 8, 2005, August 23, 2005, September 6, 2005, and April 18, 2006 are shown. Assays, conducted under ambient environmental conditions at the time of sampling (described in detail by Spears et al. 2007; Spears et al. 2008), consisted of 24-hour incubations during which SRP concentrations in the core were measured at the start and end of the 24-hour incubation, with the difference in concentration being used to estimate flux. All flux estimates were corrected for sediment surface area and expressed in units of mg FRP m^{-2} d^{-1}.

12.2.3 Estimating Long-Term Variations in Internal Loading

Long-term annual internal loadings were estimated using a water column mass balance approach. The magnitude of the May–September internal loading peak (I-Load$_{M-S}$) was corrected for lake volume to estimate the mass of P release (metric tons) required to effect the observed change in water column TP concentration, corrected for lake surface area and time, to give I-Load$_{M-S}$ load in units of mg P m^{-2} d^{-1} (Spears et al. 2006, 2008). Although we did not subtract the catchment load, since it was not measured in most years, we can confirm from the data collected in 1985, 1995, 2005, and 2015/2016 that catchment load during this period was low (Spears and May 2015). We assume that internal loading is the main source of P to the water column during this period. An estimate of I-Load$_{M-S}$ was made for each year during the *recovery period* of Loch Leven, that is, following significant catchment management as previously described (1989–2016).

12.2.4   Identifying the Long-Term Drivers of Internal Loading

The effects of water temperature, precipitation, and wind speed on the magnitude of I-Load$_{M-S}$ were assessed between 1989 and 2014 for the spring and summer means of each year. For each dataset, statistical models were developed to predict the response variables as a function of two main stressor effects and their interaction, within the framework of linear mixed effects models (LMEs). All response variables were modeled with Gaussian errors. The exact form of the model's fixed and random effects varied depending on the dataset structure. However, the full LME specification was,

$$y = b_0 + b_1x_1 + b_2x_2 + b_3x_1x_2 + S + Y + \varepsilon \qquad \text{[Eq. 12.1]}$$

where y is the response variable, x_1 and x_2 are two stressor covariates, the b_n terms are the model fixed effect coefficients and ε is the residual error. Normally distributed random effects were included for site (S) and year (Y).

All models were fitted by maximum likelihood using the lmerTest R package (Kuznetsova et al. 2017) when random effects were required, or the R lm function otherwise (R Core team 2018). Prior to model fitting, the response variables and covariates were transformed to normal distributions using Box-Cox transformations, offset by a small value to ensure that all values of the variable were greater than zero. This ensured that the models met assumptions of normality of residuals, checked by examining model residual plots.

For each dataset, a set of candidate stressor variables was identified. To identify the best combinations of stressor variables to use, all possible model combinations with up to three fixed effects were fitted and the one with the lowest Akaike Information Criteria (AIC) was selected.

From the best-fitted models, risks of the response variable exceeding the target value for in-lake TP, as set by the Loch Leven catchment management group, were evaluated across both stressor gradients and visualized as a heat map.

The heat maps were constructed by calculating exceedance probabilities from the model for a range of stressor combinations. For any values of the two stressors, the model states that observed values of the response variable are normally distributed with a mean of $b_0 + b_1x_1 + b_2x_2 + b_3x_1x_2$ and variance of $\sigma_{\frac{2}{S}} + \sigma_{\frac{2}{Y}} + \sigma_{\frac{2}{\varepsilon}}$, where $\sigma_{\frac{2}{S}}$ is the site-level random effect variance, $\sigma_{\frac{2}{Y}}$ is the year-level random effect variance, and $\sigma_{\frac{2}{\varepsilon}}$ is the residual variance.

12.3   RESULTS

12.3.1   Long-Term Variation in External P Loading and Retention Following Catchment Management

The external TP load (see Table 12.1) declined from 5.3 mg TP m^{-2} d^{-1} (25.5 t TP yr^{-1}) in 1985 to 1.4 mg TP m^{-2} d^{-1} in 1995 (7 t TP yr^{-1}), increased again to 2.1 mg TP m^{-2} d^{-1} (11.6 t TP yr^{-1}) in

2005, and decreased slightly in 2015/2016 to 2.0 mg TP m^{-2} d^{-1} (11.2 t TP yr^{-1}). The reduction between 1985 and 1995 was attributed to measures implemented within the catchment that resulted in the control of point sources of P. These included the upgrading of wastewater treatment works (WWTWs) and the controlling of an industrial point source that discharged P-rich effluent from a woolen mill. The latter, which was situated in the South Quiech sub-catchment, was identified as being rich in FURP, which is evident from the significant reduction in FURP load in the South Quiech between 1985 and 1995; the FURP load remained relatively constant thereafter. A reduction in FRP load to the lake between 1985 and 1995 is an apparent response to P-stripping measures implemented in the WWTW in the catchment.

Changes in the whole lake net gain (i.e., P in > P out) and loss (i.e., P in < P out) of P are also apparent in response to catchment management (see Figure 12.2). In 1985, during the period of high catchment loading, TP, FRP, and FURP were all retained throughout the year and at high levels relative to subsequent years, with the exception of October for TP and January for FRP. In subsequent

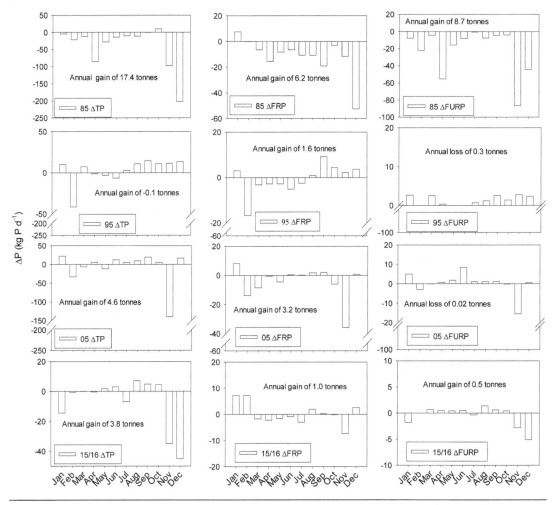

Figure 12.2   Monthly average estimates of whole lake gain (negative values) and loss (positive values) of TP, FRP, and FURP in Loch Leven in 1985, 1995, 2005, and 2015/16. The annual retention values include estimates from point sources discharging directly to the lake.

years, levels of TP retention were generally lower than in 1985, although the lake continued to retain TP on an annual basis in all years. Some seasonality is apparent in the whole lake retention of FRP and TP in 1995, 2005, and 2015/16, with net loss being more likely to occur in summer and autumn, and retention more likely to occur in winter and spring. It is likely that the whole lake net P loss in the summer and autumn of 1995, 2005, and 2015/16 was the result of internal-loading processes during a period when inflowing discharge and catchment loading were characteristically low. Conversely, the retention of P in the winter and spring was probably the result of catchment loading—this representing the period when inflows, discharges, and TP loads were characteristically high. The retention of FURP represented a significant proportion of the TP balance in 1985, with annual net P loss of FURP being reported in 1995 and 2005.

12.3.2   Water Column TP Concentrations in Response to Variations in External Load

The period of high external loading (1968–1989) was characterized by high baseline monthly water column TP concentrations (> 40 µg L^{-1}) and sporadic peaks (max. 250 µg L^{-1}), which appeared to mask seasonality in P concentrations in comparison to periods of low catchment nutrient loading (see Figure 12.3a). A change in the water column TP signal was evident post-1989, when a more regular seasonal pattern emerged that was characterized by high TP concentrations in summer and autumn, relative to winter and spring. This was attributed to the reduction in catchment loading and the onset of internal loading within the lake. The magnitude of this summer-autumn peak fluctuated between 1989 and 2017 and was lowest during 2008–2012 (40 to 60 µg L^{-1}), stabilizing in recent years at moderate levels (about 100 µg L^{-1}). A gradual decline in baseline TP concentration occurred between the 1990s (about 50 µg L^{-1}) and 2017 (about 20 µg L^{-1}).

These variations are most apparent when presented as average monthly TP values across the major management and recovery periods of the lake (see Figure 12.3b). Prior to the reduction of catchment loading (i.e., 1968–1975), water column TP concentrations were high throughout the year with little seasonality apparent. During the period of catchment management (i.e., the 1980s and 1990s), water column TP concentrations generally decreased in winter and spring and increased in summer and autumn, representing an apparent switch in the dominant source of TP to the water column from the catchment (winter-spring) to the bed sediments (summer-autumn). The intensity of this summer-autumn peak then reduced, alongside further reductions in winter and spring concentrations, as recovery proceeded (1990s to 2000s). Conditions remained relatively stable thereafter (i.e., 2006 to 2016).

12.3.3   Variation in the Intensity of Internal Loading Following a Reduction in External Load

Mass balance estimates of the rate of internal loading between May and September (I-Load$_{M-S}$) were compared with the annual TP load estimates during the recovery period (i.e. 1995, 2005, and 2015/16) (see Table 12.1). When compared to catchment (external) load estimates, I-Load$_{M-S}$ was more than double in 1995, similar in 2005, and about half in 2015/2016. Variability in I-Load$_{M-S}$ was apparent in the longer-term annual estimates during the recovery period (i.e. 1989–2016) (see Figure 12.4a) and where values were relatively high (i.e., between 2 mg TP m^{-2} d^{-1} and 3.5 mg TP m^{-2} d^{-1}) between 1989 and 1996, after which values remained below 1.75 mg TP m^{-2} d^{-1}, with the exception of 2004 to 2006 (about 2.0 mg TP m^{-2} d^{-1}). I-Load$_{M-S}$ was below 1 mg TP m^{-2} d^{-1} and stable between 2008 and 2012 and fluctuated between 0.5 mg TP m^{-2} d^{-1} and 1.75 mg TP m^{-2} d^{-1}

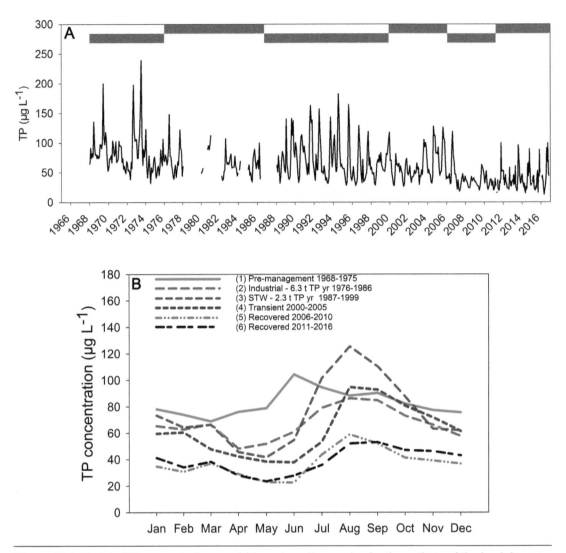

Figure 12.3   In-lake TP concentrations as (a) long-term time series for the entirety of the Loch Leven long-term monitoring period, (b) monthly averages across years within defined management and recovery periods, also indicated in (a) as staggered grey bars at the top.

between 2013 and 2016. Correlation analysis (n = 28; correlation coefficient = 0.78; p < 0.0001) confirmed that I-Load$_{M-S}$ was significantly correlated with annual mean water column TP concentration between 1989 and 2016 (see Figure 12.4b).

As a comparison with the annual I-Load$_{M-S}$ values presented in Table 12.1 and Figure 12.4 for the period 2004–2005 (i.e., about 2.0 mg FRP m^{-2} d^{-1}), we present estimates of FRP flux estimates from intact core incubations and using Fick's Law based on sediment-water FRP concentration gradients conducted in 2004 and 2005 (see Figure 12.5). Intact core flux estimates were conducted on cores collected from a single site (depth of about 3.5 m water depth) and ranged between −2.5 mg FRP m^{-2} d^{-1} and 25 mg FRP m^{-2} d^{-1} (median value about 12 mg FRP m^{-2} d^{-1}), with negative values indicating uptake of FRP from the water column to the bed sediments. Fluxes based on Fick's Law (roughly monthly monitoring frequency across the depths) indicated little variation in the median values

Table 12.1 TP, FRP, and FURP loading to Loch Leven from each of the five monitored inflows in 1985, 1995, 2005, and 2015–16. The surface area of sub-catchment drained by each inflow is shown. Sub-catchments: Ury Burn (Ua); South Queich (Sa); Pow Burn (Pb); North Queich (Na); Gairney Water (Ga). PS—point sources discharging directly to the lake that are not accounted for in the estimates of P loading from the inflowing streams. Estimates of internal load based on mass balance calculations between May and September (I-Load$_{m-s}$) for 2015–2016 taken as the average of estimates from both years; no estimate of I-Load$_{m-s}$ presented for 1985 as catchment sources were uncontrolled at this time and assumptions of summer catchment load being low were not valid in this year.

Sub-Catchment	(km²) Area	1985 (metric tons P yr⁻¹)			1995 (metric tons P yr⁻¹)			2005 (metric tons P yr⁻¹)			2015/16 (metric tons P yr⁻¹)		
		TP	FRP	FURP	TP	FRP	FURP	TP	FRP	FURP	TP	FRP	FURP
Ga	33.1	2.1	0.6	0.4	0.8	0.3	0.2	1.5	0.5	0.3	0.9	0.3	0.1
Na	41.8	6.9	2.9	0.9	2.1	1.2	0.3	3.9	1.4	0.5	2.3	0.9	0.3
Pb	10.4	1.7	0.4	0.1	0.4	0.2							
Sa	35.1	11.7	1.5	7.9	1.9	0.6	0.3	3.6	0.9	0.5	4.4	1.2	0.3
Ua	1.4	0.2	0.1	<0.1	<0.1	<0.1	<0.1						
PS	–	3.0	2.0	0.8	1.7	1.5	0.2	1.7	1.5	0.2	1.4	1.0	0.4
Catchment Load (t P yr⁻¹)	121.8	25.5	7.6	10.2	7.0	3.8	1.0	11.6	4.7	1.6	11.2	4.4	1.2
Catchment Load (mg P m⁻² d⁻¹)		5.25	1.57	2.10	1.44	0.78	0.21	2.39	0.97	0.33	2.02	0.7	0.2
I-Load$_{m-s}$ (mg P m⁻² d⁻¹)					3.57			2.14			1.15		

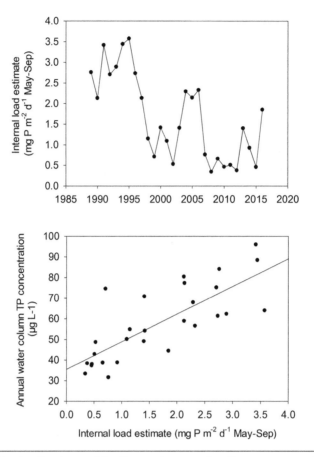

Figure 12.4   Internal load estimates as (a) a time series covering the recovery period, (b) correlated against annual average water column TP concentrations; correlation line shown (n = 28; correlation coefficient = 0.78; p < 0.0001).

(about 2.5 mg FRP m^{-2} d^{-1}) with depth. However, the range of values appeared to be lowest at intermediate water depth (3.5 m) and greatest at the deepest point (22 m). In addition, the 95th percentile values of these ranges indicated a general decrease from 2 m to 2.5 m and a general increase from 3.5 m to 22 m, with maximum flux rates of about 14 mg FRP m^{-2} d^{-1} estimated at 22 m. However, no significant variation was observed in these data in relation to depth or season.

12.3.4   Assessing the Drivers of Variation in Internal Loading

I-Load$_{M-S}$ varied markedly in the years following the reduction of external P loading, when external loading was relatively stable. Using the dredge analysis approach outlined before, we assessed the combinations of stressors that explained the highest degrees of variation in I-Load$_{M-S}$. This analysis indicated that summer precipitation and spring wind speed were the two most important stressors, in that order (see Table 12.2). No significant interaction term was reported between these two stressors and their combined effects on I-Load$_{M-S}$ are shown (see Figure 12.6). I-Load$_{M-S}$ was highest when mean summer precipitation was lowest and mean spring wind speed was highest.

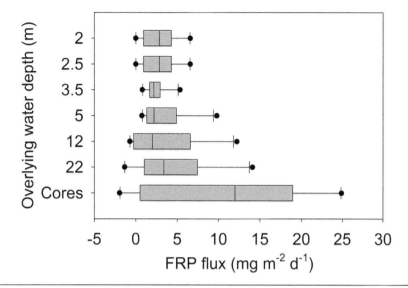

Figure 12.5   Variation in ranges of FRP fluxes estimated using Fick's Law in Loch Leven from monthly monitoring conducted April 2004 and April 2005. The range of SRP fluxes calculated from intact sediment core assays between August 2005 and April 2006 are indicated as *Cores*. Median lines are indicated within each box. The lower and upper boundaries of the box indicate the 25th and 75th percentile values, respectively. The lower and upper error bars represent the 5th and 95th percentile values, respectively. Outliers are also shown.

Table 12.2   Model outputs of the two best models: (a) describing the effects of summer precipitation and spring wind speed on internal TP load, residual standard error = 0.690 on 20 degrees of freedom, multiple R-squared = 0.567, adjusted R-squared = 0.524, F-statistic = 13.1 on 2 and 20 degrees of freedom, p-value = 0.0002 (this model was used in the construction of the heat maps); and (b) describing the effects of summer precipitation, spring wind speed, and spring water temperature on internal TP load, residual standard error = 0.649 on 19 degrees of freedom, multiple R-squared = 0.637, adjusted R-squared = 0.579, F-statistic = 11.09 on 3 and 19 degrees of freedom, p-value = 0.0002 (this model was weighted highest but was not used in heat map construction due to having 3 terms).

	Estimate	Std. Error	t value	Pr(>|t|)
Model a				
Intercept	1.96E-16	1.44E-01	0	1
Summer precipitation	−6.08E-01	1.49E-01	−4.072	0.000594
Spring wind speed	3.52E-01	1.49E-01	2.359	0.028591
Model b				
Intercept	−8.37E-17	1.35E-01	0	1
Summer precipitation	−4.44E-01	1.65E-01	−2.693	0.0144
Spring wind speed	3.01E-01	1.43E-01	2.101	0.0492
Spring water temperature	−3.20E-01	1.68E-01	−1.906	0.072

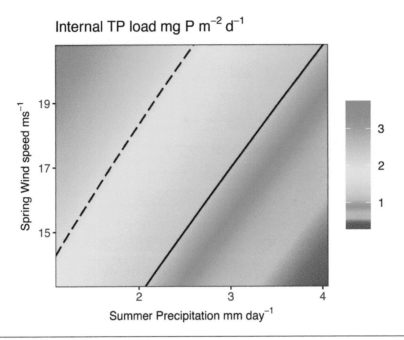

Figure 12.6 Contour plot (heat map) showing the effects of spring mean wind speed and summer mean precipitation on the expected response in internal loading. The full black line and the dotted line represent the estimated background annual load from the catchment to the lake in the early 1900s (1.08 mg P m⁻² d⁻¹) and the annual load in 2015 (2.02 mg P m⁻² d⁻¹).

However, multiple combinations of weather indicators could be used to return alternative, but less statistically robust, significant models to explain the variation in I-Load$_{M-S}$. In general, the dredge analysis indicated that water temperature could result in negative (stronger negative effect in spring, slight negative effect in summer) effects in I-Load$_{M-S}$, wind speed resulted in both positive (spring) and negative (summer) effects, and precipitation resulted in both positive (spring) and negative effects (summer).

The results of correlation analysis conducted on log$_{10}$ transformed data to assess covariation between indicators of atmospheric phenomena (data links: NAO: Jones et al. 1997, https://crudata. uea.ac.uk/cru/data/nao/; AO: Zhou et al. 2001, http://www.cpc.ncep.noaa.gov/products/precip/ CWlink/daily_ao_index/ao.shtml; and EAI: http://www.cpc.ncep.noaa.gov/data/teledoc/ea.shtml) that are known to regulate weather conditions in this region and the weather indicators discussed earlier in addition to estimates of I-Load$_{M-S}$ are reported (see Table 12.3). The primary stressor included in the model, mean summer precipitation, varied significantly and negatively with the spring average AO index, only. The secondary stressor in the model, mean spring wind speed, varied significantly and positively with the winter mean NAO index. The summer mean NAO index returned the strongest correlation, which was positive, with I-Load$_{M-S}$ although weaker correlations were also returned with the winter mean AO index (negative correlation), and the autumn and winter mean EA indices (both positive correlations). Other notable, strongly significant correlations include winter AO with winter (positive correlation) and summer (negative correlation) water temperature, and positive correlations between winter NAO index and winter precipitation, wind speed, and water temperature.

Table 12.3 Pearson's correlation coefficients between annual and seasonal average values of the EAI, the AO and the NAO index values with log₁₀ transformed values of internal-load estimates delivered between May and September of each year (I-Load$_{M-S}$; metric tons) and seasonal averages for each year between 1989 and 2013 of wind speed ('Wind'; km h⁻¹), precipitation ('Precip'; mm d⁻¹), and surface water temperature ('Temp'; °C). ***p < 0.01; **p: 0.01 to 0.049; *p: 0.05 to 0.1. Winter, spring, and summer values of wind speed, precipitation, and temperature were correlated with year-1 lag EAI, AO, and NAO index values as follows: winter versus winter, autumn y-1; spring y-1, summer y-1, spring versus spring, winter, autumn y-1; summer versus summer, spring, winter, autumn y-1; autumn versus autumn, summer, spring, winter; I-Load$_{M-S}$ versus winter, spring, summer, autumn. The values highlighted are included in the generalized linear model constructed to predict inter-annual variation in I-Load$_{M-S}$.

	Winter			Spring			Summer			Autumn			Sep–May I-Load
	Precip	Wind	Temp	Precip	Wind	Temp	Precip	Wind	Temp	Precip	Wind	Temp	
EA													
Winter	-0.01	0.05	0.55***	0.30	0.34	-0.20	-0.04	-0.01	-0.52***	-0.27	-0.36*	-0.44**	0.35*
Spring	0.07	0.11	-0.01	-0.01	0.17	0.10	0.28	0.21	-0.05	-0.14	0.20	0.27	-0.30
Summer	-0.4**	0.20	-0.05	0.23	-0.01	-0.39*	-0.14	0.09	-0.18	0.10	-0.27	0.07	0.29
Autumn	-0.02	0.25	0.18	-0.04	0.13	0.07	-0.23	-0.21	-0.34	0.01	-0.49**	0.07	0.37*
AO													
Winter	0.47**	0.33	0.05	-0.12	0.21	0.21	0.09	-0.25	0.16	0.09	-0.11	0.15	-0.31
Spring	0.22	-0.01	-0.22	0.07	-0.16	-0.10	-0.46**	-0.15	0.31	-0.02	0.19	0.36	0.13
Summer	0.39**	-0.10	-0.06	0.08	0.15	0.05	0.13	-0.04	-0.18	-0.18	-0.03	-0.21	-0.14
Autumn	0.38*	0.14	-2.3	0.32	0.08	0.06	0.22	0.15	-0.08	0.00	0.36*	-0.09	-0.05
NAO													
Winter	0.52***	0.58***	0.58***	0.13	0.54***	-0.14	-0.07	-0.44**	-0.32	-0.16	-0.18	-0.19	0.19
Spring	0.20	0.34	-0.01	0.10	0.22	0.05	0.16	0.06	-0.32	-0.15	0.04	0.00	0.17
Summer	0.00	0.01	0.05	0.15	0.28	-0.43**	-0.30	0.01	-0.07	-0.14	-0.18	-0.28	0.52***
Autumn	-0.19	0.35*	0.08	0.15	0.01	0.22	0.14	-0.20	-0.10	0.27	0.43**	0.23	0.10

12.4 DISCUSSION

12.4.1 Responses in Catchment Loading and P Retention to Catchment Management

The catchment management interventions conducted in Loch Leven were successful in reducing the overall catchment load. Reductions of between 13.9 t TP yr^{-1} and 18.5 t TP yr^{-1} (i.e., 2.5 to 3.3 mg TP m^{-2} d^{-1}, respectively), depending on which years are compared, are reported. However, the extent to which this change can be directly attributed to specific catchment interventions is questionable. The effects of the reduction in the P discharged from the woolen mill can be estimated, with relative confidence, as a result of the P being predominantly in the form of FURP and easily apportioned to the South Queich sub-catchment. The effects of the woolen mill P control measures resulted in the apparent whole-lake net FURP loss from the lake in 1995 and 2005. Spears and May (2015) reported evidence of homeostasis in FURP concentrations in Loch Leven, regardless of the load, suggesting internal processes capable of rapidly transforming high molecular weight P compounds to ortho-phosphate. Although we hypothesize that microbial communities are the main drivers here, the specific processes are not well described for lakes, but these organisms may play an important role in determining internal loading, especially in urban and shallow water bodies where FURP load may be relatively high.

The increase in FURP load between 2005 and 2015 indicates the emergence of a new, as yet unidentified, catchment source. The fact that FURP load is ubiquitous across all sub-catchments, and apparently high in the estimates of point sources, would suggest that domestic wastewater discharges may be a significant source of FURP. Domestic discharges in the Loch Leven catchment may contain high molecular weight P compounds associated with household products including, for example, cosmetics, and soap detergents (Brownlie 2014).

Variation in FURP, FRP, and TP is apparent across the sub-catchments and survey years stressing the need for long-term monitoring following catchment management. For example, following the reduction in P load from the South Queich between 1985 and 1995, FRP load increased from 0.6 to 1.2 t yr^{-1}. A similar change is apparent for the Pow Burn. This period was characterized by urbanization of both sub-catchments and although no causative evidence is available to link this land-use change to the increase in FRP load, it warrants further investigation.

In our surveys we will have captured the effects of yearly variations in weather and also changes in catchment land use, which would be expected to occur over the 30-year monitoring period. In comparison to estimates of historical catchment loading in the early 1900s (i.e., 6 t TP yr^{-1}) (May et al. 2012), the reduced load remains elevated at almost double the historical load. This is also evident in the P retention analysis where TP retention has persisted throughout the monitoring period; in 2015, about 30% of the catchment P load was retained by the lake. Also evident is the seasonality of P retention in the lake, where a large proportion of whole lake net P gain occurs in winter months when the catchment load is highest with net P loss occurring more commonly in summer and autumn, during periods when the load to the water column is predominantly from the lake-bed sediments. Nonetheless, the reduction in TP load from the catchment was significant ecologically, resulting in increased macrophyte colonization depth (May and Carvalho 2010) and reduced phytoplankton biomass (Carvalho et al. 2012). In addition, the responses presented here regarding in-lake TP demonstrate the characteristic recovery trajectory of shallow temperate lakes following catchment load reduction. This is characterized by a reduction in winter and spring TP concentrations coupled with an increase in summer and autumn concentrations. As recovery proceeds, summer and autumn concentrations slowly decline.

12.4.2.   Long-Term Responses in Internal Loading Following a Reduction in Catchment Loading

Estimates of sediment P flux representing the concentration change in the lake water between May and September indicate a general decrease following catchment load reduction from about 50% of the catchment flux estimates in 1995 to about 10% of the catchment flux in 2015/16. This decline, relative to catchment load, is the result of both increasing catchment load over this period and decreasing internal load. Given that sediment P stocks remained relatively stable between 1990 and 2004, it is unlikely that the reduction in internal loading during this period was the result of diminished mobile P in the sediments (Spears et al. 2006), although this remains untested.

Spears et al. (2007) studied in detail the processes regulating sediment P release in Loch Leven and reported that dissolved oxygen concentration was a dominant regulator of bottom-water FRP concentration, especially in warmer summer months, when P release increased during periods of anoxia. Phytoplankton detritus played a key role in returning labile P to the bed sediments and in replenishing organic P pools, which in turn were closely linked with labile P concentrations in the surface sediments. As surface-water P concentrations have declined, so has phytoplankton biomass, presumably reducing the amount of organic P reaching the bed sediments and resulting in an overall reduction in internal P loading. Using undisturbed sediment core incubations, Spears et al. (2007) demonstrated that bed sediments, high in mobile P content, would release FRP under aerobic conditions when bottom water concentrations were below 180 μg L^{-1} to 270 μg L^{-1}, and that bottom water concentrations in 2004/2005 did not exceed 100 μg L^{-1}.

The studies previously reviewed indicate the potential for continual release of FRP from the bed sediments of Loch Leven, with rates increasing during periods of anoxia, which occurs in < 5% of the surface area of the lake-bed area, during summer with autumn turnover. This is in agreement with the analysis of diffusive flux presented here. We also note that the I-Load$_{M-S}$ estimates are expressed per unit surface of lake and that release rates may be variable across specific basins depending on local conditions. We found no significant effect of season on FRP flux, and efflux from sediments was reported across all depths and seasons. Confirming the likelihood of elevated sediment P efflux in deeper sediment zones during periods of anoxia, we observed the highest flux rates in the deepest waters during summer.

Returning to the whole-lake scale, our long-term mass balance estimates of internal loading over the May-to-September period in Loch Leven indicated relatively high fluxes following catchment load reduction prior to a sharp decrease over a five-year period between 1995 and 2000. After this period, internal loading fluctuated with apparently sporadic peaks lasting up to three years in duration. The relative contribution of internal load to catchment load has decreased in recent years, mainly as a consequence of reduced rates of internal loading. However, given that catchment loading is low during summer and autumn months, that is, the growing season for primary producers, internal loading still plays a dominant role in driving water quality in the lake. This was confirmed by a strong correlation between I-Load$_{M-S}$ and annual water column TP concentration following catchment load reduction. Previous work (Spears et al. 2012) has demonstrated that current sediment P stocks are sufficient to support much higher flux rates than reported here; it is apparent that factors other than those considered previously must have been driving the long-term variation at the whole-lake scale during this period.

12.4.3   Drivers of Internal P Loading during the Recovery Period

Several studies have considered the effects of large-scale drivers on the water quality and ecology of Loch Leven in recent years. Spears and Jones (2011) proposed that increased wind-induced wave mixing could increase the magnitude of internal loading through more frequent bed disturbance leading to reduced macrophyte cover. The effects of turbulent bed disturbance can reduce the establishment of *Chara sp.* and *Potamogeton pusillus* by 91% and 45%, respectively, under laboratory conditions (Van Zuidam and Peeters 2015). Spears et al. (2012) produced a multi-stressor model associating spring wind mixing and summer air temperature with internal loading. Carvalho et al. (2012) reported a link between water temperature in the spring and *Daphnia* abundance leading to more intense clear-water conditions in spring and also proposed that increased precipitation in summer can result in increased losses of phytoplankton downstream, leading to lower in-lake chlorophyll *a* concentrations, through flushing. Taken collectively, these studies indicate that multiple weather stressors may play a key role in regulating Loch Leven water quality.

The models presented here confirm that spring wind speed, water temperature, and summer precipitation all play a significant role in regulating internal loading in Loch Leven following catchment load reduction. The underlying processes need to be further validated; however, the results are in general agreement with those hypotheses previously presented, with the exception of spring water temperature. Here we present a negative effect of spring water temperature on internal loading. Spears et al. (2007) reported a positive effect of summer water temperature on diffusive flux in both experimental incubations and analysis of field data. In our models, it is likely that spring water temperature is driving some other key process that is difficult to determine here. For example, cooler spring periods may result in fewer *Daphnia* in spring, leading to higher summer phytoplankton biomass accrual and organic P in surface sediments, along with higher internal loading in summer. Regardless, summer precipitation is clearly the dominant factor regulating internal loading in Loch Leven—most likely as a result of removal of phytoplankton biomass from the lake, although the delivery of oxygenated water to anoxic bed sediments may also be important.

12.4.4   Future Management Considerations for Loch Leven

As we consider the future management implications of the work, we are acutely aware of the need to predict or forecast water-quality changes to meet society's needs. In Loch Leven, assuming catchment P load remains relatively stable, it appears that managing the effects of weather variations and climate change should play a more prominent role in the design of management plans. With respect to short-term forecasting, we present correlations to link multiple weather indicators with large-scale teleconnection processes known to drive regional and local scale weather conditions in the east of Scotland. Of the parameters included in our models, Spears and Jones (2011) have previously demonstrated a direct link between the intensity of winter and spring wind mixing with the NAO in Loch Leven. This is confirmed in our study using a longer period of data. Summer precipitation correlates negatively with the spring phase of the AO and spring temperature correlates negatively with the summer phase in the previous year of the EA and the NAO. These responses are in general agreement with the documented relationships between these teleconnection indices and regional weather patterns as described in Section 12.3. The lagged associations open the potential for the development of forecasting models with which to predict the effects of weather on internal loading and summer water quality in Loch Leven.

By 2050, climate change is expected to result in a 1–2°C rise in annual and summer average daily temperature in the east of Scotland, at the 10% probability level and assuming a medium emissions scenario (UKCP09 SRES A1B; Nakićenović et al. 2000). Under the same scenario, summer and winter precipitation are predicted to decrease by 20%–30% and increase by 0%–10%, respectively. More frequent and intense rainfall events are also predicted. O'Reilly et al. (2015) report a contemporary warming rate of about 0.7°C per decade in Loch Leven surface waters—one of the fastest rates of warming reported for a lake in the world. We have no predictions for wind speed, either annually or seasonally for Scotland. A decrease in summer rainfall of 20%, assuming a current median precipitation value of 2.5 mm d^{-1}, would result in an increase in the annual TP load of about 0.5 mg TP m^{-2} d^{-1} to Loch Leven.

We stress that our model does not, necessarily, provide evidence to support climate change mitigation measures through further reduction in catchment load. This is because a reduction in catchment load in summer may be extremely difficult to achieve given this represents the season when the catchment load is typically lowest. One potential option for mitigation during summer is to manipulate the underlying process reported in the models, flushing rate, or wind-induced bed disturbance. If the lake was allowed to fill during the predicted wetter winter/spring conditions and this water was released in summer/autumn, then it is possible that the magnitude of internal loading would be reduced (Spears et al. 2006). For wave mixing, coastal protection structures may be considered; for example, created islands, wetlands, or dikes can be used to dissipate wave energy, perhaps also enhancing habitat for waterfowl or producing a source of renewable energy (Borsje et al. 2017). Finally, the addition of geo-engineering materials to control sediment P flux directly has been demonstrated recently in lakes (Spears et al. 2016; Lürling et al. 2016).

These potential climate change mitigation approaches require further consideration in line with a range of other measures, including the need to further reduce catchment P loads. In addition, the predicted increase in magnitude and intensity of precipitation events is likely to result in an increase in P load from the catchment in winter and spring months (May et al. 2017); hence, mitigation measures in the catchment and the lake may be necessary to meet water quality targets in the future.

12.5 CONCLUSIONS

Results from Loch Leven highlight two very important responses in lake water quality following catchment load reduction. First, improvements in water column TP concentration during the recovery period can be slow, highly variable, and at times counterintuitive. Second, the variation in the intensity of internal loading can be predictable and in Loch Leven is currently being driven by variations in weather linked to large-scale teleconnection processes, through a complex interaction between summer precipitation, spring wind speed, and spring water temperature, in order of decreasing influence. All three stressors are expected to change with climate change in this region. We present an empirical model to demonstrate the potential range of effects associated with climate change predictions in this region of Scotland in relation to internal loading in Loch Leven.

12.6 ACKNOWLEDGMENTS

We are grateful to the Loch Leven Trustees for daily outflow records. We are indebted to the Montgomery family, the owners of the loch, for their continuing support for our research. This research was funded mainly by the Natural Environment Research Council, UK, with contributions from Scottish Natural Heritage, The William Grant Foundation, the Scottish Environment Protection Agency, and Perth and Kinross Council. BMS was supported through funding from the Scottish Government RESAS Strategic Programme 2018–2019. SI was supported through a NERC funded PhD through the E3 Doctoral Training Partnership.

12.7 REFERENCES

Bailey-Watts AE and Kirika A. 1999. Poor water quality in Loch Leven (Scotland) in 1995, in spite of reduced phosphorus loading since 1985: the influences of catchment management and inter-annual weather variation. Hydrobiologia. 403:135–151.

Borsje, BW; de Vries, S; Janssen, SKH; Luijendijk, AP; and Vuik, V. 2017. Building with nature as coastal protection strategy in the Netherlands. In: Bilkovic, DN; Mitchell, MM; La Peyre, MK; and Toft, JD, (Eds.). Living Shorelines. The science and management of nature based coastal protection. CRC Press. p. 137.

Brownlie, W. 2014. Assessing the role of domestic phosphorus emissions in the human phosphorus footprint. PhD Thesis. Heriot-Watt University, School of the Built Environment. United Kingdom. p 217.

Carss, D; Spears, BM; Quinn, L; and Cooper, R. 2012 Long-term variations in waterfowl populations in Loch Leven: identifying discontinuities between local and national trends. Hydrobiologia. 681(1):85–104.

Carvalho, L; Miller, C; Spears, BM; Gunn, IDM; Bennion, H; Kirika, A; and May, L. 2012. Water quality of Loch Leven: responses to enrichment, restoration and climate change. Hydrobiologia. 681(1):35–47.

Comas-Bru, L and McDermott, F. 2014. Impacts of the EA and SCA patterns on the European twentieth century NAO—winter climate relationship. Q J Roy Meteorol Soc. 140:354–363.

Defew, LH. 2008. The influence of high-flow events on phosphorus delivery to Loch Leven, Scotland, UK. University of Edinburgh, School of Geosciences, PhD Thesis, p. 276.

Dudley, B; Gunn, I; Carvalho, L; Proctor, I; O'Hare, MT; Murphy, KJ; and Milligan, A. 2012. Changes in aquatic macrophyte communities in Loch Leven: evidence of recovery from eutrophication? Hydrobiologia. 681:49–57.

Ferguson, RI. 1986. River loads underestimated by ratings curves. Water Res. 22:74–76.

Fisher, MM and Reddy, KR. 2001. Phosphorus flux from wetland soils affected by long-term nutrient loading. J Environ Qual. 30:261–271.

Jones, PD; Jónsson, T; and Wheeler, D. 1997. Extension to the North Atlantic Oscillation using early instrumental pressure observations from Gibraltar and South-West Iceland. Int J Climatol. 17:1433–1450.

Kuznetsova, A; Brockhoff, PB; and Christensen, RHB. 2017. lmerTest Package: Tests in Linear Mixed Effects Models. J Stat Softw. 82(13):1–26.

Lerman, A. 1979. *Geochemical Processe: Water and Sediment Environments*. New York (NY): John Wiley and Sons, Inc.

Lürling, M; Mackay, E; Reitzel, K; and Spears, BM. 2016. A critical perspective on geo-engineering for eutrophication management in lakes. Water Res. 97:1–10.

May, L and Carvalho, L. 2010. Maximum growing depth of macrophytes in Loch Leven, Scotland, United Kingdom, in relation to historical changes in estimated phosphorus loading. Hydrobiologia. 646:123–131.

May, L; Defew, LH; Bennion, H; and Kirika, A. 2012. Historical changes (1905–2005) in external phosphorus loads to Loch Leven, Scotland, UK. Hydrobiologia. 681:11–21.

May, L; Moore, A; Woods, H; Bowes, M; Watt, J; Taylor, P; and Pickard, A. 2017. Loch Leven nutrient load and source apportionment study. Scottish Natural Heritage Commissioned Report; oai:nora.nerc.ac.uk:517798.

May, L and Spears, BM. 2012. Managing ecosystem services at Loch Leven, Scotland, UK: actions, impacts and unintended consequences. Hydrobiologia. 681(1):117–130.

Mehner, T; Diekmann, M; Gonsiorczyk, T; Kasprzak, P; Koschel, R; Krienitz, L; Rumpf, M; Schulz, M; and Wauer, G. 2008. Rapid recovery from eutrophication of a stratified lake by disruption of internal nutrient load. Ecosystems. 11:1142–1156.

Nakicenovic, N; Alcamo, J; Davis, G; de Vries, B; Fenhann, J; Gaffin, S; Gregory, K; Grübler, A; Yong Jung, T; Kram, T; et al. 2000. IPCC Special Report on Emission Scenarios. Cambridge University Press, UK.

O'Reilly, CM; Sharma, S; Gray, DK; Hampton, SE; Read, JS; Rowley, RJ; Schneider, P; Lenters, JD; McIntyre, PB; Kraemer, BM; et al. 2015. Rapid and highly variable warming of lake surface waters around the globe. Geophys Res Lett. 42: doi:10.1002/2015GL066235.

R Core Team, 2018. R: A Language and Environment for Statistical Computing. R Found. Stat. Comput. Vienna, Austria. URL http//www.R-project.org/.

Rust, W; Holman, I; Corstanje, R; Bloomfield, J; and Cuthbert, M. 2018. A conceptual model for climate teleconnection signal control on groundwater variability in Europe. Earth-Sci Rev. 177:164–174.

Sas, H. 1989. *Lake Restoration by Reduction of Nutrient Loading*. Academic Verlag Richarz, Germany.

Scheffer, M. 2001. *Ecology of Shallow Lakes*. London (UK): Kluwer Academic Press.

Sharpley, A; Jarvie, HP; Buda, A; May, L; Spears, BM; and Kleinman, P. 2013. Phosphorus legacy: overcoming the effects of past management practices to mitigate future water quality impairment. J Environ Qual. 42(5):1308–1326.

Sinke, AJC; Cornelese, AA; Keizer, P; Van Tongeren, AFR; and Cappenberg, TE. 1990. Mineralisation, pore water chemistry and phosphorus release form peaty sediments in the eutrophic Loosdrecht Lakes, The Netherlands. Freshwater Biol. 23:587–599.

Spears, BM; Carvalho, L; Perkins, R; Kirika, A; and Paterson, DM. 2006. Spatial and historical variation in sediment phosphorus fractions and mobility in a large shallow lake. Water Res. 40(2):383–391.

Spears, BM; Carvalho, L; Perkins, R; Kirika, A; and Paterson, DM. 2007 Sediment phosphorus cycling in a large shallow lake: spatio-temporal variation in phosphorus pools and release. Hydrobiologia. 584:37–48.

Spears, BM; Carvalho, L; Perkins, R; Kirika, A; and Paterson, DM. 2012. Long-term variation and regulation of internal phosphorus loading in Loch Leven. Hydrobiologia. 681(1):23–33.

Spears, BM; Carvalho, L; Perkins, R; and Paterson, DM. 2008. Effects of light on sediment nutrient flux and water column nutrient stoichiometry in a shallow lake. Water Res. 42(4–5):977–986.

Spears, B and Jones, I. 2010. The long-term (1979–2005) effects of the North Atlantic Oscillation on wind-induced wave mixing in Loch Leven (Scotland). Hydrobiologia. 646(1):49–59.

Spears, BM; Mackay, E; Yasseri, S; Gunn, IDMG; Waters, KE; Andrews, C; Cole, S; De Ville, M; Kelly, A; Meis, S; et al. 2016. A meta-analysis of water quality and aquatic macrophyte responses in 18 lakes treated with lanthanum bentonite (Phoslock). Water Res. 97:111–121.

Spears, BM and May, L. 2015. Long-term homeostasis of filterable un-reactive phosphorus in a shallow eutrophic lake following significant reduction in catchment load. Geoderma. 257–258:78–85.

Stevens, RJ and Smith, RV. 1978. A comparison of discrete and intensive sampling for measuring river loads of nitrogen and phosphorus in the Lough Neagh system. Water Res. 11:631–636.

Van Zuidam, BG and Peeters, ETH. 2015. Wave forces limit the establishment of submerged macrophytes in large shallow lakes. Limnol Oceanogr. 60:1536–1549.

Winfield, IJ; Adams, CE; Armstrong, JD; Gardiner, R; Kirika, A; Montgomery, J; Spears, BM; Stewart, DC; Thorpe, JE; and Wilson, W. 2012 Changes in the fish community of Loch Leven: untangling anthropogenic pressures. Hydrobiologia. 681(1):73–84.

Zhou, S; Miller, AJ; Wang, J; and Angell, JK. 2001. Trends of NAO and AO and their associations with stratospheric processes. Geophys Res Lett. 28:4107–4110.

CHAPTER 13

BARTON BROAD, UK: OVER 40 YEARS OF PHOSPHORUS DYNAMICS IN A SHALLOW LAKE SUBJECT TO CATCHMENT LOAD REDUCTION AND SEDIMENT REMOVAL

Geoff Phillips[1], Andrea Kelly[2], Jo-Anne Pitt[3], Bryan M. Spears[4]

Abstract

Barton Broad is a very shallow lowland eutrophic lake, located in the Broads National Park, UK. The lake has been monitored over the last 40 years, during which time a series of restoration measures have been undertaken. This started with the control of phosphorus (P) from wastewater treatment works (WWTW) to reduce P input to the lake, followed by the removal of the upper 30–50 cm of sediment by suction dredging to increase water depth for recreation and potentially to reduce internal P load. The catchment P load to the lake decreased from over 10 g total P (TP) m^{-2} y^{-1} to 2.5 g TP m^{-2} y^{-1} with consequent reduction in water column total P (TP) and soluble reactive P (SRP).

During the period before and after dredging, a substantial number of intact sediment cores were collected and we present an analysis of data from these, demonstrating that both the sediment TP and interstitial pore water SRP decreased in a linear pattern since the start of catchment management. In addition, the sediment exposed by dredging had lower TP and interstitial SRP concentrations, although subsequent sedimentation increased sediment TP for a period of up to eight years. P release from the sediment in the dredged area could still be demonstrated during the summer, but rates were significantly lower than those from the undredged areas. Our results suggest that during the summer, the lake still had a significant internal P load, resulting in a small annual export of P from the lake; while the lake water column achieved a new equilibrium TP concentration, that of the sediment may still be declining.

Keywords: Sediment, eutrophication, internal loading, recovery, retention, dredging.

[1] Biological and Environmental Sciences, University of Stirling, Stirling, UK, FK9 4LA. Email geoff.phillips@stir.ac.uk.

[2] The Broad's Authority, Yare House, 62-64 Thorpe Road, Norwich, Norfolk, NR1 1RY

[3] Environment Agency, Horizon House, Bristol BS1 5AH

[4] Centre for Ecology & Hydrology Edinburgh, Penicuik, Midlothian, EH26 OQB, UK

13.1   INTRODUCTION

Barton Broad is the second largest of the Norfolk Broads with an area of approximately 70 ha and an average depth of 1.3 m. Lying directly on the navigable River Ant, Barton Broad is well flushed, with an annual retention time of approximately 16 days and a summer retention time of 25 days (Phillips et al. 2005). The water body is used for navigation and is recognized for its conservation status via designation as a Site of Special Scientific Interest, a Special Area of Conservation, as a National Nature Reserve, and as a RAMSAR site.

The site has 40 years (1975 to 2015) of documented management, including various trials, to address the causes and symptoms of eutrophication (see Table 13.1). Over this period, regular water-quality monitoring has been conducted providing a comprehensive record of responses to the management activities. Of particular interest with respect to P dynamics is an assessment of the responses in sediment and water column P concentrations to both the effluent diversion and P stripping activities conducted in wastewater treatment works discharging upstream of the Broad, and sediment removal from the lake itself.

By the 1970s, it was well established that P enrichment, mainly from WWTWs, was the major cause of changes to lakes worldwide (Harper 1992; Sutcliffe and Jones 1992). Work carried out in the Broads during the 1970s confirmed eutrophication as the primary cause of the deleterious changes that had occurred in these shallow lakes (Moss 1983). While there was an initial reluctance to accept the need for controlling external nutrient loads, evidence of the success of such action elsewhere, for example in Lake Washington (Edmondson 1996), combined with a detailed study of the nutrient status of Barton Broad (Osborne and Moss 1977), resulted in the installation of an experimental reduction of P from the effluent discharged by a WWTW just upstream of Barton Broad (see Table 13.1). Shortly after this, effluent from a larger WWTW further upstream was diverted to a coastal outfall, removing a substantial amount of the P load to Barton Broad (Phillips et al. 1999).

Sediment accumulation rates were high following the onset of eutrophication in Barton Broad (Osborne and Moss 1977) and internal P loading from sediment release (Osborne and Phillips 1978; Osborne 1980) was thought likely to delay recovery. In 1995, following catchment P load reduction and investigations of sediment P release (Phillips et al. 1994), a new restoration program called *Clear Water 2000* was initiated that aimed to apply the latest scientific understanding to restore the lake using grab and suction dredging. Dredging started in 1996, working from the river inflow in discrete areas to the lake outflow. A 30–50 cm depth of silt was removed, aiming to expose a *marl* layer, deposited when the lake was dominated by submerged aquatic vegetation (Osborne and Moss

Table 13.1   Chronology of management activities influencing Barton Broad

Year	Description
1978	Experimental P removal from Stalham WWTW
1980	Diversion of effluent from North Walsham WWTW
1983	P removal from small WWTW discharging to River Ant
1996–2000	Sediment removal from Barton Broad
1996	Enhanced P removal at Stalham WWTW
1998	P removal from industrial source discharging to River Ant

1977) and likely to contain plant propagules. By December 2000, over 80% of the lake area had been dredged and bathymetric (echo-sound) surveys confirmed the volume of sediment removed. Suction dredging, although more costly and time consuming than grab dredging, was the preferred option for this site for several reasons: low turbidity creation; achieving an even bottom contour; and enabling an efficient transfer of material from the lake bed to land (Klein 1998). Suction dredging enabled direct transfer of material made up of 1:3 parts water to sediment, a distance of ~1000 m to land. Settlement lagoons were constructed on agricultural land on a slight incline, using the topsoil to form bunds. When full, and after shrinkage with drying, each lagoon had ~1 m depth of sediment. This sediment was broken up and mixed with subsoil. The topsoil was spread over the surface and the land was planted with a conditioning crop of oilseed rape or wheat before returning to commercial agricultural use.

Here we use the Barton Broad case study to examine sediment and water column TP responses to sediment removal activities and also to consider recommendations for future sediment-removal monitoring activities. The objectives of this chapter are to assess:

1. Long-term responses in catchment P loading and retention in Barton Broad and corresponding seasonal and annual responses in water column TP and SRP;
2. Seasonal dynamics of sediment pore water P and its relationship to sediment P release to the water column; and
3. Long-term responses of sediment P resulting from both reduced catchment loading and sediment removal.

13.2   METHODS

13.2.1   Estimating Catchment P Load, Water Column SRP and TP Concentrations, and Lake P Retention

Subsurface samples for analysis of water chemistry were collected from a central open water point in Barton Broad, from 1978 to 2001 at approximately fortnightly intervals during March to October and monthly during the winter, and from 2002 to 2015 at monthly intervals all year. Water samples also were collected on the same sample occasions from the lake's main inflow, the River Ant, approximately 1.8 km upstream of the Broad (Hunsett Mill) (see Figure 13.1), with the exception of 1983–1986. The analysis of water column TP and SRP followed standard procedures, as described by Osborne (1978) and further outlined by Phillips and Kerrison (1991). Data prior to this sample period were provided by Osborne (1978) and were determined using comparable methods from similar sites.

The P loads entering and leaving the Broad were estimated by taking the point concentration data for the inflow at Hunsett Mill and the Broad itself multiplied by the daily mean river discharge on the day of sampling from a gauge station at Honing Lock, 5.8 km upstream from Hunsett Mill, corrected for catchment area. River sampling was not carried out between 1983 and 1986, but for other years, the difference between inflow and outflow loads were calculated to determine relative rates of P retention or loss from the sediment, expressed as an internal P load in mg P m^{-2} d^{-1}.

TP and SRP concentration and internal load data were binned into six-year periods starting from 1975 and average monthly concentrations were compared across these periods to assess variation in seasonality.

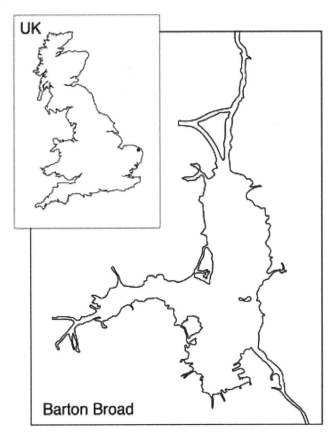

UK

Barton Broad

© Crown copyright [and database rights] 2018 OS 100021573. You are permitted to use this data solely to enable you to respond to, or interact with, the organisation that provided you with the data. You are not permitted to copy, sub-licence, distribute or sell any of this data to third parties in any form.

Figure 13.1   Left inset: location of Barton Broad (filled circle) within the UK and blow-up of Barton Broad.

13.2.2   Estimating Sediment TP Concentrations

Barton Broad has been the focus of sporadic and comprehensive sediment surveys (1987 to 2000). These surveys focused on identifying drivers and indicators of sediment P release, and involved the collection of short (typically 50 cm) sediment cores from one or more locations in the Broad using 7.0 cm diameter Perspex tubes. On most sampling occasions, 3–5 replicate cores were collected at each location sampled. Within 24 hours of collection, these cores were used to determine the interstitial pore-water SRP concentrations, and less frequently, total and different chemical fractions of P and total iron (Fe). Estimates of sediment P release rates were determined from experimental incubation of cores. Further details of the methods can be found in Phillips et al. (1994). Following the method of Davison et al. (1982), interstitial P concentrations were sampled from the fresh undisturbed sediment using hypodermic needles inserted into 3 mm holes pre-drilled in a spiral pattern at 1cm intervals. To avoid changes to Fe-bound P, the apparatus used to sample and filter the sampled pore water was flushed immediately prior to sampling with nitrogen gas. Total P concentration was determined on extruded sediment using an adaptation of the ignition method of Andersen (1976). On some cores, sequential extractions were also carried out using the method of Nürnberg (1988) to determine the mainly Fe-bound P (dithionite extraction), P bound to aluminum and metals in humic acids (sodium hydroxide extraction), and calcium bound P (hydrochloric acid extractable). Organic P content was calculated as the difference between inorganic and TP.

13.2.3   Experimental Incubation of Sediment Cores to Determine Release Rates

In 1988, 1992, 1999, and 2000, direct estimates of sediment P release were made by incubating three to five replicate cores in a growth chamber at ambient water temperature and light/dark cycle reflecting day length at time of sampling. In 1988 and 1992, the experimental approach used a flow-through system with water above the sediment being slowly exchanged (exchange time: 2.5 hours) using a peristaltic pump. In 1999 and 2000 a batch system was used with no water exchange. In both cases the overlying water was analyzed for SRP at regular intervals with experiments lasting typically 100–200 hours. P release rates were estimated from the average mass of P lost from the sediment to the overlying water; in the case of the batch method used in 1999 and 2000, the estimate was based on the change in SRP concentration in the water column during the first 24 hr of the experiment. In 1999 the interstitial water in the incubated cores was also sampled by inserting porous polymer tubes (*Rhizon* soil moisture samplers) horizontally into the sediment at several depth intervals (2, 6, 10, and 21 cm), allowing samples of interstitial water to be withdrawn during the experimental incubation.

13.2.4   Overview of Sampling Strategy and Approach to Statistical Analysis

In total, approximately 500 cores were collected from the sediment of Barton Broad between 1987 and 2000, the majority being from the summer period (May–September), but some in spring and winter. Samples were collected in 1987 and 1988 and then at approximately two-year intervals until 1994, after which samples were collected annually until 2000. From 1996, when sediment dredging started, samples were collected from both the dredged and undredged areas to allow comparisons of the effect of dredging.

All statistical analysis was carried out using R (R Development Core Team 2014). To compare the effects of dredging on both sediment P fractions and interstitial P, linear mixed effects analysis of variance using the lme function from package nlme (Pinheiro et al. 2018) were used with random intercept terms of *core* and *depth*. To quantify longer term changes in water column P, sediment TP, and interstitial P, generalized additive mixed models were fitted to all available data (1987–2000) using the generalized additive mixed model (gamm) function in the R package mgcv (Wood 2018). For the annual mean water column concentration and loading data, year was also included as an AR1 autocorrelation term. Models were fitted with thin plate regression spline smooths of *year*, for *sediment depth*, and *months since dredging*, and a cyclic cubic regression spline for *month*. Following an approach described by Simpson (2017), the fitted, smooth functions for both sediment TP and interstitial SRP with depth and month were compared between dredged and undredged sediment by fitting an additional model that used an additional ordered factor term enabling mgcv to fit smooths to the difference between a reference smooth—in this case the undredged sediment—and the dredged sediment. In the sediment models, random intercept terms were added as *site* and *core* replicates, models were fitted using maximum likelihood (ML) (Zuur et al. 2009), and the best model was selected using AIC. Comparisons of the composition of dredged and undredged sediments were made using a type II analysis of variance from the package car (Fox and Weisberg 2011) applied to a generalized least squares model, fitted using the gls function from nlme (Pinheiro et al. 2018). Initial

24-hour P release rates for the dredged and undredged sediments were made using analysis of variance. Linear models predicting lake TP from catchment and internal load were fitted using the gls function from nlme, with *year* included as an AR1 autocorrelation term; autocorrelation between covariables was checked by ensuring that the variance inflation factor (VIF) was less than 2.5 using the VIF function in package car.

13.3   RESULTS

13.3.1   Long-Term Responses in Catchment Nutrient Loading and Retention in Barton Broad

Following the introduction of P controls to effluents discharging to the River Ant and the diversion of a large WWTW effluent to a coastal outfall in 1980 (see summary in Table 13.1), the TP load to Barton Broad declined from more than 10 g P m^{-2} yr^{-1} to an average of 2.5 g P m^{-2} yr^{-1} by 2005 (see Figure 13.2). The trend of SRP concentration shows a similar relatively rapid decline in annual average concentrations reaching a minimum of 10 µg L^{-1} by 1990. The TP concentration, however, showed a more complex pattern than SRP. After an initial steep decline, the trend in the TP concentration leveled off in the late 1980s before declining again some 10 years later to reach a new minimum level by 2005 at about 70 µg P L^{-1}. The seasonal pattern of water column P remained

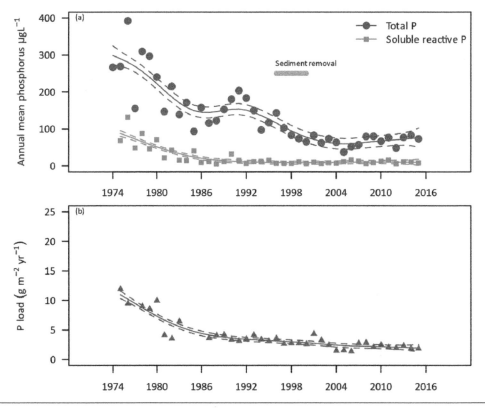

Figure 13.2   Trend in (a) annual mean total and soluble reactive phosphorus concentration (µg L^{-1}) in Barton Broad and (b) estimated upstream river load (g m^{-2} yr^{-1}) from 1975–2015, solid line fitted smooths, dotted line ± standard error.

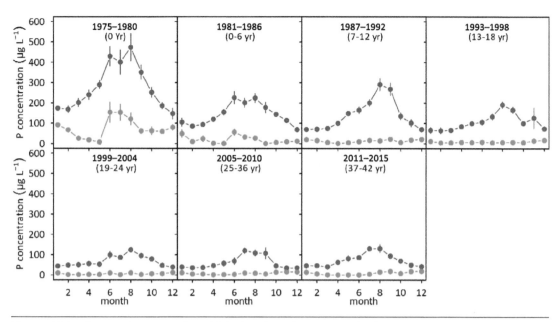

Figure 13.3 Seasonal pattern of total (blue) and soluble reactive (red) water column phosphorus concentration (μg P L⁻¹) for consecutive six-year periods from 1975–2015 (points represent monthly mean, vertical bars show ± standard error).

similar with TP peaking during the summer (see Figure 13.3). Prior to catchment control, and for the subsequent first six-year period, water column SRP was also elevated during summer (see Figure 13.3, first 2 boxes), despite substantial phytoplankton development (Phillips et al. 2005), but since 1987 has rarely been above 20 μg P L⁻¹ with an annual mean concentration of 10 μg P L⁻¹.

Mass balance estimates show that prior to catchment control of P (1975–1980), the Broad was a net P sink annually (see Table 13.2), but with a net gain to the water column (internal load caused by sediment-released P) occurring during the summer (June–September) (see Table 13.2 and Figure 13.4). For the following six-year period, net sediment P release occurred for most of the year (see Figure 13.4, 2nd box), but in subsequent periods was again restricted to the summer months. Overall, the annual net internal P load changed from net deposition prior to catchment reduction to net release. The highest rates of release occurred immediately after P control (1991–1996), but there remains a small net loss of P from the lake (see Table 13.2 and Figure 13.4, box 8).

The relative effect of this internal P load on annual water column TP is shown as a trajectory through time in Figure 13.5. Models predicting water column TP from internal load, which included a categorical variable representing different management periods (1975–1980: prior to catchment control; 1981–1986: prior to sediment removal; and 2000–2015: post sediment dredging and final catchment P reduction) showed no significant difference among the slopes for each period, demonstrating that the overall effect of the internal P load on annual mean TP did not change (see Table 13.3, Model 1). The intercept term of Model 1 represents catchment load and in comparison, the internal load had a relatively small effect on water column lake TP (see Figure 13.5). A multivariate model predicting water column mean TP from internal load, external load, and river discharge (see Table 13.3 Model 2) confirms this, as the standardized regression coefficient for internal load (0.30) is considerably lower than that for catchment load (0.77).

Table 13.2 Annual and summer (June–September) TP loads to Barton Broad from the catchment and from sediment for six-year periods before (0 Years, 1975–1980) and after catchment load reduction

		0 Years 1975–1980		+6 Years 1981–1986		+12 Years 1987–1993		+18 Years 1993–1998		+24 Years 1999–2004		+36 Years 2005–2010		+42 Years 2011–2015	
		mean	std err.	mean	std err.	mean	std err.	mean	std err.	mean	std err.	mean	std err.	mean	std err.
Catchment load (annual)	kg d⁻¹	19.0	± 1.1	9.1	± 1.7	7.1	± 0.3	6.4	± 0.4	5.5	± 0.8	4.3	± 0.5	3.9	± 0.2
	mg m⁻² d⁻¹	26.8	± 1.6	12.9	± 2.4	10.0	± 0.5	9.0	± 0.6	7.7	± 1.1	6.0	± 0.7	5.5	± 0.2
Catchment load (summer)	mg m⁻² d⁻¹	29.6	± 3.5	9.8	± 1.2	9.9	± 0.5	8.5	± 0.6	6.5	± 1.5	5.7	± 1.3	4.5	± 0.3
Internal load (annual net)	mg m⁻² d⁻¹	-1.9	± 2.1	2.6	± 1.3	1.8	± 0.3	-0.2	± 0.7	0.3	± 0.9	1.0	± 0.4	1.1	± 0.4
Internal load (summer net)	mg m⁻² d⁻¹	1.5	± 3.2	9.1	± 6.0	4.5	± 0.4	1.6	± 1.3	3.8	± 0.6	4.1	± 1.5	4.1	± 0.9

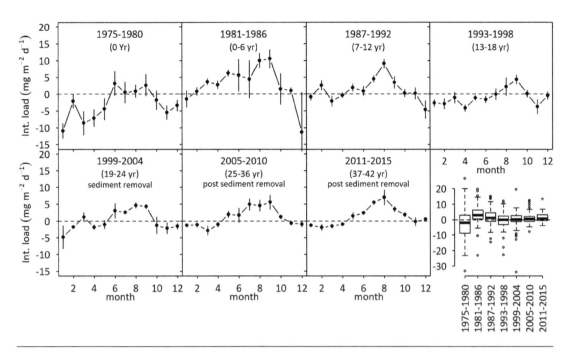

Figure 13.4 Seasonal pattern of net sediment phosphorus load (mg P m⁻² d⁻¹) determined from mass balance estimates of incoming and outgoing TP load for consecutive six-year periods from 1975–2015 (points represent monthly mean, vertical bars show ± standard error). The final box shows range of annual mean values for each period.

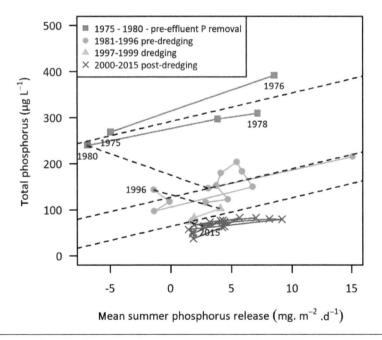

Figure 13.5 Scatter plot showing the trajectory of change (1975–2015) for annual mean water column TP concentration (μg P L⁻¹) and mean summer (May–September) internal P load (mg m⁻² d⁻¹). Lines show predicted fit from linear model, which included a categorical variable representing major periods of management actions (1975–1980: prior to catchment control; 1981–1986: prior to sediment removal; and 2000–2015 post sediment dredging and final catchment P reduction) (see Table 13.3).

Table 13.3 Model parameters for best-model predicting water column phosphorus from catchment and internal loading and river discharge. Model fitted using generalized least squares with year included as an autocorrelation AR1 term. (For Model 2, data were power transformed and centered).

	Coefficients	Std. Error	t-value	p-value
Model 1: pseudo R^2 = 0.92				
(Intercept) 1975–1980 pre-catchment P reduction	292.02	10.85	26.92	< 0.001
Summer internal load	6.19	0.76	8.11	< 0.001
Period 1981–1996 pre-dredge	−165.02	12.94	−12.75	< 0.001
Period 1997–2000 dredge	−227.82	17.12	−13.31	< 0.001
Period 2001–2015 post-dredge	−249.49	13.21	−18.88	< 0.001
Model 2: pseudo R^2 = 0.89 (standardized)				
(Intercept)	−0.02	0.04	−0.50	0.619
River discharge	−0.41	0.04	−9.04	< 0.001
Catchment TP load	0.77	0.05	15.38	< 0.001
Internal TP load	0.30	0.06	4.63	< 0.001
Interaction between catchment: internal TP loads	−0.11	0.06	−2.023	0.052

13.3.2   Seasonal Dynamics of Sediment Pore-Water P and Its Relationship to Sediment P Release

The upper sediment of Barton Broad is highly fluid, with a 90% water content and, in comparison to the water column (median 7.5 µg P L^{-1}), has a high SRP concentration (median 0.54 mg P L^{-1} interquartile range 0.15–1.5 mg P L^{-1}). Relatively frequent sampling carried out in 1987–1988 showed that the SRP concentration generally increased with sediment depth, although high concentrations typically occurred closer to the sediment-water interface in autumn (September–October) (see Figure 13.6). The lowest SRP concentrations occurred in June and July, the period when SRP peaks were present in the overlying water and when mass balances showed net sediment P release was greatest. This is surprising if it is assumed that diffusion is the primary mechanism of release. Experimental estimates of gross P release, using cores incubated at ambient conditions also showed maximum release rates in the same period and demonstrated a significant negative relationship with both the interstitial SRP gradient at the sediment-water interface (r^2 = 0.51; $p < 0.001$) and the mean interstitial SRP concentration (r^2 = 0.40; $p < 0.001$) (see Figure 13.7). The measured gross P release rates are much higher than the mass balance net estimates. The use of a flow-through system may, by continually removing P, maximize release rates, but the results suggest the potential for a substantial loss of P from the sediment that cannot be accounted for by diffusion.

13.3.3   Changes in Sediment P—the Relative Effects of Sediment Removal and Catchment Controls

Results from the analysis of sediment cores collected from dredged and undredged areas during the dredging period of 1996–2000 show that following sediment removal, there was a significant change in the TP and Fe content of the upper 20 cm of sediment. Phosphorus was lower, and Fe

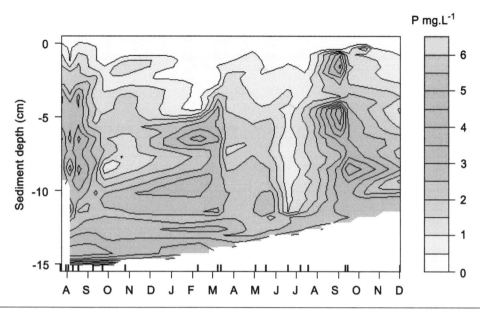

Figure 13.6   Heat map showing concentration of SRP (mg P L^{-1}) in the interstitial pore water of the surface sediment in Barton Broad (August 1987–December 1988); sampling occasions are shown as inner tick marks.

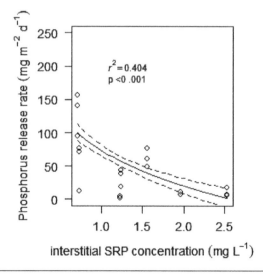

Figure 13.7   Relationship between experimental assessment of gross P release rate (mg m^{-2} d^{-1}) and mean interstitial pore-water SRP concentration (mg L^{-1}) of the upper 20 cm of sediment (1989 and 1992); the line shows log regression fit.

higher, resulting in an increase in the molar ratio of total Fe to TP from 7.6 to 18.9 (see Table 13.4) in sediment collected from the dredged areas. The TP changes resulted from a significant reduction in the organic P and a smaller reduction in the humic/metal-bound fractions (81.5% to 55% and 11% to 8%, respectively). The Fe and calcium-bound fractions did not change. There was a slight but

Table 13.4 Comparison of soluble and particulate forms of P and iron in the upper 20 cm of dredged and undredged sediment collected from Barton Broad during the dredging period of 1996–2000. (Analysis of deviance shows a significant difference between means from a mixed linear model with core identity and depth as nested random variables).

Metric	Dredged		Un-dredged		Analysis of Deviance	
	mean	std error	mean	std error	Chi Square	p
Interstitial SRP (mg L^{-1})	0.31	0.03	0.47	0.05	5.11	0.0238
Slope interstitial SRP at sediment water interface	−0.01	0.00	-0.02	0.00	9.40	0.0022
Interstitial Fe(II) (mg L^{-1})	0.58	0.05	0.58	0.05	0.06	0.8143
Excess SRP: Fe μmol	−3.80	1.40	15.78	5.34	4.51	0.0338
TP (mg g^{-1})	0.82	0.01	1.07	0.02	76.67	< 0.0001
Total Fe (mg g^{-1})	20.39	0.25	14.56	0.53	27.60	< 0.0001
Total Fe:P (molar ratio)	18.94	1.96	7.55	0.29	7.35	0.0067
Organic P (mg g^{-1})	0.55	0.03	0.81	0.03	7.77	0.0053
Fe-bound P (mg g^{-1})	0.20	0.02	0.18	0.02	0.25	0.6182
Calcium-bound (mg g^{-1})	0.09	0.01	0.09	0.01	0.09	0.7664
Aluminum- and humic-bound P (mg g^{-1})	0.08	0.01	0.11	0.01	5.85	0.0156

significant decrease in the SRP concentration of the interstitial pore water of the dredged sediment areas (0.47 mg L^{-1} to 0.31 mg L^{-1}) and a significant decrease in its concentration gradient at the sediment water interface. Unlike total Fe, the concentration of the pore-water ferrous Fe did not change, but the reduction of SRP concentration resulted in a significant reduction in the excess of pore-water SRP relative to soluble Fe. Sediment from the undredged areas had more SRP than Fe (15.8 μmol), whereas the reverse was true from the dredged areas (−3.8 μmol). Phosphorus flux, estimated from incubations of intact cores at ambient conditions, from dredged sediment was lower than from the undredged sediment, although only cores collected in June 1999 showed statistically significant differences (see Table 13.5). This P flux was significantly related to the excess interstitial pore-water SRP concentration relative to Fe in the upper 5 cm of sediment ($r^2 = 0.74$; $p < 0.001$).

Longer-term changes in sediment TP and interstitial pore-water SRP are shown in Figure 13.8 and Figure 13.9. The shape of the depth profiles and seasonal variation of the dredged and undredged sediments were similar, but the undredged sediment had more pronounced seasonal variation. We tested the differences in these trends by fitting an additional GAM model; significant differences are indicated where the confidence intervals do not overlap the zero line. The right-hand boxes of Figure 13.8 show these differences, from which it can be seen that the dredged sediment had a

Table 13.5 SRP release rates (mean ± standard error) from cores collected from undredged and dredged areas of Barton Broad during the summers of 1999 and 2000, and the relationship between SRP release rate and excess concentration of SRP relative to soluble Fe in the upper 5 cm of sediment and a factor representing dredged/undredged sediment. Dates are in format of month/day/year.

	P Release Rate (mg m^{-2} d^{-1})						F	p
	Dredged			Un-dredged				
05/11/1999	0.05	±	0.36	0.56	±	0.29	1.19	0.306
06/07/1999	0.66	±	0.11	1.95	±	0.35	12.66	0.007
06/21/1999	1.43	±	0.18	5.51	±	1.17	11.83	0.009
08/16/1999	7.01	±	1.63	12.70	±	6.82	0.66	0.440
06/27/2000	−0.38	±	0.61	2.17	±	1.25	3.38	0.103
08/08/2000	−0.05	±	0.05	1.84	±	0.84	5.07	0.054

	Coefficients	Std. Error	t value	Pr(> |t|)
Intercept (Dredged)	3.58	0.86	4.16	0.000212
Excess SRP:Fe (μmol)	222.82	23.05	9.67	3.75E-11
Un-dredged	0.10	1.22	0.08	0.936362

Multiple R-squared: 0.7542, Adjusted R-squared: 0.7393
F-statistic: 50.64 on 2 and 33 DF, p-value: 8.777e-11

slightly steeper depth gradient and less pronounced seasonal variation. There was also a clear linear decrease in the TP concentration over the whole period of monitoring from 1987–2007 (see Figure 13.8, box 7). By re-running the GAM model with *year* as a parametric rather than smoother term, we can estimate the slope (-0.02 ± 0.006, p = 0.0003); we calculate that, overall, the sediment TP content of the surface sediment of Barton Broad has been declining by approximately 0.02 mg P g^{-1} dry weight (dw) per year. It is also clear that while dredging exposes deeper sediment, with lower P content, over time this difference flattens such that after 8 years (96 months), the effect of dredging on the sediment TP content is reduced (see Figure 13.8, box 8).

Similar differences are seen in the pore-water SRP profiles of dredged and undredged sediments (see Figure 13.9). The most marked differences are in the depth profile (see Figure 13.9, boxes 1–3). SRP concentration in both dredged and undredged sediment increases with depth, but there is a much shallower SRP gradient at the sediment-water interface and almost uniform concentrations below 10 cm in the dredged sediment. As was the case for the sediment TP, the SRP in the undredged sediments varied significantly with the season, being high in spring and autumn and low in summer and winter, while in the dredged sediment the same pattern can be detected, but is barely significant (see Figure 13.9, box 5). There is also an overall linear decline in the average concentration of SRP (parametric model slope -0.099 ± 0.012, p < 0.0001) of approximately 0.01 mg L^{-1} per year (box 7), again showing a similar pattern to the TP.

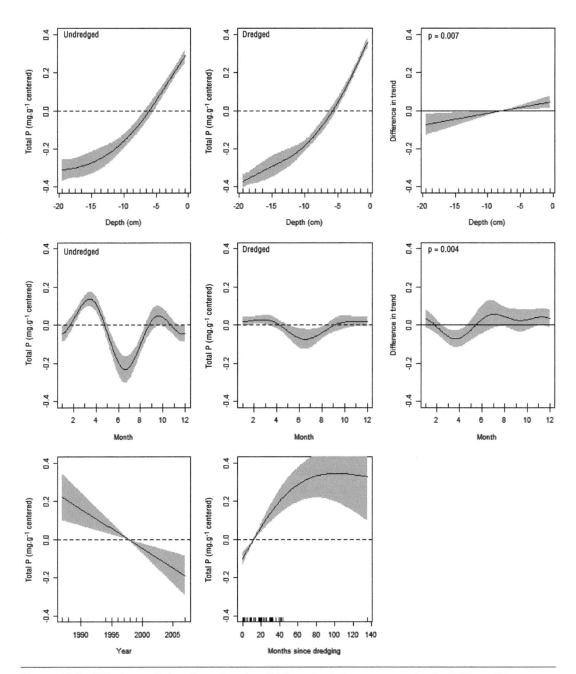

Figure 13.8 Fitted smooth functions (trend: solid line; shaded area: approximate 95% confidence intervals) of TP content of the sediment of Barton Broad (mg g^{-1} dw), and for depth and month the difference between the smooths fitted to dredged and undredged sediments. Top row compares trend with sediment depth for undredged and dredged sediment, middle row compares seasonal trends, bottom row shows overall temporal trend 1987–2007 and changes in dredged sediment since time of dredging (adjusted R^2, overall deviance explained by model = 0.51).

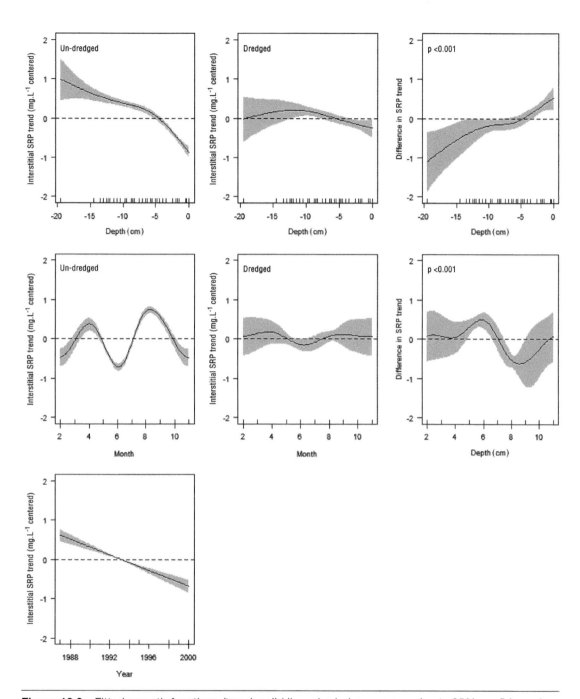

Figure 13.9 Fitted smooth functions (trend: solid line; shaded area: approximate 95% confidence intervals) of interstitial SRP content for the sediment of Barton Broad (mg L^{-1}) and, for depth and month, the difference between the smooths fitted to dredged and undredged sediments. Top row compares trend with sediment depth for undredged and dredged sediment, middle row compares seasonal trends, bottom box shows overall temporal trend 1987–2000 (adjusted R^2, overall deviance explained by model = 0.49).

13.4 DISCUSSION

13.4.1 Responses to Changes in Catchment Loading and Long-Term Responses in TP and SRP Concentration in the Water Column

It is now well established that the recovery of very shallow lakes from eutrophication is a slow process, typically requiring at least 15 years to achieve a new equilibrium condition (Jeppesen et al. 1991). Despite being a highly flushed riverine lake, Barton Broad is no exception to this as it was not until 2004 (24 years since catchment controls were largely completed) that it had reached a lower, stable TP concentration. Winter and spring TP concentrations declined more rapidly, with consequent changes in the seasonal pattern of phytoplankton (Phillips et al. 2005), but summer TP, although substantially lowered, remained relatively high as a consequence of internal P loading. Soluble reactive phosphorus concentrations changed more rapidly than TP. The substantially elevated concentrations seen during the summer, which were evidence of supply exceeding the uptake requirements of the phytoplankton (Osborne and Phillips 1978), disappeared within 12 years. This is likely a consequence of reduced availability of SRP from reduced catchment input, but also potentially an indication of reduced internal P loading.

13.4.2 Changes in P Retention

Osborne and Moss (1977) demonstrated that Barton Broad was a major P sink, accumulating P in the surface sediment, not only through deposition from the water column but also via a *bed load*, as sediment enriched with P from the river slowly entered the lake. Such a *bed load* almost certainly continues today, assisted by resuspension of sediment by frequent boat traffic (Hilton and Phillips 1982). In the slow-moving inflowing river, suspended inorganic particulate P, unavailable for phytoplankton growth, enters the Broad and is deposited within the lake. In contrast, the particulate P leaving the lake is almost entirely contained within phytoplankton cells, whose growth is supplied largely by SRP that is released from the sediment during the summer months (Osborne 1981). This makes interpreting simple mass-balance calculations as an indicator of internal P load difficult, but they provide an estimate of overall net retention and give indications of periods when the sediment is acting as a P source to the water column.

An earlier review of the long-term monitoring data for Barton Broad (Phillips et al. 2005) concluded from a mass balance that following the reduction of catchment P load, the lake had, after an initial period of net P export, again returned to being a net P sink, although at much lower rates than it had prior to catchment controls. They estimated that during the period of export, the lake had potentially lost (over a 10-year period) 2.3 mg P m^{-2}, equivalent to a reduction of 0.15 mg P g^{-1} dw of sediment in the upper 10 cm of the bed (0.015 mg P g^{-1} dw y^{-1}). More recent data, however, suggest that the lake has continued to be a slight net annual exporter of P, with a net summer internal P load exceeding deposition at other times of the year. Our analysis of the long-term TP content of the surface sediment suggests a linear decline of 0.02 mg P g^{-1} y^{-1}, which is slightly higher than the estimate of Phillips et al. (2005). Using our rate of decline and taking the overall mean sediment TP for the upper 20 cm of sediment (0.91 mg P g^{-1} dw) as indicative of conditions in 1998 (the intersection of the year trend line with 0 in Figure 13.8) allows us to calculate that in 1975, prior to catchment controls, the TP content of the upper sediment in the lake was approximately 1.4 mg P g^{-1} dw, which is within the range reported by Moss (1980) for cores collected between 1975 and 1977 at similar locations (1–2 mg P g^{-1} dw). Thus, it seems plausible that the sediment of Barton Broad has been continuing to release P to the overlying water, generating a net export of P throughout this period which has resulted in a measurable reduction in surface sediment P content.

13.4.3   Evidence of Reduced Internal Loading Following Sediment Removal

There is mixed evidence from the literature concerning the effectiveness of sediment removal on the control of internal P loading. Short-term studies generally show reduced P release rates (Kleeberg and Kohl 1999; Reddy et al. 2007), but there are fewer long-term studies with which the effectiveness of this approach can be assessed with confidence. It has been suggested that to successfully reduce internal P loading, sediment removal should expose a *new* sediment surface with a high P-binding capacity and low *release-sensitive* sediment P content (Søndergaard et al. 2007). Our data suggest that after the dredging in Barton Broad, deeper sediments with lower P and higher Fe content were exposed, resulting in the dredged areas having a surface sediment Fe:P ratio of 18.9 by mole (33 by mass) in comparison to 7.4 by mole (13 by mass) for the undredged areas. Both values are higher than the value of 8 by mole proposed by Jensen et al. (1992) to control P release, but the higher values in the dredged sediment might reduce the release even further due to the increased availability of Fe. However, Phillips et al. (1994) showed that in other very shallow lakes the highest rates of internal P loading were associated with high sediment total Fe:P ratios and that release occurred when there was an excess of soluble P relative to soluble Fe, a condition that occurred when hydrogen sulphide adsorbed the available Fe. There was no difference in the amount of Fe-bound P in the dredged and undredged sediment. Rather it was the organic P, the most abundant P fraction, which was reduced by dredging. There are insufficient data to explore changes in the different fractions of P with time, but our analysis shows that while the newly dredged areas had markedly lower P content, over time the fresh sediment deposition reduces the difference.

Bjork et al. (2010) and Liu et al. (2016) point out that for long-term benefits, sediment removal needs to be accompanied by reduction of external P loading, otherwise benefits are typically limited to a few years. Thus, it seems unlikely that dredging increased the binding capacity of sediment; rather, it was the removal of historically deposited P that lowered the overall TP content of the surface layers. Others have found varying evidence for increased binding capacity. Yu et al. (2017) found that the equilibrium P concentration (EPC_0) of dredged and undredged sediment from Lake Taihu was not significantly different, while Steinman and Ogdahl (2016) showed that the EPC_0 decreased following dredging of a re-wetted wetland area that had previously been used for agriculture, although due to changes in composition, the dredged sediments had a low P-binding capacity (Oldenborg and Steinman 2019). Thus, the benefits of sediment removal are perhaps more likely to be related to the removal of significant reservoirs of P rather than through the exposure of older sediment with higher P-binding capacity.

Sas (1990) demonstrated that sediment TP concentration can be used as an indicator of the likely extended recovery time, as a result of sediment P release, following external load reduction. Sas (1990) classified the sediment TP concentration in the upper 15 cm of lake sediments in relation to the estimated recovery time following reduction of external P loading. At concentrations of less than 1 mg TP g^{-1} dw, internal loading is expected to be negligible with moderate summer sediment P release events. At concentrations between 1 mg TP g^{-1} dw and 2.5 mg TP g^{-1} dw, net annual sediment P release will be high initially, with recovery expected within a five-year period; a high summer release event would be expected to occur that will be affected by pH, temperature, dissolved oxygen, and microbial activity. At concentrations in excess of 2.5 mg TP g^{-1} dw, net annual sediment

P release will occur for more than five years; in this situation, sediment P release is expected all year round and will be greatly influenced by pH, dissolved oxygen, and microbial activity.

Prior to catchment controls, the sediment of Barton Broad was likely approaching Sas's upper category, suggesting recovery times in excess of five years. Thus, by the late 1990s, after more than 15 years of catchment control, when dredging started, it is perhaps not surprising that the sediment P had already declined to about 1 mg TP g^{-1} dw, entering the lowest risk category. Sediment dredging further reduced this by about 0.25 mg g^{-1} dw, a value that by our calculations would have otherwise taken an additional 10–15 years.

Like other studies (Kleeberg and Kohl 1999; Reddy et al. 2007; Chen et al. 2018; Oldenborg and Steinman 2019), we show lower P release rates in sediments from dredged areas. In addition, further evidence that dredging reduced the rate of sediment P release is suggested by changes in the SRP pore-water profiles with depth and season. The dredged sediment had significantly lower SRP concentrations close to the sediment-water interface, reducing the potential for diffusive release, but perhaps more importantly, the spring and autumn peaks and mid-summer dip in sediment SRP concentration are no longer observed, which we suggest is indicative of sediment P release.

13.5 CONCLUSIONS

Reduction of the P loading from the catchment to Barton Broad has been an incremental process. The largest changes were those initially made between 1977 and 1980, but this was followed by a series of smaller interventions, the last of which was concurrent with the onset of the sediment dredging program in 1996. This makes it difficult to unequivocally relate responses in water column P to specific management actions. However, what emerges from our analysis are the different rates of change of catchment loading, and both water column and sediment P.

Our analysis suggests that an internal P load from the sediment continues to occur during the summer in Barton Broad. This has an influence on the seasonal pattern of TP in the lake, accounting for the summer TP peak values. Mass balance estimates suggest that the annual net effect of the internal P load has not significantly changed over the 40 years of monitoring. The annual mean TP concentration in the lake and, therefore its overall nutrient status, is mostly dependent on the catchment load. We had expected to find evidence of a slowing rate of sediment P loss as conditions in the lake reached a new equilibrium. There was some evidence of this, with lower interstitial SRP concentration gradients and less marked seasonal variation, but modeling of both the interstitial SRP and the sediment TP suggested a continued linear decline in P. Thus, while the water column appears to have reached a new steady state with respect to TP, the sediment has not.

Sediment removal has benefits in addition to the control of internal P load. For example, in Barton Broad it has increased water depth, and thus the potential for water-based recreational activity in this very shallow lake. However, despite significant reductions in catchment load prior to the dredging, P release from the sediment has continued, and for now, we conclude only that the recovery time for the lake may have been reduced.

13.6 REFERENCES

Andersen, JM. 1976. An ignition method for determination of total phosphorus in lake sediments. Water Res. 10(4):329–331.

Bjork, S; Pokorny, J; and Hauser, V. 2010. "Restoration of lakes through sediment removal, with case studies from Lakes Trummen, Sweden and Vajgar, Czech Republic." In: Eiseltova M (Ed.). *Restoration of Lakes, Streams, Floodplains, and Bogs in Europe: Principles and case studies.* Dordrecht: Springer; pp. 101–122.

Chen, M; Cui, J; Lin, J; Ding, S; Gong, M; Ren, M; and Tsang, DCW. 2018. Successful control of internal phosphorus loading after sediment dredging for 6 years: A field assessment using high-resolution sampling techniques. Sci Total Environ. 616–617:927–936.

Davison, W; Woof, C; and Turner, DR. 1982. Handling and measurement techniques for anoxic interstitial waters. Nature. 295(5850):582–583.

Edmondson, WT. 1996. *Uses of Ecology: Lake Washington and Beyond*. Seattle (WA): University of Washington Press.

Fox, J; and Weisberg, S. 2011. *An R Companion to Applied Regression, Second Edition*. Thousand Oaks CA: Sage: Sage; [accessed September 8, 2018]. http://socserv.socsci.mcmaster.ca/jfox/Books/Companion.

Harper, D. 1992. Eutrophication of freshwaters—principles, problems and restoration. Dordrecht: Springer.

Hilton, J and Phillips, GL. 1982. The effect of boat activity on turbidity in a shallow broadland river. J Appl Ecol. 19(1):143–150.

Jensen, HS; Kristensen, P; Jeppesen, E; and Skytthe, A. 1992. Iron-phosphorus ratio in surface sediment as an indicator of phosphate release from aerobic sediments in shallow lakes. Hydrobiologia. 235:731–743.

Jeppesen, E; Kristensen, P; Jensen, JP; Sondergaard, M; Mortensen, E; and Lauridsen, T. 1991. Recovery resilience following a reduction in external phosphorus loading of shallow, eutrophic Danish lakes: duration, regulating factors and methods for overcoming resilience. Memorie dell'Istituto Italiano di Idrobiologia Dott Marco de Marchi. 48:127–148.

Kleeberg, A and Kohl, J-G. 1999. Assessment of the long-term effectiveness of sediment dredging to reduce benthic phosphorus release in shallow Lake Müggelsee (Germany). Hydrobiologia. 394(0):153–161.

Klein, J. 1998. Sediment dredging and macrophyte harvest as lake restoration techniques. Land and Water. 42(3):10–12.

Liu, C; Zhong, JC; Wang, JJ; Zhang, L; and Fan, CX. 2016. Fifteen-year study of environmental dredging effect on variation of nitrogen and phosphorus exchange across the sediment-water interface of an urban lake. Environ Pollut. 219:639–648.

Moss, B. 1980. Further studies on the paleolimnology and changes in the phosphorus budget of Barton Broad, Norfolk. Freshw Biol. 10(3):261–279.

Moss, B. 1983. The Norfolk Broadland: Experiments in the restoration of a complex wetland. Biol Rev. 58(4):521–561.

Nurnberg, GK. 1988. Prediction of phosphorus release rates from total and reductant soluble phosphorus in anoxic lake sediments. Can J Fish Aquat Sci. 45(3):453–462.

Oldenborg, KA and Steinman AD. 2019. Impact of sediment dredging on sediment phosphorus flux in a restored riparian wetland. Sci Total Environ. 650:1969–1979.

Osborne, PL. 1978. Relationship between the phytoplankton and nutrients in the River Ant and Barton, Sutton, and Stalham Broads, Norfolk. Norwich. PhD thesis, University of East Anglia.

Osborne, PL. 1980. Prediction of phosphorus and nitrogen concentrations in lakes from both internal and external loading rates. Hydrobiologia. 69(3):229–233.

Osborne, PL. 1981. Phosphorus and nitrogen budgets of Barton Broad and predicted effects of a reduction in nutrient loading on phytoplankton biomass in Barton, Sutton, and Stalham Broads, Norfolk, United Kingdom. Internat Rev Gesamten Hydrobiol. 66(2):171–202.

Osborne, PL and Moss, B. 1977. Paleolimnology and trends in the phosphorus and iron budgets of an old man-made lake, Barton Broad, Norfolk. Freshwater Biol. 7(3):213–233.

Osborne, PL and Phillips, GL. 1978. Evidence for nutrient release from the sediments of two shallow and productive lakes. SIL Proceedings, 1922–2010. 20(1):654–658.

Phillips, G; Bramwell, A; Pitt, J; Stansfield, J; and Perrow, M. 1999. Practical application of 25 years' research into the management of shallow lakes. Hydrobiologia. 395–396(0):61–76.

Phillips, G; Jackson, R; Bennett, C; and Chilvers, A. 1994. The importance of sediment phosphorus release in the restoration of very shallow lakes (the Norfolk-Broads, England) and implications for biomanipulation. Hydrobiologia. 275:445–456.

Phillips, G; Kelly, A; Pitt, J-A; Sanderson, R; and Taylor, E. 2005. The recovery of a very shallow eutrophic lake, 20 years after the control of effluent derived phosphorus. Freshwater Biol. 50(10):1628–1638.

Phillips, G and Kerrison, P. 1991. The restoration of the Norfolk Broads: The role of biomanipulation. Memorie dell'Istituto Italiano di Idrobiologia Dott Marco de Marchi. 48:75–97.

Pinheiro, J; Bates, D; DebRoy, S; Sarkar, D;and Team RDC. 2018. Nlme: Linear and nonlinear mixed effects models. R package version 3.1–137. [accessed September 7, 2018]. https://CRAN.R-project.org/package=nlme.

R Development Core Team. 2014. R: A Language and Environment for Statistical Computing. R Foundation for Statistical Computing. Vienna, Austria.

Reddy, KR; Fisher, MM; Wang, Y; White, JR; and James, RT. 2007. Potential effects of sediment dredging on internal phosphorus loading in a shallow, subtropical lake. Lake Reserv Manag. 23(1):27–38.

Sas, H. 1990. Lake restoration by reduction of nutrient loading: expectations, experiences, extrapolations. Verh Internat Verein Limnol. 24:247–251.

Simpson, G. 2017. Comparing smooths in factor-smooth interactions ii, ordered factors. [accessed September 16, 2018]. https://www.fromthebottomoftheheap.net/2017/12/14/difference-splines-ii/.

Søndergaard, M; Jeppesen, E; Lauridsen, TL; Skov, C; Van Nes, EH; Roijackers, R; Lammens, E; and Portielje, R. 2007. Lake restoration: successes, failures and long-term effects. J Appl Ecol.44:1095–1105.

Steinman, AD; Ogdahl, ME. 2016. From wetland to farm and back again: phosphorus dynamics of a proposed restoration project. Environ Sci Pollut Res. 23(22):22596–22605.

Sutcliffe, DW and Jones, JG. 1992. Eutrophication research and application to water supply. Freshwater Biological Association. 6–29.

Wood, S. 2018. Mgcv v1.8–24: Mixed GAM Computation Vehicle with Automatic Smoothness Estimation. [accessed September 7, 2018]. https://www.rdocumentation.org/packages/mgcv.

Yu, JH; Ding, SM; Zhong, JC; Fan, CX; Chen, QW; Yin, HB; Zhang, L; and Zhang, YL. 2017. Evaluation of simulated dredging to control internal phosphorus release from sediments: Focused on phosphorus transfer and resupply across the sediment-water interface. Sci Total Environ. 592:662–673.

Zuur, AF; Leno, EN; Walker, N; Saveliev, AA; and Smith, GM. 2009. Mixed Effects Models and Extensions in Ecology with R. New York: Springer-Verlag.

CHAPTER 14

INTERNAL PHOSPHORUS LOADING IN ESTHWAITE WATER, UNITED KINGDOM: CONSIDERING THE ROLE OF WEATHER AND CLIMATE

Eleanor B. Mackay[1] and Ian D. Jones[2]

Abstract

Esthwaite Water is a small, stratifying lake in the northwest of the United Kingdom. It has experienced decades of nutrient enrichment from its catchment, including both point and diffuse sources, and as a result is meso-eutrophic. Various management actions have been undertaken in the past to address the catchment nutrient sources and while some improvement in surface-water phosphorus (P) concentrations has occurred, significant internal P loading from the bed sediments takes place during the summer. This study considers internal loading of soluble reactive P (SRP) between the hypolimnion and epilimnion. It combines data from a previous investigation of internal SRP loading (Mackay et al. 2014a) with long-term monitoring data to consider how variability in physical forcing leads to intra- and inter-annual variation in SRP fluxes and how the influence of climate change might affect these fluxes over longer timescales. Inter-annual variability in the hypolimnetic flux was found to drive changes in the overall SRP budget. The variation in this flux is associated with changes in anoxia and stratification drivers, which in turn are influenced by inter-annual differences in weather conditions. Over longer timescales, increasing strength and length of stratification is occurring alongside an increase in hypolimnetic anoxia, which is likely to have implications for the duration of sediment P release and the size of the hypolimnetic SRP pool. Better constraining of our estimates of internal SRP fluxes and their role in promoting poor water quality in the lake is therefore essential when deciding on whether active management of internal loading is justified.

Key Words: Eutrophication, SRP, entrainment, physical mixing, hypolimnion

[1] Lake Ecosystems Group, Centre for Ecology and Hydrology, Library Avenue, Bailrigg, Lancaster LA1 4AP, UK

[2] Earth Observation and Geotechnologies, Stirling University, Stirling, FK9 4LA, UK. E-mail for correspondence: ellcka@ceh.ac.uk

14.1   INTRODUCTION

Esthwaite Water is a small (surface area: 1 km^2), natural lake with a glacial origin, situated in the temperate climate of the English Lake District in the northwest of the United Kingdom (UK) (54°21′N, 3°0′W) (see Figure 14.1). The lake volume is 6.7×10^6 m^3, the maximum depth is 16.0 m, and the mean depth is 6.9 m (Mackay et al. 2012). The lake is usually monomictic and stratifies seasonally during the summer. Esthwaite Water has a catchment area of 17 km^2, the majority of which is agricultural grassland (52%), with some woodland (32%) and arable (7.5%) land uses (Fuller et al. 2000), and a long-term average hydraulic residence time of 100 days. It has most recently been classified as eutrophic on the basis of its chlorophyll *a* concentration and Secchi depth, and mesotrophic for its total phosphorus (TP) concentration (Maberly et al. 2016).

The lake provides a range of ecosystem services and has a number of conservation designations. It is a RAMSAR wetland site designated for the lake habitats that it supports and the presence of a diverse assemblage of aquatic plants including the nationally rare species *Najas flexilis*. The Lake District is a key tourist destination for the UK, providing cultural and amenity services to national and international tourists and was recently designated a United Nations Educational, Scientific, and Cultural Organization world heritage site. Esthwaite Water contributes to the cultural and amenity services of the national park with its close links to the world famous writers William Wordsworth and Beatrix Potter, along with the recreational trout fishery and osprey tours available on the lake.

The growth in tourism, human population, and changing agricultural practices has been responsible for the gradual eutrophication of the lake. Palaeolimnological studies of Esthwaite Water reveal that nutrient enrichment of the sediments began during the 19th Century as the expansion of the railway to the nearby village of Windermere brought more visitors to the area (Bennion et al. 2000; Dong et al. 2012; McGowan et al. 2012). The rapid acceleration of eutrophication in the lake started in the 1970s, when the village of Hawkshead to the north of the lake was added to a main sewerage system. There was a small treatment works constructed at that time that discharged into Black Beck, the main lake inflow, just upstream of where the stream joins the lake. At this time, the changing agricultural practices in the Lake District, including the intensification of sheep farming, which had begun after the Second World War, also accelerated. This coincided with the UK joining the European common market (later the European Union or EU) (Bennion and Winchester 2010; McGowan et al. 2012), which also likely increased nutrient loads to the lake. The declining water quality and large algal blooms occurring during the late 1970s prompted action to address the enrichment problems. During the 1980s, the first attempts at managing P input to Esthwaite Water began with the introduction of P stripping at the sewage treatment works in 1986. Earlier in this decade, a license was granted for a fish farm on the lake and the stocking of non-native rainbow trout (*Oncorhynchus mykiss*) for a recreational fishery. This effectively swapped one nutrient point source for another and winter time SRP concentrations continued to increase (Maberly et al. 2011). As a result, initial improvements in water quality were not sustained and by the late 2000s additional action was taken to target the main P sources to the lake. An increase in storm-water storage and an upgrade of the sewage treatment works was carried out in 2013, the fish farm and fish stocking license was revoked, and the farm was bought out and closed at the end of 2009, with only a reduced level of native brown trout (*Salmo trutta*) stocking allowed with the continuation of the recreational fishery. In tandem with this approach to point sources, efforts to target diffuse nutrient sources have been carried out using mechanisms from the EU Common Agricultural Policy cross compliance and agri-environment schemes.

Although reduced, the present day catchment sources of nutrients for Esthwaite Water remain, with the legacy of nutrient loading resulting in P-enriched sediments (Mackay et al. 2012) and

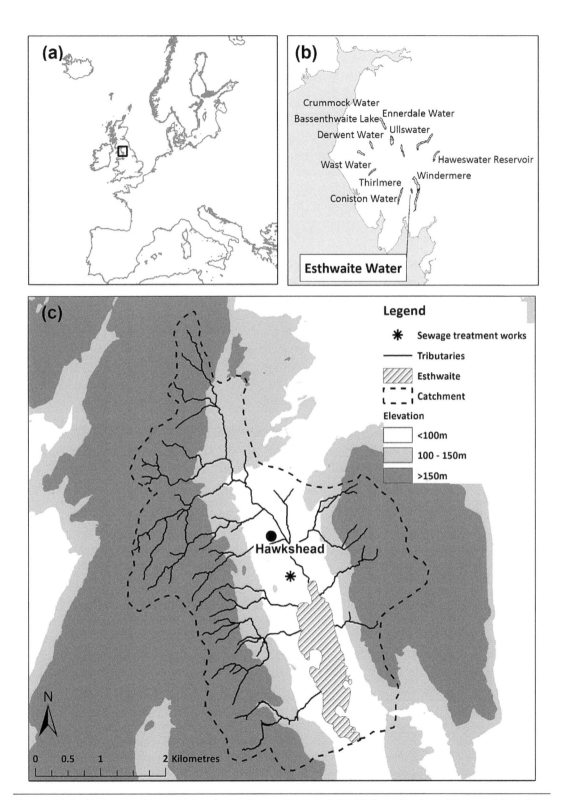

Figure 14.1   Esthwaite Water, (a) located in the northwest of the United Kingdom, (b) one of the lakes making up the English Lake District, (c) the lake catchment area and drainage network including the location of the main settlement at Hawkshead and sewage treatment works.

evidence for internal loading. The dominant catchment P sources continue to be the sewage treatment works, agricultural diffuse sources from pastoral farming, and rural septic tanks. The total burden from these sources, particularly the sewage treatment works, are much reduced from the high historical levels seen in previous decades—and declines in the TP concentration of the water have been seen (Maberly et al. 2011, 2016). The internal sediment source of P in Esthwaite Water has been recognized for many years, particularly during the summer, when it is thought to be an important source of P to the lake (Hall et al. 2000; Miller 2008; Mackay et al. 2014a). The legacy of decades of enrichment from activities in the catchment suggests that this is likely to be the case for many years in the future.

There is also growing evidence that changes associated with climate change are altering the rainfall patterns (Malby et al. 2007) and the thermal regime of the lake (Maberly et al. 2011), which may have implications for external and internal P loads in the future. The direct impacts of climate change on the lake include an earlier onset and later overturn of the thermal stratification, increasing the period of time that the hypolimnion is anoxic, which in turn is likely to have implications for an increased time window for potential P release from the sediment. In addition, regional changes in rainfall with increasing storm intensity are likely to increase the pulses of P that are flow-dependant, such as those from diffuse agricultural sources. The timing of these storm *pulses* will be important in the context of the eutrophication of the lake in the delivery of nutrients, particularly when set in the context of the predictions for average seasonal changes in river flow. Future changes in river flow predicted for the region (Fowler and Kilsby 2007), suggest that in general, lower flows are expected during the summer period, which will increase the lake residence time and reduce the loss of SRP from the lake via the outflow, enhancing algal growth.

Esthwaite Water is one of the seminal sites for the study of internal P loading in stratified lakes. Early work on redox-mediated sediment release of P and other dissolved metals and nutrients carried out during the 1940s by Clifford Mortimer (Mortimer 1941, 1942) has been cited over 1000 times and represents one of the classic studies that first explored internal nutrient loading (see Chapter 1). More recently, the relative importance of different mechanisms controlling P release has expanded beyond the iron hydroxy-oxide reduction under anoxic conditions proposed in the Mortimer papers, revealing the complexity of interactions that may contribute to internal P release in any given lake (Golterman 2001; Hupfer and Lewandowski 2008; see Chapter 4). Esthwaite Water littoral sediments have also been studied to examine the potential for the release of P under oxic conditions. Drake and Heaney's work in the 1980s (Drake and Heaney 1987) showed how elevated pH at the sediment-water interface can lead to high rates and rapid release of dissolved P into the overlying water. More recently, the role of physical drivers of mixing in promoting the movement of SRP, the most bioavailable fraction, from the hypolimnion across the thermocline to the epilimnion, has been quantified in Esthwaite Water to provide a new perspective on understanding the links between internal P loading and its ecological impacts (Miller 2008; Mackay et al. 2014a), which is beginning to be tested in algal community models (Page et al. 2017, 2018).

Reductions in catchment P inputs to Esthwaite Water and a delayed recovery in water quality indicate that internal loading of P from bed sediments is likely to represent an increasingly important component of the bioavailable P budget for this lake in the future. Changes in weather and climate are likely to impact the relative importance of this source at inter-annual and decadal scales timescales; therefore, the question we will address in this chapter is, "How does variation in stratification and physical mixing affect internal supply of bioavailable P to the epilimnion in a small stratified lake?" To address this question, we have the following objectives:

1. Quantify the SRP supply from the hypolimnion across the thermocline during the summer stratified period over two years, based on vertical diffusion and entrainment calculations,

illustrating how controls on physical mixing contribute to variation in the relative and total contribution of the hypolimnetic internal load to the summertime total. We contrast these estimates with epilimnetic sediment SRP release based on published values and external SRP loads for the summer period.

2. Explore the implications of long-term changes in stratification and anoxia in Esthwaite Water on the potential magnitude of internal loading fluxes.

3. Examine the sensitivities of vertical flux estimates with an illustration of how small variations in the calculation of diffusive mixing may result in large uncertainties of this P flux.

4. Discuss, with the help of a conceptual model, the role of stratification and mixing in determining the size of the hypolimnetic bioavailable P pool as the potential supply, and how the timing of mixing events during the summer period can affect the final bioavailable P loading reaching the epilimnion.

14.2 METHODS

14.2.1 Data Collection

The majority of data used in this analysis are taken from the study of Mackay et al. (2014a) and more detailed information on methods is provided in that publication. Briefly, the measurement of meteorological and physical lake conditions was carried out by the Centre for Ecology and Hydrology (CEH) Automatic Water Quality Monitoring Station in the north basin of the lake and a meteorological station on the shore. Wind speed and solar radiation were measured over the lake surface as hourly averages and a series of 12 platinum resistance thermometers (PRT) were deployed at meter intervals through the water column to a depth of 11.5 m, providing water temperature at two-minute intervals averaged to hourly values. Bathymetric information on the area and volume of different water layers was provided by Mackay et al. (2012). Daily discharge data were provided by an Environment Agency for England and Wales gauging station ~1 km downstream of the lake—adjusted to reflect the catchment at the lake outflow. Total inflow to the lake was assumed to equal the total outflow on the seasonal and inter-annual timescales used in this study.

Vertical profiles of dissolved oxygen concentrations at 1 m intervals and discrete samples of SRP at 0.5 m, 11 m, and 14 m were taken on a weekly basis from June to October 2008, September to October 2009, and fortnightly from June to August 2009. Measurement of the external SRP load was carried out at fortnightly intervals from June to October 2008 and 2009 on the main-lake inflow, which represents ~50% of the total discharge to the lake and drains ~80% of the total catchment area. Water samples for SRP analysis were filtered in the field using GF/C filters in a closed filtration unit to minimise air ingress, placed in pre-acid washed bottles, and analyzed on the same day using the molybdenum blue method of Murphy and Riley (1962) following Stephens (1963).

Historical data on changes to stratification in Esthwaite Water are provided by the CEH Cumbrian Lakes Long-Term Monitoring Programme (https://auth-ceh.axis12.com/our-science/monitoring-site/lake-observatories). Vertical temperature and dissolved oxygen profiles from the deepest site in the lake were collected at weekly or fortnightly intervals between 1957 and 2010. The temperature data were linearly interpolated to daily values, converted to water density, and the mixed-layer depth calculated as the depth at which a density difference of 0.1 kg m^{-3} from the surface was exceeded. The timing of the onset and overturn of continuous stratification was identified as the first or last day when a mixed-layer depth different from the full-lake depth was detected in the lake for a time interval greater than 14 days during the summer. The dissolved oxygen data were also linearly interpolated to daily values and used to quantify the number of days and area of sediment that were anoxic (< 1 mg L^{-1} dissolved oxygen) in each year and allow the computation of the annual Anoxic Factor according to Nürnberg (1995) (see Chapter 3 in this book).

14.2.2   P Loading Calculations

Internal loading of SRP from the hypolimnion to the epilimnion is the focus of this analysis; we differentiate between the entrainment of SRP through the deepening and shoaling of the epilimnion bringing high SRP concentration water to the surface versus the incremental vertical diffusion of SRP across the thermocline from the hypolimnion (Mackay et al. 2014a). A detailed description of the methods used to calculate these fluxes is provided in Mackay et al. (2014a), therefore only a brief description follows. The discrete SRP concentration measurements were initially linearly interpolated between the bottom of the mixed layer (assumed to be the same concentration as that measured at 0.5 m) and the 11 m measurement, and then again between the 11 m and 14 m measurements. These interpolations were then corrected with a 27% reduction in concentrations between the bottom of the mixed layer and 11 m, and a 10% increase between 11 m and 14 m, to more accurately reflect the shape of the concentration gradients measured in a previous, more highly vertically resolved study by Miller (2008). Detraining (loss of SRP) or entraining (gaining of SRP) events in the epilimnion were defined by a change in the mixed depth of 1 meter or more. The daily flux due to entrainment, *Hent* (mg day^{-1}), was then calculated as:

$$Hent = \frac{\Sigma C_i' v_i}{\Delta t},$$ [Eq. 14.1]

where Δt is change in time (day), v_i is volume of each meter of depth layer that is entrained or detrained each day (m^3), and C_i' is the previous day's SRP concentration (mg m^{-3}). Daily values were averaged to weekly values and converted to kilograms.

The vertical diffusive flux of SRP *Hvd* (mg day^{-1}) was calculated as:

$$Hvd = K_z^m \frac{\Delta C}{\Delta z} A_m,$$ [Eq. 14.2]

where $\Delta C / \Delta z$ is the concentration gradient of SRP calculated between the center of the epilimnion and center of the hypolimnion (mg m^{-3}), A_m is the surface area at the depth of the center of the metalimnion (m), and K_z^m is the daily vertical eddy diffusivity (m^2 day^{-1}) calculated using the heat flux method of Jassby and Powell (1975). The K_z^m values were calculated as monthly averages from the hourly temperature data at a mid metalimnion depth, also calculated as a monthly average. Daily values of *Hvd* were averaged to weekly values and converted to kilograms. The error estimate for both calculations is taken from Mackay et al. (2014a).

To provide a context for the internal hypolimnetic load, an epilimnetic load (I_{epi}) was calculated for the same period. A value for the epilimnetic sediment area (S_j, m^2) was calculated for each day (j) based on the area of sediment above the depth of the mixed layer, and an estimate of the SRP loading (K_{sed}) of 0.46 mg m^{-2} day^{-1} taken from the study by Steinman et al. (2009), which was summed over the number of days in the study period (m):

$$I_{epi} = \sum\nolimits_{j=1}^{m} K_{sed} S_j.$$ [Eq. 14.3]

The external SRP load (E, mg) coming from the lake catchment is taken from the calculations in Mackay et al. (2014a) and represents the sum of the point and diffuse sources calculated for the summertime period in 2008 and 2009. Although these estimates were based on relatively low frequency sampling of SRP concentrations compared to that expected from storm input, the authors addressed this by increasing the external load by 50% based on the findings of a study that compared a high-frequency to a low-frequency sampling effort (Cassidy and Jordan 2011).

14.2.3   The Influence of Long-Term Changes in Stratification on the Potential for Internal Loading

The physical influences of stratification and mixing on the variability of summertime loading reported in Mackay et al. (2014a) suggest that changes to stratification due to climate change are likely to influence the size and timing of hypolimnetic loading in the future. To examine the potential influence of these changes, we first examined the statistical association between changes in epilimnion depth and changes in the late summer hypolimnetic SRP pool using breakpoint detection from the strucchange package in R. This analysis assesses deviations from a standard single linear fit of data to identify the optimal location or locations (breakpoints) of a change in the model coefficients, estimated by minimizing the residual sum of squares. Confidence intervals are calculated for each breakpoint and where these overlap between the datasets, we infer that the changes in epilimnion depth and hypolimnetic SRP are statistically associated. Using data from the more stable stratified year considered here (2008), models with differing numbers of breakpoints were applied to the hypolimnetic SRP and Schmidt stability data for the latter half of the summer. Akaike information criteria (AIC) values were used to compare between candidate models to select the most parsimonious model for each data set.

To examine the long-term changes and potential influence of stratification on anoxia and internal loading, we quantify historical patterns in stratification strength, length, and onset, as well as overturn timing and the changes in the number of anoxic days and anoxic factor for Esthwaite Water using generalized least squares models that include an autoregressive structure to account for temporal autocorrelation in the times series data using the nlme package in R (Pinheiro et al. 2018). Combining information from the statistical trends in stratification and anoxia, we then calculated a simple projection of the potential change in the hypolimnetic P pool in 2008 that could occur with an earlier onset in the accumulation of P in the hypolimnion due to earlier onset of anoxia and stratification associated with climate warming.

14.2.4   Sensitivities in Physical Factors Influencing Hypolimnetic Loading Estimates

A number of components of the hypolimnetic SRP load calculations are difficult to quantify accurately on realistic timescales and are therefore uncertain. In particular, estimates for physical mixing and the vertical diffusivity term K_z are highly variable in nature, are lake specific, and can span many orders of magnitude in size (10^{-2} to 1.4×10^{-7}). This large potential variability will therefore have consequences for the estimation of the P flux transported across the thermocline, if all other variables are held equal. To illustrate the potential impact that this variation may have on the vertical diffusive load estimates relative to differences in the SRP gradient across the thermocline, a sensitivity analysis of *Hvd* was carried out using the range of K_z values observed in the Mackay et al. (2014a) study to contextualize the SRP flux and total summertime internal hypolimnetic load calculated for 2008 and 2009—years with contrasting vertical gradients in SRP.

14.3   RESULTS

14.3.1   Inter- and Intra-Annual Variability in SRP Fluxes

Inter-annual variability in internal hypolimnetic loading is larger than the variability seen in the other calculated fluxes. Total SRP loads into the lake differed by just over 25% between years in the summer period (June to October) (see Table 14.1). In contrast, variation in the internal hypolimnetic load between years was much larger, with nearly a 60% reduction in load from 2008 to 2009. This meant that in 2008 the hypolimnetic load was comparable to the external load to the lake—45% compared to 49% of the total, respectively. In 2009, the hypolimnetic load represented only 27% of the total. The majority of this inter-annual variation occurred due to differences in the entrainment flux (see Table 14.1 and Figure 14.2).

Hypolimnetic loading over the summer period is associated with large temporal variability (see Figure 14.2). In 2008 and 2009, the majority of the vertical entrainment flux occurred during the 30-day period prior to the lake overturn in early October (day 280). This was particularly marked in 2008 where over 60% of the flux occurred during this period. Early in the summer, both the entrainment and diffusive fluxes are very low and only begin to increase during the midsummer period. In both years, and particularly 2009, this early period was mostly dominated by the diffusive flux.

14.3.2   Drivers in Temporal Variation of the Fluxes

Physical drivers of stratification, the mixing energy and heat content, and metrics of stratification and anoxia, a control on hypolimnetic P accumulation, differed between years over the course of the summer (see Figure 14.3). To compare the two summer periods, it is useful to divide the time into three periods characterized by the mixing conditions; the first coincides with relatively high levels of stratification stability and shallow mixed depths in each year, the midsummer is intermediate, and the third period is characterized by low stability and deep mixing. Although in both years the lake displayed these same general characteristics of seasonal change; shorter-term variations overlaid on the seasonal pattern resulted in some clear differences between years in these drivers. The mixing energy was much higher during the midsummer period in 2009 compared to 2008 (see Figure 14.3a) and the lake had a negative heat balance, so had begun to cool (see Figure 14.3b). As a result, the lake stability over this period was lower in 2009 compared to 2008, and the epilimnion depth was on average deeper (see Figure 14.3c, d). In contrast, both Schmidt stability and epilimnion depth in 2008 remained at similar levels to the early summer period. The changes in the epilimnion depth and stability were echoed by the oxygen dynamics of the hypolimnion during the midsummer period, where the area of sediment overlain by anoxic water began to decline in 2009, but increased during 2008 from the early summer area (see Figure 14.3e). This pattern of midsummer anoxia was repeated for the volume of the lake water that was anoxic, with the difference between years being 39% on average at this time. The dynamics in these physical drivers of mixing and the response of the deep water anoxia both have the potential to contribute to the variability in the hypolimnetic flux through the upward transport of SRP across the thermocline and control over the accumulation of hypolimnetic SRP.

The accumulation of SRP in the hypolimnion—one of the key source terms of the hypolimnetic flux—showed large variations over the summer and between years (see Figure 14.4). In 2008,

Table 14.1   Comparison of the different SRP fluxes to Esthwaite Water between 2008 and 2009 over the summer period (June to October); bold rows indicate the total contributions from external, hypolimnetic, and epilimnetic SRP sources.

Source	2008		2009	
	kg SRP for the summer period	mg SRP m^{-2} d^{-1} for the summer period	kg SRP for the summer period	mg SRP m^{-2} d^{-1} for the summer period
Total External	**235 ± 43**	**1.92**	**229 ± 41**	**1.87**
Internal—Entrainment	151 ± 60	1.23	51 ± 20	0.42
Internal—Diffusion	62 ± 16	0.51	41 ± 11	0.33
Total Hypolimnetic	**213**	**1.74**	**92**	**0.75**
Epilimnetic	**23 ± 12**	**0.19**	**24 ± 12**	**0.20**
Total	**471**	**3.85**	**345**	**2.82**

Based on Mackay et al. (2014a)

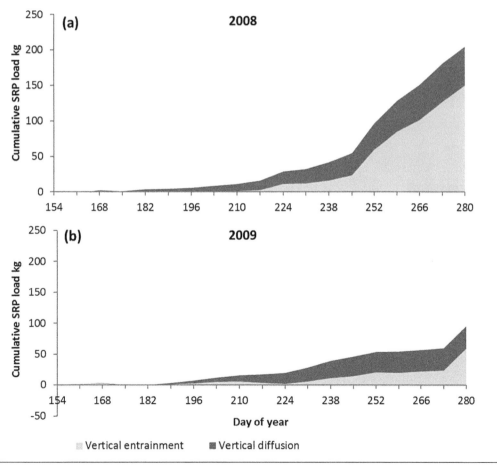

Figure 14.2   Intra-annual variation in the hypolimnetic SRP fluxes in Esthwaite Water for (a) 2008 and (b) 2009. Based on Mackay et al. (2014a).

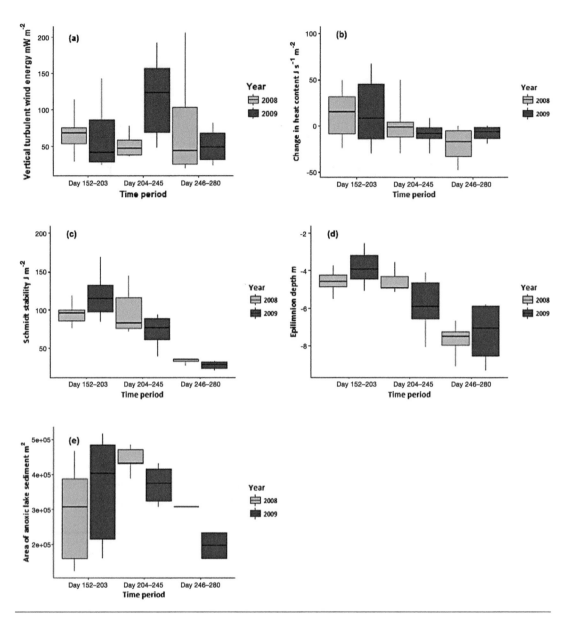

Figure 14.3 A comparison of changes in physical drivers of stratification: (a) vertical turbulent wind energy, (b) change in heat content, (c) Schmidt stability, (d) epilimnion depth, and (e) area of anoxic sediment in 2008 and 2009 over the summer period in Esthwaite Water.

the average concentration of SRP in the hypolimnion > 10 m deep showed a quasi-linear increase between days 154 and 245, peaking at over 140 mg m^{-3}. In 2009, while the concentration of SRP increased over this period, the gradient was shallower and the concentration peaked earlier in the summer at a lower value (74 mg m^{-3}). Thus, the available pool of hypolimnetic SRP differed substantially between years.

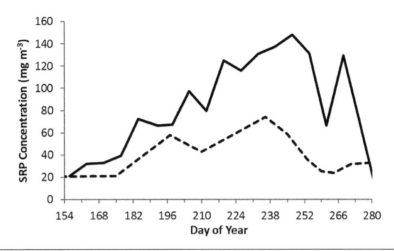

Figure 14.4   Average hypolimnetic > 10 m deep water concentration of SRP in Esthwaite Water in 2008 (solid line) and 2009 (dashed line). Based on Mackay et al. (2014a).

14.3.3   Linking Stratification and Hypolimnetic SRP to Examine the Implications of Long-Term Changes in Physical Conditions on Hypolimnetic Loading

Comparing mid to late summer epilimnion depth, the surface *mixed layer*, and hypolimnetic SRP in 2008 using the breakpoint analysis, suggests that changes to the depth of the *mixed layer* are at times closely associated with changes in the hypolimnetic SRP pool (see Figure 14.5). The optimal SRP model had five breakpoints at days 221, 231, 249, 260, and 270, while breakpoints were identified at days 247 and 264 for the epilimnion depth model with the lowest AIC value. The timings of these breakpoints and their confidence intervals indicate that correspondence between changes in the two datasets varies over this time period. However, the breakpoint associated with the main decline in hypolimnetic SRP occurred just two days after the breakpoint in the epilimnion depth data and their confidence intervals overlapped (see Figure 14.5). At this time, the mixed layer deepened from around 5 m to 8 m, representing a change in around 21% of the lake water volume and the hypolimnetic SRP pool declined by around 50%.

Long-term changes in stratification have been occurring in Esthwaite Water over the last five decades (see Figure 14.6). The length of the continuously stratified period in the summer has increased by 39 days on average, from 152 to 191 days, between the 1960s and the 2000s. The onset of stratification now occurs, on average, 23.5 days earlier and the overturn is 15.6 days later. Stability of stratification has also increased during the summer period (June–September) in this time, with significant ($p < 0.05$) increases in average, maximum, and minimum values of Schmidt stability. This change in stratification has also been accompanied by a similar significant increase ($p < 0.05$) in the number of days in which anoxia occurs in the lake and consequently the lake anoxic factor (see Figure 14.7).

Stronger and longer stratification in Esthwaite Water may promote a larger accumulation of hypolimnetic SRP in the late summer due to the prolonged suppression of mixing that would act to remove SRP or add dissolved oxygen to the deeper waters. Owing to the approximately linear

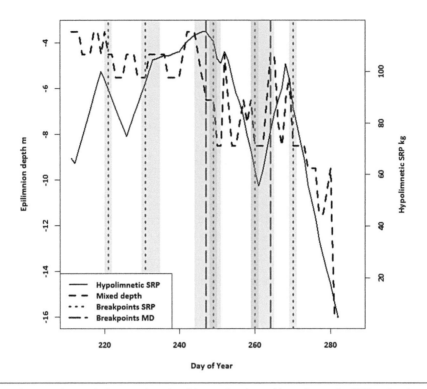

Figure 14.5   Associations of changes in mixed layer depth and hypolimnetic SRP using breakpoint analysis in the mid to late summer of 2008 in Esthwaite Water. The grey shaded areas represent the confidence intervals around the predicted break points.

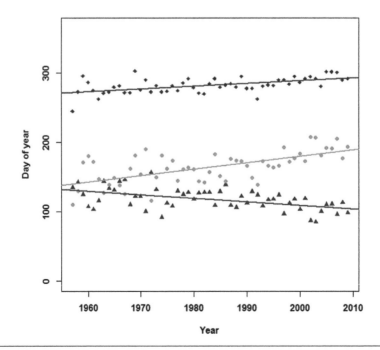

Figure 14.6   Long-term changes in the onset (green triangles), end (blue diamonds), and duration (yellow circles) of stratification in Esthwaite Water. Lines denote the gls model fit for each data set.

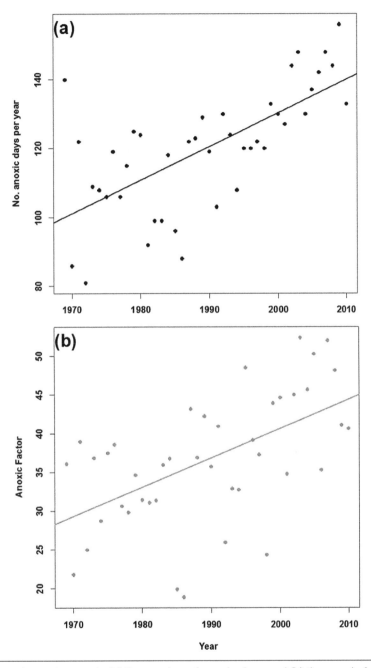

Figure 14.7   Long-term changes in (a) the number of anoxic days and (b) the anoxic factor in Esthwaite Water. The lines represent the gls model fit to the data.

increase in hypolimnetic SRP seen during the early to midsummer period, it is possible to make a simple projection of the impact of a continuation of an earlier onset in SRP accumulation, suggested by the stratification and anoxia changes. By altering the intercept of the linear increase in hypolimnetic SRP in 2008 to reflect an earlier onset of 14 or 30 days reveals that, assuming similar physical, chemical and biological conditions, there may be the potential for higher peak concentrations of

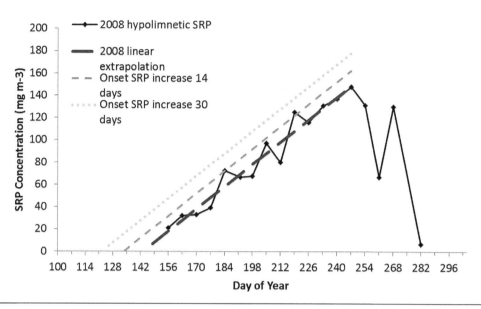

Figure 14.8 A projection of the peak concentration of hypolimnetic SRP in Esthwaite Water with an earlier onset in the accumulation of SRP.

SRP (increase of ~20 mg m^{-3}) by the late summer (see Figure 14.8), simply due to the longer period of time available for the P to accumulate in the hypolimnion.

14.3.4 Sensitivities in Diffusive Flux Estimates

All calculations of P fluxes will have varying levels of uncertainty associated with the different terms of the equations due to limitations in measurements or accuracy of methods, for example. Quantifying the impact of these different uncertainties on the final value of the flux enables the most sensitive parameters to be identified and, potentially, where more effort in future studies should be applied. We illustrate this using the example of the vertical diffusive flux, which is made up of the mixing term K_z and a concentration gradient of SRP across an area at the depth of the thermocline. K_z values are highly variable in time and space and are very difficult to quantify accurately on appropriate timescales. The values used in this study were calculated as monthly averages of hourly calculations of the heat flux method from the PRT temperature data, which smoothed some of the noise within the results, but may dampen some of the contribution to the epilimnetic SRP load made by vertical diffusion. Taking the highest and lowest calculated K_z from the summers of 2008 and 2009, 1.77 × 10^{-6} m^2 s^{-1} and 2.02 × 10^{-7} m^2 s^{-1}, respectively, we recalculated the SRP flux from vertical diffusion in both years (see Figure 14.9). The range in the K_zs, which was a difference of over an order of magnitude in Esthwaite Water over the two summer periods, has the potential to have a large influence on the calculated flux from this source. These differences resulted in a 59–67% increase in the calculated vertical diffusive flux in both years for the maximum and an 80–81% decrease in flux for the minimum value. This level of variation is comparable to the differences seen in the measured SRP gradient over the course of the summer, where surface and deep-water SRP concentrations can differ by around two orders of magnitude between the start of the stratified period and peak in hypolimnetic SRP accumulation.

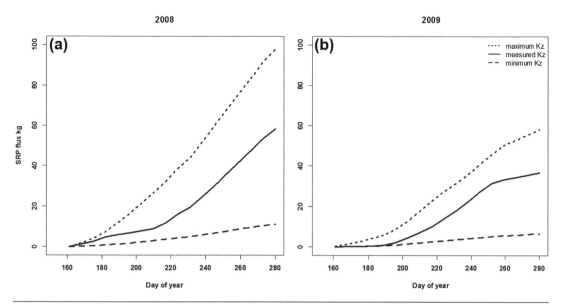

Figure 14.9   The response of the cumulative vertical diffusive flux to changes in the K_z value used for (a) 2008 and (b) 2009 in Esthwaite Water.

14.4   DISCUSSION

14.4.1   Factors Influencing Magnitude and Frequency of Internal Loading

14.4.1.1   Weather

Variability in the inter- and intra-annual summertime SRP load to the epilimnion of Esthwaite Water can be strongly influenced by short-term changes in weather conditions that go on to influence physical mixing, anoxic conditions, and the accumulation of hypolimnetic P. These changes can be relatively subtle yet have large impacts, as illustrated in this study, where only a small change in wind speed has a disproportionate effect on the mixing energy since it is the cube of the wind speed. The timing of variations in weather conditions is also likely to be crucial in terms of the actual impact it has on internal P loading. The data from 2008 and 2009 show a contrast between a year that is relatively stable with shallow epilimnion depths until the late summer and one where mixing energy is higher and heating is lower earlier in the summer resulting in lower stability and deeper epilimnion depth during the midsummer. As a consequence, hypolimnetic anoxia is greater in the more stably stratified year compared to the more weakly stratified year, due to less dissolved oxygen being mixed downward, increasing the area of the sediment that becomes anoxic. As a result, hypolimnetic SRP accumulation is larger, with larger vertical fluxes to the epilimnion.

Physical factors such as surface heating and wind mixing are important in determining the start and end of stratification, which both directly and indirectly influence the onset of hypolimnetic SRP accumulation and the vertical transport of SRP. In addition, the balance of these factors also plays a role in determining the size of this flux through the control of mixing during the middle of the summer stratified period. Using the three periods in the summer that were previously defined, it is possible to conceptualize a model of hypolimnetic SRP loading where the timing of mixing is crucial in how it impacts the overall hypolimnetic load (see Figure 14.10). During the early summer following

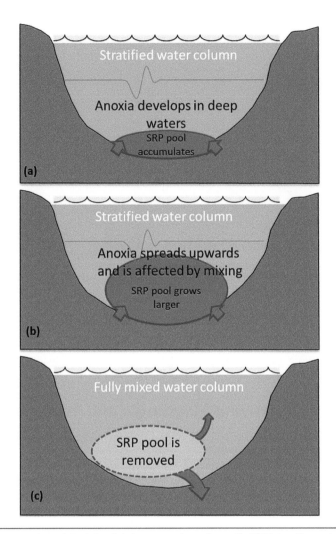

Figure 14.10   Conceptual model of the link between hypolimnetic SRP loading, stratification, and mixing in the (a) early summer, (b) midsummer and (c) late summer; the grey line indicates the mixed layer depth.

stratification onset, anoxia develops in the deep water of the hypolimnion and redox changes at the water-sediment boundary result in biological and chemical-mediated SRP release into the overlying water. At this time the SRP pool is relatively small and is unaffected by direct mixing across the thermocline (see Figure 14.10a). As the summer progresses, water column anoxia affects larger volumes of the hypolimnion and hypolimnetic sediment area and continued SRP release from sediments or remineralization in the water column results in the hypolimnetic SRP pool growing in size. Mixing events occurring at this time have the potential to impact the overall size of this pool and therefore help to determine the total available hypolimnetic SRP that can be mixed across the thermocline at the end of the summer (see Figure 14.10b). The largest vertical fluxes of SRP occur when the lake starts to overturn in the late summer and a proportion of the hypolimnetic SRP pool will be mixed upward to the epilimnion via vertical entrainment or diffusion, and the remainder will return to the sediment (see Figure 14.10c).

14.4.1.2   Climate

Using the framework of the conceptual model (see Figure 14.10) it is also possible to consider the potential influences of the long-term changes seen in stratification and anoxia on internal loading of SRP from the hypolimnion to the epilimnion. In early summer, the earlier onset of stratification may be expected to lead to an inhibition of mixing and the earlier and more rapid formation of deep water anoxia, with associated redox changes resulting in a prolonged period of time when SRP loading from the sediments can occur and therefore the potential for a larger peak in the hypolimnetic SRP pool. Foley et al. (2012) found that rates of oxygen depletion and the extent of deep water anoxia were associated with long-term changes in stratification, and that these rates were particularly sensitive to lake thermal conditions during the spring stratification onset period. In midsummer, the potential for shallower mixed layers and more stable stratification may increase the anoxic sediment extent through the increased inhibition of dissolved oxygen mixing downward from the surface and cause a reduction in the vertical transport of SRP into the epilimnion due to lower vertical diffusivity and fewer mixing events. The potential consequence of these changes will be to increase the hypolimnetic SRP pool but decrease the upward transport of SRP to the epilimnion. During the late summer, longer, more stable stratification and prolonged anoxia may extend the period of time that the hypolimnetic SRP pool can accumulate, resulting in an overall larger pool size.

The influence of physical mixing processes on the hypolimnetic SRP flux to the epilimnion, in terms of its size and potential source for algal growth, is a complex balance between the influence of stratifying and mixing forces. The former promotes the sedimentation of epilimnetic P and the inhibition of the supply of dissolved oxygen to the hypolimnion, while the latter promotes the upward mixing of SRP to the surface mixed layer where it can be accessed by algae. The relative importance of these different processes will vary within and between years with the result that the hypolimnetic contribution to the overall SRP load to the lake will differ in response to changes in the drivers of physical mixing, as illustrated by the differences seen in 2008 and 2009.

14.4.2   Factors Influencing the Relative Importance of Internal Versus External Loading

Both weather and the influence of climate change have the potential to influence the relative importance of internal versus external P loading to the epilimnion through the transport of catchment nutrients and internal vertical mixing during storm events and the reduction in the export of nutrients from lakes during low flow periods. In the two summer periods considered in this analysis, both of which had higher than average discharge levels, the relative importance of internal versus external loads differed on an inter-annual basis. In 2008, internal and external loads were comparable, but external loads dominated in 2009, largely due to a smaller internal SRP flux. Mackay et al. (2014a) recalculated these external loads in both years based on long-term average discharge levels, which revealed that the internal load has the potential to be the dominant flux for Esthwaite Water in years when flows are at or below average levels. The relative importance of flow-dependent (diffuse sources) and flow-independent (point and internal) sources of nutrients for algal growth is clearly important in the context of climate change where changing rainfall and discharge patterns will alter lake residence times at different times of the year (Jones et al. 2011). Reductions in discharge during the summer growing season are likely to increase the importance of point or internal nutrient sources through a reduction in the dilution of incoming nutrients and by reducing the nutrient losses at the outflow, with an overall increase in nutrient concentrations in the lake and consequently, the potential growth of algae (Jones et al. 2011).

14.4.3   Uncertainties and Sensitivities in the Flux Calculations

The sensitivity analysis of the K_z calculation illustrates that if this measurement is poorly constrained, our estimates of SRP transport to the epilimnion could be off by at least an order of magnitude. This level of error is comparable to the difference in the size of the SRP gradient between the early summer and peak of SRP accumulation in the hypolimnion, when fluxes vary from being negligible to representing an important contribution to the SRP budget. It also indicates that well-established uncertainties in external loading due to low-resolution sampling, which may not adequately capture storm delivery of P (Johnes 2007; Cassidy and Jordan 2011; Defew et al. 2013), are equally relevant for internal P fluxes. The estimates presented in this paper are based on a simplified calculation, which excludes the effect of sediment heating. Excluding this effect can lead to additional errors in the estimation of K_z. It is therefore clear that to obtain realistic calculations for the upward flux of hypolimnetic SRP, which will contribute to more accurate P budgets that can be used to examine the importance of this P source for promoting algal blooms, more effort should be made to improve the characterization of internal-lake physical mixing processes.

Values of K_z also vary among lakes since the strength of vertical mixing is associated with exposure to wind-mixing energy, with larger values found in lakes and inland seas with a large surface area. Salas de León (2016) report values from a large number of lakes that show variation from 10^{-3} to values around the speed of the molecular diffusion of water (1.4×10^{-7}). This large variation clearly has implications for the relative importance of this transport process in different lake types and under different climatic conditions. In larger or wind-exposed lakes, the role of the diffusive transport of SRP into the epilimnion is likely to be more important than for smaller, wind-sheltered lakes.

14.4.4   Possible or Active Management Strategies to Address Internal Loading in Esthwaite Water

The legacy of historical nutrient loading to Esthwaite Water is evident in the elevated P concentrations of the sediments. Mackay et al. (2012) found that the surface sediments, particularly in the deep-water areas of the lake, were highly enriched in TP, up to 6 mg g^{-1} dry weight (dw) compared to concentrations found in oligotrophic and mesotrophic lakes of 1.5–2.2 mg g^{-1} dw (Carey and Rydin 2011). The current management strategies for this lake have focused on reducing the external P sources, with natural recovery being adopted for the internal sediment sources of P, allowing for either natural burial by new sediments or for sediments to slowly equilibrate over a number of years. Active internal load management, such as the use of Phoslock®, is not currently being considered in the lake; however, a hypothetical estimate of the costs of the material required, excluding application, can be made with reference to the P content of the sediment. The cost of Phoslock equates to £245 (approximately $300 US) per kg of P, based on assumptions from Mackay et al. (2014b) that 100 kg of Phoslock will bind 1 kg of P and 1 metric ton of Phoslock costs £2454 (approximately $3000 US). Utilizing the P depth distribution data from Mackay et al. (2012), which considers only TP and is likely to overestimate the mobile fraction, and the top 2 cm of lake sediment, which is likely to be an underestimate of the depth of mobile P in the sediment (Meis et al. 2013), there is approximately 6223 kg of P in Esthwaite Water. The cost of Phoslock required to treat this quantity of P would, therefore, be around £1.5 million ($1.8 million US). In addition, the ongoing sedimentation of P being brought in from the catchment, estimated by Mackay et al. (2012) to be c. 1000 kg per year, implies an ongoing dosing cost of ~£250,000 per year (approximately $303,000 US), assuming all the P would eventually be released. These costs suggest that addressing catchment P sources should remain the priority for Esthwaite Water over the short term.

14.5   CONCLUSIONS

Intra- and inter-annual variability in summertime SRP loading to Esthwaite Water is an important feature of the seasonal budget. Understanding this variability is essential since it determines when bioavailable P is supplied to the lake and is likely to have consequences for how and when the nutrient may be used for algal growth. The influence of physical changes in the weather conditions experienced over the lake affects the lake heat budget and thermal stratification, which in turn can affect the extent of hypolimnetic anoxia and the potential accumulation of SRP in the lake hypolimnion. This indicates that both short-term weather variations and longer-term climate change are likely to play an important role in how much SRP is available to be transported to the epilimnion. In addition, changes in the mixed-layer depth are also associated with changes in hypolimnetic SRP at the end of the summer period, which is when the largest upward fluxes of bioavailable P occur, at a potentially important time during the growing season. Improving our understanding and quantification of the mixing processes that are responsible for this upward transport of SRP is essential given that current limitations may result in large differences in calculated and actual supply. Constraining these values is important for the development of process-based models of internal P loading that can be used to inform future management of the lake's water quality.

14.6   ACKNOWLEDGMENTS

The authors would like to thank Ruth and Tom Hansard, Elizabeth Hurrell, Daniel Wright, Fanghua Li, Rebecca Messham, Rebecca Jackson, and Jennifer Carrie for assistance with fieldwork; and Jack Kelly for technical assistance with the meteorological data that formed the data from the Mackay et al. (2014a) paper. We would also like to acknowledge the staff of the Freshwater Biological Association, Institute for Freshwater Ecology, and Centre for Ecology and Hydrology who have collected and maintained the invaluable long-term data set for Esthwaite Water. We are also grateful to the Environment Agency for England and Wales for provision of the discharge data used to calculate the external SRP loads. The project was funded by NERC project NEC06415 Cumbrian Lakes monitoring and a PhD studentship awarded to E. B. Mackay by the Faculty of Science and Technology at Lancaster University.

14.7   REFERENCES

Bennion, H; Monteith, D; and Appleby, P. 2000. Temporal and geographical variation in lake trophic status in the English Lake District: evidence from (sub)fossil diatoms and aquatic macrophytes. Freshwater Biol. 45:394–412.

Bennion, H and Winchester, AJL. 2010. Linking Historical Land-Use Change with Palaeolimnological Records of Nutrient Change in Loweswater, Cumbria. Lancaster University, Report produced for the Loweswater Care Project (LCP). Available at: http://www. lancaster. ac. uk/fass/projects/loweswater/research.htm.

Carey, CC and Rydin, E. 2011. Lake trophic status can be determined by the depth distribution of sediment phosphorus. Limnol Oceanogr. 56(6):2051–2063.

Cassidy, R and Jordan, P. 2011. Limitations of instantaneous water quality sampling in surface water catchments: Comparison with near-continuous phosphorus time-series data. J Hydrol. 405:182–193.

Defew, LH; May, L; and Heal, KV. 2013. Uncertainties in estimated phosphorus loads as a function of different sampling frequencies and common calculation methods. Mar Freshwater Res. 64(5):373–386.

Dong, X; Bennion, H; Battarbee, RW; and Sayer CD. 2012. A multiproxy palaeolimnological study of climate and nutrient impacts on Esthwaite Water, England over the past 1200 years. The Holocene. 22(1):107–118.

Drake, JC; and Heaney, SI. 1987. Occurrence of phosphorus and its potential remobilization in the littoral sediments of a productive English lake. Freshwater Biol. 17:513–523.

Foley, B; Jones, ID; Maberly, SC; and Rippey, B. 2012. Long-term changes in oxygen depletion in a small temperate lake: effects of climate change and eutrophication. Freshwater Biol. 57(2):278–289.

Fowler, HJ and Kilsby, CG. 2007. Using regional climate model data to simulate historical and future river flows in northwest England. Climatic Change. 80(3–4):337–367.

Fuller, RM; Smith, GM; Sanderson, JM; Hill, RA; Thomson, AG; and Hall, MW. 2000. Land Cover Map 2000, Huntington, Cambridgeshire.

Golterman, HL. 2001. Phosphate release from anoxic sediments or "What did Mortimer really write?" Hydrobiologia. 450:99–106.

Hall, GH; Maberly, SC; Reynolds, CS; Winfield, IJ; James, BJ; Parker, JE; Dent, MM; Fletcher, JM; Simon, BM; and Smith, E. 2000. Feasibility study on the restoration of three Cumbrian lakes. Ambleside: Centre for Ecology and Hydrology, UK.

Hupfer, M and Lewandowski, J. 2008. Oxygen controls the phosphorus release from lake sediments—a long-lasting paradigm in limnology. Int Rev Hydrobiol. 93(4–5):415–432.

Jassby, A and Powell T. 1975. Vertical patterns of eddy diffusion during stratification in Castle Lake, California. Limnol Oceanogr. 20(4):530–543.

Johnes, PJ. 2007. Uncertainties in annual riverine phosphorus load estimation: Impact of load estimation methodology, sampling frequency, baseflow index and catchment population density. J Hydrol. 332:241–258.

Jones, ID; Page, T; Elliott, JA; Thackeray, SJ; and Heathwaite, AL. 2011. Increases in lake phytoplankton biomass caused by future climate-driven changes to seasonal river flow. Global Change Biol. 17:1809–1820.

Maberly, S; De Ville, MM; Feuchtmayr, H; Jones, ID; Mackay, EB; May, L; Thackeray, SJ; and Winfield, IJ. 2011. The limnology of Esthwaite Water: historical change and its causes, current state and prospects for the future. A Report to Natural England.

Maberly, SC; De Ville, MM; Thackeray, SJ; Ciar, D; Clarke, M; Fletcher, JM; James, JB; Keenan, P; Mackay, EB; Patel, M; et al. 2016. A survey of the lakes of the English Lake District: The Lakes Tour 2015. A report to United Utilities.

Mackay, EB; Folkard, AM; and Jones, ID. 2014a. Interannual variations in atmospheric forcing determine trajectories of hypolimnetic soluble reactive phosphorus supply in a eutrophic lake. Freshwater Biol. 59(8):1646–1658.

Mackay, EB; Jones, ID; Folkard, AM; and Barker, P. 2012. Contribution of sediment focusing to heterogeneity of organic carbon and phosphorus burial in small lakes. Freshwater Biol. 57:290–304.

Mackay, EB; Maberly, SC; Pan, G; Reitzel, K; Bruere, A; Corker, N; Douglas, G; Egemose, S; Hamilton, D; Hatton-Ellis, T; et al. 2014b. Geoengineering in lakes: welcome attraction or fatal distraction? Inland Waters. 4(4):349–356.

Malby, AR; Whyatt, JD; Timmis, RJ; Wilby, RL; and Orr, HG. 2007. Long-term variations in orographic rainfall: analysis and implications for upland catchments. Hydrolog Sci J. 52(2):276–291.

McGowan, S; Barker, P; Haworth, EY; Leavitt, PR; Maberly, SC; and Pates, J. 2012. Humans and climate as drivers of algal community change in Windermere since 1850. Freshwater Biol. 57(2):260–277.

Meis, S; Spears, BM; Maberly, SC; and Perkins, RG. 2013. Assessing the mode of action of Phoslock® in the control of phosphorus release from the bed sediments in a shallow lake (Loch Flemington, UK). Water Res. 47:4460–4473.

Miller, HJ. 2008. Investigation into mechanisms for the internal supply of phosphorus to the epilimnion of a eutrophic lake. PhD Thesis. Lancaster: Lancaster University.

Mortimer, CH. 1941. The exchange of dissolved substances between mud and water in lakes. J Ecol. 29:280–329.

Mortimer, CH. 1942. The exchange of dissolved substances between mud and water in lakes. J Ecol. 30:147–201.

Murphy, JA and Riley, JP. 1962. A modified single solution method for the determination of phosphate in natural waters. Analytica Chimica Acta. 27:31–36.

Nürnberg, GK. 1995. Quantifying anoxia in lakes. Limnol Oceanogr. 40(6):1100–1111.

Page, T; Smith, PJ; Beven, KJ; Jones, ID; Elliott, JA; Maberly, SC; Mackay, EB; De Ville, M; and Feuchtmayr, H. 2017. Constraining uncertainty and process-representation in an algal community lake model using high frequency in-lake observations. Ecol Mod. 357:1–13.

Page, T; Smith, PJ; Beven, KJ; Jones, ID; Elliot, JA; Maberly, SC; Mackay, EB; De Ville, M; and Feuchtmayr, H. 2018. Adaptive forecasting of phytoplankton communities. Water Res. 134:74–85.

Pinheiro, JC; Bates, D; Debroy, S; and Sarkar, D. 2018. nlme: Linear and Nonlinear Mixed Effects Models. Available at: https://cran.r-project.org/package=nlme.

Salas De León, DA; Alcocer, J; Gloria, VA; and Quiroz-Martinez, B. 2016. Estimation of the eddy diffusivity coefficient in a warm monomictic tropical lake. J Limnol. 75(1S):161–168.

Steinman, AD; Chu, X; and Ogdahl, M. 2009. Spatial and temporal variability of internal and external phosphorus loads in Mona Lake, Michigan. Aquat Ecol. 43:1–18.

Stephens, K. 1963. Determination of low phosphate concentrations in lake and marine waters. Limnol Oceanogr. doi.org/10.4319/lo.1963.8.3.0361

LAKE SØBYGAARD, DENMARK: PHOSPHORUS DYNAMICS DURING THE FIRST 35 YEARS AFTER AN EXTERNAL LOADING REDUCTION

Martin Søndergaard and Erik Jeppesen[1]

Abstract

Lake Søbygaard (Denmark) is a shallow and highly eutrophic lake that was subjected to very high levels of external phosphorus (P) loading from a nearby town until 1982. Since then, the lake has suffered from an internal loading of P, which during the first 10–15 years after external load reduction led to summer total P (TP) concentrations being three to four times higher than the inlet concentrations. During the same period, mean summer TP in the lake was between 0.3 mg L^{-1} and 0.8 mg L^{-1}, chlorophyll a (Chl a) between 120 µg L^{-1} and 900 µg L^{-1}, and the net sediment release between 5 g P m^{-2} and 15 g P m^{-2} annually. As of 2017, 35 years after the external loading reduction, summer P concentrations remain about two times higher than the inlet concentrations for three to four months during summer and Chl a is close to 100 µg L^{-1}. The yearly mean inlet concentrations of P are still lower than the yearly mean lake-water concentrations, suggesting that the lake continues to act as a P source. P originates from a very extensive sediment pool accumulated at depths down to 20–25 cm. Most of the sediment P is bound to iron (extractable with NaOH), but during summer a considerable part is transformed to easily releasable P (NH_4Cl extractable), probably facilitated by high pH in the water column and the top sediment layers.

Key words: Sediment, internal loading, phosphorus fractions, seasonal variation, chlorophyll a

[1] Department of Bioscience, Aarhus University, Denmark

15.1 INTRODUCTION

15.1.1 Description of Lake Søbygaard and Its Management History

Lake Søbygaard is a 40 ha shallow (mean depth = 1.0 m, maximum depth = 1.9 m) lake located in the central part of Jutland, Denmark (see Figure 15.1 and Table 15.1). The lake has one main inlet representing about 90% of the total hydraulic loading to the lake, and one main outlet. The average hydraulic retention is two to four weeks and varies from 15–20 days in winter to 25–30 days in summer (Søndergaard et al. 1993). The catchment area of Lake Søbygaard is 11.6 km² and is dominated by agriculture (44%), urban areas (28%), and forest (17%) (Rolighed et al. 2016).

For many years, Lake Søbygaard was heavily loaded with wastewater from the nearby town of Hammel, including wastewater from a slaughterhouse. A paleolimnological study on chironomid subfossil remains in the lake indicated a succession from a *naturally* eutrophic state to a hypereutrophic state during recent centuries and that the sediment accumulation rate had doubled in recent decades in the 20th century (Brodersen et al. 2001).

In 1969, mechanical treatment of the wastewater was introduced, and in 1976 biological treatment was added. In the 1970s and early 1980s, the annual loading was about 30 g P m^{-2} yr^{-1} (Søndergaard 1988). Chemical treatment of the wastewater for P removal was introduced in autumn 1982 and this reduced the P loading by 80–90% (to 6.7 g P m^{-2} yr^{-1} in 1983 and 4.9 g P m^{-2} yr^{-1} in 1984). In 1983–1984, the net P retention in the lake was about -8 g P m^{-2} yr^{-1} (Søndergaard 1988). In 1987, the slaughterhouse was closed, and in 1996 nitrogen (N) removal was implemented at the wastewater plant. During the first 15 years after the external loading reduction (1984–1999), the P retention in the lake ranged from -5 to -15 g P m^{-2} yr^{-1} (Jeppesen et al. 2007). In 2006, all wastewater was diverted from the lake, but it still receives storm water. In 2007–2010, the external P loading was estimated to be 2.7–3.8 g P m^{-2} yr^{-1} and the external N loading to be 35–45 g N m^{-2} yr^{-1} (Rolighed et al. 2016).

The lake experienced frequent fish kills in the 1970s, and in the 1980s short-term collapses in the phytoplankton biomass occurred, which had significant effects on the internal P loading (Jeppesen

Table 15.1 Morphological characteristics of Lake Søbygaard and chemical data from shortly (1983–1990) and then 25 years (2007–2014) after the main external nutrient loading reduction in 1982 (presented as minimum, mean, and maximum summer (May 1 to September 30) means of the two eight-year periods).

Area (ha)	40					
Mean depth (m)	1.0					
Maximum depth (m)	1.9					
Hyd. ret. time (weeks)	2–4					
	1983–1990			**2007–2014**		
	Min	**Mean**	**Max**	**Min**	**Mean**	**Max**
Mean summer TP (µg L^{-1})	510	743	1046	227	419	768
Mean summer PO$_4$ (µg L^{-1})	113	321	518	57	222	439
Mean summer TN (mg L^{-1})	2.07	3.08	4.94	0.61	1.14	1.52
Mean summer NO$_3$ (mg L^{-1})	0.19	0.79	1.52	0.01*	0.03*	0.06
Mean summer Chl-a (µg L^{-1})	124	385	901	70	107	180

*In four out of eight years, summer mean NO$_3$ was below the detection limit of 0.01 mg L^{-1}. In this case, 0.01 mg L^{-1} was used to calculate the mean value.

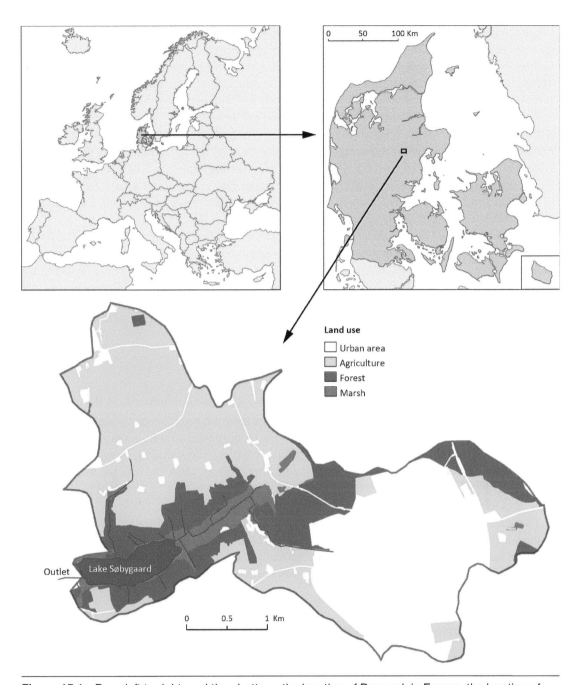

Figure 15.1   From left to right, and then bottom, the location of Denmark in Europe, the location of Lake Søbygaard (9°48′36″ E, 56°15′20″ N) in Denmark, and Lake Søbygaard surrounded by its catchment areas. Modified from Rolighed et al. (2016).

et al. 1990; Søndergaard et al. 1990; see Chapter 4). In the 1970s, Chl *a* levels in the lake were low despite very high nutrient concentrations, and this has been ascribed to the frequent occurrence of fish kills and the subsequently high zooplankton biomass (Søndergaard 1987; Jeppesen et al. 1998). Until the late 1990s, when the percentage of piscivorous fish (mainly perch, *Perca fluviatilis*) increased, the fish community was totally dominated by cyprinids (mainly roach, *Rutilus rutilus*, and rudd,

Scardinius erythropthalmus) with a high biomass—as determined from catches in multi-mesh-sized gillnets (Jeppesen et al. 2007). A study comprising fish community data from 1990 to 2010 found 11 species, with perch, roach, rudd, and zander (*Stizostedion lucioperca*) as the dominants, but a decreasing trend in fish biomass was observed as a consequence of reduced nutrient concentrations and declining mean planktivorous fish size due to increased temperatures (Gutierrez et al. 2016).

Today, Lake Søbygaard is still highly eutrophic and exhibits high Chl concentrations. The water is turbid with a Secchi depth around 0.5 m most of the year and submerged macrophytes are absent. According to the latest Danish water management plans involving implementation of the European Water Framework Directive, the lake is in a bad ecological state (based on Chl levels) (The Danish Environmental Protection Agency 2013). Further descriptions and studies on Lake Søbygaard can be found in Søndergaard (1989, 1990), Søndergaard et al. (1993), Jensen et al. (1992), Jeppesen et al. (1996, 1998), Liboriussen and Jeppesen (2003), and Sørensen et al. (2011).

The aim of this chapter is to describe how the internal P loading has changed in Lake Søbygaard in the years following the external nutrient loading reduction based on monitoring data comprised during the years 1971 to 2017. We focused on the long-term internal P loading as reflected in the lake-water nutrient concentrations on a yearly/seasonal scale, supplemented with the changes in P concentrations recorded in the sediment profile.

15.2 METHODS

15.2.1 Water Sampling and Analyses

Water was sampled from a central position in the lake and in the inlet 100 m from the entrance to the lake as surface samples. The sampling frequency varied from only one or two samples in summer (May 1–September 30) in the early and mid-1970s to more than 20 summer samples from 1985 to 1994 (distributed throughout the season). From 1995–2007, the sampling frequency was twice a month during summer and once a month during winter. Since 2008, the lake has been sampled once a month. Ice cover during winter ranged from a few weeks to several months, but if possible, samplings were collected through the ice.

The water samples were analyzed for TP from the lake (TP_lake) and the inlet (TP_inlet), and in the lake for orthophosphate (PO_4), total nitrogen (TN), nitrate+nitrite (NO_3), pH, and Chl. All chemical variables were measured using standard analytical procedures (see also Søndergaard et al. 2005). Yearly means or summer means (May 1 to September 30) were calculated as the mean value of monthly means.

15.2.2 Sediment Sampling and Analyses

Sediment was sampled with a Kajak bottom sampler or a piston corer. The inner core diameter was 52 mm and the core length varied between 30 and 50 cm. Sampling was conducted at the same position in the center of the lake, representing the whole lake since sediment is accumulated in most of the lake area (Andreasen et al. 1984). The sediment samples were cut into sections at 1 or 2 cm intervals and kept dark and cool until analyses. Each analysis included a composite sample of three cores.

Sediment dry weight (dw) was determined by drying at 105°C for 24 hours, and loss on ignition (LOI) was subsequently estimated by drying to constant weight at 550°C. TP in the sediment (TP_sed) was analyzed spectrophotometrically as molybdate reactive P after extraction of ash-free sediment with 1 M HCl (modified from Andersen (1976)). Phosphorus fractionation was conducted according to the method of Hieltjes and Lijklema (1980), defining the following fractions: NH_4Cl-P,

NaOH-P, and HCl-P. Residual P (Res-P) was calculated as TP_sed − (NH$_4$Cl-P + NaOH-P + HCl-P). Sediment analyses were carried out in 1985, 1991, 1998, and 2004 (only TP_sed). Volumetric concentrations of TP_sed were calculated using TP_sed, dw, and LOI, assuming an inorganic matter density of 2.6 g cm^{-3} and an organic matter density of 1.05 g cm^{-3}. Sediment profiles and the sediment surface after 1985 were adjusted to take into account a net sediment increase of 0.5–0.6 cm yr^{-1} (Søndergaard et al. 1993).

15.2.3   Methods to Describe the Internal Loading

The internal P loading of Lake Søbygaard was described and illustrated in different ways using:

- *Yearly TP_lake and TP_inlet*: comparison of lake and inlet TP concentrations to illustrate how lake water TP is influenced by P release from the sediment on a yearly scale;
- *Seasonal TP_lake*: comparison of winter and summer TP in the lake to demonstrate the importance of season;
- *Sediment profile changes*: comparison of the profiles of TP_sed and P fractions in the upper 30 cm of the sediment for up to 20 years after the external loading reduction to show the long-term impact from the sediment;
- *Seasonal sediment changes in NH$_4$Cl-P and NaOH-P*: to display the interaction between the sediment and water on a seasonal scale; and
- *Yearly mass balance studies (based on previous analyses and calculations)*: to identify the long-term impact of the sediment acting as P source.

15.3   RESULTS

15.3.1   Yearly TP and Chl

In the 1970s and early 1980s, TP concentrations were very high (> 0.5 mg L^{-1}) in both the inlet to Lake Søbygaard and in the lake itself—even up to 5 mg L^{-1} in the inlet and 2 mg L^{-1} in the lake (see Figure 15.2). After the introduction of chemical treatment, TP concentrations decreased but remained at a high level; TP_lake varied between 0.16 and 0.71 mg L^{-1} and TP_inlet varied between 0.09 and 0.37 mg L^{-1} in the period 1983–2017. With the exception of 2005 and 2006, the yearly mean inlet concentrations were always lower than the yearly mean lake TP. Chl concentrations were relatively low in the 1970s but very high (up to 900 μg L^{-1}) in the mid-1980s. Since 1987, summer mean Chl has fluctuated between 70 and 298 μg L^{-1}, displaying a decreasing trend.

15.3.2   Seasonal TP_Lake

Seasonally, TP_lake has varied by a factor of two or more between winter and summer months (see Figure 15.3). Particularly in the first years after the loading reduction (1983–1996), TP_lake was much higher in summer, with the median being four times higher during July and August than during January–March. Maximum TP_lake occurred in July with mean concentrations up to 1.53 mg L^{-1} (1987). In the two last seven-year periods (2004–2010 and 2011–2017), summer concentrations were still higher than in winter, but the difference has been declining. Maximum TP_lake occurred in August with mean concentrations up to 1.02 mg L^{-1} (2009). In the first two seven-year periods after the loading reduction (1983–1996), mean TP_lake exceeded TP_inlet in all months from April to November, whereas in the last two seven-year periods, the months when TP_lake exceeded TP_inlet has declined to May/June through October.

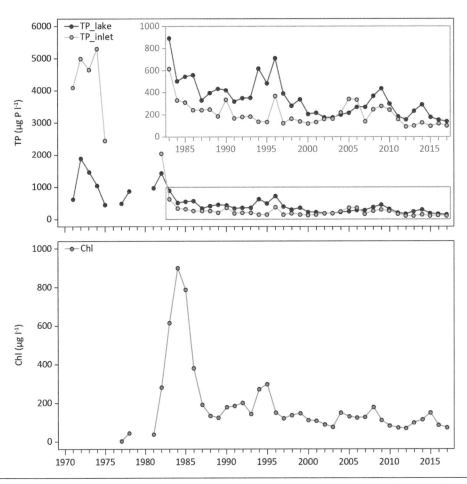

Figure 15.2 *Upper*: Yearly mean concentrations of TP_lake and TP_inlet in Lake Søbygaard from 1971 to 2017 (1983 to 2017 in the enlarged figure inserted in upper right corner); based on mean monthly data. Number of months in 1971: 2, 1972: 1, 1973: 1, 1974: 2–3, 1975: 2, 1977: 2 (lake only), 1978: 7 (lake only), 1981: 7 (lake only), 1982: 8–9, 1983: 4–7, 1984–2014: 12, 2015–2017: 9–13. *Lower*: summer mean concentrations of Chl in the lake.

15.3.3 Lake-Water and Sediment P Fractions Seasonality in 1985

Detailed seasonal measurements are available from both the sediment and lake water during 1985 (i.e., three years after the major external loading reduction). This is also one of the years when a collapse in the phytoplankton biomass was recorded (the last two weeks of July) (see Chapter 4). Apart from the ice-covered period during January and February when Chl was lower than 10 μg L^{-1} and during the phytoplankton collapse in July (17 μg L^{-1}), Chl was high throughout the year and reached a maximum of 1268 μg L^{-1} in August (see Figure 15.4). Mirroring the high Chl, pH increased from near neutral in winter up to a maximum of 10.9 in early July. During or shortly after the phytoplankton collapse, pH decreased to 9.2 before it reached 10.5 in September.

TP_lake was less than 0.20 mg L^{-1} until late March when the ice disappeared and then increased steadily until early June, reaching a maximum of 1.00 mg L^{-1} (see Figure 15.4). Another maximum of 1.23 mg L^{-1} was reached in August and TP_lake remained high (around 0.9 mg L^{-1}) during the

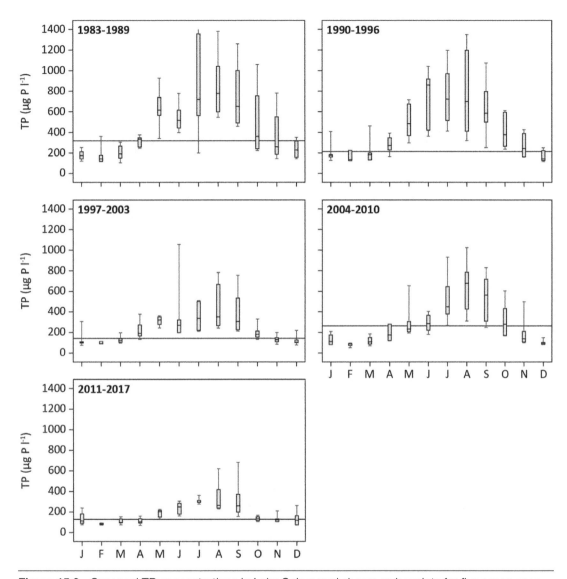

Figure 15.3   Seasonal TP concentrations in Lake Søbygaard shown as boxplots for five seven-year periods from 1983 to 2017. Mean inlet TP concentrations are shown as reference lines for the same periods. The boxes show 10%, 25%, 75%, and 90% percentiles.

entire autumn period. PO_4 was always higher than 0.02 mg L^{-1}, and during late August and autumn it was much higher with concentrations up to 0.70 mg L^{-1}.

Data from the uppermost 1 cm of the sediment showed seasonal variations in the fractions of NH_4Cl-P and NaOH-P, starting with relatively low NH_4Cl-P concentrations in May of about 0.4 mg P g^{-1} dw and ending up with a maximum of 1.9 mg P g^{-1} dw in August (see Figure 15.4). In the same period, NaOH-P decreased from about 4.5 mg P g^{-1} dw to around 2.7 mg P g^{-1} dw. During and shortly after the phytoplankton collapse, NH_4Cl-P declined and NaOH-P increased. The collapse in phytoplankton biomass caused an increase in TP_lake of about 1 mg L^{-1} during a 10-day period due to the continued high gross release rate of P, but the sedimentation rate was low (see Chapter 4). This

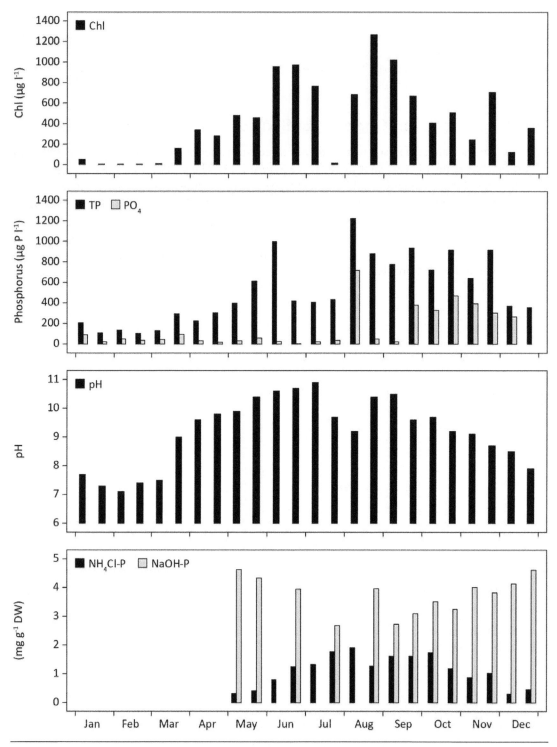

Figure 15.4 Chl *a*, P (TP, PO₄), pH, and the amount of NH₄Cl-P and NaOH-P in the upper 0–1 cm of the sediment during the season of 1985 in Lake Søbygaard (fortnightly data, if available). From mid-January to mid-March, the lake was covered with ice. No data on P fractions in the sediment before May. The collapse in Chl *a* in July is also described in Chapter 4.

corresponds to a sediment release of about 1 g P m^{-2} or 100 mg m^{-2} day^{-1}. Assuming an average dw of 5%, an LOI of 25%, and a TP_sed of 8 mg g^{-1} dw (see Figure 15.5), this 10-day net P release corresponds to approximately 10% of the TP_sed present in the upper 1 cm of the sediment. NaOH-P + NH$_4$Cl-P comprises 60–70% of TP_sed in the upper sediment; the remaining part consists mainly of organic-bound P (Søndergaard 1988).

15.3.4   Long-Term Changes in Sediment TP and P Fractions

Sediment profile data on TP_sed were gathered three years after the loading reduction in 1985 and then measured again 19 years later in 2004 (see Figure 15.5). The profile from 1985 showed a very marked TP_sed maximum around 15 cm depth. Assuming a sediment accumulation of 0.6 cm yr^{-1} in the period with high sedimentation rates (Søndergaard et al. 1993) and no transport disrupting the overall sediment concentrations, this corresponds to a maximum P loading/P retention around 1960. In 2004, a TP_sed maximum was present at the same depths, but all depths down to 30 cm had lower TP_sed than the 1984 profile. Expressed as P per sediment volume, P increased in the upper 4 cm relative to the 1984 profile.

Changes in the sediment profiles of P fractionation were compared using profile data from 1985, 1991, and 1998 (see Figure 15.6). In the upper 20 cm (1985 profile), the majority of the sediment P was extracted as NaOH-P, which comprised 60–80% of TP_sed. HCl-P was relatively constant throughout the sediment and comprised up to 10% of TP, whereas Res-P (organic-bound) comprised 10–25%. Res-P was highest in the upper 5 cm of the sediment. In the period from 1985 to 1998, NaOH-P declined dramatically between depths of 15 cm and 25 cm, whereas there were no clear changes in HCl-P; however, a reduction in Res-P was observed in the upper 5 cm.

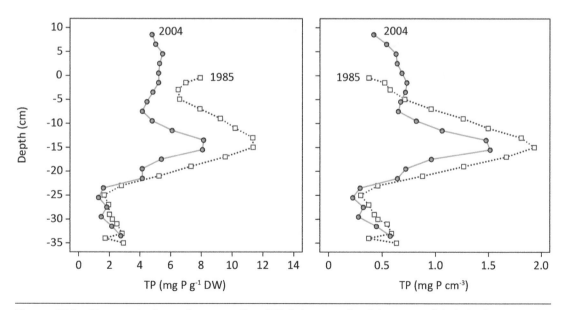

Figure 15.5   Changes in the sediment profile of TP (left: per g dw, right: per cm^3) in Lake Søbygaard from 1985 to 2004. The y-axis scale corresponds to 1985 data, the 2004 sediment surface starting 9.5 cm above the 1985 sediment surface in accordance with an assumed annual net sediment growth of 0.5 cm (Søndergaard et al. 1993). Modified from Søndergaard et al. (1999).

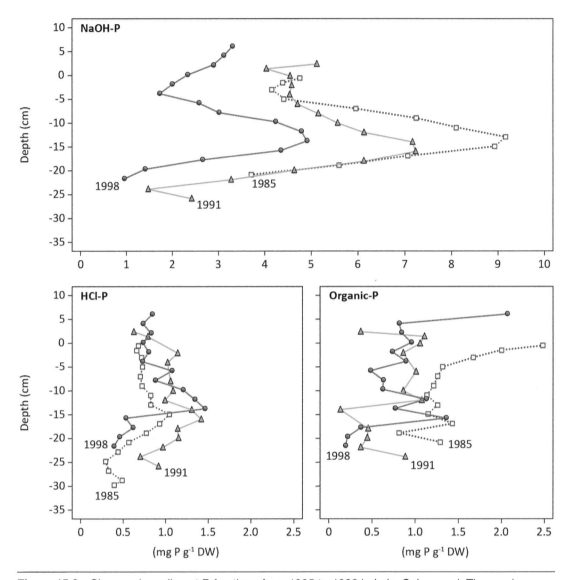

Figure 15.6 Changes in sediment P fractions from 1985 to 1998 in Lake Søbygaard. The y-axis scales represent 1985 data and the 1991 and 1998 sediment surface values with adjustment in accordance with an annual net sediment growth of 0.6 cm (Søndergaard et al. 1993). Modified from Søndergaard et al. (2003). Note different scales for the x-axis.

15.4   DISCUSSION

The highly eutrophic conditions in Lake Søbygaard have resulted in various extremes that have not been observed so far in most other lakes that are under study. For example, TP_inlet concentrations were as high as 5 mg L^{-1}, summer mean TP_lake above 1 mg L^{-1}, summer mean Chl up to 900 µg L^{-1}, pH up to 10.9, and TP_sed up to 12 mg P g^{-1} dw. In addition, net sediment P release rates of 100 mg P m^{-2} day^{-1} were measured. The drivers behind this condition were the very high external loading until 1982, followed by a high release of P from the lake plus high external loading of N until 1996.

It can be argued that the results on nutrient dynamics from such a highly eutrophic lake as Lake Søbygaard cannot be readily transferred to other lakes; however, the extreme conditions prevailing in this lake may reveal some mechanisms that are not easily detected yet relevant also for other lakes. One such mechanism is the long-term net annual release of P from the sediment, which induced changes in the sediment profile.

The active part of a lake's sediment or the depth from which P can be transported to the water column is often debated and is an important parameter when estimating the total pool of mobile sediment P, for example in relation to determining the amount of aluminum that needs to be applied to prevent internal sediment P loading (James 2005; Wang and Jiang 2016). The active layer has been defined as 10 cm (Boström et al. 1982) or is set in relation to the water content of the sediment (Kapanen 2012), but few data are available to provide a robust definition. Usually, the net P release from lake sediment is small compared with the total amount of P present in the sediment, and this makes it difficult to determine the depth from where the P is actually released. However, this was possible for Lake Søbygaard. Based on the TP_sed profile, the active layer is clearly deeper than 10 cm—at least 20 cm or maybe even 25 cm (Søndergaard et al. 1999). A marked maximum pool at 15 cm depth, originating from a period with maximum external loading, was slowly diminished in the years after the loading reductions.

Mass balance calculations have shown that in the period 1985–1998, a total of 39 g P m^{-2} was released from the sediment (Søndergaard et al. 1999). However, sediment profile data from the same period indicate that 57 g P m^{-2} was released in total; at some depths TP_sed was reduced by up to 21 g P m^{-2}. Although the changes recorded in the sediment profile and the mass balance calculation do not perfectly correspond, the results of the two methods seem comparable considering the relatively high uncertainty of both estimates.

We do not know how well the findings from Lake Søbygaard represent other lakes; long-term studies on sediment profiles are rare and changes are difficult to detect because annual release rates are usually very small compared to the total amount of P that is present in the sediment. Furthermore, lake sediments and transport mechanisms may vary significantly between lakes and probably also within a lake, making generalizations problematic. However, our findings underline the importance of considering depths below 10 cm as active sediment layers when investigating the potential long-term internal loading, and under some conditions at least, the upper 20 cm should also be included.

Another clear effect of the internal P loading in Lake Søbygaard is the marked seasonality with high summer and low winter concentrations, which was particularly notable in the first 15 years after the external loading reduction. However, even 35 years after the reduction, this seasonality is still pronounced, with summer concentrations being about twice as high as in winter. A similar seasonal pattern with persistently high summer concentrations has been recorded in other Danish lakes, implying positive retention during winter and negative retention during summer (Søndergaard et al. 2013, 2017a). This may be a general phenomenon in shallow eutrophic lakes in the temperate zone, which means that P availability is higher during summer, even though a lake, on a yearly basis, still may exhibit net positive P retention. A further implication is that nutrient limitation is likely to change over the season, with N most probably being the limiting nutrient during summer and P in spring (Søndergaard et al. 2017b; Ding et al. 2018). Relatively high PO$_4$ concentrations but low NO$_3$ concentrations during the growing season suggest that Lake Søbygaard is more N-limited than P-limited for most of the summer, as suggested also from model simulations (Rolighed et al. 2016).

Finally, Lake Søbygaard illustrates the highly dynamic coupling between sediment and water. The dramatic increase in TP_lake, which occurred rapidly during the phytoplankton collapse in

1985 (see Chapter 4), underlines that the net release of P, affecting lake-water P concentrations, is usually a result of the difference between two large counteracting transport rates: downward sedimentation and upward gross sediment release. Another coupling is between the photosynthetically elevated pH in the lake water and the sediment, constituting a positive loop between high P concentrations (Xie et al. 2003). Thus, high primary production creates high pH in the water and top sediment layer. This entails a higher concentration of more easily releasable P, and if this is released, primary production may be stimulated even further. Pore-water profiles from Lake Søbygaard in July 1985—the same year in which seasonal changes in NH_4Cl-P were observed—showed enhanced pH down to 10 cm depth (Søndergaard 1990). In the upper 5 cm, pH ranged between 9 and 10, being as high as 9.7 in the uppermost 1 cm. Thus, NH_4Cl-P fluctuated, in particular in the upper 10 cm of the sediment. High pH may also be a reason for the transformation of NaOH-P to NH_4Cl-P during summer, as P mobilization and release of NaOH-P are favored under alkaline conditions (Gao 2012).

Three main aspects presented in Chapter 4 of this book are considered important drivers of internal loading: (1) the presence of mobile sediment P; (2) mobilization processes; and (3) transport. They are all discernable in Lake Søbygaard. First, P concentrations in the sediment are very high and P mainly occurs in its mobile form as NaOH-P, likely constituting Fe-bound P that is redox sensitive and, therefore, potentially mobile. Second, a number of mobilization processes are at work, including the aforementioned effects of pH and mineralization of organic matter and organic-bound P, which are traceable in the changes recorded in Res-P during the years following the loading reduction. It is also probable that redox processes associated with low oxygen concentrations and the absence of nitrate during summer are of strong significance for the release of P from the large pool of NaOH-P. Third, several transport mechanisms may be in operation in Lake Søbygaard, including high pore-water concentrations in the upper part of the sediment, with a typical phosphate gradient of 0.5–1.0 mg P L^{-1} cm^{-1} (Søndergaard 1990). This creates an upward transport of PO_4, which is probably further boosted by gas ebullition from the fermentation of organic-rich sediment in deeper parts of the sediment. The importance of wind resuspension has been studied in Lake Søbygaard and it has revealed that strong winds can resuspend up to 1 cm of the sediment and thereby provoke major changes in the pore-water profile down to 6–8 cm depth (Kristensen and Jensen 1987; Søndergaard 1990).

The longevity of internal P loading is an important issue in the management of lakes in which external P loading has been reduced, but where no major improvements in lake-water quality have occurred. The duration of internal P loading after an external loading reduction depends on a number of factors (as described in Chapter 4), but it generally lasts for 10–15 years (Jeppesen et al. 2005). In Lake Søbygaard, internal loading has been significant for more than 25 years, and even today (2017), the yearly TP_inlet is lower than the yearly mean TP_lake, which indicates that Lake Søbygaard, 35 years after the reduction of the external loading, is still not in equilibrium. The reason why the internal loading has continued for a longer period in Lake Søbygaard than in most other lakes is probably that the multiple years of extremely high loading resulted in a very significant P legacy, as well as the lake's iron-rich sediment, which is known to increase the P retention capacity (Søndergaard et al. 1996).

Very few long-term case studies on internal P loading are available; one example is, though, Finnish Lake Tuusulanjärvi, which over a 40-year study period showed no decrease in internal loading despite numerous within-lake management efforts. This is likely due to the potential of the sediment to release P for decades even after external loading is reduced to acceptable levels (Horppila et al. 2017).

15.5   CONCLUSIONS

Lake Søbygaard is an example of how a previously heavily P-loaded lake is severely impacted by internal loading for many years, even after the external loading has been effectively curbed. Today, 35 years after the loading reduction, the mean lake-water P concentrations in the lake are generally still higher than the mean inlet concentrations. Particularly during summer, the P concentrations are much higher than the inlet concentrations, indicating a significant impact of internal loading in Lake Søbygaard whose hydraulic retention time is only a few weeks. During winter, the inlet concentrations are higher than the lake-water concentration, an indication of positive retention of P.

P originates from a pool of mainly NaOH-extractable P and from sediment depths down to at least 20 cm. During summer, photosynthetically elevated pH in the water column helps liberate loosely sorbed P in the surface sediment and is likely one of the mechanisms behind high internal loading during summer. A significant proportion of organic-bound P, which eventually is decomposed and potentially released to the water column, represents another important pathway for the internal cycling of P.

15.6   ACKNOWLEDGMENTS

The project was supported by the European Union (EU) project MARS (Managing Aquatic ecosystems and water Resources under multiple Stress) funded under the 7th EU Framework Programme and is an AU Water Technology Centre (WATEC.au.dk) publication. Anne Mette Poulsen and Juana Jacobsen are acknowledged for editorial and layout assistance.

15.7   REFERENCES

Andersen, JM. 1976. An ignition method for determination of total phosphorus in lake sediments. Water Res. 10:329–331.

Andreasen, K; Søndergaard, M; and Schierup, HH. 1984. En karakteristik af forureningstilstanden i Søbygaard Sø—samt en undersøgelse af forskellige restaureringsmetoders anvendelighed til en begrænsning af den interne belastning (in Danish). MSc project, Botanical Institute, Aarhus University, p. 164.

Boström, B; Jansson, M; and Forsberg, C. 1982. Phosphorus release from lake sediments. Arch Hydrobiol Ergeb Limnol. 18:5–59.

Brodersen, KP; Odgaard, B; Vestergaard, O; and Anderson, NJ. 2001. Chironomid stratigraphy in the shallow and eutrophic Lake Søbygaard, Denmark: chironomid—macrophyte co-occurrence. Freshwater Biol. 46:253–267.

Ding, SM; Chen, MS; Gong, MD; Fan, XF; Qin, BQ; Xu, H; Gao, SS; Jin, ZF; Tsang, DCW; and Zhang, CS. 2018. Internal phosphorus loading from sediments causes seasonal nitrogen limitation for harmful algal blooms. Sci Total Environ. 625:872e884.

Gao, L. 2012. Phosphorus release from the sediments in Rongcheng Swan Lake under different pH conditions. Proc Environ Sci. 13:2077–2084.

Gutierrez, MF; Devercelli, M; Brucet, S; Lauridsen, TL; Søndergaard, M; Jeppesen, E. 2016. Is recovery of large-bodied zooplankton after nutrient loading reduction hampered by climate warming? A long-term study of shallow hypertrophic Lake Søbygaard, Denmark. Water. 8:341.

Hieltjes, AHM and Lijklema, L. 1980. Fractionation of inorganic phosphates in calcareous sediments. J Environ Qual. 9:405–407.

Horppila, J; Holmroos, H; Niemistö, J; Massa, I; Nygrén, NA; Schönach, P; Tapio, P; and Tammeorg, O. 2017. Variations of internal phosphorus loading and water quality in a hypertrophic lake during 40 years of different management efforts. Ecol Eng. 103:264–274.

James, WF. 2005. Alum: Redox-sensitive phosphorus ratio considerations and uncertainties in the estimation of alum dosage to control sediment phosphorus. Lake Reserv Manage. 21:159–164.

Jensen, JP; Jeppesen, E; Kristensen, P; Christensen, PB; and Søndergaard, M. 1992. Nitrogen loss and denitrification as studied in relation to reductions in nitrogen loading in a shallow, hypertrophic lake (Lake
Søbygaard, Denmark). Int Rev Ges Hydrobiol. 77:29–42.

Jeppesen, E; Søndergaard, M; Jensen, JP; Havens, KE; Anneville, O; Carvalho, L; Coveney, MF; Deneke, R;
Dokulil, M; Foy, B; et al. 2005. Lake responses to reduced nutrient loading—an analysis of contemporary
long-term data from 35 case studies. Freshwater Biol. 50:1747–1771.

Jeppesen, E; Søndergaard, M; Jensen, JP; Mortensen, E; Hansen, A; and Jørgensen, T. 1998. Cascading trophic
interactions from fish to bacteria and nutrients after reduced sewage loading: An 18-year study of a shallow hypertrophic lake. Ecosystems. 1:250–267.

Jeppesen, E; Søndergaard, M; Jensen, JP; Mortensen, E; and Sortkjær, O. 1996. Fish-induced changes in zooplankton grazing on phytoplankton and bacterioplankton: a long-term study in shallow hypertrophic
Lake Søbygaard. J Plankton Res. 18:1605–1625.

Jeppesen, E; Søndergaard, M; Lauridsen, TL; Kronvang, B; Beklioglu, M; Lammens, E; Jensen, HS; Kohler, J;
Ventelä, A-M; Tarvainen, M; et al. 2007. Danish and other European experiences in managing shallow
lakes. Lake Reserv Manage. 23:439–451.

Jeppesen, E; Søndergaard, M; Sortkjær, E; Mortensen, E; and Kristensen, P. 1990. Interactions between phytoplankton, zooplankton and fish in a shallow, hypertrophic lake: a study of phytoplankton collapses in Lake
Søbygaard, Denmark. Hydrobiologia. 191:149–164.

Kapanen, G. 2012. Pool of mobile and immobile phosphorus in sediments of the large, shallow Lake Peipsi over
the last 100 years. Environ Monit Assess. 184:6749–6763.

Kristensen, P and Jensen, P. 1987. Sedimentation and resuspension in Søbygaard Sø. Report from Botanical
Institute, University of Aarhus/The Freshwater Laboratory. p. 150 (in Danish).

Liboriussen, L and Jeppesen, E. 2003. Temporal dynamics in epipelic, pelagic and epiphytic algal production in
a clear and a turbid shallow lake. Freshwater Biol. 48:418–431.

Rolighed, J; Jeppesen, E; Søndergaard, M; Bjerring, R; Janse, JH; Mooij, WM;and Trolle, D. 2016. Climate
change makes recovery from eutrophication more difficult in shallow Danish Lake Søbygaard. Water.
8:459.

Søndergaard, M. 1987. Lake Søbygaard: A shallow lake in recovery after a reduction in phosphorus loading.
GeoJournal. 14(3):381–384.

Søndergaard, M. 1988. Seasonal variations in the loosely sorbed phosphorus fraction of the sediment of a shallow and hypereutrophic lake. Environ Geol Water Sci. 11:115–121.

Søndergaard, M. 1989. Phosphorus release from a hypertrophic lake sediment—Experiments with intact sediment cores in a continuous-flow system. Arch Hydrobiol. 116:45–59.

Søndergaard, M. 1990. Porewater dynamics in the sediment of a shallow and hypertrophic lake. Hydrobiologia.
192:247–258.

Søndergaard, M; Bjerring R; and Jeppesen, E. 2013. Persistent internal phosphorus loading during summer in
shallow eutrophic lakes. Hydrobiologia. 710:95–107.

Søndergaard, M; Jensen, JP; and Jeppesen E. 1999. Internal phosphorus loading in shallow Danish lakes. Hydrobiologia. 408/409:145–152.

Søndergaard, M; Jensen, JP; and Jeppesen, E. 2003. Role of sediment and internal loading of phosphorus in
shallow lakes. Hydrobiologia. 506–509: 135–145.

Søndergaard, M; Jeppesen, E; Jensen, JP; and Amsinck, SL. 2005. Water Framework Directive:
ecological classification of Danish lakes. Appl Ecol. 42:616–629.

Søndergaard, M; Jeppesen, E; Kristensen, P; and Sortkjær, O. 1990. Interactions between sediment and water
in a shallow and hypertrophic lake—A study on phytoplankton collapses in Lake Søbygaard, Denmark.
Hydrobiologia. 191:139–148.

Søndergaard, M; Kristensen, P; and Jeppesen, E. 1993. 8 years of internal phosphorus loading and changes in
the sediment phosphorus profile of Lake Søbygaard, Denmark. Hydrobiologia. 253:345–356.

Søndergaard, M; Lauridsen, TL; Johansson, LS; and Jeppesen, E. 2017a. Repeated fish removal to restore lakes:
Case study Lake Væng, Denmark—Two biomanipulations during 30 years of monitoring. Water. 9:43.

Søndergaard, M; Lauridsen, TL; Johansson, LS; and Jeppesen, E. 2017b. Nitrogen or phosphorus limitation
in lakes and its impact on phytoplankton biomass and submerged macrophyte cover. Hydrobiologia.
795:35–48.

Søndergaard, M; Windolf, J; and Jeppesen, E. 1996. Phosphorus fractions and profiles in the sediment of shallow Danish lakes as related to phosphorus load, sediment composition and lake chemistry. Water Res. 30: 992–1102.

Sørensen, T; Mulderij, G; Søndergaard, M; Lauridsen, TL; Liboriussen, L; Brucet, S; and Jeppesen, E. 2011. Winter ecology of shallow lakes: strongest effect of fish on water clarity at high nutrient levels. Hydrobiologia. 664:147–162.

The Danish Environmental Protection Agency, 2013. Basis analyse for vandområdeplaner 2015–2021 (in Danish) http://miljoegis.mim.dk/spatialmap?&profile=vandrammedirektiv2basis2013.

Wang, C and Jiang, H-L. 2016. Chemicals used for in situ immobilization to reduce the internal phosphorus loading from lake sediments for eutrophication control. Crit Rev Environ Sci Technol. 46(10):947–997.

Xie, L; Xie, QP; and Tang, HJ. 2003. Enhancement of dissolved phosphorus release from sediment to lake water by Microcystis blooms—an enclosure experiment in a hyper-eutrophic, subtropical Chinese lake. Environ Pollut. 122:391–399.

CHAPTER 16

ACCUMULATION OF LEGACY SEDIMENT PHOSPHORUS IN LAKE HJÄLMAREN, SWEDEN: CONSEQUENCES FOR LAKE RESTORATION

Brian J. Huser[1], Mikael Malmaeus[2], Ernst Witter[3],
Anders Wilander[1], and Emil Rydin[4]

Abstract

Lake Hjälmaren, the 4th largest lake in Sweden, currently suffers from poor water-quality conditions due to excess inputs of nutrients to the system. It is a complex system, consisting of four major basins, two of which are shallow and well mixed, and two that are deeper and intermittently stratified. In addition to external input of nutrients and other pollutants, the water level of the lake was lowered in the late 1800s by approximately two meters to increase the area of farmland surrounding the lake. The current state of Lake Hjälmaren is considered to be eutrophic to hypereutrophic (depending on the basin). Historic (pre-2000) average total phosphorus (TP) concentrations during summer (June–September) ranged from 41 µg L^{-1} to 324 µg L^{-1} in the different basins. In the most recent five years (2012–2016), this range has dropped to between 41 µg L^{-1} and 75 µg L^{-1}. Algal productivity was elevated as well, with chlorophyll *a* concentrations ranging from 22 µg L^{-1} to 97 µg L^{-1} during 2012–2016. Water clarity has changed little since measurements began in the 1960s. Although external loading has decreased significantly, lake-water quality remained essentially unchanged (or slightly worse) in the deeper basins where sediment accumulation bottoms exist. In these areas, over 4000 metric tons of potentially mobile (i.e.,

continued

---

[1] Department of Aquatic Sciences and Assessment. Swedish University of Agricultural Sciences, Box 7050. 750 07 Uppsala, Sweden. E-mail for correspondence: brian.huser@slu.se.

[2] IVL Svenska Miljöinstitutet AB, Box 210 60,100 31 Stockholm, Sweden.

[3] County Administrative Board of Örebro, 701 86 Örebro, Sweden.

[4] Naturvatten i Roslagen AB, Norra Malmavägen 33, 761 73, Norrtälje, Sweden.

legacy) phosphorus (P) has accumulated in the sediment due to elevated external P inputs, driving internal P loading in these parts of the lake. Herein, we discuss the past and current conditions of the lake, along with how continued management of nutrients may return the lake to a less productive, more natural state.

Key words: Phosphorus, internal loading, water quality, sediment, eutrophication, Sweden.

16.1   INTRODUCTION

Lake Hjälmaren is the 4th largest lake in Sweden (483 km²) and one of the 25 largest lakes in Europe. The lake is 58 km long, up to 18 km wide, and is divided into four sub-basins: Hemfjärden, Mellanfjärden, Storhjälmaren (including Södra Hjälmaren), and Östra Hjälmaren (see Figure 16.1). Overall Lake Hjälmaren is considered shallow (polymictic), but the sub-basins represent different types of limnic systems. Hemfjärden and Mellanfjärden are shallow and well mixed, whereas Storhjälmaren occasionally stratifies and Östra Hjälmaren stratifies regularly during the growing season (see Table 16.1). Because of the large fetch, however, none of the sub-basins would be considered dimictic, as the entire water column can recirculate during the growing season. Reports from as early as the 1700s indicated the lake already suffered from excess nutrient inputs and associated problems with eutrophication (Willén 2001). From 1878 to 1887, the surface level was lowered by between 1.3 m and 2 m in one of the largest lake-draining projects in Sweden, reducing the lake volume by 28%. The outlet of the lake (Eskilstuna River) flows to Lake Mälaren and then on to the Baltic Sea. Water level is now regulated via the outlet where a number of hydropower stations are located. Lake Hjälmaren is jointly governed by a number of municipalities, a local farmer's co-op, power utilities, and a collective that represents new land areas that were exposed during the lowering of the lake level in the late 1800s. The lake is an important resource and provides numerous ecosystem services, including fisheries, drinking water, transport, and recreation.

The bulk of flow enters Lake Hjälmaren via the river Svartån through Hemfjärden, but as much as 35% enters Storhjälmaren from the south, with a main contribution from River Täljeån (see Table

Table 16.1   Morphological and hydrological data for Lake Hjälmaren and its sub-basins (Håkanson 1978, 1981)

Basin	Area km²	Max Depth m	Mean Depth M	Volume 10⁶ m³	Tributary Flow %	Hydraulic Residence Time Year
Hemfjärden	25	2.6	1.0	24	55	0.05
Mellanfjärden	40	3.2	1.8	73	1.5	0.15
Storhjälmaren	377	22	6.9	2617	35	3.3
Östra Hjälmaren	36	24	5.0	178	8	0.22
Entire lake	478	24	6.1	2893	100	3.4

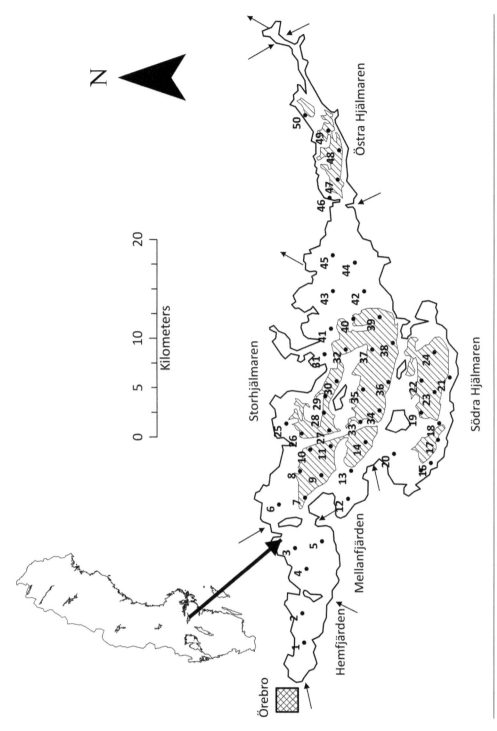

Figure 16.1 Location of sediment sampling stations in Lake Hjälmaren with basins, from left to right, Hemfjärden, Mellanfjärden, Storhjälmaren, Södra Hjälmaren (often included as a part of Storhjälmaren), and Östra Hjälmaren. Diagonal lines represent accumulation bottoms (modified from Håkanson 1981). Small arrows pointing to or from the lake represent major inflows and outflows, respectively. The square with cross-hatched lines indicates the location of Örebro city.

Table 16.2 External P input amounts and percent of contribution of the total P load from major tributaries to Lake Hjälmaren (SLU 2014)

Source	Receiving Basin	P Load	Total Input*	Agriculture	Wastewater**
		Ton year⁻¹		%	
River Svartån	Hemfjärden	21.0	36.0	65	3
River Täljeån	Storhjälmaren	15.0	25.7	74	2 (11)
Other small rivers	Storhjälmaren, Östra Hjälmaren	8.0	13.7	Not calculated	
Deposition	All	3.0	5.1	NA	NA
Additional diffuse input	All	11.4	19.5	NA	NA
Total		58.4			

* Total P input is further split into percentages coming from agriculture and wastewater for each source.
** Number in parenthesis includes contribution from on-site, individual septic systems.

16.2). Lake Hjälmaren's catchment area is 3800 km², of which currently, nearly half (44%) is forest, 35% is agricultural land, and 15% is water (including Lake Hjälmaren itself). Numerous smaller towns and cities surround the lake, the largest being Örebro with a population of 129,703 (2007). Nevertheless, urban land use comprises less than 3% of the catchment area. There are no reliable data to allow a numerical comparison between current and historic land use, but an ocular comparison of current land-use maps and historic maps (1864–1873) suggests that the amount of woodland has not changed significantly. Nonetheless, up until the 1950s (in contrast to now), most woodlands were dual purpose and also used for grazing of livestock. Numbers of livestock (cattle and sheep) were in fact more than twice as numerous in the latter half of the 19th century as they are now. A significant change in land use during the mid-19th century was the drainage of wetlands (previously used for grazing or hay production) and lakes to make more land suitable for arable crops. The lowering of the water level of Lake Hjälmaren between 1878 and 1887 can be seen as one such project that created nearly 20,000 ha of new land that is now mainly used for agriculture. Significant portions of the rivers Täljeån and Svartån, as well as their contributories, were also canalized, reducing the hydraulic residence time in the catchment.

To reach water-quality goals for Lake Hjälmaren, the average, lake-wide total phosphorus (TP) concentration must be reduced by over 50% to 22 μg L⁻¹, which would mean a further reduction of total loading by 50% to 60%. It is unlikely that external measures alone will be adequate for achieving P loading reduction considering the amount of legacy phosphorus (P) contained in the sediment (due to historical P loading) that contributes to internal P cycling. Because of the varying morphologies of the sub-basins, internal P loading is likely driven by different processes. During stratification in deeper areas, low redox conditions may promote the reduction of iron (Fe) and subsequent release of historically accumulated legacy P in the sediment. In shallower bays, resuspension and degradation of organic matter likely contribute the bulk of the internal load. Because of this, Lake Hjälmaren is a complex system, requiring a different focus on measures to restore water quality in the different basins. To determine the potential for internal loading of P across the lake, a sediment-sampling campaign was carried out and a model was developed using both sediment and external and internal water-quality monitoring data. The results of these projects, along with potential management scenarios to improve water quality in Lake Hjälmaren, are presented.

16.2   METHODS

16.2.1   Water Chemistry

Consistent sampling of Lake Hjälmaren began in the 1960s and data are stored by both the Hjälmarens vattenvårdsförbund (http://www.vattenorganisationer.se/hjalmaren/index.php) and the Department for Aquatic Sciences and Assessment at the Swedish University of Agricultural Sciences (https://www.slu.se/institutioner/vatten-miljo/). Analytical methods have varied over the years, and information on these can also be found at the aforementioned links. Although water chemistry and quality have been monitored regularly since the 1960s, samples were most often collected at the surface (0.5 m) only in late winter/early spring and summer (usually August). In addition, bottom water in the deeper basins (generally between 1 m and 2 m from the sediment surface) was not always sampled (or analyzed for P) during each sampling event. Basic statistics were conducted using Sigmaplot (Version 11.0.1) and time-series data were analyzed using the Mann-Kendall trend test (Loftis et al. 1991). Significant monotonic trends ($p \leq 0.05$) were determined and Sen's-slope (Sen 1968) was used for trend estimates.

16.2.2   Sediment Sampling and Analysis

Sediment sampling was carried out in May 2015 and 50 sediment cores were collected across Lake Hjälmaren. The sampling stations were selected to represent both transport and accumulation bottoms in the different sub-basins of the lake (see Figure 16.1).

   All sediment cores were sliced on-site (0–1, 1–2, 3–4, 6–7, 9–10, 14–15, 24–25 and 34–35 cm layers) and analyzed for water and organic content, P fractions (Psenner et al. 1988), and TP. All fractions were analyzed as soluble reactive P (SRP), except for the NaOH extraction step where both SRP and TP were analyzed in order to determine both Al-P and organic P. Potentially releasable, mobile P was quantified by subtracting the mean P concentration in deeper sediment layers, representing the background concentration. The difference is considered to be the sum of the P that may be released over time (Rydin et al. 2011), and consists of organic forms together with Fe-bound, loosely bound, and pore-water P (i.e., legacy P). P content in the layers that were not analyzed was linearly interpolated, and potentially mobile P in each sediment layer was converted to mass (per m^2) and summed for the upper 15 cm of sediment. The average mobile P mass for the different sub-basins was multiplied by the accumulation and transport bottom area (Håkanson 1981) to determine the total amount of legacy P potentially available for release to the water column in these areas. Additional sediment cores from four representative stations (4, 23, 37, and 40) were sliced at 2 cm intervals (down to 36 cm) and the activity of ^{137}Cs was measured using a gamma detector to determine the approximate age of the sediment.

   Because water-quality data were not sufficient to calculate P release rates from the sediment, a model was used to estimate potential internal loading rates (Pilgrim et al. 2007). The mobile (pore water, loosely sorbed, and Fe-bound) P mass in surficial sediment was used to estimate the amount of P that can be released to the water column. Potential internal P loading rates were calculated based on these fractions for the 50 sediment cores (mentioned previously) collected from Lake Hjälmaren in 2015.

16.2.3   Mass Balance Modeling

The Lake Eutrophication, Effect, Dose, Sensitivity (LEEDS) model (Malmaeus and Hakanson 2004) is a one-dimensional, mass-balance model that calculates the concentration of P in the water column of a lake by numerically solving a system of ordinary differential equations resulting from the parameterization of the fluxes (see Figure 16.2). To apply the model for the four basins of Lake Hjälmaren, four sub-systems are connected (see Figure 16.3). In the model, accumulation bottoms (A-sediment) are separated from erosion and transport bottoms (ET-sediment). The distribution of different bottom types was defined using empirical data (Håkanson 1981). The wave base and thermocline were determined from statistical models in combination with measured temperature profiles.

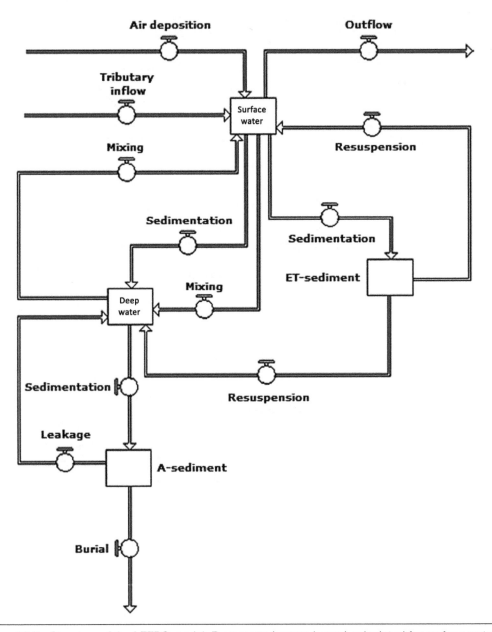

Figure 16.2 Structure of the LEEDS model. P amounts (square boxes) calculated for surface water, deep (bottom) water, and internal sediment release from erosion and transport (ET) and accumulation (A) bottoms. Transport between compartments is indicated by valves (circles).

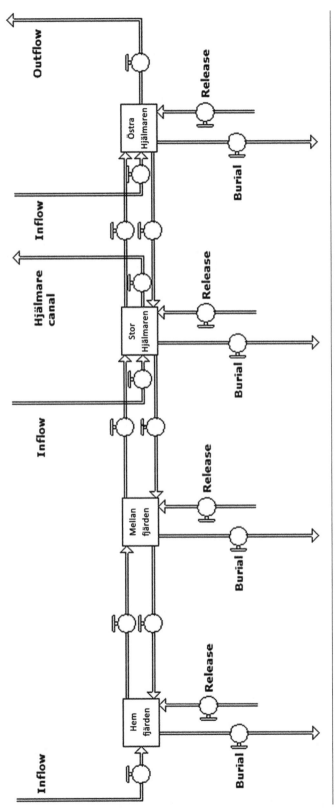

Figure 16.3 Overview of Lake Hjälmaren as a system with four separate basins. Transport between basins and between sediment and water are indicated by valves.

In addition to water fluxes, the model simulates the vertical mixing between surface water and deep water, sedimentation, resuspension, burial, and release of P from sediment. The model is driven by morphological and hydrological data with calibrated algorithms for internal processes (sedimentation, resuspension, release, mixing), which in turn, are driven by oxygen concentration and temperature in surface and deep waters. With this model, effects on P dynamics of the system driven by reduced exchange between water and sediment, as well as from reduced external P loading, were simulated.

16.3 RESULTS

16.3.1 Surface Water Quality

Lake Hjälmaren has been and continues to be affected by excess amounts of external loading. Over time, this has led to an accumulation of legacy P in sediment that continues to negatively affect water quality, even as external loading has been reduced. The effects of this can be clearly seen when comparing water quality in the different sub-basins (see Figure 16.4).

Substantial declines in TP concentration were seen until the year 2000 in both Hemfjärden (4.5 µg L^{-1} yr^{-1}, p ≤ 0.00001) and Mellanfjärden (2.5 µg L^{-1} yr^{-1}, p ≤ 0.0001) after initiation of advanced wastewater treatment in the 1970s. In Storhjälmaren, however, little change occurred and

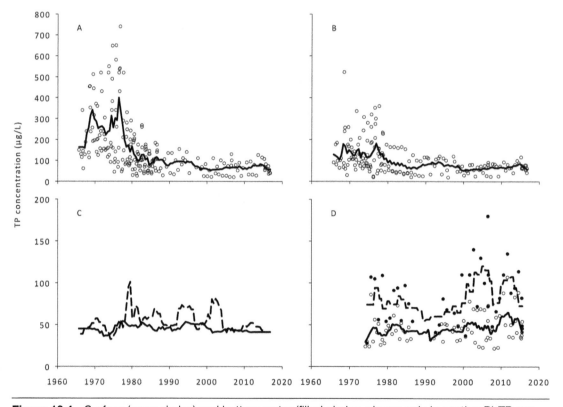

Figure 16.4 Surface (open circles) and bottom water (filled circles, shown only in section D) TP concentrations in (A) Hemfjärden, (B) Mellanfjärden, (C) Storhjälmaren, and (D) Östra Hjälmaren. Running averages shown for surface water (solid line) and deep water (dashed line). (Data points excluded from C to improve readability.)

TP actually increased in Östra Hjälmaren during this period ($0.2\ \mu g\ L^{-1}\ yr^{-1}$, $p \leq 0.05$). Surface water TP concentration in Östra Hjälmaren was also significantly greater ($p \leq 0.01$) during more recent years (2000–2016) compared to historical data (pre-2000). This increase may be due to elevated internal P loading (see Figure 16.4D), as increasing amounts of legacy P accumulated in the sediment, but with the low frequency of water chemical data available, it was not possible to show this quantitatively. Even though substantial declines in TP were seen from 1965–1999 in some areas of the lake, no significant trends were detected for water transparency (as Secchi depth), except for a slight increase in Storhjälmaren ($1.7\ cm\ yr^{-1}$, $p \leq 0.01$).

16.3.2   Legacy Sediment P in Lake Hjälmaren

The amount of mobile sediment P mass was generally greatest in the sediment of Storhjälmaren, slightly lower in the southern (Södra Hjälmaren) and eastern (Östra Hjälmaren) basins, and significantly lower in Mellanfjärden. In Hemfjärden, no potentially mobile P was detected (see Table 16.3). Potential sediment release rates, using loosely bound pore water and Fe-P, were calculated based on the upper 5 cm of sediment (Huser and Pilgrim 2014; Pilgrim et al. 2007). Mean sediment release rates for each basin ranged from $0\ mg\ P\ m^{-2}\ d^{-1}$ to $8.8\ mg\ P\ m^{-2}\ d^{-1}$ across the lake, with higher potential loading rates in Storhjälmaren and Östra Hjälmaren, and generally very low or no internal loading in Hemfjärden and Mellanfjärden (see Table 16.3).

The average burial (or background) concentration of sediment P ($700\ \mu g\ g^{-1}\ dw$) and a sediment accumulation rate of 3–6 mm per year indicates that between 48 and 170 tons of P are buried each year in Lake Hjälmaren sediment, of which 48–96 tons exist in accumulation bottom and 36–96 tons in transport bottom sediment. However, we can assume that sediment accumulation is less in transport bottom areas, so a burial rate in the order of 100 tons yr^{-1} is a reasonable estimate.

In total, approximately 4,300 tons of sediment P are considered potentially available for release to the water column in Lake Hjälmaren. In general, the P contained in the top 15 cm of sediment in the profile has accumulated during the previous 20–30 years, given the estimated sediment increase according to sediment dating (^{137}Cs, Malmaeus and Rydin 2015). Assuming that the system is currently in *equilibrium*, about 140–210 tons of P should, therefore, be released to the water column every year, 100–150 tons of which would come from sediment in Storhjälmaren. Organic P mineralized below the upper 15 cm of sediment, however, might diffuse toward the sediment surface, adding to the mobile P pool in the surficial sediment. Thus, the previous calculations may underestimate the effect of P in older sediment on internal loading in Lake Hjälmaren.

Table 16.3   Mobile P in the different basins of Lake Hjälmaren, including Fe-P and organic P in surficial sediment collected in 2015. Station refers to the location of each sediment core shown in Figure 16.1. Potential internal loading rates (Li; Pilgrim et al. 2007) are shown as well.

Basin	Station	Mobile P			Li	
		Total	Fe-P	Org-P	Range	Mean
		g m^{-2}			mg m^{-2} d^{-1}	
Hemfjärden	1–2	0	0	0	0.6–0.7	0.7
Mellanfjärden	3–5	6	1	5	0.2–0.4	0.3
Storhjälmaren	6–14; 25–45	12	6	6	4.6–13	8.8
Södra Hjälmaren	15–24	11	5	5	6.3–9.1	7.1
Östra Hjälmaren	46–50	9	5	4	6.0–8.6	7.7

The use of a background sediment P concentration (average of the 24–25 cm and 34–35 cm layer concentrations) to identify potentially mobile P implies that the higher P concentrations in surficial sediment layers are expected to decrease via diffusion toward the sediment surface (Carey and Rydin 2011). In lakes where the composition of deposited material has undergone diagenesis, P burial rate and concentration can be affected and cause under- or over-estimation of true burial rates. The reduction of water level in Lake Hjälmaren in the late 1800s resulted in an increased transport of eroded material to the lake (Malmaeus and Karlsson 2015). The proportion of clay material decreased during the time period represented in the cores (i.e., the clay proportion was significantly larger 3 decimeters down compared to the younger, surficial layers). Thus, it is possible that the background concentration was underestimated because clay generally has lower P content compared to organic-rich sediment. Nonetheless, the aforementioned results indicate that measures to reduce internal P loading would decrease P concentrations in the water column by increasing permanent burial in sediment.

16.3.3   Modeling—Phosphorus Dynamics of Lake Hjälmaren

P was modeled in Lake Hjälmaren with help of the LEEDS model (Malmaeus and Karlsson 2015) and calibrated using sediment and external and in-lake monitoring data. Phosphorus dynamics—including fluxes to, from, and within the system; exchange between water and sediment; and exchange between the basins were calculated. Management scenario analyses were then conducted and different measures (both external and internal) were evaluated. Model results are shown together with measurements (mean of data from 2004–2013) in Hemfjärden and Mellanfjärden surface water and in the surface and deep water of Storhjälmaren and Östra Hjälmaren (see Figure 16.5).

According to measurements, the annual average external P load to the lake was 52 tons during 2004–2013. The measured outflow during the same period averaged 40 tons per year, giving a net retention of 12 tons of P per year (~25% of the total inflow mass). P concentrations in the water column increased during summer, indicating internal loading, and then gradually decreased from late autumn (see Figure 16.5). The modeled output, based on the LEEDS mass balance model and measured values, agreed reasonably well with measured values (see Table 16.4). The model does overestimate retention slightly (15 tons), suggesting that internal sources are somewhat underestimated as well. Even so, as indicated by the modeled exchange of P between sediment and the water column in Storhjälmaren (see Table 16.4), internal release of P is approximately 10 times the external inflow and outflow to this basin, which may be explained by the large area of accumulation sediment and mass of mobile P in this basin. There is also an exchange between ET sediments and the water column (resuspension and sedimentation). The large fluxes between water and sediment reflect considerable exchange; although the annual net fluxes between water and sediment may be small, imbalances between these fluxes during the year may cause considerable changes in TP, especially during the growing season when sediment P release is greatest.

Flow through Lake Hjälmaren is generally considered to run west to east, from Hemfjärden to Östra Hjälmaren basin. Transport of P within the lake, however, can occur both in the *upstream* and *downstream* directions (see Figure 16.3 and Table 16.4). Due to the size and morphology of the lake, it was modeled as four, distinct basins (basically interconnected lakes). Thus, P is transported to the lake from the watershed, but it can also move between basins within the lake. For example, the flow from the *upstream* basin (e.g., from Hemfjärden to Mellanfjärden with a P source of 29 kg to Mellanfjärden) corresponds with the outflow from the upstream Hemfjärden (a P outflow of 29 kg)

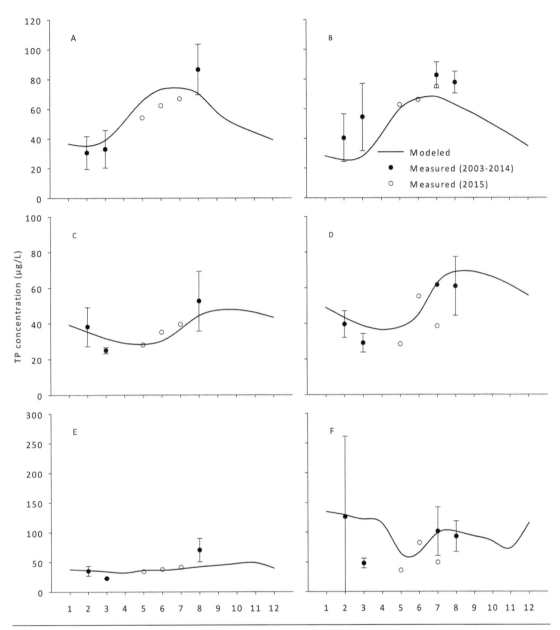

Figure 16.5   Modeled and measured means by month for TP from 2004–2013 (filled circles) or single measured values for 2015 (open circles) in (A) Hemfjärden, (B) Mellanfjärden, (C—surface and D—bottom water) Storhjälmaren, and (E—surface and F—bottom water) Östra Hjämaren. Standard deviations shown for data from 2004–2013.

(see Table 16.4). The exception to this is a small fraction (8%) of the outflow from Storhjälmaren that exits the lake through Hjälmare canal and Östra Hjälmaren where the main outlet (Eskilstuna River) exists. Fluxes between basins are significant, and in the two western basins (Hemfjärden and Mellanfjärden), fluxes from the deeper basins (Storhjälmaren and Östra Hjälmaren) are substantial.

Table 16.4 Modeled P mass balance for Lake Hjälmaren (ton yr⁻¹). Fluxes are illustrated in Figures 16.2 and 16.3. Note that three tons exit the lake via outflow through the Hjälmaren canal, whereas the remainder (35 tons) exits the lake via Östra Hjälmaren.

	Hemfjärden	Mellanfjärden	Storhjälmaren	Östra Hjälmaren	Total
Sources					
External inflows	21	0	27	5	53
From upstream basin		29	30	29	
From downstream basin	8	11	0		
Resuspension	61	125	666	15	867
Sediment release	0	1	255	8	264
Sum	90	166	978	57	1184
Sinks					
Outflow	29	30	31	35	38
Export to upstream basin	0	8	11	0	
Sedimentation ET-bottoms	61	125	666	15	867
Sedimentation A-bottoms	0	3	267	6	276
Sum	90	166	975	57	1180
Sediment burial	0	0	80	6	86

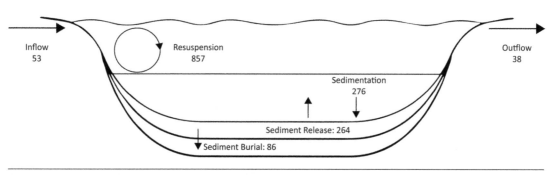

Figure 16.6 Mass balance of P in Lake Hjälmaren (ton yr⁻¹)

That is, P from the deeper basins can be transported *upstream* and thus internal P loading in the deeper basins can also affect water quality in shallower areas of the lake. The burial flux is the differences between inflows and outflows and is sustained by a pool of mobile, legacy P available in the sediment (see Figure 16.6).

In summary, the results of both modeling and sediment P content indicate the importance of internal loading for P dynamics in Lake Hjälmaren, especially in the central and eastern basins. During the summer there is a release of P from sediment in these basins that can be explained by higher temperature throughout the water column and low oxygen concentrations during stratification in deeper parts of the lake. Accordingly, P concentration in the entire water column is generally elevated during the summer relative to the winter months. While Hemfjärden and Mellanfjärden show the most pronounced increase in TP during the summer (see Figure 16.5A and B), it is mostly explained by resuspension, rather than release of mobile P from the sediment in these basins.

16.3.4   Modeling—Management Scenario Analysis

Evaluating scenarios further illustrates the relative importance of external and internal fluxes for management of P dynamics in the lake. Malmaeus and Karlsson (2015) evaluated the following management scenarios:

- 50% reduction of external P loads from rivers to all basins;
- Complete reduction of internal P loading in 50% of the accumulation area in Storhjälmaren (no other basins treated);
- A combination of the two previous measures.

By treatment of sediment accumulation areas in Storhjälmaren only, it was assumed that all mobile P in the treated sediments was permanently immobilized and hence would not contribute to P flux to the water column. It was not specified what method would be used to achieve this, but the addition of mineral binders (e.g., Al salts) or complete removal of the surficial sediment (top 15 cm) would produce these results.

The effect on P concentration in the epilimnion of the four basins based on these three scenarios was compared with current conditions (see Figure 16.7). In these simulations, a reduction of external P loading would benefit primarily Hemfjärden and Mellanfjärden, whereas the effect in the two eastern basins (Storhjälmaren and Östra Hjälmaren) would be minor. Treatment of sediment (50% of the accumulation bottom) in Storhjälmaren produced a substantially larger effect both in Storhjälmaren and in Östra Hjälmaren, but surface TP would also be reduced in Mellanfjärden due to the interaction between the basins. Good chemical status (indicated by the black lines in Figure 16.7) would be achieved in three of the four basins with either one or a combination of these measures, except in Östra Hjälmaren. It is likely that reduction of internal sources of P in Östra Hjälmaren, and/or a greater reduction of internal P loading in Storhälmaren, would reduce surface water P levels sufficiently to meet water quality goals in this basin as well.

Taken together, the model simulations indicate that external loading has a greater influence on epilimnetic P concentrations in Hemfjärden and Mellanfjärden, whereas internal processes largely drive P concentrations in Storhjälmaren and Östra Hjälmaren, with sediment playing a key role. This is apparent from the scenarios where reductions of external loading generated substantial effects in the two western basins while P concentration in the central and eastern basins remained largely unaffected. This is also in accordance with historical evidence where further improvements in wastewater treatment in the 1970s resulted in substantial declines in surface water P concentration in Hemfjärden and Mellanfjärden. Phosphorus concentrations in Storhjälmaren and Östra Hjälmaren were largely unaffected, and since 2000, TP concentrations have actually increased in Östra Hjälmaren, most likely due to increased internal P loading (see Figure 16.4).

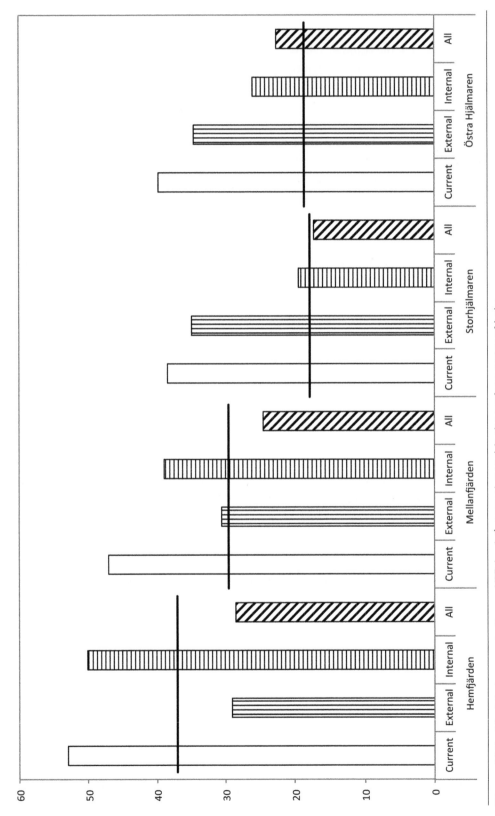

Figure 16.7 Modeled concentrations of TP (μg L⁻¹, annual mean) in the surface water of Lake Hjälmaren under current loading (open), reduced external loading (vertical lines), reduced internal loading (horizontal lines), and a combination of these measures (diagonal). The horizontal lines (solid) represent the limits for good water quality status in the respective basins.

16.4   DISCUSSION

16.4.1   Surface-Water P Dynamics

Clear reductions have been seen in surface water TP since the initiation of advanced wastewater treatment in the 1970s in Lake Hjälmaren (see Figure 16.4); however, these reductions have been limited to the western, shallow basins of the lake (Hemfjärden and Mellanfjärden). In Östra Hjälmaren, epilimnetic TP has significantly increased since the year 2000 (see Figure 16.4). The reason why a reduction of the external P loading generates little to no effect on TP concentration in the central and eastern basins (Storhjälmaren and Östra Hjälmaren, respectively) is that there is a major P flux from a pool of mobile, legacy P present in the surficial sediment. This pool of legacy P may emanate from P inputs to the lake that occurred before advanced wastewater treatment was implemented. However, it seems unusual that inputs that occurred 30 to 40 years ago still affect the current ecological status of the lake. In other large lakes in central Sweden (including Lake Vänern, Lake Vättern, and Lake Mälaren), the recovery after reduction of external nutrient loads has been significantly faster (Wilander and Persson 2001). This could be due to the shallow, polymictic nature of Lake Hjälmaren, as well as the smaller portion of accumulation areas in Lake Vänern, Vättern, and Mälaren, where permanent burial of P may occur. Stratification in deeper, more strongly stratified lakes will generally limit the transfer of P from bottom to surface waters (diffusion dominates during summer), and a smaller accumulation area will increase the burial rate.

16.4.2   Sediment P and Internal Loading in Lake Hjälmaren

Much of the P transported to lakes is incorporated in the bottom sediment of the deeper areas where fine-particle material either settles or is transported from erosion and transport bottoms. Mineralization of newly deposited organic material, such as phytoplankton, is a primary source of dissolved phosphate in sediment (Xie 2006). The release of mineralized P (as phosphate) is often controlled by the reduction-oxidation (redox) potential at the sediment water interface. When oxygen deficiency (anoxia) occurs, bacteria begin to use other elements as electron acceptors for energy production, including Fe. Once Fe is reduced from its ferric (Fe^{3+}) to ferrous form (Fe^{2+}), it releases previously bound P to the sediment pore water, which in turn can cause increases in water-column P via diffusion or mechanical disturbance of the sediment. However, P release can also occur in oxygenated sediments due to degradation of organic matter, elevated pH (> 9) that solubilizes metals such as Fe and Al, or lack of sufficient sediment binding capacity (e.g., Fe, Al, and Ca). Over time, the capacity of the sediment to retain P upon final burial will influence the amount of P released to the water column. Estimating this pool of P is crucial for a number of reasons, such as predicting quantities of ecologically relevant P stored in a lake and potential for transport downstream over time. It also forms the basis for remediation measures (e.g., application of Al salts) to increase P binding capacity in the sediment.

The mass of legacy sediment P in Lake Hjälmaren (averaging up to 12 g m^{-2} in Storhjälmaren) is considered high compared to other large water bodies in and around Sweden. Average levels of mobile P in Baltic Sea archipelago areas, where eutrophication is a substantial problem, have been estimated to range from 2.5 g m^{-2} (Malmaeus et al. 2012) to 3.5 g m^{-2} (Puttonen et al. 2014). The mass of legacy P in surficial sediment of Lake Erken (25 km^2) averages 5 g m^{-2} in accumulation bottoms (Rydin 2000). Potential internal loading rates, reaching 13 mg m^{-2} d^{-1} in Storhjälmaren (based on Pilgrim et al. 2007), are also considered high and indicative of eutrophic to hyper-eutrophic conditions (Nürnberg and LaZerte 2004). The modeled rates may, however, overestimate internal P loading due to polymictic conditions in Lake Hjälmaren that can result in higher dissolved oxygen at the sediment water interface during some years. But they may also underestimate internal loading because the model does not incorporate mineralization and release of P from organic matter. Because water temperatures can reach 20°C during summer throughout the water column, and approximately 50% of the potentially mobile P is in the organic form, mineralization likely contributes substantially to P release from the sediment during the growing season.

According to the model, sediment burial of P is 86 tons per year, which is close to the 100 tons estimated from sediment analysis (Malmaeus and Rydin 2015). Annual P release from sediment amounts to 264 tons, with 255 tons coming from sediment in Storhjälmaren, which means that the existing amount of potentially mobile, legacy P (approximately 2,900 tons in Storhjälmaren alone) would be flushed from the system in approximately 11 years, assuming a constant release rate. If the system was in balance and met reference targets for P, about 140–210 tons of sediment P would be released to the water column every year, of which 100–150 tons would come from Storhjälmaren. There are, of course, uncertainties in the sediment measurements and modeling, so this difference in itself is not remarkable. However, it is likely that the lake is not in balance (i.e., excess internal P loading), and thus sediment P release is larger than burial. This further suggests that measures to bind legacy sediment P would reduce P in the water column.

16.4.3   Other Factors That May Affect Internal Loading of P

Fish kills have occurred in Lake Hjälmaren, mainly in the western portions of the lake. The latest occurrences of these were in 2007 and 2011. While the reason(s) behind these are not certain, no toxins were found in the dead fish. It has been hypothesized that nitrogen poisoning (cf. Mortonson and Brooks 1980; Wuertz et al. 2013) during fall 2007 when conversion of ammonium to nitrate was limited by lower temperatures, likely caused the fish kill in 2007 (SRK 2007). Although information is lacking, degradation of water quality and fish habitat likely led to more frequent fish kills in the past before reductions in external loading were made in the 1970s. Not unexpectedly, the fish community in Lake Hjälmaren is currently dominated by fish typically found in eutrophic, polymictic lakes. From the 1950s to 1970s, noticeable declines in perch and the European whitefish (Salmonidae family) occurred while the number of bream and roach increased. Catches from 2002–2016 included 24 species—but perch, white bream, roach, bream, and pike-perch were most common. Species in the cyprinid family were, on average, 56% of the catch (by weight). The wels catfish (sheatfish), chub, brown trout, and river lamprey were found in the lake in the past but are no longer present (Degerman et al. 2001).

The diversity of macrophytes is also poor in Lake Hjälmaren and resembles the poor diversity often found in other eutrophic lakes. According to a survey done between 1970 and 1973, aquatic vegetation was dominated by the common reed (*Phragmites australis*), which composed 86% of the vegetated area (Andersson and Eriksson 1974). Due to a combination of low water clarity and wave action, the depth limit for macrophyte growth was only 1.5 m in Hemfjärden and Mellanfjärden and 2.0–2.5 m in Storhjälmaren and Östra Hjälmaren. Unfortunately, no similar surveys have been conducted since the 1970s, but aerial photos taken more recently do not indicate a substantial increase in vegetation coverage area in the lake.

Both burrowing activity by benthic feeding fish species and limited macrophyte coverage can lead to increased turbidity and release of sediment P. Carp were shown to increase the sediment mixing depth in a shallow lake in the US, resulting in a doubling of potentially mobile, legacy P that could be released to the water column (Huser et al. 2016a). Although previous reporting on macrophytes in Lake Hjälmaren attributed the low diversity and coverage to nutrient-related turbidity and wind-driven resuspension, benthic feeding fish can also affect macrophytes negatively by both reducing water clarity (via sediment resuspension) and by physically burrowing in the sediment and uprooting macrophytes in search of food (Bajer et al. 2009; Lougheed et al. 1998). Rooted macrophytes stabilize sediment, leading to less resuspension (Scheffer 1990). They can also supply oxygen to sediment via the root system, minimizing or preventing the release of reductant soluble P (Horppila and Nurminen 2005); however, increased pH within the plant bed itself may stimulate P release (James et al. 1996).

16.4.4   Management—Past and Present

Wastewater input coming from cities such as Örebro, runoff from agricultural areas, and input from industrial activities (among other sources) all contributed to the store of legacy P that exists in Lake Hjälmaren sediment today. Substantial amounts of organic matter (up to 520 kg of cellulose fiber per day from one paper mill alone) were also discharged to the lake, with reported massive algal blooms in connection to these pollutant releases. Hemfjärden and Mellanfjärden basically functioned as settling and oxygenation basins for the incoming organic pollutant loads. Treatment plants were constructed in the 1950s and 1960s to treat wastewater using mechanical and biological methods. By the 1970s, precipitation using metal salts (e.g., Al) further decreased P input through river Svartån (approximately 50 to 150 tons per year after initiation of tertiary treatment). The heavy, long-lasting algal blooms that occurred during summer have been substantially reduced since the 1970s (Willén 2001); however, frequent blooms still occurred in shallow basins and near towns (Persson 1996). A cardboard manufacturing plant was closed in 2010 and ongoing projects to reduce P from other external sources continue to be implemented. Given the approximate 50 tons of P that enter the lake each year, substantial reductions will still be necessary to reach a further reduction of 25 tons (Willén 2001). This reduction will likely result in a substantial decrease in surface water P and phytoplankton growth in the two recipient basins (Hemfjärden and Mellanfjärden), but further measures to reduce internal loading in the easternmost basins of the lake will be required to meet water quality goals.

16.4.5   Development of Internal P Loading in the Lake

Release of legacy sediment P was a problem in all basins before substantial reductions in external loading were made in the mid-1970s, and often P retention in the lake was negative (Wilander and Persson 2001). More recently, retention has been generally positive, but there was negative retention (−8 tons) as recent as 2010 (Malmaeus and Karlsson 2015). Due to short residence times and the transitory nature of Hemfjärden and Mellanfjärden lake bottoms, legacy sediment P was likely transported rather quickly from these basins to deeper areas of the lake after the previous reductions to external loading. Since then, release of legacy sediment P remains a problem in the eastern basins of the lake, despite there now being an average net retention of approximately 12 tons per year (mean 2004–2013). Although water quality data do not cover the growing season consistently (generally only spring and late summer, except for 2015), intermittent stratification has been detected in the deeper basins (Storhjälmaren and Östra Hjälmaren). Under these polymictic conditions, P concentrations do not reach levels often seen in the bottom water of eutrophic or hyper-eutrophic dimictic lakes, but the P that is released is quickly mixed into the photic zone. This process can happen multiple times during the growing season, creating a cycle of P release wherein organic matter production (algae) increases then settles to the sediment. This presumably adds to the sediment oxygen demand that promotes further release of sediment P.

16.4.6   Future Management Options

According to modeling results, using both sediment and water-chemical data, Lake Hjälmaren will require substantial reductions to both external and internal loading in order to meet water quality goals set out in the European Union Water Framework Directive (EUWFD 2000). A reduction of external loading by 50% will be required for all basins to meet water quality goals; however, Storhjälmaren and Östra Hjälmaren (the two deeper basins) will also require at least a 50% reduction of internal loading. Because this chapter focuses on internal loading of P, external loading measures are not discussed in detail. But projects designed to reduce P input to Lake Hjälmaren are underway or being planned. Most of these projects focus on reducing nutrients in storm water, individual sewer systems, and agricultural runoff since these are some of the largest sources of P to the lake.

Reducing sediment P release in Lake Hjälmaren will require major efforts. Planning efforts are underway and pilot studies will soon be conducted to determine the best method(s) to reduce internal loading in the lake. Two of the most likely options for managing internal P loading in the lake include dredging and increasing sediment P-binding capacity via the addition of P-binding minerals such as gibbsite (Al-hydroxide). The scale of such a project, however, makes it difficult to accurately estimate the cost of different management measures. The size of such a project may reduce overall cost due to economy of scale, but the large size may also make certain parts of a project more difficult logistically (i.e., pumping sediment across kilometers of lake bed or travel time to add Al to the lake bottom).

The best way to calculate and compare cost effectiveness is to determine the cost per unit of P removed or inactivated in a lake. There are many data available for Al treatment of lake sediment because it has been used since the 1960s to reduce internal loading and hundreds of lakes have been treated around the world (Huser et al. 2016b; Welch and Cooke 1999). Cost efficiency of Al treatment is quite low, ranging from 13 to 82 euro per kg P inactivated by Al (2014 Euro) (Huser et al. 2016d). Most lakes that have been treated, however, are a fraction of the size of Lake Hjälmaren. Thus, we used treatment data only for lakes that were larger than 150 ha in a recent study conducted for the Swedish Environmental Protection Agency (EPA) (Huser et al. 2016d). Unfortunately, most dredging projects reviewed in the study lacked sediment P data, and P fractions were not analyzed in any of the projects. Thus, a comparison was made using the average cost per unit area of lake surface (see Table 16.5).

Table 16.5 Cost estimates for Al treatment of sediment and sediment removal. Costs (2014 basis) are based on treating or dredging 50% of the accumulation areas in Storhjälmaren and Östra Hjälmaren and include cost per unit area (Euro ha^{-1}) and total cost (Euro) for the two management methods. Note that total cost is in millions of euros.

Basin	Accumulation Bottom		Legacy P	Al Treatment*		Dredging**	
	%	km²	Tons	Euro ha^{-1}	Euro	Euro ha^{-1}	Euro
Hemfjärden	0	0	NA	Not considered for treatment			
Mellanfjärden	0	0	NA	Not considered for treatment			
Storhjälmaren	49.5	187	2213	1123	10.5	8516	79.5
Östra Hjälmaren	60.4	21.7	196	1123	1.2	8516	9.3

* Huser et al. 2016c, Huser et al. 2016d
** Huser et al. 2016d, Iowa DNR 2017

Clearly the costs for managing internal loading are substantial, irrespective of the method, but it should also be noted that estimating cost by area of lake surface, and not by how much legacy P is in the sediment, increases the uncertainty of these calculations. If, for example, one lake has double the legacy P in surficial sediment compared to another, cost for Al treatment would be nearly double, whereas cost for dredging would remain the same. One must also take into account the effectiveness of the method. Assuming, however, that the same amount of legacy P is either permanently inactivated (Al treatment) or removed (dredging), the longevity of these two management methods would be the same and only depend on continued input (or lack thereof) of excess external P loads. Another important factor when considering project cost is the revenue generation that can be created after restoration (e.g., increased tourism and improved fishery), which can often pay for the restoration project in a few years (Iowa DNR 2017; Isely et al. 2018). Revenue may be generated by repurposing of the sediment, but even though the content of potentially available P in the sediment of Lake Hjälmaren is relatively high, it is much lower (mean = 1.2 mg g^{-1}, max = 2.1 mg g^{-1} total P, not potentially available P) than that found in commercial fertilizer. In addition, other pollutants such as heavy metals and organic toxins are also often present in eutrophic lake sediment, meaning removal of these would be necessary depending on the type of reuse. But as mineral P deposits become harder to find and more costly to mine, reuse of dredged sediment as fertilizer, for example, may generate revenue in the future. And if sediment pollutants are low and agricultural areas are close to the lake, pumping dredge spoils directly on to land also may be feasible from an economic perspective.

Although modeling was based on reducing only 50% of the legacy sediment P in Storhjälmaren, Östra Hjälmaren also has a substantial amount of legacy P that contributes to internal P loading in the lake. Ten tons of legacy P also remain in the sediment of Mellanfjärden, and this P will eventually be released to the water and/or transported along the sediment bottom to deeper areas (Malmaeus and Rydin 2015). Targeting legacy sediment P with P binding minerals (e.g., Al) in areas with transport bottoms has been shown to be a cost effective method for reducing mobile P input to, and internal loading in, deeper areas of lakes. This is due to the transitory nature of sediment in shallow areas, and targeting this internal P loading source as it moves naturally to deeper areas has resulted in greatly improved binding efficiency and lower internal loading rates in untreated, deeper lake areas (Huser 2017). Thus, management of legacy sediment P, other than the 50% of the accumulation bottom in Storhjälmaren, should be considered to help meet water quality goals.

16.4.7   Innovative Methods

There are two innovative treatment methods that have recently been developed for reducing internal P loading in Swedish lakes. The first is a sediment injection method that pumps the amorphous Al mineral to a specific depth range in the surficial sediment. Al is also added to the water column in lakes where bottom-water P is elevated during the treatment. The reasons for developing the injection were twofold. Sediment injection should reduce horizontal transport across the lake bottom that focuses the Al mineral floc in deeper areas, thereby increasing treatment effectiveness (Schütz et al. 2017) and binding efficiency of the added Al (de Vicente et al. 2008; Huser 2012). The first lake treated with this method was Lake Flaten in Stockholm (treated in 2000), which continues to have good water quality and very low TP concentrations in the hypolimnion throughout the summer stratification period (ca. 15 µg L^{-1}) (Schütz et al. 2017). The binding efficiency, however, was similar to that found in lakes where Al was applied to the water and the amorphous mineral was allowed to settle to the sediment. Because of the longer time it takes to treat a lake using the injection method, the cost has been greater compared to traditional water-column precipitation treatments. But only a small number of lakes have been treated using this method to date, and thus results on cost efficiency have a high degree of uncertainty.

The second method is a low-flow, hydraulic dredging system that removes surficial sediment that generally contains greater amounts of legacy P. The unit is designed to move along the sediment bed via remote control, pumping sediment back to shore for treatment. This method is still in the pilot-stage testing phase, so no data on cost effectiveness were available to include in this chapter.

16.4.8 Resuspension

Resuspension of sediment also contributes to TP measured in the epilimnion, but much of this is likely due to the morphology of the lake, especially in the shallow, westernmost basins Hemfjärden and Mellanfjärden. The effects of resuspension can be clearly seen when comparing P concentration in Hemfjärden in late winter (when the lake is normally ice covered) with those in late summer. On average, summer P concentrations are twice as high as those in late winter. This difference is not related to any differences in P concentrations in the main inflow Svartån, which shows no such seasonal pattern. An interesting question is: what role does the P that is released by resuspension play? Given that the sediment data clearly show that virtually none of the sediment P collected from Hemfjärden and Mellanfjärden is potentially mobile, it is likely that very little is bioavailable if resuspended.

Water-quality guidelines have been developed to attempt to take natural resuspension of sediment into account (Huser and Fölster 2013), but differentiating turbidity between resuspended sediment and excess algal production is difficult. Reduction of low-density organic matter (e.g., algae) may reduce resuspension potential, but given the large wind fetch, resuspension of sediment will likely continue (Håkanson and Jansson 1983). Reduction of benthic feeding fish would likely reduce both sediment resuspension in shallower areas and the amount of legacy P available for release from sediment in deeper areas (Huser et al. 2016a). Costs for this method were not calculated due to uncertainty about the current biomass density and how effective such a treatment would be in a lake as large as Hjälmaren.

16.5 CONCLUSIONS

Lake Hjälmaren is a complex system that has suffered the effects of anthropogenic eutrophication for at least 300 years. Although external P loading has been reduced significantly, nutrient inputs continue to negatively affect water quality and contribute to legacy P accumulation in the sediment. Management of the lake will require multiple measures in order to meet water-quality goals. External loading and sediment resuspension control water quality in the two shallow western basins, whereas internal loading of legacy sediment P sustains poor water quality in the two deeper basins to the east. Using sediment and water chemical data, lake modeling suggested that a reduction of 50% to both external and internal loading is needed to achieve water quality goals. Additional management of legacy P in sediment bottoms in Östra Hjälmaren and Mellanfjärden will likely improve the chance for successful restoration of Lake Hjälmaren.

16.6 ACKNOWLEDGMENTS

The authors are grateful to Lake Hjälmaren's vattenvårdsförbund and the Dept. of Aquatic Sciences and Assessment at SLU for providing water quality data. We are also grateful to the editors Bryan Spears and Alan Steinman for their helpful comments on the manuscript.

16.7   REFERENCES

Andersson, B and Eriksson, S. 1974. Aquatic vegetation in Lake Hjälmaren 1970–1973 (in Swedish). Report SNV PM 461 Swedish EPA limnological investigations 75. Uppsala, Sweden.

Bajer, PG; Sullivan, G; and Sorensen, PW. 2009. Effects of a rapidly increasing population of common carp on vegetative cover and waterfowl in a recently restored midwestern shallow lake. Hydrobiologia. 632(1): 235–245.

Carey, CC and Rydin, E. 2011. Lake trophic status can be determined by the depth distribution of sediment phosphorus. Limnol Oceanogr. 56(6):2051–2063.

de Vicente, I; Huang, P; Andersen, FO; and Jensen, HS. 2008. Phosphate adsorption by fresh and aged aluminum hydroxide. Consequences for lake restoration. Environ Sci Technol. 42(17):6650–6655.

Degerman, E; Hammar, J; Nyberg, P; and Svardson, G. 2001. Human impact on the fish diversity in the four largest lakes of Sweden. Ambio. 30(8):522–528.

EUWFD. 2000. Directive 2000/60/ec of the european parliament and of the council of 23 october 2000 establishing a framework for community action in the field of water policy. European Parliament, Council No. L 372.

Håkanson, L. 1978. Hjälmaren—en naturgeografisk beskrivning. Swedish EPA. Report SNV PM 1079.

Håkanson, L. 1981. Sjösedimenten i recipientkontrollen—principer, processer och praktiska exempel. Swedish EPA. Report SNV PM 1398.

Håkanson, L and Jansson, M. 1983. *Principles of Lake Sedimentology*. Berlin: Springer-Verlag.

Horppila, J and Nurminen, L. 2005. Effects of different macrophyte growth forms on sediment and p resuspension in a shallow lake. Hydrobiologia. 545:167–175.

Huser, BJ. 2012. Variability in phosphorus binding by aluminum in alum treated lakes explained by lake morphology and aluminum dose. Water Res. 46(15):4697–4704.

Huser, BJ. 2017. Aluminum application to restore water quality in eutrophic lakes: maximizing binding efficiency between aluminum and phosphorus. Lake Reserv Manage. 33(2):143–151.

Huser, BJ; Bajer, PG; Chizinski, CJ; and Sorensen, PW. 2016a. Effects of common carp (*Cyprinus carpio*) on sediment mixing depth and mobile phosphorus mass in the active sediment layer of a shallow lake. Hydrobiologia. 763(1):23–33.

Huser, BJ; Egemose, S; Harper, H; Hupfer, M; Jensen, H; Pilgrim, KM; Reitzel, K; Rydin, E; and Futter M. 2016b. Longevity and effectiveness of aluminum addition to reduce sediment phosphorus release and restore lake water quality. Water Res. 97:122–132.

Huser, BJ and Folster, J. 2013. Prediction of reference phosphorus concentrations in Swedish lakes. Environ Sci Technol. 47(4):1809–1815.

Huser, BJ; Futter, M; Lee, JT; and Perniel, M. 2016c. In-lake measures for phosphorus control: the most feasible and cost-effective solution for long-term management of water quality in urban lakes. Water Res. 97:142–152.

Huser, BJ; Löfgren, S; and Markensten, H. 2016d. Internal loading of phosphorus in Swedish lakes and coastal areas—an overview of knowledge and recommendations for measures for water management (in swedish). Uppsala, Sweden: Sveriges lantbruksuniversitet.

Huser, BJ and Pilgrim, KM. 2014. A simple model for predicting aluminum bound phosphorus formation and internal loading reduction in lakes after aluminum addition to lake sediment. Water Res. 53(0):378–385.

Iowa DNR. 2017. Lake Restoration Program 2017 Report and 2018 Plan. Iowa Department of Natural resources. Des Moines, Iowa. http://www.iowadnr.gov/Portals/idnr/uploads/fish/programs/files/17report.pdf.

Isely, P; Isely, ES; Hause, C; and Steinman, AD. 2018. A socioeconomic analysis of habitat restoration in the muskegon lake area of concern. J Great Lakes Res. 44(2):330–339.

James, WF; Barko, JW; and Field, SJ. 1996. Phosphorus mobilization from littoral sediments of an inlet region in lake delavan, wisconsin. Arch Hydrobiol. 138(2):245–257.

Loftis, JC; Taylor, CH; Newell, AD; Chapman, PL. 1991. Multivariate trend testing of lake water-quality. Water Resour Bull. 27(3):461–473.

Lougheed, VL; Crosbie, B; and Chow-Fraser, P. 1998. Predictions on the effect of common carp (*Cyprinus carpio*) exclusion on water quality, zooplankton, and submergent macrophytes in a great lakes wetland. Can J Fish Aquat Sci. 55(5):1189–1197.

Malmaeus, JM and Hakanson, L. 2004. Development of a lake eutrophication model. Ecol Model. 171(1–2):35–63.

Malmaeus, M and Karlsson, M. 2015. Phosphorus dynamics in Lake Hjälmaren. Result of Simulations (in Swedish). Report C72. IVL Svenska Miljöinstitutet. Stockholm, Sweden.

Malmaeus, M and Rydin, E. 2015. Sediment study in Lake Hjälmaren. Results from sampling in May 2015 (in Swedish). Report C136. IVL Svenska Miljöinstitutet. Stockholm, Sweden.

Malmaeus, JM; Rydin, E; Jonsson, P; Lindgren, D; and Karlsson, OM. 2012. Estimating the amount of mobile phosphorus in baltic coastal soft sediments of central Sweden. Boreal Environ Res. 17(6):425–436.

Mortonson, JA and Brooks, AS. 1980. Occurrence of a deep nitrite maximum in Lake Michigan. Can J Fish Aquat Sci. 37(6):1025–1027.

Nürnberg, GK and LaZerte, BD. 2004. Modeling the effect of development on internal phosphorus load in nutrient-poor lakes. Water Resour Res. 40(1):W01105.

Persson, G. (Ed.). 1996. Lake Hjälmaren during 29 years of study (In Swedish). Report 4535. Swedish Environmental Protection Agency. Stockholm, Sweden.

Pilgrim, KM; Huser, BJ; and Brezonik, PL. 2007. A method for comparative evaluation of whole-lake and inflow alum treatment. Water Res. 41(6):1215–1224.

Psenner, R; Boström, B; Dinka, M; Pettersson, K; Puckso, R; and Sager, M. 1988. Fractionation of phosphorus in suspended matter and sediment. Arch Hydrobiol Suppl. 30:98–103.

Puttonen, I; Mattila, J; Jonsson, P; Karlsson, OM; Kohonen, T; Kotilainen, A; Lukkari, K; Malmaeus, JM; and Rydin, E. 2014. Distribution and estimated release of sediment phosphorus in the northern Baltic Sea archipelagos. Estuar Coast Shelf S. 145:9–21.

Rydin, E. 2000. Potentially mobile phosphorus in Lake Erken sediment. Water Res. 34(7):2037–2042.

Rydin, E; Malmaeus, JM; Karlsson, OM; and Jonsson P. 2011. Phosphorus release from coastal Baltic Sea sediments as estimated from sediment profiles. Estuar Coast Shelf S. 92(1):111–117.

Scheffer, M. 1990. Multiplicity of stable states in fresh-water systems. Hydrobiologia. 200:475–486.

Schütz, J; Rydin, E; and Huser, BJ. 2017. A newly developed injection method for aluminum treatment in eutrophic lakes: effects on water quality and phosphorus binding efficiency. Lake Reserv Manage. 33(2):152–162.

Sen, PK. 1968. Estimates of the regression coefficient based on kendall's tau. J Am Stat Assoc. 63(324):1379–1389.

SLU 2014. Swedish recipient control for the Eskilstunaåns watershed (in Swedish). Report 2014:10. 43p. Inst. för vatten och miljö, Sveriges lantbruksuniversitet. Uppsala, Sweden.

SRK 2007. Swedish recipient control for the Eskilstunaåns watershed (in Swedish). Hjälmarens Vattenvårdsförbund. https://www.vattenorganisationer.se/hjalmaren/downloads/50/Hjlmarens_vvf_Rapport_2007.pdf.

Welch, EB, Cooke GD. 1999. Effectiveness and longevity of phosphorus inactivation with alum. Lake Reserv Manage. 15(1):5–27.

Wilander, A and Persson G. 2001. Recovery from eutrophication: experiences of reduced phosphorus input to the four largest lakes of Sweden. Ambio. 30(8):475–485.

Willén, E. 2001. Four decades of research on the Swedish large lakes Malaren, Hjalmaren, Vattern and Vanern: the significance of monitoring and remedial measures for a sustainable society. Ambio. 30(8):458–466.

Wuertz, S; Schulze, SGE; Eberhardt, U; Schulz, C; and Schroeder, JP. 2013. Acute and chronic nitrite toxicity in juvenile pike-perch (Sander lucioperca) and its compensation by chloride. Comp Biochem Phys C. 157(4):352–360.

Xie, P. 2006. Biological mechanisms driving the seasonal changes in the internal loading of phosphorus in shallow lakes. Sci China Ser D. 49:14–27.

CHAPTER 17

LIMITED ROLE OF INTERNAL LOADING IN A FORMERLY HYPERTROPHIC SHALLOW LAKE IN THE NETHERLANDS

Ruurd Noordhuis[1], Gerlinde Roskam, and Leonard Osté

Abstract

Lake Eemmeer in the central part of The Netherlands is a medium-sized, shallow, buffered lake. It was a hypertrophic lake with heavy algal blooms during the 1970s and 1980s, fed by large amounts of nutrients (phosphorus (P) up to 45 g m^{-2} yr^{-1}) mainly from the catchment of the River Eem.

The external loading gradually decreased due to measures taken in the 1980s in the catchment of the River Eem; the decrease was more dramatic in 1995/1996 due to a reduction in the contribution of P coming from treatment plants. Extremely high discharge in two consecutive winters (1993/94 and 1994/95) may have caused unintentional flushing, followed by a period of exceptionally low discharges that further decreased the P load from the River Eem. Following the reduction of external loading, in-lake total P (TP) concentrations remained high for only about one year before declining.

Moreover, due to increased fishery pressure for bream, the physically induced internal loading due to resuspension of P-rich sediments was reduced resulting in increased transparency. An increase in dreissenids and macrophytes, covering sediment and collecting suspended matter, may have amplified the effect.

The reduction of the P loading led to decreasing P concentrations in the sediment. The Fe:P ratio in the sediment and the P concentrations in the pore water both suggest that internal loading currently does not play a significant role in this lake.

A reduction of the water-column TP concentration from the present value of 0.15 mg L^{-1} to 0.09 mg L^{-1}, the *good ecological potential* (GEP) value for this lake, requires a reduction of the external loading by another 50%. Other lakes in the area have met this target

continued

---

[1] Deltares, PO Box 85467, 3508 AL Utrecht, The Netherlands E-mail: Ruurd.Noordhuis@deltares.nl

value; however, they had a much longer hydrologic residence time. Due to its strong riverine input, Lake Eemmeer is not likely to meet the GEP-value without extensive measures reducing (mainly) diffuse sources of P. Moreover, a reduction in the P concentration may not have a positive effect on the lake's ecological condition if it results in a slowdown in the growth of dreissenids, thereby reducing the removal of phytoplankton by filtration.

Keywords: Lake Eemmeer, phosphorus, internal loading, The Netherlands

17.1   INTRODUCTION

17.1.1   Key Water-Body Features

Lake Eemmeer is a 15 km² lake with an average depth of 2.1 m. It shares its history with Lake IJsselmeer and Lake Markermeer as the remaining compartments of a former brackish, inland sea of 5,900 km², which was closed by a dam creating a freshwater system in 1932. In later years, about half of the enclosed area was reclaimed, and in this process, Lake Eemmeer was constructed as part of a chain of Borderlakes around the reclaimed polders in 1957. To the east, Lake Eemmeer has an open connection to channel-like *Lake Nijkerkernauw* and to the west, to Lake Gooimeer, together forming the southern part of the chain. Other Borderlakes mentioned in this chapter are the central Lake Veluwemeer and Lake Wolderwijd, and the northern Lake Ketelmeer (see Figure 17.1), which receives water directly from a River Rhine branch (River IJssel) and has a residence time of only a few days. The residence time of Lake Eemmeer is also relatively short, due to the seasonal pattern in discharge of its main source of water, the River Eem, and ranges between 13 days in winter and 40 days in summer. The lake became hypertrophic in the 1970s due to input of highly eutrophic water from the River Eem. External nutrient loading, however, has been strongly reduced over the last two decades. Since 1995, the water has become progressively clearer, and macrophytes and mussel densities have been increasing.

17.1.2   Challenges

External nutrient loading of Lake Eemmeer has decreased but still remains relatively high due to diffuse sources in the valley of the River Eem; variables such as total phosphorus (TP) and chlorophyll *a* (Chl *a*) still do not meet the requirements of the European Union Water Framework Directive (EUWFD). Average Chl *a* concentrations have decreased but blooms of cyanobacteria still occur locally. Also, in recent years, excessive growth of pondweeds and filamentous algae forming extensive *floating algae beds* (flab) caused problems regarding the recreational use of the lake. While a further reduction of external P loading becomes more and more difficult, this study focuses on the role of internal P loading in the lake. Due to strong retention of nutrients in past decades, internal loading may slow down the process of re-oligotrophication of the lakes. This study consists of three parts:

1. A historical analysis of changes in water quality and ecology;
2. The results of a field campaign of sediment sampling carried out in autumn 2017; and
3. An estimate of fluxes and P-cycling within Lake Eemmeer.

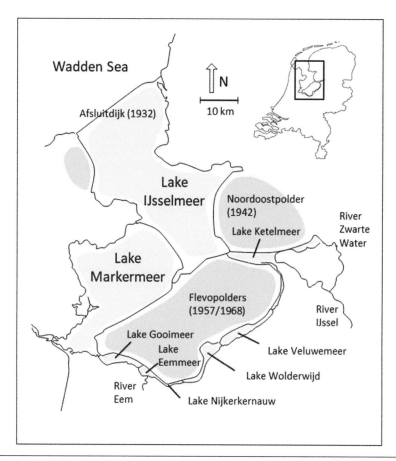

Figure 17.1   Lake IJsselmeer area with locations of the lakes mentioned in the text.

The key questions of this chapter are:

1. Is it currently internal loading that keeps the lakes from meeting WFD targets now that external loading has been substantially reduced?
2. And, if so, what are the drivers of internal loading and can they be manipulated in order to meet the targets?

17.2   METHODS

17.2.1   Long-Term Monitoring of Surface Water

A country-wide monitoring program of national water bodies provided a basis for historical analysis of past changes. This program includes monthly sampling of water quality at one central location in Lake Eemmeer (phosphorus (P), nitrogen (N), Chl *a*, algal species composition, suspended matter, transparency, and many other variables), that has been carried out since around 1970. Every two to four years, macrophyte densities and species composition, dreissenid mussel densities, and the composition of the fish community have been surveyed.

17.2.2   Sediment Sampling

In early September 2017, bed sediments were sampled at 13 locations in Lake Eemmeer and the surrounding Borderlakes to provide insight into spatial patterns of the variables to assess the role of sediment P release. The sediment was collected with an Uwitec sediment core sampler; the top 10 cm of three to five cores were combined to form one sample per location. The sediment was analyzed by TGA (Thermogravimetric analysis) and the element content was determined by ICP-OES after destruction with a mixture of $HClO_4$, HF, and HNO_3. Pore water was sampled from the mixed sediment by using a rhizon with a mean pore size of 0.15 µm. The pH, dissolved organic carbon (DOC), ortho-P, and element concentrations (by ICP-OES) of the pore water were determined.

17.2.3   Calculation of External and Internal Loading: a Mass Balance

A lot of research has been done to quantify nutrient transfer across the sediment-water interface in shallow lakes (Böstrom 1988; Van der Molen and Boers 1994). This is nevertheless a complicated process. First, the water-sediment interface in shallow lakes has a steep redox gradient. Within mm to cm, the conditions can change from fully aerobic to anaerobic (Penn et al. 2000). This has a huge effect on the mobility of P. Iron-bound P mobilizes in anaerobic sediment, but once it enters the surface water it is bound by iron again. Second, P is a very dynamic element. The P-cycle via algae, decomposition of (settled) detritus, and the interaction with inorganic suspended matter (erosion and settling) makes it very difficult to distinguish the contribution of the sediment from other processes. For a proper evaluation of sediment as a P source, other load sources and internal dynamics also need to be considered. This section will first describe a number of methods to estimate the P flux from (or to) sediments. After the focus on sediments, flux will be considered as a part of the P-cycle, which clearly shows the importance of the sediment compared to other processes.

Internal loading cannot be evaluated in its entire context without considering the external loads. In the longer term, if external loads are more or less constant, the internal loading will adjust to the external loading, resulting in a stationary situation. Lake restoration by reducing the internal load can speed up the recovery as long as the internal loading is not proportional to the external loading (Spears et al. 2013). Such an evaluation starts by creating a mass balance that takes both external loading and the internal processes into account. Our approach to construct such a mass balance is shown in Figure 17.2.

First, six fluxes are distinguished for particulate and dissolved P:

1. The external loads via surface water and groundwater;
2. The outflow of P via surface water and groundwater;
3. The gross internal load (release);
4. Algae production turning dissolved P into particulate P;
5. Storage in sediment and biota; and
6. Sorption, both adsorption and desorption.

This section describes the stepwise approach that is used to come to this mass balance.

External Loads

External loads are all determined individually, based on monitoring data (µg P L^{-1}) and the inflow (m^3 water yr^{-1}). Two P-forms are distinguished: $P_{dissolved}$ (measured) and $P_{particulate}$ (calculated: $P_{total} - P_{dissolved}$). This is a conservative estimate, as it does not account for other possible external

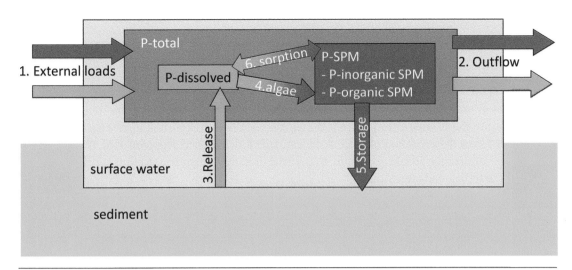

Figure. 17.2   General approach to compose a mass balance that includes internal processes. SPM = suspended particulate matter, P = phosphorus.

sources, such as atmospheric deposition, groundwater discharge, or fecal material of roosting geese and swans that forage on land. Given the load from the River Eem, these sources are assumed to be too small to justify quantification.

P-Outflow

Similar to the inflow, the outflow is calculated, but now the concentration of the lake water is used. The outflow is equal to the total inflow, corrected for rainfall and evaporation, but these are minor factors if lakes have a relatively short residence time. Other reasons for *outflow* of P can be the harvesting of biota (e.g., fish, vegetation). We calculated the contribution of fishing, but it appeared to be negligible on the total mass balance.

Gross Internal Loading

Common methods to determine gross internal loading are: core incubation; a flux chamber; or pore-water profiles (Orihel et al. 2017; see Chapter 2). Core and flux-chamber experiments result in a flux that is valid for the specific condition of the experiments. Pore-water profiles, or even total concentrations in the top layer, require a calculation of a concentration into a flux. These calculation methods can be validated for several locations to generalize their applicability. Van der Molen and Boers (1994) analyzed 49 shallow lakes and derived a general formula to estimate the internal P loading based on the total concentrations of P and Fe in sediments (Table 3 in Van der Molen and Boers 1994):

$$\text{Log } L_{int} = 1.29 \times \log (P_{sed}/Fe_{sed}) + 1.86 \qquad \text{[Eq. 17.1]}$$

in which:

L_{int} = internal loading (g m^{-2} yr^{-1})
P_{sed}/Fe_{sed} = the P/Fe-ratio in sediment (g g^{-1})

This can be rewritten to a flux in mg P m^{-2} day^{-1}:

$$L_{int} = 198 \times (P_{sed}/Fe_{sed})^{1.29} \qquad \text{[Eq. 17.2]}$$

Hin et al. (2010) developed a guideline for sediment assessment for practical water management. The proposed method is valid if P is the element that hampers the ecological objectives. If N is the limiting factor, the N concentrations in sediments generally decrease soon after external loads are reduced, whereas the reduction of sediment P can take much longer; Jeppesen et al. (2005) argued, on the basis of data for 35 lakes, that due to internal loading P oligotrophication is delayed 5 to 10 years after external load reduction.

The following rules of thumb are used to determine whether or not the sediment is an important source with respect to eutrophication. The first check is whether Fe/S (g g^{-1}) < 1. The Fe/S ratio indicates whether the iron in the sediment is able to bind P sufficiently. If the Fe/S ratio < 1, there is a possibility that all the Fe in the sediment is present as iron sulphides, resulting in a lack of binding sites for P. In that case the sediment has a high potential for the release of P. If the Fe/S ratio > 1, not all Fe is bound as iron sulphides, and is therefore available to bind with P and retain it in the sediment.

If Fe/S > 1, the P-binding capacity is estimated by evaluating the Fe/P ratio, which indirectly determines the P availability. If the Fe/P ratio is very high, it is assumed that a major part of P is bound to Fe and thus not available. If the Fe/P ratio is low, the potential for P release from sediments is high. Criteria are set to distinguish low P release (Fe/P > 20), moderate P release (Fe/P between 10 and 20), and high P release (Fe/P < 10). The practical use of the Fe/P ratio leads to an additional criterion: if the TP concentration in sediment is below 0.5 g kg^{-1}, there is no risk for internal eutrophication (Implementatieteam Besluit Bodemkwaliteit 2010).

More recently, sediment fluxes were measured in 29 sites in the Netherlands (STOWA 2012). They derived the best relationship between the average pore-water concentration in the top 5 cm and the P flux to surface water as measured in sediment core incubation experiments. The relationship between the pore-water concentration and internal loading is:

$$L_{int,15°C} \text{ (mg P m}^{-2}\text{ day}^{-1}) = 0.81 \times P_{pw} - 0.29 \text{ (R}^2 = 0.74) \qquad \text{[Eq. 17.3]}$$

in which:

$L_{int,15°C}$ = internal loading as determined in experiments at 15°C
P_{pw} = P concentration in pore water (mg L^{-1})

As stated before, experimental results are valid only for the experimental conditions. STOWA (2012) derived a generic temperature correction to transfer experimental results at 15°C to a different temperature to account for seasonal variation.

$$L_{int,T} = L_{int,15°C} \times [(0.0543 \times T) + 0.193] \qquad \text{[Eq. 17.4]}$$

in which:

$L_{int,T}$ = internal loading of the lake for average temperatures during the considered period
T = the temperature in °C (yearly average is 12°C)

Algae Production Turning Dissolved P into Particulate P

Detritus production by algae is estimated according to Los (2009). The P concentration in algae (P-algae) is derived from the Chl *a* concentration in Lake Eemmeer:

$$\text{P-algae (mg P L}^{-1}) = 0.001 \times \text{chlorophyll (µg L}^{-1}) / 1.65 \qquad \text{[Eq. 17.5]}$$

Consequently, the formation of detritus P (in Figure 17.2, part of P-organic or SPM) is calculated according to:

$$\text{P-detritus (g P m}^{-3} \text{ day}^{-1}) = 0.65 \times 0.04 \times 1.083^{T} \times \text{P-algae (mg P L}^{-1}) \qquad \text{[Eq. 17.6]}$$

in which:

T = the water temperature in °C

Storage in Sediment and Biota

Fluxes 5 (storage) and 6 (sorption) (see Figure 17.2) are factors to complete the balance. First, we fit the total P-balance. The gap between total loads and total outflow is considered as storage (or release) in the system (see Equation 17.7). The storage mechanism can be net settling of particulate P or uptake by vegetation, mussels, or other biota:

$$\text{P-storage} = \text{TP load (1)} + \text{P release (3)} - \text{P outflow (2)} \qquad \text{[Eq. 17.7]}$$

The numbers in the equation refer to Figure 17.2.

Sorption—Both Adsorption and Desorption

Once the TP balance is in equilibrium, we distinguish between particulate and dissolved P. Internal processes often contribute to chemical changes of P in the system. Particulate P in sediment is mineralized to dissolved P, which diffuses to the surface water, which in turn can be taken up by algae and transformed to organic P. These two processes have been quantified independently (see fluxes 3 and 4 in Figure 17.2), but if all P fluxes are considered, there might be a net shift from dissolved to particulate P or *vice versa*. This shift is fitted by a sorption flux (see Equation 17.8):

$$P_{\text{sorption}} = P_{\text{outflow,dissolved}} + P_{\text{algae}} - P_{\text{external,dissolved}} - P_{\text{release}} \qquad \text{[Eq. 17.8]}$$

If the result of Equation 17.8 is positive, particulate P desorbs into solution and if the result is negative, adsorption takes place.

17.3 RESULTS

17.3.1 Historical Overview of Loading and Water Quality

External P loading of Lake Eemmeer was extremely high up to 1988 (ca. 40–45 g m^{-2} yr^{-1}). Annual TP concentrations in the lake were on average 1.1 mg L^{-1} (1976–1988). For comparison, these concentrations are twice as high as during a turbid, cyanobacteria-dominated state in Lake Veluwemeer (one of the other Borderlakes) in the 1970s. TP concentrations in the River Eem almost reached

2 mg L^{-1} in 1980. The concentrations in the only other significant source to Lake Eemmeer, the neighboring Lake Wolderwijd, were almost ten times lower (see Figure 17.3). Since the mid-1980s, the concentrations have gradually decreased due to measures taken to reduce the P load in the catchment of the River Eem. In 1995–1996, the P load coming from the River Eem declined to a greater degree due to improved P removal at one of the wastewater treatment plants and the closure of another, leading to a strong decrease in the concentrations in 1995. Record high discharge values in the winters of 1993/94 and 1994/95 temporarily increased loading, but also may have caused unintentional flushing of the lake, leading to declining P concentrations. Exceptionally low discharge volumes in the winter of 1995/96 and the summer of 1996 reinforced the decrease in P loading of the lake (see Figure 17.4).

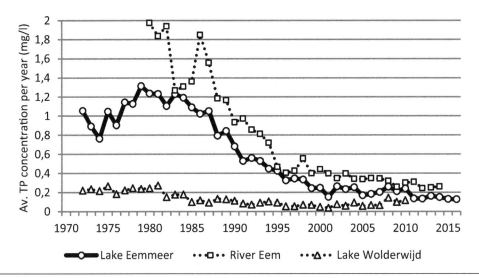

Figure 17.3 Average TP concentrations per year in the River Eem, Lake Eemmeer, and Lake Wolderwijd.

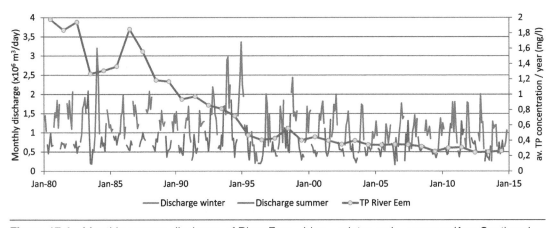

Figure 17.4 Monthly average discharge of River Eem—blue = winter, red = summer (Apr–Sept) and average TP concentration per year in the River Eem.

The TP concentrations in the lake were always lower than those in the River Eem, but both decreased in parallel. This coherent decline in TP concentrations suggests that the role of internal loading was limited to about one year in response to changes in external loading. The largest drop in external loading, which took place in 1987 and 1988, was matched by a similar drop of in-lake concentrations with a delay of a little more than a year. The decrease in 1995 and 1996 was also followed by a decrease in concentrations in Lake Eemmeer after about one year. Present summer TP concentrations are 0.15 mg L^{-1} (average April–September 2011–2016).

The decrease in P concentrations in the lake was eventually followed by ecological improvements. Concentrations of planktonic Chl a in the water column decreased and transparency increased (see Figure 17.5a), followed by an increase in coverage by aquatic macrophytes (see Figure 17.5b). Lake

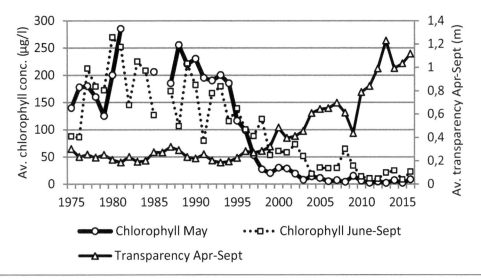

Figure 17.5a   Changes in (planktonic) Chl a concentrations and transparency in Lake Eemmeer.

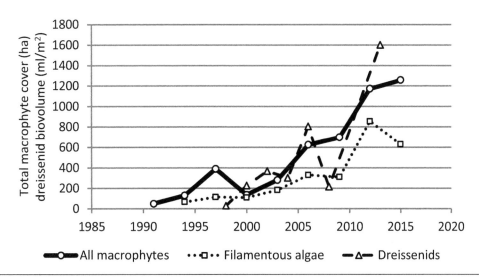

Figure 17.5b   Increase of total aquatic vegetation (including filamentous algae), of filamentous algae separately, and of dreissenid mussels in Lake Eemmeer.

Eemmeer seems to have shifted to a clear-water, macrophyte-dominated state. The extent of macrophyte recovery was unexpected because the nutrient concentrations are still high compared to the values at which macrophyte dominance becomes likely according to multi-lake studies (Kosten et al. 2009). The increase of macrophytes is accompanied by a high relative abundance of filamentous macro-algae that formed floating algae beds over large areas (Noordhuis et al. 2016). In spite of decreased Chl *a* concentrations, episodic blooms of cyanobacteria are still reported almost every year.

The early transition to a clear water state is probably linked to the strong influence of the River Eem. Apart from a short residence time (see Figure 17.6a), the lake has several other riverine characteristics, such as very low concentrations of Chl *a* in winter (see Figure 17.6b), high percentages of dissolved nutrients (see Figure 17.6c, d), and high densities of dreissenid mussels. After 1995, zebra mussel (*Dreissena polymorpha*) densities increased in Lake Eemmeer (see Figure 17.5b), and the community reached even higher densities after this species was joined (and eventually largely replaced) by quagga mussels (*D. bugensis*). Mussel densities, as well as average shell lengths, in the Lake IJsselmeer area are highest in the lakes with direct input of nutrients with river water: around the mouths of the River Eem in Lake Eemmeer and of the River IJssel in Lake Ketelmeer (see Figure 17.7).

Dissolved P represents a large proportion of TP during most of the year (see Figure 17.6c, d), suggesting that, in general, P is not limiting algal growth. After two decades of light limitation, SRP percentages did decrease with the decrease of TP around 1990, particularly in early spring. However, around 1995 a clear-water phase developed in May, indicative of zooplankton grazing, and SRP percentages started to increase in summer. In later years, an increase in the population of the

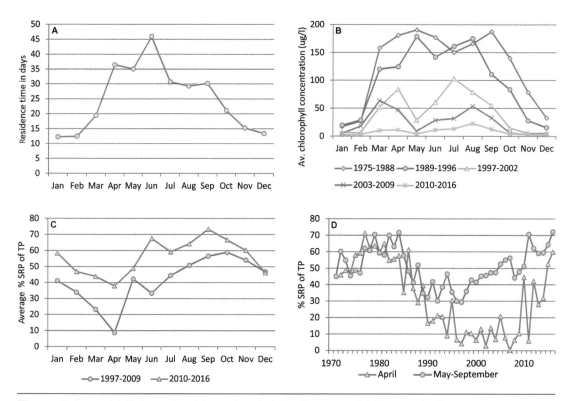

Figure 17.6a-d   (a) Estimated residence time in Lake Eemmeer; (b) Changes in the seasonal pattern of Chl *a*; (c) Seasonal pattern of % SRP (soluble reactive phosphorus) of TP in the water column of Lake Eemmeer; and (d) Changes in the % SRP in April and in summer (average per year 1971–2016).

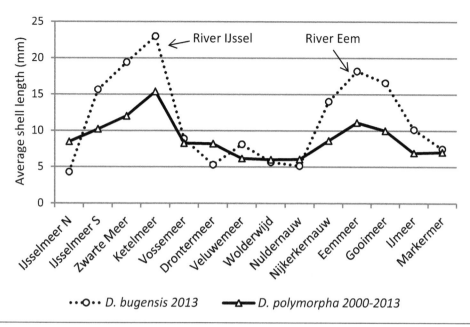

Figure 17.7 Average shell lengths of zebra mussels *Dreissena polymorpha* (2000–2013) and quagga mussels *D. bugensis* (2013) in the chain of lakes in the Lake IJsselmeer area. The lakes are sequenced from north to south, starting in Lake IJsselmeer, through the chain of Borderlakes and into Lake Markermeer, with indications of the location of the two main rivers (main nutrient sources).

filter-feeding zebra mussels took place. In combination with the increase of the dissolved fraction of TP, this suggests increasing removal of phytoplankton biomass by filtration, resulting in an increase in transparency. Around 2010, the percentage of SRP showed a strong increase, even in April, concurrent with a further increase of dreissenids, due to the invasion of quagga mussels.

Physical loading through resuspension of P-rich sediment is likely to have decreased over the past 30 years. The increase in dreissenids and macrophytes, covering sediment and collecting suspended matter, are relevant in this context. However, changes in the population of benthivorous fish, particularly bream (*Abramis brama*) also played a role. Bream biomass decreased in the second half of the 1980s, but biomass was further reduced in the adjacent eastern Borderlakes by an increase in commercial fishing. In the southern Borderlakes, bream biomass remained relatively high at ca. 50 kg ha^{-1}, but according to a survey in the winter of 1995/96, the population largely consisted of small, planktivorous bream. Only 3% of the biomass consisted of fish larger than 25 cm, as compared to 34.5% in the winter of 1993/94. In 2012, the percentage of large fish increased to 90%, but the total biomass decreased to 26 kg ha^{-1} in Lake Eemmeer.

17.3.2   Results of Sediment Sampling

In September 2017, 13 locations in Lake Eemmeer and Lake Gooimeer were sampled for sediment and pore water (see Figure 17.8 and Table 17.1).

The highest P-content in the sediment was 20 times higher than the lowest (see Table 17.1). However, in the presence of Fe, the availability of phosphate may be limited due to the binding of phosphate to iron hydroxides. The P concentrations in pore water ranged from 0.27 mg L^{-1} to 2.21 mg L^{-1}, which is a factor of eight between the highest and lowest value. This is in the middle of

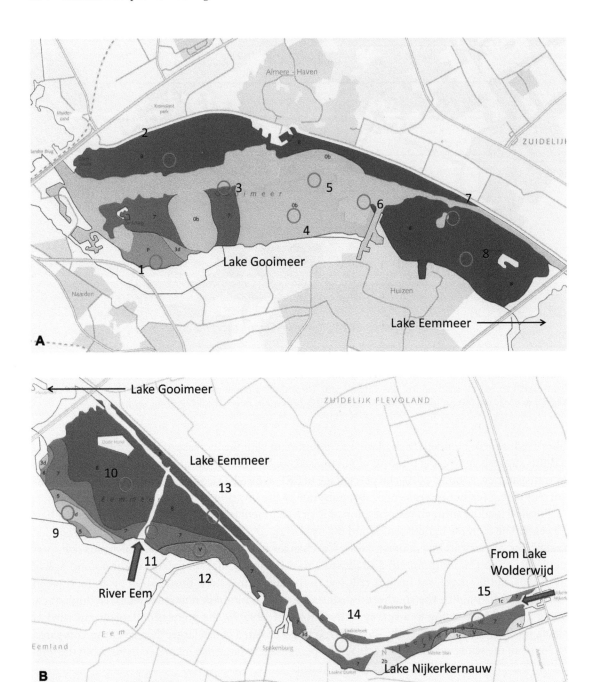

Figures 17.8a and 17.8b   Locations of the field campaign of September 2017 (a) Lake Gooimeer (locations 1–8); and (b) Lake Eemmeer (locations 9–15). The maps show the different sediment types: yellow and orange for sandy material, the green colors represent clay, and the purple areas indicate the presence of peat. Locations 5 and 6 were not sampled due to bad weather.

Table 17.1 Sediment characteristics and concentrations of Fe, S, and P in sediment and pore water in Lake Gooimeer, Lake Eemmeer, and Nijkerkernauw (NN)

Lake	loc	Sediment characteristics[a]			Sediment			Pore water		
		OM	Al	CaCO$_3$	Fe	S	P[b]	Fe	S	P
		%	g kg^{-1}	%	g kg^{-1}	g kg^{-1}	g kg^{-1}	mg L^{-1}	mg L^{-1}	mg L^{-1}
Lake Gooimeer	1	0.4	10.5	1.7	1.97	0.43	0.06	< 0.01	14.4	0.43
	2	5.6	55.5	10.3	33.23	9.65	0.54	0.1	8.6	0.37
	3	10.5	48.9	10.2	32.95	6.59	1.21	1.03	6.8	1.45
	4	0.5	12	8.7	2.37	0.41	0.07	< 0.01	14.2	0.5
	7	4.2	40.4	13.9	23.95	8.74	0.51	0.15	6.5	1.05
	8	5.6	44.5	11.2	29.18	12	0.51	0.44	5.6	0.86
Lake Eemmeer	9	3	36.4	13	17.37	3.3	0.47	0.51	2.7	0.58
	10	4.7	39.3	22.6	22.35	5.56	0.54	0.32	3.5	0.96
	11	4.6	43.1	9.9	25.34	6.53	0.66	0.68	2.8	2.21
	12	30.8	46.7	4.1	28.82	20.68	0.49	0.04	7.3	0.27
	13	7.3	46.4	9	26.61	7.52	0.54	0.32	9.3	1.26
NN[b]	14	2.4	18.6	4.3	5.74	1.79	0.23	0.05	19.2	2.02
	15	59.2	4.4	5.6	6.38	15.39	0.33	< 0.01	22	0.51

[a]Peat at locations 12 and 15; sandy material (low in Al) at locations 1 and 4; clay at the other locations. Location 3 is on the slope of a deep pit. The water depth was 14 m, the water content of the material was 80% (for comparison, the other clayey sediments had a water content of 30–60%).
[b]NN = Nijkerkernauw, the channel between Lake Wolderwijd and Lake Eemmeer.

the range found in the Netherlands: ~0.003 mg L^{-1} to 19 mg P L^{-1} (STOWA 2012). Because we could not attribute the variation to certain properties (e.g., geographic gradient, relation to soil type, etc.), we used an average value to calculate fluxes from the sediment to surface water.

In the case of Lake Eemmeer, these data can be added to a time series of TP and Fe concentrations in sediment. TP samples have been taken at only one location (in the sampling campaign number 13) in Lake Eemmeer and always in autumn in the years 1995–2000, 2010, and during the more extensive campaign in 2017 (see Figure 17.9a). In general, the concentrations show a decrease from 1.4 g kg^{-1} dry weight (dw) and 1.1 g kg^{-1} dw in 1995 and 1996, respectively, to 0.71–0.82 g kg^{-1} dw in 1997–2000 and finally to 0.54 g kg^{-1} dw at the same location in 2017. The 2010 value of 1.27 g kg^{-1} dw seems to be an outlier, as all the lakes in the Lake IJsselmeer area show a more or less continuous decrease (see Figure 17.9a). The decrease from 1995/1996 to 1997 may be influenced by an unusually long period of low river discharge from January–July 1996 and consequent flushing caused by the high river discharge in November and December 1996. Due to the decreased P content, the estimated Fe/P ratio increased to above 20 in 1997, suggesting that the role of internal loading was limited (see Table 17.2 and Figure 17.9b).

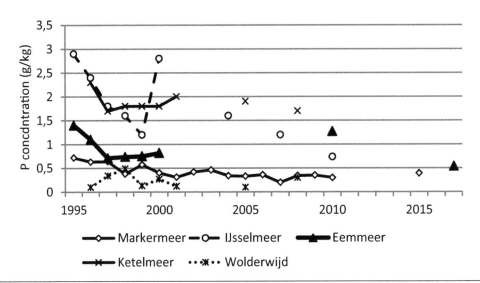

Figure 17.9 (a)   TP concentrations in the top (0–10 cm) sediment layer of Lake Eemmeer and of four other lakes in the IJsselmeer area. One sample in each year, October/November.

Table 17.2   Calculated ratios and fluxes. Most numbers exceed criteria set for the specific ratio: Fe/S should be larger than 1 (g g^{-1}), Fe/P should be larger than 20 (g g^{-1}).

		Ratios in sediment		P-flux: Sediment to surface water	
	loc	Fe S$^{-1 c}$ g g^{-1}	Fe P$^{-1 d}$ g g^{-1}	Van der Molen and Boers (1994) mg m^{-2} day^{-1}	Stowa (2012) mg m^{-2} day^{-1}
Lake Gooimeer	1*	4.6	30.9	2.37	0.05
	2	3.4	61.9	0.97	0.01
	3	5	27.2	2.79	0.75
	4*	5.7	35.4	1.99	0.10
	7	2.7	46.6	1.39	0.47
	8	2.4	57.4	1.07	0.34
Lake Eemmeer	9*	5.3	37.2	1.86	0.15
	10	4	41.5	1.62	0.41
	11	3.9	38.5	1.78	1.27
	12*	1.4	59.3	1.02	−0.06
	13	3.5	49.3	1.30	0.62
NN	14*	3.2	24.8	3.15	1.14
	15*	0.4	19.4	4.32	0.10
Average Lake Eemmeer		3.62	45.16	1.52	0.48

* No need to check the Fe/P-ratio if the P-concentration in sediment is below 0.5 g kg^{-1} (Implementatieteam Besluit Bodemkwaliteit 2010).

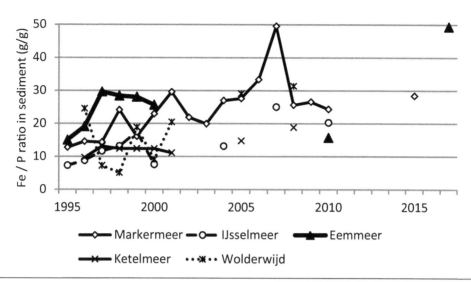

Figure 17.9 (b)   Fe/P ratios in the top sediment layer of Lake Eemmeer and of four other lakes in the IJsselmeer area. Fe and TP often sampled in different years, in those cases shown in the year of TP sampling, using average Fe concentrations.

17.3.3   External and Internal Fluxes

As mentioned in the Methods section, we constructed a mass balance showing external and internal fluxes in six steps.

The External P Loads via Surface Water and Groundwater

The dominant external P load of Lake Eemmeer is the River Eem. Based on total and dissolved P concentrations in location *Eemdijk*, a load of 17.2 mg P m^{-2} day^{-1} was estimated. About 25% consists of dissolved P. A small contribution comes from Lake Wolderwijd, east of Lake Eemmeer, resulting in a total external load of 20.4 mg P m^{-2} day^{-1}. As noted before, this is a conservative estimate because it does not account for other possible external sources, such as atmospheric deposition, groundwater discharge, or faecal deposition from roosting geese and swans.

The Outflow

The concentrations at monitoring location *Eemmeer Midden* (number 13 in Figure 17.8) proved to be representative for Lake Eemmeer since values of the main parameters in surface water samples during the 2017 field campaign were close to the average values of all locations in the lake. This concentration multiplied by the outflow discharge estimate (equal to the inflow of the River Eem and Lake Wolderwijd) results in 15.3 mg P m^{-2} day^{-1} being transported to Lake Gooimeer. Whereas dissolved P in the river Eem composed 25% of TP, the ratio was 50% in Lake Eemmeer.

Internal Loading

Table 17.2 shows the elemental ratios calculated for Lake Gooimeer (locations 1–8), Lake Eemmeer (locations 9–13), and Nijkerkernauw (locations 14 and 15). All locations meet the eutrophication criteria except location 15. Location 15 is a peat sediment (60% organic matter) with a low P concentration (0.33 g P kg^{-1} dw). It is unlikely that high P fluxes are released from this location. Another sample (location 3) shows an elevated P content (1.21 g P kg^{-1}) (see Table 17.1) suggesting some risk of release of phosphate; however, the Fe-based ratios are low, so the overall risk of P release is limited.

Using Equation 17.2, the internal loading ranges from 1.0 to 2.8 mg m^{-2} day^{-1}. The average value for Lake Eemmeer is 1.52 mg m^{-2} day^{-1}. If Equations 17.3 and 17.4 are used, and assuming an average yearly temperature of 12°C, the internal loading ranges from 0 to 1.5 mg P m^{-2} day^{-1}, with an average for Lake Eemmeer of 0.48 mg m^{-2} day^{-1}. Although the approach using pore-water concentrations has been developed more recently and gives a better correlation, we used the average from both equations in the mass balance: 1.0 mg m^{-2} day^{-1}.

P Uptake by Algae

Equations 17.5 and 17.6 are used to determine the detritus production that settled to the bottom. The calculation was done for four quarters resulting in 0.27 mg m^{-2} day^{-1} in winter (Q1, Q4) up to 2.5 mg m^{-2} day^{-1} in Q3. The yearly average in the mass balance is 1.0 mg m^{-2} day^{-1}.

P Storage

The storage in the system can be calculated based on the total load (21.4 mg m^{-2} day^{-1}) including the flux from the sediment, minus the outlet (15.3 mg m^{-2} day^{-1}). Settling of SPM will be an important factor because the contribution of suspended matter to the TP concentration in surface water decreases in Lake Eemmeer compared to the River Eem. However, storage in biota is probably an important factor because the abundance of mussels and vegetation in Lake Eemmeer has increased strongly in the last decade (see Figure 17.5b). In any case, storage is assumed to be particulate P, either adsorbed to the sediment or taken up by biota.

Sorption

After step 5 the total mass balance was balanced, as the inflow equaled the outflow, assuming step 6 was in equilibrium (see Figure 17.2). However, if the mass balance is split into dissolved and particulate P, there are inconsistencies. According to Equation 17.8, the amount of dissolved P leaving the system (dissolved P$_{outlflow}$ and detritus production) is larger than the amount entering the system (dissolved P$_{external}$ and release from sediment). This is most likely due to desorption of P from the suspended particles from the River Eem. The calculated desorption is 3.2 mg m^{-2} day^{-1}.

17.3.4   The P-Balance and Some Concluding Remarks

Figure 17.10 shows all P fluxes in Lake Eemmeer including primary production, but it does not include other biota such as vegetation, mussels, and fish, explicitly. The water quality of the lake is still governed by the external P load from the River Eem, even though that load has strongly decreased (see Figure 17.3). Almost 30% of the TP entering Lake Eemmeer is stored in the system. This is partly due to settling, but also to net uptake by biota. The latter is supported by the increase of mussels during the last decade. Another conclusion is that desorption from suspended matter is likely occurring, resulting in an increased concentration of dissolved P in Lake Eemmeer. This fits well with the observation that the sediment top layer of Lake Eemmeer does not show high P content. Finally, the P balance also shows that the release of P from the sediment is a minor factor, irrespective of the method that is used to determine it.

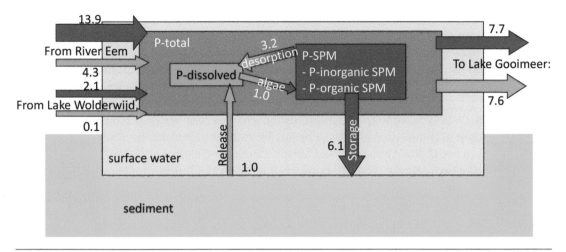

Figure 17.10   P balance for Lake Eemmeer showing external and internal fluxes in mg m^{-2} day^{-1}. The purple area is P in surface water; the green arrows represent transport of dissolved P; the brown arrows = transport of particulate P; and SPM = suspended particulate matter.

17.4 DISCUSSION

17.4.1 Historical Analysis

Even for an estuarine lake, the low sediment fluxes are surprising considering the extremely high external loading during the 1970s and 1980s. Earlier work had shown that the amount of nutrients leaving these Borderlakes is smaller than the amount entering, so retention was taking place (Portielje and Oostinga 2003). Studies in other lakes show that ecological recovery of eutrophic lakes may be delayed for decades after the external loading has been reduced due to internal loading (Søndergaard et al. 2001). Based on comparison of the patterns of external P loading and in-lake TP concentrations, combined with the low fluxes determined in 2017, we conclude that in this case, recovery was relatively quick, with delays limited to one or two years at the most, and affected the pace of recovery only somewhat around 1996.

Within the Lake IJsselmeer area, other lakes also showed short delays between decreases in external P loading and decreases in in-lake P concentrations. This is the case in Lake Ketelmeer, with a residence time of only a few days, but also in Lake IJsselmeer, with a residence time of three to four months. In Lake Veluwemeer, eutrophication was treated in 1979 by installing a P-removal facility in a wastewater treatment plant that was responsible for supplying 80% of the nutrients to the lake. The expected delay in decrease of P concentrations in the lake water due to internal loading was addressed by flushing the lake with water from the adjacent polders, which were low in P. The P removal combined with a reduction of the residence time from 1.5 years to three months caused a sudden decrease of the in-lake P concentrations by 75% (Hosper 1997). In contrast, the decrease of in-lake concentrations in Lake Markermeer was delayed for many years after a strong decrease in external P loading in the mid-1980s; however, the lag time before observing a decrease in concentration was linked to a reduced water supply and a strong increase in residence time. The natural residence time in Lake Eemmeer, particularly in winter, is much shorter than the residence times of Lake Markermeer before 1985 and in Lake Veluwemeer after the flushing started. A high winter discharge of the River Eem after the decrease of the TP concentrations, such as in the winter of

1994/95, may have increased flushing effects, reducing internal P loading, once the concentrations had started decreasing.

Internal P loading may be affected by changes in environmental conditions, such as the biomass of benthivorous fish and abundance of dreissenid mussels. Benthic feeding of bream *Abramis brama* may have increased the release of nutrients from the sediment during the years in which bream biomass was high and also contributed to high turbidity. Removal of 80% of bream biomass in Lake Wolderwijd in 1990/91 failed to trigger a shift of the lake to a clear-water, macrophyte-dominated state. An increase of commercial interest in bream around five years later, however, did contribute to such a shift in both Lake Wolderwijd and Lake Veluwemeer (Ibelings et al. 2007), approximately 17 years after the two main measures were taken (P removal and flushing). This shift more or less coincided with the transition in Lake Eemmeer that is described in this chapter, and is partly connected to it since a limited amount of water generally flows from Lake Wolderwijd to Lake Eemmeer. Gradual changes like those connected to the implementation of program measures in the catchment area of the river Eem (P removal at wastewater treatment plants, rerouting of wastewater and closure of other treatment plants, enforcement of regulations on agricultural use of fertilizers) and the increase of commercial interest in bream have led to abrupt changes in lake ecology, nutrient dynamics, and water quality. This acceleration of change also seems to have been influenced by unusual climatic conditions, in this case acting through variations in discharge, although the severe winters of 1996 and 1997 may also have added to the decrease of bream biomass. While other drivers of change differ across systems, the shifts in the eastern Borderlakes on the one hand and the southern Borderlakes on the other may have been synchronized by climatic fluctuations.

17.4.2 Present Situation

The data presented in this chapter clearly show that currently, internal P loading in Lake Eemmeer (and Lake Gooimeer) is limited and at least an order of magnitude lower than external loading. However, our analysis shows that almost half of the dissolved P is supplied by desorption of P from suspended matter, most likely originating from the more eutrophic River Eem. To prove this phenomenon, it would be necessary to better assess the availability of P from suspended particulate matter being delivered to the lakes by the River Eem; this could be done by analysis of the P fractions, conditions under which release occurs, and its bioavailability. As a rule of thumb, the contribution of internal P loading in lakes with a residence time below one month during at least 90% of the year is expected to be limited, and thus the chance of not meeting the WFD P target due to internal loading is low. The residence time of Lake Eemmeer is longer than one month in summer; however, the short residence times in winter compensate for summer, resulting in a system with annual residence times shorter than one month for 90% of the year. The strong seasonal pattern in water column P poses questions about possible risks of internal loading over the year. More important to assessing the role of internal P loading than the exact residence time, however, are other clues of riverine or estuarine functioning including very low Chl *a* concentrations in winter, large sizes and densities of dreissenids, and a high percentage of SRP in summer TP. Assessed in this way, Lake Gooimeer (the lake downstream of Lake Eemmeer) is also in the low-risk range with respect to internal P loading. Both lakes appear to function as an *estuarine* lake as opposed to *stagnant* lakes like most other lakes in the IJsselmeer area, which are lacking riverine influences. Apart from that, the large contribution of SRP to TP during summer suggests that P is not the limiting factor for algal growth; the recent increase of the SRP:TP ratio suggests that filtration by dreissenids controls algal growth. Dreissenids in these lakes are among the largest in the area, together with the mussels in the lakes around the mouth of the River IJssel. Most likely their growth rates are linked to high external P loading, but the mussels may be benefiting from high nutrient concentrations in the sediment as well.

17.4.3 WFD Targets

From this analysis, it is clear that in recent years internal P loading has played a minor role in the southern Borderlakes and was not a contributing factor in these systems not meeting their WFD targets. External loading has been strongly reduced by P-removal at the wastewater treatment plants in the catchment of the River Eem, but this does not appear to have resulted in an increased relative contribution of internal loading to water column TP. After the initial decrease of external P loading in the years 1985–95, further decreases have been slow. Most effective measures have included P removal at wastewater treatment plants discharging into the river or the lakes, and eliminating point sources by rerouting their waters to the treatment plants. The remaining P loading is largely due to diffuse sources, already contributing an estimated 75% in 2003 (Portielje and Oostinga 2003). Measures taken to reduce diffuse loading will take longer to show results.

Around 30 years have passed since the first measures were taken. The difficulties to meet the targets are in contrast with the ecological developments, which include strong increases in aquatic macrophytes and transparency, as well as a general increase in species diversity. The high transparency in recent years is connected to the invasion of *Dreissena bugensis*, and in turn, the high density and large size of dreissenid mussels are connected to relatively high P loading. To a certain extent, this is to be considered natural in these estuarine lakes. In the present Dutch WFD typology, the southern Borderlakes are considered a single water body of type M14: medium-sized, shallow, buffered lakes. GEP for P is set at 0.09 mg L^{-1}, while the present value is approximately 0.15 mg L^{-1} (April–September 2011–2016). The eastern Borderlakes (Lakes Veluwemeer and Wolderwijd) are also combined and considered an M14 water body, but these lakes are not connected to a river, have longer residence times, and do not show the characteristics of an estuarine system. The GEP value for these lakes is also 0.09 mg L^{-1} and this target has been met since 1996; the present average value is 0.02 mg L^{-1}. It could be argued that a higher target is sufficient for an estuarine M14 water body. However, the present average TP concentration in the River Eem itself, approximately 0.27 mg L^{-1}, is still well above the background range of 0.02–0.14 mg L^{-1} for TP concentrations in rivers, as estimated by Laane (2005). Reaching this range would mean a further reduction of external loading to Lake Eemmeer by another 50%. Even then, it is anticipated that the contribution of internal P loading will be insignificant; the reduction will probably result in meeting the TP target, but it may also slow down the growth of dreissenids and affect their filtration control of phytoplankton.

17.5 CONCLUSIONS

In 1995/96, external loading of the southern Borderlakes was strongly reduced in response to measures taken in the catchment of the River Eem and reduced loading from the eastern Borderlakes. An unusually long period of low discharge values from January to July 1996 quickened this reduction. Physically induced internal P loading declined as a result of a decrease in benthivorous fish due to increased commercial fishery for bream (which was also one of the main triggers of the shift to the clear-water, macrophyte-dominated state of the eastern Borderlakes) and of sediment covering and filtering mussels. The in-lake TP concentrations declined quickly, approximately one to two years following the decrease in external loading, suggesting a minor role of internal loading.

The P concentrations in the sediment have also decreased. Iron:P ratios have increased to values that suggest an insignificant role of internal P loading in the ecological functioning of the lake. External P loading has strongly decreased but is, nevertheless, still relatively high. The total internal P loading as estimated from pore-water concentrations is at least an order of magnitude smaller than the external P load. The whole-lake P balance shows a net storage of 25% of the total external P load, but at the same time a shift from particulate into dissolved P.

17.6   ACKNOWLEDGMENTS

The study presented in this chapter was funded by Rijkswaterstaat, an agency of the Ministry of Infrastructure and Water Management in The Netherlands, which is also responsible for the monitoring program that supplied the long-term data on water quality and biota.

17.7   REFERENCES

Böstrom, B; Andersen, JM; Fleischer, S; and Jansson, M. 1988. Exchange of phosphorous across the sediment-water interface. Hydrobiologia. 170:229–244.

Hin, JA; Osté, LA; and Schmidt, C. 2010. Guidance Document for Sediment Assessment. Ministry of Infrastructure and the Environment—Directorate General Water. Available on: https://www.helpdeskwater.nl/secundaire-navigatie/english/sediment/guidance-document/

Hosper, SH. 1997. Clearing lakes: an ecosystem approach to the restoration and management of shallow lakes in The Netherlands. Dissertation, Wageningen University.

Ibelings, BW; Portielje, R; Lammens, EHRR; Noordhuis, R; Van den Berg, MS; Joosse, W; and Meijer, M-L. 2007. Resilience of alternative stable states during the recovery of shallow lakes from eutrophication: Lake Veluwe as a case study. Ecosystems. 10:4–16.

Implementatieteam Besluit Bodemkwaliteit. 2010. Handreiking voor het herinrichten van diepe plassen (in Dutch) (Guidance Document for reshaping deep waterlogged former sand pits). Ministry of Infrastructure and the Environment. Available on: https://www.bodemplus.nl/onderwerpen/wet-regelgeving/bbk/grond-bagger/diepe-plassen/.

Jeppesen, E; Søndergaard, M; Jensen, JP; Havens, KE; Anneville, O; Carvalho, L; Coveney, M; Deneke, R; Dokulil, MT; Foy, B; et al. 2005. Lake responses to reduced nutrient loading—an analysis of contemporary long-term data from 35 case studies. Freshwater Biol. 50:1747–1771.

Kosten, S; Kamarainen, A; Jeppesen, E; Van Nes, EH; Peeters, EHM; Mazzeo, N; Sass, L; Hauxwell, J; Hansel-Welch, N; Lauridsen, TL; et al. 2009. Climate-related differences in the dominance of submerged macrophytes in shallow lakes. Global Change Biol. 15:2503–2517.

Laane, RWPM; Brockmann, U; Van Liere, L; and Bovelandera, R. 2005. Immission targets for nutrients (N and P) in catchments and coastal zones: a North Sea assessment. Estuar Coast Shelf Sci. 62:495–505.

Los, FJ. 2009. Eco-hydrodynamic modeling of primary production in coastal waters and lakes using BLOOM. PhD-thesis, Wageningen University.

Noordhuis, R; Van Zuidam, BG; Peeters, ETHM; and Van Geest, GJ. 2016. Further improvements in water quality of the Dutch Borderlakes; combined Bream reduction and increased Dreissenid filtration as a mechanism of change. Aquat Ecol. 50:521–539.

Orihel, DM; Baulch, HM; Casson, NJ; North, RL; Parsons, CT; Seckar, DCM; and Venkiteswaran, JJ. 2017. Internal phosphorus loading in Canadian fresh waters: a critical review and data analysis. Can J Fish Aquat Sci. 74:2005–2029.

Penn, MR; Auer, MT; Doerr, SM; Driscoll, CT; Brooks, CM; and Effler, SW. 2000. Seasonality in phosphorus release rates from the sediments of a hypereutrophic lake under a matrix of pH and redox conditions. Can J Fish Aquat Sci. 57:1033–1041.

Portielje, R and Oostinga, K. 2003. De waterkwaliteit en ecologie van het Eem- en Gooimeer. H_2O. 3:22–24.

Søndergaard, M; Jensen, JP; and Jeppesen, E. 2001. Retention and internal loading of phosphorus in shallow, eutrophic lakes. Sci World J. 1:427–442.

Spears, BM; Meis, S; Anderson, A; and Kellou, M. 2013. Comparison of phosphorus (P) removal properties of materials proposed for the control of sediment p release in UK lakes. Sci Total Environ. 442:103–110.

STOWA. 2012. BaggerNut. Maatregelen Baggeren en Nutriënten (in Dutch) (Measures for eutrophic sediments). STOWA-report 2012-40.

Van der Molen, DT and Boers, PCM. 1994. Influence of internal loading on phosphorus concentration in shallow lakes before and after reduction of the external loading. Hydrobiologia. 275/276:379–389.

CHAPTER 18

A REVIEW OF INTERNAL PHOSPHORUS LOADING EVIDENCE IN SÄKYLÄN PYHÄJÄRVI, FINLAND

Anne-Mari Ventelä[1], Petri Ekholm[2], Teija Kirkkala[1], Jouni Lehtoranta[2], Gertrud Nürnberg[3], Marjo Tarvainen[4], and Jouko Sarvala[5]

Abstract

Säkylän Pyhäjärvi (61°00′N, 22°18′E, 45 m.a.s.l.) in southern Finland is a large lake (155 km^2) that has been affected by human activities in its catchment and by lake fishery for centuries. The lake is shallow both in absolute terms (mean depth 5.5 m, maximum depth 26 m) and relative to its surface area; the openness of the basin enhances water movement and wave action. The catchment area is 431 km^2. The Pyhäjärvi Restoration Program was voluntarily established 20 years ago by local municipalities, companies, and associations after the lake had shown visible symptoms of eutrophication such as cyanobacterial blooms; monitoring data had also indicated that the external nutrient load was too high to maintain the oligotrophic state that was observed before the 1980s. Here we review four earlier studies on internal phosphorus (P) load in Pyhäjärvi: two mass balance and model-based studies, and two studies based on laboratory experiments. The results indicate that the role of internal P load is important and it forms a significant nutrient source for new summer time production. However, as the monitoring has not included sediment-related variables and most of the research has dealt with pelagic processes, the mechanisms, magnitude, and dynamics of internal P load in Pyhäjärvi require focused, additional research.

Key Words: shallow lakes, phosphorus, internal loading, climate change, lake restoration, biomanipulation

[1] Pyhäjärvi Institute, Sepäntie 7, 27500 Kauttua, Finland. Email for correspondence: anne-mari.ventela@pji.fi

[2] Finnish Environment Institute, Latokartanonkaari 11, 00790 Helsinki, Finland

[3] Freshwater Research, 3421 Hwy, 117, RR1, Baysville, Ontario, P0B 1A0, Canada

[4] Southwest Finland's Centre for Economic Development, Transport and the Environment, PL 236, 20101 Turku, Finland

[5] Department of Biology, University of Turku, 20014 Turku, Finland

18.1 INTRODUCTION

Pyhäjärvi is a lowland lake in SW Finland (see Figure 18.1). It is one of the country's most extensively studied lakes due to its important role in supporting local ecosystem services. The lake serves as a water source for households along with paper and food industries. Its fisheries have been important for centuries and the commercial fishery has currently become increasingly important for the local economy. The lake is used for recreational activities including swimming and boating. More than 1,000 summer cottages are situated on its shoreline and several tourist companies are linked to the lake.

The lake is shallow both in absolute terms (mean depth: 5.5 m; maximum depth: 26 m) as well as relative to its large surface area (155 km^2), as more than 90% of the area is < 7 m deep; the openness of the basin enhances water movement and wave action. The catchment area is 431 km^2, giving a ratio of catchment area: lake area = 2.8, and the main rivers entering the lake are the Yläneenjoki from the south and the Pyhäjoki from the east (catchment areas 234 and 78 km^2, respectively). Most of the lake bottom is subjected to erosion and sediment transport, whereas permanent deposition is limited to a narrow deep basin on the western side of the lake (8% of total lake area).

During 1980–2017, the average annual precipitation in the area was 614 mm, the annual mean temperature was 4.9°C, winter (January–March) mean air temperature in the area was −4.4°C, and the lake was covered by ice for 110 days on average. The catchment is usually covered with snow in winter, although warmer periods with several melting events have become common phenomena. The warmest month is generally July—when the average air temperature was 16.8°C during the period 1980–2017.

Cultivated fields cover 22% of the catchment area, with the rest composed of forests (50%), peatlands (20%), and built-up areas. Land use of the area changed dramatically in the 1850s when the lake water level was lowered more than 2 m and large areas of former lake bottom were turned into agricultural land.

Southwestern Finland is an important food production area, and the intensity of the agriculture and forestry sectors has clearly increased after World War II, which also can be seen in the lake sediment (Räsänen et al. 1992). In addition, the municipalities developed sewage systems and flush toilets became increasingly common from the 1950s onward. Currently, all industrial and municipal wastewaters are treated in wastewater treatment plants and lead downstream to the Pyhäjärvi outlet Eurajoki.

The water level of Pyhäjärvi is regulated. The present regulation amplitude between the permitted lower and upper water levels is only 58 cm; continuous monitoring of river flows and improved hydrological forecasts have made it possible to mostly fulfil the authorized water-level targets (Marttunen et al. 2000).

Pyhäjärvi has been the focus of a large-scale restoration program since 1995 (Kirkkala 2014). A multitude of measures have been implemented in the catchment in order to reduce the external phosphorus (P) and nitrogen (N) loads (Kirkkala et al. 2012; Kirkkala 2014). Pyhäjärvi has also been intensively biomanipulated for decades by removing planktivorous fish by the commercial fishery. Fish removal has traditionally been undertaken in winter and the annual harvest rate of vendace (*Coregonus albula*) has approached their total production (Sarvala et al. 1999). The harvest of commercially unwanted fish has been subsidized since 1995 by the Pyhäjärvi Restoration Program and other actors. This fishery was especially intensive in 2002–2004 and may have caused the observed water-quality improvement, together with a lower external nutrient load (Ventelä et al. 2007).

Since the early 1990s, the effects of climate change on the lake and its catchment have become evident, including increases in winter precipitation and nutrient load (Ventelä et al. 2011), lake

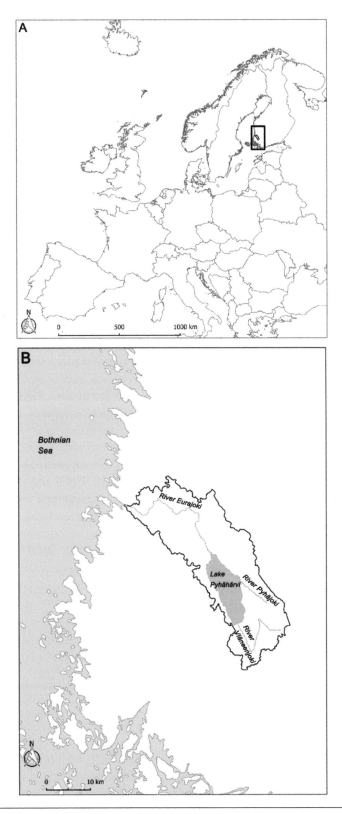

Figure 18.1a,b   Maps of Pyhäjärvi location in (a) Europe and (b) its catchment.

temperature (Jeppesen et al. 2012), and changes in phytoplankton assemblage structure, such as the increasing dominance of cyanobacteria (Deng et al. 2016).

Processes in the bottom sediments have an important role in controlling water quality in all lakes, but especially in shallow lakes with a small volume relative to bottom area. The bottom of a lake may act both as a sink and source of nutrients during the seasonal cycle. A large amount of P settles in particulate form on the lake bottom. Then, processes in the bottom sediments often transform the particulate P to a dissolved form that can be released back into the water. This release of P from the bottom is often termed *internal loading*, even if it originated from external inputs. The objective of this study is to review the variation in magnitude and frequency of internal loading and their drivers in Pyhäjärvi from earlier studies (Ekholm et al. 1997; Lehtoranta and Gran 2002; Tarvainen et al. 2010; Nürnberg et al. 2012).

18.2 METHODS

18.2.1 Data and Basic Monitoring

The data used in this study originate from national, regional, and project-funded monitoring programs; nonetheless, the monitoring protocol and methods are comparable. The water chemistry and hydrology of Pyhäjärvi have been monitored since the 1960s as part of a national program. Since 1971, monitoring has been more intensive and involves the use of automated instruments continuously measuring the river flows (Ekholm et al. 1997). The nutrient concentrations in the lake and in the two main incoming rivers, the Yläneenjoki and the Pyhäjoki, have been determined since 1980, first as part of a statutory monitoring program (Sarvala and Jumppanen 1988), and by the regional authorities at approximately a 2–3 week interval during the open-water season since 1992.

Water samples for nutrient analyses were taken from several depths covering the whole water column (0, 1, 5, 10, 15, 20, and 24 m) at a single sampling station situated in the deepest area of the lake. For chlorophyll *a*, composite samples from 0–2 or 0–5 m were used. The methods of analyses followed Finnish standard laboratory procedures (Ekholm et al. 1997) and are accepted by FINAS (Finnish Accreditation Service). All of these monitoring data are available in the Finnish Environment Institute's Open Data service (http://www.syke.fi/opendata).

18.2.1.1 Mass Balance-Based Calculations and Models for Internal P Load

Ekholm et al. (1997) estimated the internal P load from nutrient balances using sedimentation measurements and bioassays. Two P mass budgets were estimated: a long-term budget for the period 1980–1992 and a more detailed one for the year 1992.

The annual total P (TP) and total N (TN) loads/fluxes from the two major rivers were calculated based on TP and TN concentrations in water samples and continuously measured stream flow data. Atmospheric deposition of TP, TN, and inorganic N directly into the lake was estimated from the bulk deposition measurements made at three stations adjacent to the lake (data from the air-protection register, maintained by the Finnish Environment Institute). The mean amount of nutrients removed from the lake with the fish catch (mainly vendace and whitefish, *Coregonus lavaretus*) was estimated by multiplying the nutrient contents of the main fish species (Jumppanen 1983) by the estimated mean catch in 1980–1992.

To determine the effect of external loading on the level of TP and TN in the lake, the daily concentrations were calculated with a mass-balance model from measured external load and outflow and calibrated net sedimentation. If internal loading does not markedly affect the concentrations in the lake, the calculated and observed concentrations should be close to those observed.

For the ice-free period of 1992, the internal loads of TP, TN, and total suspended solids (TSS) were estimated from sedimentation measurements (Huttula 1994) by substituting the measured gross sedimentation for the sedimentation term and using mean seasonal values for input and outflow. The mean net sedimentation was calculated by subtracting storage change in water and outflow from the catchment and atmospheric input.

Nürnberg et al. (2012) (see also Chapters 3 and 6) used three methods to estimate internal load in Pyhäjärvi:

- *Partial net estimates of internal load from in situ summer increases*—the determination of partial net internal load L_{int_1} by increases of *in situ* P concentration throughout the summer was done for years 1990, 1992, 1999, and 2003 when the loads from the catchment were low.
- *Net estimates from complete P budgets (mass-balance approach)*—the mass balance calculations yield net estimates of internal load, L_{int_2} and are based on the sedimentation and release flux of P via the theory of P retention. Measured P retention (R_{meas}) is the proportion of P retained in the whole lake and should include all input and output fluxes; in the case of Pyhäjärvi, this includes also the P content of the fish catch.
- *Gross estimates from anoxia and anoxic sediment P release rates*—gross internal load (L_{int_3}) is determined for summer and winter, and then summed for an annual estimate. The release rate of P from sediment surfaces (RR) was estimated from models developed for oligotrophic and mesotrophic, iron-controlled lakes with a comparable geochemistry and trophic state to Pyhäjärvi (Nürnberg 1988). The model included the anoxic factor describing the length of time that an area similar to the lake surface area is actively releasing P. pH was not considered to affect the release or oxygen reduction rates as pH values were circumneutral and rarely above 8. Winter internal load was considered to be small but was modelled to capture annual variability. Steady-state annual average P concentration (P_{ann}) in the water column was modelled according to Nürnberg (1998, 2005), except that the term F_{out} was added to predicted retention to account for the large export as fish catch. Detailed description of the modelling work is given by Nürnberg et al. (2012). All P budget components were determined according to Ventelä et al. (2007) for the calendar years 1980–2005.

18.2.1.2   *Measurements of Nutrient Release from the Bed Sediments*

Lehtoranta and Gran (2002) conducted sediment core incubations during the period of September 1997 through August 1998 to investigate the release of nutrients at the accumulation and transport bottoms. The sediment cores taken from the lake depth of 25 m represent accumulation bottoms and the shallower sampling location (6 m) represents areas where sediment accumulates temporarily (i.e., the surface layers are occasionally resuspended by wind-induced currents and transported elsewhere). The sampling dates covered the seasonal cycle (i.e., early autumn, winter, early spring, early summer, and late summer). To measure pore-water nutrient concentrations, the sediment cores were sliced to 0–1, 1–2, 2–3, 3–5, 5–7, and 7–10 cm slices.

The release of nutrients from the sediment to the water was measured in a temperature-controlled laboratory from triplicate undisturbed sediment tubes equipped with flow-through nylon plugs. Near-bottom water from the sampling site was circulated between a test-water container and the headspace inside the sediment tube. Before the actual incubation, the test water was circulated for 60 minutes through the sediment tube at a rate of 8 mL min^{-1} to ensure similar initial concentrations of the water above the sediment in the parallel tubes at the start of the actual incubation. The sediment cores were incubated in the dark for 12–24 hours with a water flow of 2 mL min^{-1}; the total volume of the test water varied from 200 to 400 mL between the sampling times. The concentrations of dissolved reactive phosphorus (DRP) and dissolved total phosphorus (DP) were measured from filtered samples (0.2 μm cellulose-acetate membrane). The release rates of nutrients were calculated on the basis of incubation time, the area of sediment sample, and the concentration difference in the beginning and the end of incubation.

Tarvainen et al. (2005) estimated the effects of ruffe (*Gymnocephalus cernuus* (L.)) on water quality on the basis of their nutrient release in laboratory experiments. The average total weight-specific P release rate of 1.4 μg g^{-1} h^{-1} was assumed to represent the mean release of the ruffe stock during the open-water season. The P release rates of vendace (*Coregonus albula* (L.)), roach (*Rutilus rutilus* (L.)), perch (*Perca fluviatilis* (L.)), and bream (*Abramis brama* (L.)) were taken from the literature.

The species-specific P release rates of smelt (*Osmerus eperlanus* (L.)), bleak (*Alburnus alburnus* (L.)), and whitefish (*Coregonus lavaretus* (L.)) were assumed to be the same as for vendace. The stocks of species were estimated for autumn 2002 by using data on the fish species in seine catches for winter 2002–2003 combined with the stock size of vendace estimated with DeLury's method (Helminen et al. 1993) (for description of the method, see Tarvainen et al. 2008). Total P release was estimated for 150 days during the summer (water temperature > 10°C), when P loads from the catchment are commonly low and P release from fish may have significant impact on water quality.

18.3 RESULTS

The estimates of internal P load vary markedly because different methods were applied for different years and different aspects of sediment release were assumed as internal loading. Study periods overlap from 1980 to 1992, and the data from the year of 1992 were included both in Ekholm et al. (1997) and in Nürnberg et al. (2012). Both studies suggested that P released from the bottom may temporarily exceed the load from the catchment, even at the annual level.

The mass-balance model of Ekholm et al. (1997) showed that the mean annual P concentration in the lake was regulated mainly by internal loading, whereas the loads from the catchments played a minor role. They reported that the internal loading of TP during the open-water season of 1992 in the whole Lake Pyhäjärvi was 6060 kg day^{-1}, which is equivalent to 39 mg m^{-2} day^{-1} or 5811 mg m^{-2} for the ice-free period (see Table 18.1). Because the gross sedimentation rates that were used to derive this value are more than an order of magnitude lower in winter under the ice cover (Niemistö and Horppila 2007), it cannot be extrapolated to the whole year. During the ice-free period, the variations in P concentrations were controlled almost solely by internal processes, perhaps by resuspension of inorganic and organic bottom matter. The release rate of DRP estimated by Ekholm et al. (1997) was 2.2–54.8 mg m^{-2} yr^{-1}, or more than two orders of magnitude lower than the internal load of TP. Although the internal load of bioavailable P is significant, the mechanism by which it was released to the water was not identified.

In Nürnberg et al. (2012), the long-term averages of three annual P release estimates were comparable with each other (see Table 18.1) and were about 60% of the mean annual load from the catchment, estimated as 105 mg m^{-2} yr^{-1}. Internal load estimates were larger than external loads in years with high water temperature and low discharge. Regression analysis revealed that external load was decreasing ($R^2 = 0.22$, $p < 0.02$) but internal load was increasing over time ($R^2 = 0.46$, $p < 0.0001$, 1980–2005), as was lake water temperature ($R^2 = 0.17$, $p < 0.05$). Internal load also correlated negatively with discharge expressed as annual water load ($R^2 = 0.23$, $p < 0.02$). The interaction of climate variables with internal loading was further established by a significant improvement (14%) of the relationship with time when using *date of ice-out* as an additional predictor (negative correlation with a partial-p = 0.023). This means that with the longer periods of stratification, the warmer water temperature coupled with low flows create conditions that enhance internal P loading in Pyhäjärvi.

Table 18.1 Internal loading of total phosphorus, mg P m^{-2} yr^{-1} unless otherwise indicated

Study	Ekholm et al. 1997	Nürnberg et al. 2012			Tarvainen et al. 2010
Years	1992	1980–2005	average of 1990, 1992, 1999 and 2003	1992	2002
Time period of experiment	May 4 to Sept 30	whole year	June–September	whole year	May–Sept
Mass balance model, includes resuspension	5811				
In situ summer internal load			63	44	
Mass balance method, gross		62		66	
Gross estimates from anoxia and anoxic sediment P release rates.		61		46	
TP release by fish					11

DRP release rates of laboratory incubation experiments ranged between -0.06 and 0.7 mg m^{-2} yr^{-1} (mean 0.22 mg m^{-2} yr^{-1}) in the accumulation bottom, and between -0.1 and 0.2 mg m^{-2} yr^{-1} (mean 0.06 mg m^{-2} yr^{-1}) in transport bottom sediments at varying oxygen concentrations (Lehtoranta and Gran 2002). These rates imply annual DRP release of 80 mg m^{-2} yr^{-1} for the accumulation bottoms, and 22 mg m^{-2} yr^{-1} for the transport bottoms. The bottom water at the site of accumulation was oxic except for January and April, when bottom water was hypoxic and almost anoxic, respectively. At the transport bottom, the water was well oxidized throughout the annual cycle. Most of the time, only minor amounts of DRP were released from the sediment but at the end of the summer and beginning of autumn the release of DRP increased. The sediment Fe:P ratios in the deep basin were 18–20, suggesting that the sediment in the deep parts of the lake was likely able to adsorb a considerable amount of DP to iron oxides when the water at the sediment-water interface was oxic.

Tarvainen et al. (2005) measured the role of ruffe (Gymnocephalus cernuus) and benthic chironomid larvae for resuspension in laboratory experiments. Both ruffe and chironomids increased nutrient concentrations and turbidity of water either via nutrient release (ruffe) or bioturbation (or both) (see Table 18.1).

18.4   DISCUSSION

The research of internal load in Pyhäjärvi has been closely linked to lake management. The first internal load estimations (Ekholm et al. 1997) were made to form the basis of the long-term management plan for the Pyhäjärvi Restoration Program, which started in 1995.

The monitoring of Pyhäjärvi has mainly been funded by public resources and until recently, it was versatile and extensive, including both lake and catchment. During the last decade, the entire national monitoring system has been reduced and an increasing share of monitoring has been covered by different projects of the Pyhäjärvi Restoration Program.

Most of the research and monitoring has focused on the pelagic system, which is relatively well described and understood (Sarvala and Jumppanen 1988; Sarvala et al. 1998; Ventelä et al. 2007, 2011, 2016; Deng et al. 2016). Sediments have received less attention—the estimates being based on mass balances, modeling, or a few laboratory experiments, except for some initial studies on the importance of groundwater in Pyhäjärvi (Rautio and Korkka-Niemi 2011, 2015). The interactions between the groundwater, sediment pore waters, and the surface-water systems will be an important topic for future research.

Most internal load estimates presented in this chapter were in general agreement, although there was some variation due to different methods used and processes included. An exception was the internal load estimate for TP from Ekholm et al. (1997), which was more than two orders of magnitude higher than the values obtained with other methods. In contrast, their estimates for the bioavailable P were rather consistent with the other studies. This discrepancy arises from differences in the methodological approach. Ekholm et al. (1997) calculated the internal loading of TP from a mass balance equation using gross sedimentation values measured in the field. Net sedimentation was calculated by subtracting the outflow and storage change in water from the external loading, and then the internal loading was obtained as the difference between gross and net sedimentation.

However, in shallow lakes the estimation of internal loading via gross sedimentation is problematic because of the dominant role of resuspension, leading to very high sedimentation rates in the prevailing turbulent conditions. In these lakes, particles at the sediment surface are suspended into the water even under a moderate wind stress, settle back during calm periods, but are resuspended when wind increases. Thus, the gross sedimentation values largely depend on the same particles circulating between water and sediment. The long-term contribution of resuspended material to the total flux of particulate matter in lakes has been estimated at 85% (Evans 1994). The long wind fetch and shallowness (over 90% of lake surface area is shallower than 7 m) of Pyhäjärvi make it sensitive to sediment resuspension.

Moreover, all kinds of sediment traps alter the surrounding flow field producing a hydrodynamic bias for collecting settling particulate matter (Buesseler 1991), although sedimentation cylinders with high aspect ratio usually yield acceptable sedimentation rates (Bloesch and Burns 1980).

Most important, under oxic conditions and close to neutral pH—as in Pyhäjärvi—the resuspended matter will release very little, if any, P; in fact, the suspended matter may even adsorb P from water and transport it to the sediment (Koski-Vähälä and Hartikainen 2000). The internal load estimates derived using gross sedimentation measurements are thus not very informative, because only a small and seasonally variable fraction of the particle-bound TP is involved in the biological processes. Therefore, estimates of the bioavailable part of the P load given by Ekholm et al. (1997) were more meaningful. For most of the summer, release of P from the resuspended particles is negligible, but significant desorption can occur in late summer if pH is high (Holmroos et al. 2009). The late summer increase in resuspension-related particle-bound internal load may also include a redox-related dissolved component, when sediment pore water P concentrations increase due to low redox and then resuspension mixes that pore water into lake water. This was likely the case even in Pyhäjärvi, where the magnitude and timing of internal load were mainly due to resuspension (Ekholm et al. 1997).

In addition, changes in precipitation, discharge, regulated water level, nutrient load, light, duration of the ice cover, and different ecological processes resulted in considerable variation among years. The TP concentration of Pyhäjärvi is relatively low (long-term summer average below 20 µg L^{-1}), which means that even a small natural variation can produce large proportional differences. This severely reduces the accuracy of the fluxes and mass balance components (Nürnberg et al. 2012). The role of anoxia in Pyhäjärvi is not clear. Being shallow and open (there are a few islands close to shoreline), it is fully mixed most of the time. Nürnberg et al. (2012) concluded that Pyhäjärvi shows many signs of redox-related internal load despite a general lack of stratification. These indications include increased P concentration at the end of summer and in early fall in the surface and bottom water, and an increase in P from the central location to the outlet (visible in annual averages). Thus, it is likely that the sediment-water interface is anoxic (or at a minimum, the oxic surface layer of the sediment is very thin) in the late summer and supports sediment P release in Pyhäjärvi despite oxygen-rich water in the mixed water layer. Similar internal loading has been reported in other mesotrophic mixed lakes (Steinman and Ogdahl 2015; Nürnberg and LaZerte 2016). More detailed data on sediment chemistry at different depths, including sediment fractionation analysis, may support this conclusion.

Based on the current monitoring, *in situ* internal load from stratified periods at the deep basin can be determined only occasionally. The deep and occasionally stratified area is only < 8% of lake area and is located around a relatively small deep hole down to 26 m, with a small volume compared to the entire water volume of the lake. However, more detailed monitoring and information on oxygen concentrations at the summer sediment-water interface and under ice would help to assess the role of P release year-round.

Since Pyhäjärvi is a shallow and clear water lake (Secchi depth 2–5 m), most of the water column is productive and P-limited (Sarvala and Jumppanen 1988). The DRP in water is efficiently taken up by algae and partly transferred to zooplankton and fish. This efficient removal of dissolved P is supported by observations (Ekholm et al. 1997; Lehtoranta and Gran 2002) showing that DRP in the lake was predominantly below the detection limit or at very low concentrations.

The benthic communities and food webs are often neglected in lake ecosystem studies (Vadeboncoeur et al. 2002). This is the case also in Pyhäjärvi, where most of the research has focused on pelagic food webs. The role of benthic algae has not been studied in Pyhäjärvi. According to Søndergaard et al. (2003), the sediment of shallow lakes may be in direct contact with the photic zone during the whole season, which, together with a higher sediment surface per volume of water as water depth decreases, increases the importance of sediment-water interactions, particularly in shallow lakes. The regular mixing regime in shallow lakes guarantees stable and near optimum conditions for primary production (Nixdorf and Deneke 1995). In Pyhäjärvi, the area of the photic zone is large especially in years of low water level, and the biological activity of benthic algae can be significant in summer months. Also, filamentous macroalgae can contribute to the use of nutrients in Pyhäjärvi; in recent years, there have been signs that water has been very clear and filamentous macroalgae, such as *Enteromorpha spp.* have increased (Ventelä, unpublished). Similarly, other players in the benthic food web can contribute to resuspension and internal P load. Tarvainen et al. (2005, 2010) showed that the role of benthic fish can be significant. The ecological role of abundant crayfish (*Pacifastacus leniusculus*) populations has been studied only preliminarily in Pyhäjärvi, but the studies made in other Finnish lakes show that crayfish have an important intermediary role between benthic and pelagic food webs (Ruokonen 2012).

The reviewed studies do not consider the role of macrophytes, which cover most of the shallow (< 3 m) bottom of Pyhäjärvi (Leka 2010). Submerged macrophytes are the most abundant macrophyte group; *Elodea canadensis* covered 141 ha (0.9%) of the water area in 2010. The area of emergent and floating leaved macrophytes was 64 ha and 23 ha, respectively. Macrophyte abundance in Pyhäjärvi seems to be closely linked to variation in water level (Kalpa 1999; Leka 2010; Ventelä et al. 2016). The influence of vegetation on sediment resuspension and its effects on nutrients have been studied in large and shallow Lake Taihu (China; see Chapter 11). The occurrence of macrophytes reduced the sediment resuspension rate up to 29-fold (Zhu et al. 2015). A similar experimental study has also been implemented in Lake Pyhäjärvi; and although the results are not as dramatic as in Taihu, they show a similar trend, as the average resuspension rate was 40% less in the macrophyte site compared to the open water (Nurminen et al. unpublished).

18.4.1   Management

The major management challenge of Pyhäjärvi is still the high nutrient load from the catchment. Climate change has already challenged many of the implemented water-protection measures (Ventelä et al. 2011, 2016) and now the current management strategy includes not only load-reduction measures such as constructed wetlands and filters, but most of all the agricultural practices.

Another cornerstone of the management of Pyhäjärvi is biomanipulation. Water quality of Pyhäjärvi is closely linked to both the commercial fishery and the quality and quantity of fish stocks (Ventelä et al. 2007, 2016). First, the most important mechanism to improve water quality is the top-down control of phytoplankton via the planktivorous fish—zooplankton—phytoplankton food chain since the most important commercial catch species are planktivores. Second, TP export via fish catch improves the P cycle. Between 1995 and 2016, more than 15 million kg of fish were removed as total fish catch of Pyhäjärvi and with this, about 100,000 kg of TP were removed from the system (Ventelä unpublished). Based on the P budget, 25–30% of the incoming P is removed annually from the system in the fish catch. Third, internal nutrient loading is enhanced by benthic fish, so an active fishery that removes these species is beneficial for lake water quality (Tarvainen et al. 2005, 2010; see Chapter 1). Nürnberg et al. (2012) determined that the P content of the annually removed fish catch equals approximately the estimated annual internal P load. Since 2008, both the fishery and fish stocks have been dramatically affected by climate change (loss of ice, collapse of cold water species) and the fishery has been forced to adapt to new conditions. While the fishery of the previously dominant commercial species—vendace and whitefish—has suffered, an increasing share of the other species in the catch has become commercially valuable as the value of the clean local food is increasing in Europe. This may ultimately improve the financial possibilities to implement biomanipulation in Pyhäjärvi.

18.5   CONCLUSIONS

In spite of the meritorious research history of Pyhäjärvi, the complex mechanisms of internal load in this shallow, large, and mixed lake are not well monitored, let alone understood. There is a need for more specific sediment measurements, along with studies on benthic microbial processes and the role of bioturbation at the sediment-water interface.

In the future, climate change is expected to alter many sediment-related processes. As the ice cover has already decreased by one month (Ventelä et al. 2011, and unpublished records), the wind exposure, light conditions, mixing processes, etc. in Pyhäjärvi are facing dramatic changes (Wu et al. 2018). These changes affect the nutrient cycles of the catchment, the sediment, and the pelagic system, inviting the re-evaluation of previous data and the revision of established models. The monitoring program in Pyhäjärvi will also need updating, as more components from catchment to lake, and from sediment to surface, have to be included, as well as links to the groundwater system. In a world of diminishing resources for environmental monitoring, this represents a challenge, but is necessary for the successful adaptation of management activities to climate change.

18.6   ACKNOWLEDGMENTS

Data collection and analyses were funded from many different sources, in particular a series of grants from Academy of Finland (including 09/109, 09/074, 1071004, 1071149, 1071255, 35619, 44130, 52271, 104483, 201414, 256240), Finnish Environment Institute, Southwest Finland's Centre for Economic Development, Transport and the Environment and Pyhäjärvi Protection Fund. We thank Elisa Mikkilä for providing maps.

18.7   REFERENCES

Bloesch, J and Burns, NM. 1980. A critical review of sedimentation trap technique. Schweiz Z Hydrol. 42: 15–55.
Buesseler, KO. 1991. Do upper-ocean sediment traps provide an accurate record of particle flux? Nature. 353:420–423.
Deng, J; Qin, B; Sarvala, J; Salmaso, N; Zhu, G; Ventelä, A-M; Zhang, Y; Gao, G, Nurminen, L; Kirkkala, T; et al. 2016. Phytoplankton assemblages respond differently to climate warming and eutrophication: a case study from Pyhäjärvi and Taihu. J Great Lakes Res. 42(2):386–396. DOI: 10.1016/j.jglr.2015.12.008.
Ekholm, P; Malve, O; and Kirkkala, T. 1997. Internal and external loading as regulators of nutrient concentrations in the agriculturally loaded lake Pyhäjärvi (southwest Finland). Hydrobiologia. 345:3–14.
Evans, RD. 1994. Empirical evidence of the importance of sediment resuspension in lakes. Hydrobiologia. 248:5–12.
Helminen, H; Auvinen, H; Hirvonen, A; Sarvala, J; and Toivonen, J. 1993. Year-class fluctuations of vendace (Coregonus albula) in Lake Pyhäjärvi, southwest Finland, in 1971–1990. Can J Fish Aquat Sci. 50:925–931.
Holmroos, H; Niemistö, J; Weckström, K; and Horppila, J. 2009. Seasonal variation of resuspension-mediated aerobic release of phosphorus. Boreal Environment Res. 14:937–946.

Huttula, T. 1994. Suspended sediment transport in Lake Säkylän Pyhäjärvi. Aqua Fennica. 24:171–185.

Jeppesen, E; Mehner, T; Winfield, IJ; Kangur, K; Sarvala, J; Gerdeaux, D; Rask, M; Malmquist, HJ; Holmgren, K; Volta, P; et al. 2012. Impacts of climate warming on lake fish assemblages: evidence from 24 European long-term data series. Hydrobiologia. 694:1–39.

Jumppanen, K, (Ed.). 1983. Säkylän Pyhäjärven ainetaseet, vedenlaatu ja kasviplanktonin perustuotanto v. 1982 ja kolmivuotiskautena 1980–1982. [Material balances, water quality and the phytoplankton production in Lake Pyhäjärvi in 1982 and 1980–1982.] Lounais-Suomen vesiensuojeluyhdistys r.y., Finland 55: 1–67. (in Finnish).

Kalpa, A. 1999. Säkylän Pyhäjärven rantakasvillisuus kesällä 1997. [Littoral macrophyte vegetation in Säkylän Pyhäjärvi in summer 1997.] Lounais-Suomen ympäristökeskus. Alueelliset ympäristöjulkaisut 121. p. 58 (in Finnish).

Kirkkala, T. 2014. Long-term nutrient load management and lake restoration: case of Säkylän Pyhäjärvi (SW Finland). Annales Universitatis Turkuensis. A II:286.

Kirkkala, T; Ventelä, A-M; and Tarvainen, M. 2012. Long-term field-scale experiment on using lime filters in an agricultural catchment. J Environ Qual. 41:409–410.

Koski-Vähälä, J and Hartikainen, H. 2000. Resuspended sediment as a source and sink for soluble phosphorus. Verh Int Ver Limnol. 27:3141–3147.

Lehtoranta, J and Gran, V. 2002. Sedimentistä veteen vapautuvat ravinteet Säkylän Pyhäjärvellä [Nutrient flux from sediment into water in Lake Säkylän Pyhäjärvi]. Finnish Environment Institute Mimeographs.

Leka, J. 2010. Säkylän Pyhäjärven vesikasvillisuus vuonna 2010. Pyhäjärvi-instituutin julkaisuja nro. 22. ISBN 978-952-9682-66-9 (in Finnish).

Marttunen, M; Hellsten, S; Rotko, P; Faehnle, M; and Visuri, M. 2000. Selvitys Säkylän Pyhäjärven sää nnöstelyn vaikutuksista ja kehittämismahdollisuuksista. (The effects and the development of the regulation of Lake Pyhäjärvi.) Finnish Environment Institute, Alueelliset ympäristöjulkaisut 166. p. 43. ISBN 952-111-0731-6. ISSN 1238-8610 (in Finnish).

Niemistö, JP and Horppila, J. 2007. The contribution of ice cover to sediment resuspension in a shallow temperate lake: possible effects of climate change on internal nutrient loading. J Environ Qual. 36:1318–1323.

Nixdorf, B and Deneke, R. 1995. Why 'very shallow' lakes are more successful opposing reduced nutrient loads. Hydrobiologia. 342/343:269–284.

Nürnberg, GK. 1988. Prediction of phosphorus release rates from total and reductant-soluble phosphorus in anoxic lake sediments. Can J Fish Aquat Sci. 45:453–462.

Nürnberg, GK. 1998. Prediction of annual and seasonal phosphorus concentrations in stratified and polymictic lakes. Limnol Oceanogr. 43:1544–1552.

Nürnberg, GK. 2005. Quantification of internal phosphorus loading in polymictic lakes. Verh Internat Verein Limnol. 29:623–626.

Nürnberg, GK and LaZerte, BD. 2016. More than 20 years of estimated internal phosphorus loading in polymictic, eutrophic Lake Winnipeg, Manitoba. J Great Lakes Res. 42:18–27.

Nürnberg, GK; Tarvainen, M; Ventelä, AM; and Sarvala, J. 2012. Internal phosphorus load estimation during biomanipulation in a large polymictic and mesotrophic lake. Inland Waters. 2(3):147–162.

Rautio, A and Korkka-Niemi K. 2011. Characterization of groundwater—lake water interactions at Pyhäjärvi, a lake in SW Finland. Bor Environ Res. 16(5):363–380.

Rautio, A and Korkka-Niemi, K. 2015. Chemical and isotopic tracers indicating groundwater/surface-water interaction within a boreal lake catchment in Finland. Hydrogeol J. 23(4):687–705.

Räsänen, M; Salonen, VP; Salo, J; Walls, M; and Sarvala, J. 1992. Recent history of sedimentation and biotic communities in Lake Pyhäjärvi, SW Finland. J Paleolimn. 7:107–126.

Ruokonen, T. 2012. Ecological impacts of invasive signal crayfish in large boreal lakes. Jyväskylä Studies in Biological and Environmental Science 244. p. 40. Dissertation, University of Jyväskylä. ISBN 978-951-39-4832-0 (PDF).

Sarvala, J; Helminen, H; and Auvinen, H. 1999. Portrait of a flourishing freshwater fishery: Pyhäjärvi, a lake in SW-Finland. Bor Environ Res. 3:329–345.

Sarvala, J; Helminen, H; Saarikari, V; Salonen, S; and Vuorio, K. 1998. Relations between planktivorous fish abundance, zooplankton and phytoplankton in three lakes of differing productivity. Hydrobiologia. 363: 81–95.

Sarvala, J and Jumppanen, K. 1988. Nutrients and planktivorous fish as regulators of productivity in Lake Pyhäjärvi, SW Finland. Aqua Fennica. 18:137–155.

Søndergaard, M; Jensen, JP; and Jeppesen, E. 2003. Role of sediment and internal loading of phosphorus in shallow lakes. Hydrobiologia. 506-509:135–145.

Steinman, AD and Ogdahl, ME. 2015. TMDL reevaluation: reconciling internal phosphorus load reductions in a eutrophic lake. Lake Reserv Manage. 31:115–26.

Tarvainen, M; Ventelä, A-M; Helminen, H; and Sarvala, J. 2005. Nutrient release and resuspension generated by ruffe (Gymnocephalus cernuus) and chironomids. Freshwater Biol. 50:447–458.

Tarvainen, M; Ventelä, A-M; Helminen, H; and Sarvala, J. 2010. Selective removal fishing—water quality and practical viewpoints. International Association of Theoretical and Applied Limnology Proceedings 30: 1653–1656.

Tarvainen, M; Vuorio, K; and Sarvala, J. 2008. The diet of ruffe Gymnocephalus cernuus (L.) in northern lakes: new insights from stable isotope analyses. J Fish Biol. 72:1720–1735.

Vadeboncoeur, Y; Vander Zanden, MJ; and Lodge, DM. 2002. Putting the lake back together: reintegrating benthic pathways into lake food web models: lake ecologists tend to focus their research on pelagic energy pathways, but, from algae to fish, benthic organisms form an integral part of lake food webs. BioScience. 52(1):44–54.

Ventelä, A-M; Amsinck, SL; Kauppila, T; Johansson, LS; Jeppesen, E; Kirkkala, T; Søndergaard, Weckström, J; and Sarvala J. 2016. Ecosystem change in the large and shallow Lake Säkylän Pyhäjärvi, Finland, during the past ~400 years: implications for management. Hydrobiologia. 778(1):273–294.

Ventelä, A-M; Kirkkala, T; Lendasse, A; Tarvainen, M; Helminen, H; Sarvala, J. 2011. Climate-related challenges in long-term management of Säkylän Pyhäjärvi (SW Finland). Hydrobiologia. 660:49–58.

Ventelä, A-M; Tarvainen, M; Helminen, H; and Sarvala, J. 2007. Long-term management of Pyhäjärvi (southwest Finland): eutrophication, restoration–recovery? Lake Reserv Manage. 23:428–438.

Wu, T; Qin, B; Zhu, G; Huttula, T; Lindfors, A; Ventelä, A-M; and Sheng, Y. 2018. The contribution of wind wave changes on diminishing ice period in Lake Pyhäjärvi during the last half-century. Environ Sci Pollut Res. 25: 24895. https://doi.org/10.1007/s11356-018-2552-7

Zhu, M; Zhu, G; Nurminen, L; Wu, T; Deng, J; Zhang, Y; Qin, B; Ventelä, A-M. 2015. The influence of macrophytes on sediment resuspension and the effect of associated nutrients in a shallow and large lake (Lake Taihu, China). PlosOne https://doi.org/10.1371/journal.pone.0127915

INTERNAL LOADING OF PHOSPHORUS TO LAKE ERIE: SIGNIFICANCE, MEASUREMENT METHODS, AND AVAILABLE DATA

Eliza M. Kaltenberg[1] and Gerald Matisoff[2]

Abstract

Lake Erie is the shallowest and most heavily populated lake of the Laurentian Great Lakes. Due to the large volume of wastewater treatment effluents it receives and a large percentage of agricultural land contributing bioavailable phosphorus (P) from fertilizers, the lake has experienced severe anthropogenic eutrophication since the 1960s. The problems manifest through harmful algal blooms (HABs) in the western basin and hypoxia in the central basin. Introduction of P reduction programs that were designed to meet the Great Lakes Water Quality Agreement (GLWQA) guidelines in the 1970s resulted in significant improvement in water quality (IJC 1978). Unfortunately in the mid-1990s, the water quality started decreasing again despite lower external total P inputs that met or exceeded the GLWQA guidelines. In response to this apparent decoupling between external loading and water quality, work was conducted to better understand P loadings and P cycling in the lake, including obtaining estimates of internal P loading. This chapter compiles experimental and modeled internal P loading data for Lake Erie and reveals significant gaps in the data for the central and eastern basins. In the western basin, for which the largest quantity of data are available, the basin-wide annual internal P input is calculated and compared to total P target loads for Lake Erie. The results indicate a small, but not negligible, contribution of internal loading to the total P budget in western Lake Erie.

Key words: Lake Erie, Great Lakes, internal P loading, P flux, Fickian diffusion

[1] Battelle, 141 Longwater Dr., Suite 202, Norwell, MA 02061; elizakaltenberg@gmail.com

[2] Case Western Reserve University, 112 A. W. Smith Bldg., 10900 Euclid Ave, Cleveland, OH 44106; gxm4@case.edu

19.1 INTRODUCTION

19.1.1 Key Water-Body Features

Lake Erie is the smallest (by volume) and the shallowest of the North American Laurentian Great Lakes (see Figure 19.1). The lake is elongated in the east-west direction and its shores are bounded by Canada on the north and by the United States (US) on the south. The major inlet of water to the lake is the Detroit River, which drains Lake Huron through Lake St. Clair, located at the west end of Lake Erie. The outlet for Lake Erie, the Niagara River, is located at its east end. The lake has a surface area of 25,700 km², an average depth of 19 m, with the deepest point reaching 64 m, and a volume of 484 km³ (US EPA and Government of Canada 1995).

Lake Erie is divided into three basins that behave almost as separate lakes: the western, central, and eastern basins. The western basin is the shallowest, with an average depth of 7.4 m and a maximum depth of 19 m. It receives the bulk of the water inflow to the lake from the Detroit River and the majority of the external P loading to the lake from the Maumee River (Maccoux et al. 2016). Although the western basin does stratify thermally, it is sufficiently shallow that wind events usually destroy the stratification within a few days (Bridgeman et al. 2006). The central basin is characterized by a very uniform depth, averaging ~18 m, with a maximum depth of 25 m. This basin thermally stratifies seasonally and the relatively thin hypolimnion goes anoxic prior to fall overturn. The eastern basin is the deepest and displays strong bottom elevation gradients. The average depth is 24 m, the maximum depth is 64 m, and, like the central basin, the basin stratifies thermally, but the much thicker hypolimnion contains sufficient oxygen that it does not go anoxic (see Table 19.1).

The Lake Erie watershed encompasses 78,000 km² (US EPA and Government of Canada 1995), of which about two-thirds fall within the US (Myers et al. 2000). The watershed is heavily populated, with 11.6 million people living within its borders (International Joint Commission 2014). The predominant land use is agricultural (of the US part of the watershed, about 67% is agriculture; Richards et al. 2010), followed by urban areas, and, to a lesser extent, industrial uses. The lake itself

Table 19.1 Summary of Lake Erie characteristics

Surface area	25,700 km²
Water depth	Average: 19 m, Max: 64 m
Volume	484 km3
Western basin depth	Average: 7.4 m, Max: 19 m
Central basin depth	Average: ~18 m, Max: 25 m
Eastern basin depth	Average: 24 m, Max: 64 m
Main water inlet and outlet	Inlet: Detroit River, Outlet: Niagara River
Main external point sources of P	Maumee River (1st) and Detroit River (2nd)
Water residence time	2.6 years
Watershed size	78,000 km²
Watershed population	11.6 million
Predominant land use	Agricultural, urban, industrial
Socioeconomic importance	Drinking water, tourism, recreational, fisheries
Predominant environmental challenge	Western basin—harmful algal blooms, Central basin—hypoxia

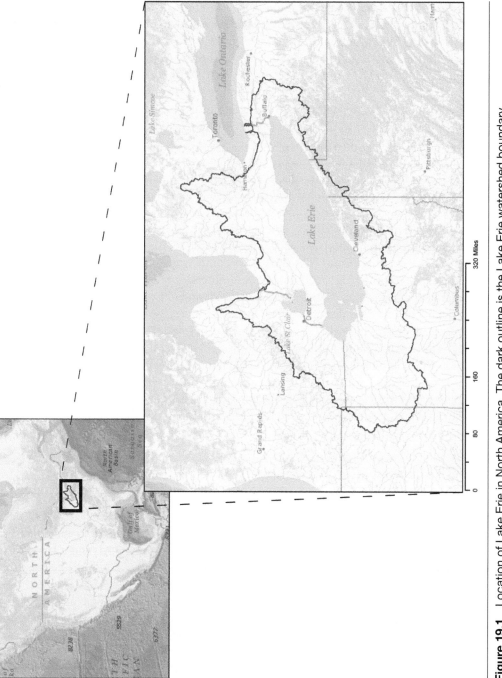

Figure 19.1   Location of Lake Erie in North America. The dark outline is the Lake Erie watershed boundary.
Source: http://www.globalgreatlakes.org/

serves as a drinking water source for 11 million people and provides recreational value for tourists, anglers, boaters, birdwatchers, and visitors of the local parks (Allan et al. 2015).

Intensive row-crop agriculture and urbanization within the Lake Erie watershed after World War II led to significant increases in nutrient delivery to the lake from nonpoint source runoff from fields and municipal wastewater treatment plants, respectively, that created water quality problems starting in the 1960s (Ohio EPA 2010). The western and central basins experienced severe eutrophication, development of HABs, hypolimnion hypoxia, and fish kills, all of which were linked to excess phosphorus (P). The annual loading of total P (TP) to Lake Erie reached a high of 28,000 metric tons in 1968 (Ohio EPA 2010). The Great Lakes Water Quality Agreement (GLWQA) signed by the International Joint Commission (IJC) in 1972 and the consequent implementation of P reduction programs resulted in remarkable improvement of water quality in Lake Erie. Improved efficiency of wastewater treatment plants and removal of P in detergents resulted in successful reduction of TP loads to below the target 11,000 metric tons per annum (mta) (Lake Erie LaMP 2011). However, despite P loadings at or below the GLWQA target, HABs and hypoxia started occurring again in the mid-1990s (Matisoff and Ciborowski 2005; Scavia et al. 2016), possibly as a consequence of increased internal P loading (Ohio EPA 2010). To combat the again worsening water quality, Annex 4 of the most recent 2012 GLWQA update recommended reducing TP loading to the lake by 40% to a target load of 6,000 mta. Matisoff et al. (2016) identified five potential causes of the worsening water quality in Lake Erie despite stable P loads: (1) increased internal loading of P, possibly mediated by invasive quagga mussels; (2) underestimation of some external P inputs (e.g., combined sewer overflows); (3) changes in nutrient uptake mechanisms and nutrient balance in the lake; (4) increases in the bioavailable fraction of P while maintaining relatively constant concentrations of TP; and (5) weather/climate-induced changes affecting water levels and temperature, the timing of nutrient pulses, and wind events that can affect nutrient release through sediment resuspension. Of these factors, internal loading has received relatively little attention and the existing data are very limited, both spatially and temporally.

19.1.2 Challenges and Stressors

Even though Lake Erie is the smallest (by volume) and shallowest of the Great Lakes, about one third of the total population of the Great Lakes region lives in the Lake Erie watershed, making it the most densely populated and most susceptible to anthropogenic changes in the Great Lakes (Richards et al. 2010). The watershed is heavily industrialized and urbanized, and a large portion of the watershed consists of agricultural land. Together these land uses result in the delivery to the lake of significant quantities of effluents from point sources such as wastewater treatment plants, and from nonpoint source pollutants such as farming activities. Both wastewater treatment plants and farmland fertilizers/manure deliver significant amounts of readily bioavailable soluble reactive P (SRP) to Lake Erie (LaMP 2011). The combination of a high nutrient delivery and a shallow water depth or volume of the lake makes Lake Erie the most biologically productive of the five Great Lakes. The major challenge in the shallow western basin is that this high biological productivity generates excessive and harmful algal blooms, which decrease both oxygen concentrations in the water column and light penetration depths. More important, some of the algal blooms contain cyanobacteria (commonly known as green-blue algae) that potentially produce toxins, such as the HAB-forming species *Microcystis aeruginosa*. Development of HABs in the western basin provides a threat to the health of the ecosystem, impacts the quality of city water derived from the lake, and decreases the recreational value and beneficial uses of the lake. Several major algal blooms have occurred in recent years, including the 2014 bloom that shut down Toledo (Ohio, U.S.) city water. In the central basin, the main problem related to eutrophication is bottom water hypoxia, which can directly affect the basin's fauna and the taste and odor of drinking water, and may indirectly contribute to further increasing P concentrations through anoxic release of P from sediments.

With a relatively short 2.6-year water residence time, Lake Erie is very responsive to climate change drivers such as increased water temperature, decreased snow and ice cover, and a higher frequency of high-intensity rain events. Based on global climate change models, by the end of the century the Great Lakes region is expected to experience a 2–6°C rise in temperature and a 20–30% increase in the winter and springtime precipitation due to a higher frequency of storm events (Cousino et al. 2015 and literature cited therein). High precipitation during the spring has been shown to significantly correlate with the intensity of summertime algal blooms (Stumpf et al. 2012; Obenour et al. 2014; Ho and Michalak 2017), presumably because increased runoff will deliver more agricultural P to the lake. Therefore, an increase in spring precipitation due to climate change, especially after application of fertilizer, is likely to cause further problems with controlling eutrophication in Lake Erie.

19.1.3   Key Questions to Be Addressed

Measuring internal loading of P is very challenging and multiple approaches have been used to try to estimate it (see Chapter 5). The current chapter compiles and compares the advantages and limitations of the different field, laboratory, and modeling methods that have been used in Lake Erie. Field methods include bottom water chamber incubations that determine the flux by monitoring concentration changes over a fixed surface area and volume of water over time. Laboratory methods include core incubations that are conducted similarly to the bottom chamber incubations except that they are conducted on a sediment core in the laboratory. Alternatively, some laboratory methods are based on Fickian diffusion using concentration gradients established by expressing (separating from a bulk sediment sample) or electrode probing of pore waters, or in a more innovative approach by utilizing diffusive equilibrium in thin films (DET). Modeling methods use complex mass balances where the internal loading is an adjustable parameter or incorporates Fickian diffusion-based modeled concentration gradients. Finally, this chapter provides a synthesis of the available internal P loading estimates for Lake Erie. The hypothesis to be addressed is: "Internal loading of P provides a significant contribution to the total water-column P and therefore contributes to HABs and hypoxia."

19.2   METHODS

This chapter outlines the methods used for estimating internal P loading to Lake Erie. Field and laboratory methods may estimate either a gross P flux or a net P flux, and although they are not necessarily the same we, here, report them together and note the possible differences where appropriate. Bottom chamber and core incubations determine net fluxes; Fickian calculations based on pore-water profiles yield gross fluxes. Removal processes at the sediment-water interface and in the water column will lead to a lower apparent flux in the incubations; hence incubations determine a net flux. Conversely, the Fickian calculation from a concentration gradient does not account for any removal mechanisms; hence it is a gross flux. Internal loadings may be estimated from three main approaches: (1) direct net flux measurements; (2) Fickian calculations of the gross diffusive P flux based on its concentration gradient between the overlying water and pore water; and (3) modeling approaches.

19.2.1   Direct Net Flux Measurements

19.2.1.1   Sediment Core Incubations

The most typical direct experimental measurement of P flux consists of a sediment core that is incubated at desired temperature and redox conditions (James 2012; Gibbons 2015). Redox is particularly important when measuring P flux because oxic conditions are well known to significantly inhibit release of P from sediment through sorption of P onto iron oxyhydroxides. A shift of the redox conditions toward anoxia causes dissolution of the iron oxyhydroxides and a subsequent release of sorbed P. Surficial oxic conditions can be maintained by pumping air into the water in the sediment core tube, whereas anoxic conditions are achieved by flushing nitrogen or other inert gas (sometimes mixed with CO_2 to buffer the pH) through the water overlying the sediment. The net flux of P from sediment to the water column in both aerobic and anaerobic incubations can be calculated based on time-series measurements of the overlying water concentration following the equation:

$$F = \frac{dC}{dt} \cdot \frac{V}{A}$$ [Eq. 19.1]

where F is the flux (mg m^{-2} day^{-1}), $\frac{dC}{dt}$ is the time rate of change of the P concentration in the water (mg m^{-3} day^{-1}), V is the volume of the overlying water in the core tube (m^3) at the time of sampling, and A is the surface area of the core (m^2). Fluxes for Lake Erie determined from core incubations in both oxic and anoxic conditions are reported by James (2012), Gibbons (2015), and Matisoff et al. (2016).

19.2.1.2   Bottom Chambers

Net flux of P from sediment to the water column can be measured in the field (*in situ*) using bottom chambers (Tenberg et al. 2004). Although much less common, this method uses a similar approach to the one used in core incubations—the flux is estimated using Equation 19.1 from the time-rate of change of the P concentration in the contained water volume. This approach was used by Matisoff et al. (2016). They deployed plastic bottom chambers that encapsulated a water volume of about 42.5 L over a surface area of 2,232 cm^2. The water inside the chamber was recirculated at 1.5 L min^{-1} with a Dynamax DM100vv DC pump powered by a car battery. The enclosed water was sampled in time series by SCUBA divers using plastic syringes to withdraw 60–65 mL aliquots per sample.

19.2.2   Fickian Diffusion Flux Calculations

This group of methods provides a calculated diffusive gross flux of SRP across the sediment-water interface based on the concentration gradient at the sediment-water interface. The SRP concentrations in the surface water just above the sediment-water interface and in the sediment pore water are measured and the flux is calculated using Fick's First Law of diffusion. In the sediment, Fick's First Law of diffusion says that the flux of solute is proportional to its concentration gradient at the sediment-water interface:

$$F = -D_s \frac{\partial c}{\partial x}$$ [Eq. 19.2]

where D_s is the P diffusion coefficient in the sediment and $\frac{\partial c}{\partial x}$ is the solute concentration gradient at the sediment-water interface. In the sediment, the gradient is calculated in the depletion layer (see Figure 19.2). Because in sediment, diffusive transport of the solute occurs only through pore water and sediment porosity has to be taken into consideration. D_s can be calculated by applying porosity and tortuosity corrections to the diffusion coefficient in pure water (D_0), and hence Equation 19.3 takes the following form (Boudreau 1996):

$$F = -\frac{\phi \cdot D_0}{\theta^2} \frac{\partial c}{\partial z}$$
[Eq. 19.3]

where ϕ is the sediment porosity (dimensionless), θ is the sediment tortuosity (dimensionless), and $\frac{\partial c}{\partial z}$ is the concentration gradient in the depletion layer of the thickness z (g cm^{-4}). Tortuosity can be estimated from the relationship (Boudreau 1996):

$$\theta^2 = 1 - \ln(\phi^2)$$
[Eq. 19.4]

If porosity is close to 1, $\theta^2 \approx 1/\phi^2$ (Boudreau 1996). This approximation was used by Paytan et al. (2017).

It can be assumed that the mass of P coming out from the sediment pore water through the sediment-water interface is equal to the mass of P that is entering the water column from the underlying sediment (conservation of mass). Therefore, the flux can be calculated also by using the

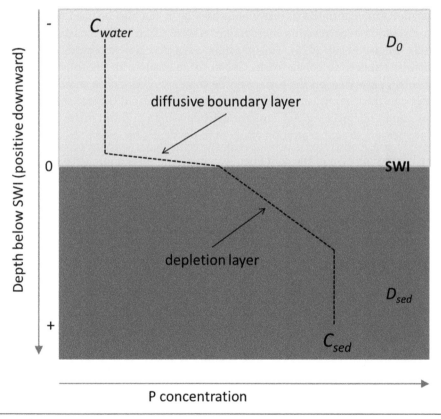

Figure 19.2 Schematic P concentration profile near the sediment-water interface (SWI). C_{water}—concentration of P in water; C_{sed}—concentration of P in sediment pore water; D_0—diffusivity (diffusion coefficient) of P in water; D_{sed}—diffusivity of P in sediment pore water. The P flux from sediment to the overlying water can be calculated using Equations 19.3 and 19.4 and by using a linear approximation of the gradient in the sediment in the depletion layer or in the water using the diffusive boundary layer.

diffusion coefficient in water (D_0), and ϕ and $\theta = 1$ (their values in water), and considering the concentration gradient in the diffusive boundary layer in the water column (see Figure 19.2). However, with typical thickness of the diffusive boundary layer ranging between 50 and 500 μm (Apell et al. 2015), determination of the concentration gradient is challenging.

Fickian diffusion calculations of P flux from sediment to the overlying water based on concentration gradient data obtained using (a) expressed pore water, (b) P-sensitive microelectrodes, and (c) P-DET gel, were reported for Lake Erie by Matisoff et al. (2016) and in this chapter from additional new results.

19.2.2.1   Expressed Pore Water

Pore water is expressed from a sediment core at 1 cm or 2 cm intervals, ideally in a nitrogen-filled glove box to avoid oxidation of dissolved iron (II) and subsequent loss of P due to sorption to iron oxyhydroxides. The pore water is then forced through a 1 μm-pore-sized membrane filter and collected into a bottle. SRP can be then measured using a standard spectrophotometric method (e.g., Strickland and Parsons 1972). This method was used by Matisoff et al. (2016) for their analyses in the western basin and by Paytan et al. (2017) for their core from the central basin.

19.2.2.2   Phosphate Microelectrode

The concentration gradient in the diffusive boundary layer can also be used to calculate diffusive fluxes. High-resolution concentration measurements were taken using a custom made phosphate microelectrode (Matisoff et al. 2016). This phosphate-sensitive potentiometric microelectrode utilizes a cobalt wire oxidized to cobalt oxide. Cobalt oxide reacts with phosphate ions causing precipitation of cobalt phosphate on the surface of the cobalt wire. Concentration of phosphate in the sampled water can then be measured by monitoring the potential difference between the cobalt microelectrode and a reference Ag/AgCl microelectrode using a millivolt meter. The microelectrode can be used to measure the concentration gradient in the depletion layer or in the diffusive boundary layer, assuming that the layer is thicker than the vertical resolution of the electrode (0.1 cm). Additional details of the phosphate microelectrode construction and flux calculation are given in Ding et al. (2015) and Matisoff et al. (2016).

19.2.2.3 Phosphate DET

Another way of obtaining high spatial resolution phosphate concentration data is through a combination of DET hydrogels and colorimetry (Matisoff et al. 2016; Kaltenberg 2016). This method enables two-dimensional measurements of phosphate concentrations in pore water and surface water with a spatial resolution of about 0.01 cm. The technology could be used both *in situ* and *ex situ*, although only the *ex situ* approach was used by Matisoff et al. (2016). Details of the preparation, sampling, and analysis of the method are presented in Kaltenberg (2016). Briefly, sheets of hydrogel are prepared by casting a polyacrylamide solution (Jézéquel et al. 2007) between two parallel pieces of glass separated by spacers (gel casters, GE Health Sciences). The prepared hydrogel sheets are then soaked in deionized water (DI) to remove any unpolymerized reagents and stored in DI until needed. Water sampling is conducted by inserting a piece of hydrogel vertically across the sediment-water interface into an undisturbed sediment core. The sampling gel is left in the sampled sediment and overlying water for about 18 hours and is then retrieved for analysis. The sampling gel is placed on a piece of glass and overlaid by a staining gel, which is the same type of hydrogel but presoaked in colorimetric reagents for two hours before use. The hydrogels are covered with a transparency sheet to minimize evaporation. After 20 minutes of color development, the gels are scanned using a conventional flatbed scanner alongside a set of standards. The standards are prepared from small pieces of hydrogel soaked in standard solutions with varying concentration of phosphate, which are colorized with staining gels, just like the sample gels. Calibrated images allow high-resolution determination of the concentration gradients across the sediment-water interface, as well as provide insight into two-dimensional changes of phosphate concentration throughout the sampled region (Matisoff et al 2016; Kaltenberg 2016).

19.2.3   Modeling Approach

Several modeling approaches of varying complexities have been used for estimating internal P loading to Lake Erie. Estimates of the P flux may be determined by mass balance differences (mass balance models) or by Fickian calculations of computed P concentration profiles (diagenetic models).

19.2.3.1   Mass Balance Model

Di Toro (2001) presents a model that assumes a steady-state balance between immobilization of P by sediment particles through mineral precipitation and/or adsorption and remobilization of P through dissolution of those solids upon burial. Matisoff et al. (2016) adapt this mass balance model in which the diagenetic production of SRP is calculated as the difference between the depositional flux of organic P and the burial flux of organic P. The flux of P from sediment to the water column is then calculated by subtracting the burial P flux from the diagenetic P production.

Verhamme et al. (2016b) also use the Di Toro (2001) model in their calculation of the P flux. The Western Lake Erie Ecosystem Model (WLEEM) (Verhamme et al. 2016b) was developed by Limno-Tech (Ann Arbor, Michigan, USA) to predict water quality conditions in western Lake Erie based on a large number of input environmental stressors including hydrology, sediment and nutrient loads, wind, and other climate-related stressors. This three-dimensional, fine-scale, process-based eutrophication model incorporates four sub-models designed to simulate the wind-wave relationships, hydrodynamics and temperature, sediment transport, and water quality. The combination of these four sub-models allows resolution, in space and time, of numerous processes such as hydrometeorological conditions, nutrient and suspended solids transport, nutrient concentrations, cycling and mass budgets, phytoplankton dynamics, dissolved oxygen, and light penetration depths. An advantage of this model is that the model can calculate the seasonal variation in P flux. Those results have not yet been published but key internal loading calculations have been presented by Verhamme et al. (2016a). An estimate of the annual internal loading to the western basin of Lake Erie in 2014 calculated by the model was included in Matisoff et al. (2016) based on personal communication with the WLEEM authors—and those results are discussed in this chapter.

19.2.3.2   Diagenetic Model

A diagenetic model (Dittrich et al. 2009; Reichert 1994) was used to calculate P flux from sediments. The model is integrated into version 2.1e of the computer program Aquasim. The diagenetic model is a 1-dimensional, non-steady-state transport reactive model designed to model diagenesis of solids and solutes in sediment. It incorporates solute transport via diffusion, biodiffusion, and bioirrigation and transport of solids due to compaction, burial, and particle diffusion, as well as the transformation rates of both the dissolved and particulate phases. The P flux from sediments is then calculated using Fickian diffusion of computed SRP profiles in sediments.

19.2.4   Statistical Analyses

All statistical analyses included in this chapter have been conducted using Minitab 18 statistical software. All datasets were tested for normality before performing any further analyses. When needed, the Box-Cox procedure was used to normalize the data. Mean fluxes under oxic and anoxic conditions were determined using on-sample t-test assuming 0.05-significance level.

19.3   RESULTS

This chapter provides a summary of currently available estimates of internal P loading in Lake Erie. Most of the data included in this chapter are from published studies including: Matisoff et al. (2016); Verhamme et al. (2016b); and Paytan et al. (2017). Included also are unpublished data (Verhamme et al. (2016a), 22 new experimental results from G. Matisoff (Case Western Reserve University, Cleveland, Ohio, USA) and one central basin diagenetic modeling result from M. Dittrich (University of Toronto Scarborough, Canada). The authors were able to identify 119 published and unpublished internal P estimates (measured and modeled) for Lake Erie, of which 111 were for the western basin. These numbers include both laboratory and field replicate measurements, so the actual number of sampling stations for which estimates exist is much lower. A total of 21 sampling stations in Lake Erie have been sampled, with 19 stations located in the western basin and only two in the central basin. No reported data were identified for the eastern basin.

The majority of the internal P loading estimates were obtained by aerobic and anaerobic incubations (see Table 19.2), followed by P-DET measurements (Matisoff et al. 2016). In general, fluxes under aerobic conditions are lower than those obtained under anaerobic conditions, consistent with the general understanding of P retention in sediments through sorption onto iron oxyhydroxides in oxidized waters (Mortimer 1941). Comparing the two methods for which a large number of data were collected, aerobic core incubations and P-DET, show that, on average, fluxes obtained from P-DET measurements are higher than those obtained using aerobic core incubations. However, the P-DET results are close to the estimate obtained from the diagenetic model, and on the low side compared to that obtained from the mass balance model. It is also possible that the difference is because core incubations measure a net flux and results determined from P-DET are gross fluxes.

The different methods yield P flux estimates that are within about a factor of three of each other. Undoubtedly, some of the variation is because not all methods were conducted at the same locations in the lake, and even within the same sampling station, variability is expected because of sediment heterogeneity. Additionally, not all samples used in flux calculations presented in this chapter were collected during the same period of time (sampling times span late spring and/or summer months of 2012, 2013, and 2014), which could lead to different fluxes due to differences in ambient temperature. The differences may also result from the fact that some methods yield net fluxes whereas other methods yield gross fluxes. While the methods used for internal P loading estimation in Lake Erie do not agree with each other perfectly and the determination of which method is the most accurate still has not been made, the results compiled in this chapter can be used to estimate the average basin-wide internal loading of P to Lake Erie. Following the method described by Matisoff et al. (2016), we tabulated all available data, calculated mean values for replicate measurements conducted at each sampling station using any particular method, then used the results to calculate both the mean aerobic and the mean anaerobic fluxes for each station (average from all aerobic or all anaerobic methods, respectively). Finally, we averaged the aerobic and anaerobic results from all the stations within each basin to calculate the basin-wide internal P loading. The mean internal loading of P to the western basin under oxic conditions was 1.40 mg m^{-2} day^{-1} with a 95% confidence interval of 0.92 to 1.89 mg m^{-2} day^{-1}, and the mean anaerobic internal

Table 19.2   Summary of SRP fluxes from sediment to the water column in Lake Erie. Data presented by basin and by method.

Method of Flux Estimation	n	SRP Flux (mg m^{-2} d^{-1})			
		Mean	SE	Minimum	Maximum
Western Basin					
Aerobic					
Experimental					
Core incubations	18	1.07	0.18	0.24	2.99
Bottom chambers	7	0.93	0.39	0.14	2.87
P microelectrode	2	0.95	0.94	0.01	1.89
P-DET	10	2.72	0.44	0.88	5.33
Expressed porewater	1	0.79	N/A	N/A	N/A
Models					
Mass balance model	1	4.93	N/A	N/A	N/A
Diagenetic model	1	2.38	N/A	N/A	N/A
Anaerobic					
Core incubations	18	4.21	1.16	0.17	18.76
P microelectrode	3	1.75	0.50	0.79	2.50
Central Basin					
Aerobic					
Experimental					
Core incubations	2	0.95	0.50	0.45	1.45
Expressed porewater	2	0.38	0.25	0.13	0.63
Models					
Mass balance model	1	0.95	N/A	N/A	N/A
Anaerobic					
Core incubations	2	0.98	0.73	0.25	1.71
Eastern Basin—no data					

n = number of locations with existing experimental or modeled dissolved P flux data
SE = standard error
SRP = soluble reactive phosphorus

loading was 2.73 mg m^{-2} day^{-1} with a 95% confidence interval of 1.52 to 4.58 mg m^{-2} day^{-1}. The aerobic average is in excellent agreement with the 1.35 mg m^{-2} day^{-1} calculated by Matisoff et al. (2016), which means that the additional data included in this chapter were in the same range as the previously published data. The result also agrees with Verhamme et al. (2016a) who used WLEEM and reported a seasonal P flux that ranged from 0.06 mg m^{-2} day^{-1} in February 2014 to 1.25 mg m^{-2} day^{-1} in August 2014. The anaerobic estimate was about 30% lower than the previous average of 6.01 mg m^{-2} day^{-1} (Matisoff et al. 2016), which is in good agreement considering the significantly higher variation of the anaerobic core incubation results compared to aerobic incubations (standard error (SE) of 1.16 mg m^{-2} day^{-1} and 0.18 mg m^{-2} day^{-1}, respectively). For the central basin, the average oxic internal loading was 0.65 mg m^{-2} day^{-1} and the anoxic loading was 0.98 mg m^{-2} day^{-1}. However, these numbers are based on only four experimental measurements plus one modeling result for oxic conditions, two experimental results for anoxic conditions, and the total of two sampling stations, so this average value should be viewed with caution (calculated 95% confidence intervals on these results are −3.92 to 5.22 mg m⁻² day⁻¹ for the flux under oxic conditions, and −8.30 to 10.26 mg m⁻² day⁻¹ for the flux under anoxic conditions).

The internal loading estimates described before can be used to calculate an annual basin-wide P contribution from internal loading. However, several assumptions are necessary. First, it is very important to acknowledge that all internal loading measurements for Lake Erie summarized in this chapter were conducted using late spring and summer months measurements (from May to August) and in a narrow span of time (2012–2014). Due to a limited number or years measured and the majority of the data having been collected in 2014, it is impossible to predict how the values presented in this chapter relate to longer timescale averages. When it comes to seasonal changes in P flux from sediment to the overlying water, somewhat contrasting information can be found in the literature. For example, North et al. (2015) found that the winter internal P loading values were about 35% higher than the summertime values, while previous studies suggested low positive or even negative flux of P from sediments to the water column in various lakes during winter months (Holdren and Armstrong 1980; Søndergaard et al. 1999). Considering the complex nature of P cycling that encompasses a large number of physical, chemical, and biological processes, it is not possible to accurately predict the behavior of P in Lake Erie during winter months based on data collected in other lakes. However, it is clear that the flux will be affected by time of year because water temperature affects the diffusion coefficients, which are directly proportional to flux (see Equation 19.2). Verhamme et al. (2016b) reported seasonal variations that are based on the temperature dependency of the diffusion coefficient. A temperature correction of the diffusion coefficient can be conducted using the Stokes-Einstein relationship, where the diffusion coefficient at 25°C is 2.19 times greater than the diffusion coefficient at 0°C (Li and Gregory 1974). Alternatively, the calculations of DiToro (2001) and Testa et al. (2013) allow the diffusion coefficient at any temperature to be calculated by multiplying the diffusion coefficient at 20°C by $1.08^{(T-20)}$, where T is the temperature in °C. For the well-mixed western basin, which is likely to be oxidized for most of the year (MacIsaac et al. 1992; MacIsaac et al. 1999; Zhang et al. 2011), it seems safe to base the calculation on the oxic flux only. Using the two temperature corrections mentioned previously and assuming an average annual temperature of the sediment surface in Lake Erie of 10°C (Matisoff et al. 2016), the calculated 95% confidence interval on the mean annual flux to the western basin is 0.62 to 1.27 mg m⁻² day⁻¹ or 0.29 to 0.6 mg m⁻² day⁻¹, depending on which temperature correction is used. Multiplying these numbers by the surface area of the western basin (3,284 km²), the 95% confidence interval of the total annual internal P load to the basin is 739 mta to 1,518 mta or 353 mta to 725 mta, depending on the temperature correction. These results are in good agreement with the prediction of WLEEM of 484 mta. The authors of this chapter refrain from attempting to make a similar calculation for the central basin for the following reasons: (1) only two stations were sampled in the central basin and the data are limited to seven experimental measurements and one modeled value, therefore extrapolation would carry huge uncertainty; (2) the hypolimnion in a portion of the central basin goes anoxic during part of the year, but it is not clear for what percentage of time or how that affects the P flux; and (3) the basin thermally stratifies, which has been recently proposed to increase the internal loading of P in some lakes through a mechanism called the *nutrient pump* (Orihel et al. 2015).

19.4 DISCUSSION

A wide array of methods has been used for assessing internal P fluxes from sediments to the water column in Lake Erie. These include direct *in situ* (bottom chambers) and *ex situ* (core incubations) net P flux determinations, Fickian diffusion calculations to determine the gross P flux, and models of varying complexity that determine either the net or the gross P flux. However, the *true* value of the flux remains unknown. Because P can be rapidly utilized by microbes, direct measurements of the net P released from sediment to the overlying water (bottom chambers and core incubations) carry potential bias related to biological loss of P released from the sediment—particularly when the experiments are not conducted in darkness. These measurements are also subject to sorption of SRP to iron oxides, the core liner, or chamber material (Holdren and Armstrong 1980). On the other hand, Fickian calculations of gross diffusive flux of SRP rely on the measurement of the concentration gradient near the sediment-water interface. The biggest uncertainty of these measurements is related to the resolution of the measurement (e.g., pore water expressed at 1 or 2 cm intervals may not provide sufficient resolution to determine the depth of the depletion layer or the thickness of the diffusive boundary layer), and fulfillment of the Fick's First Law of diffusion assumptions (steady-state system). While development of high-resolution tools such as a P microelectrode may help improve the vertical resolution of the measurement, this method measures only the concentration gradient along one specific profile, and therefore many measurements would have to be taken to obtain spatially representative results. Among the Fickian calculation methods, phosphate DETs seem to offer the best balance between high-resolution of the measurement and spatial representation of the data that is accomplished through integration of the vertical profiles across the entire width of the DET (Matisoff et al. 2016). Both the direct measurements and Fickian calculations are affected by strong spatial heterogeneity of sediment P conditions. The data presented by Matisoff et al. (2016) showed that replicate measurements of internal P flux at the same sampling station and using the same method varied by a factor of up to four. It is possible that the flux of P from sediments under aerobic conditions is controlled, in part, by the presence of iron oxyhydroxides in the surficial sediment. Since iron oxyhydroxides are the major constituents in sediment responsible for SRP binding, it was hypothesized that DET images of the dissolved iron (II) will be correlated to SRP distribution because reduction of strongly sorptive iron precipitates should release bound phosphorus. However, no such correlation was observed (Kaltenberg 2016).

Limited data are available on internal loading of P to Lake Erie, with no published data available for the eastern basin, and only a few estimates for the central basin. This limited number of internal P loading measurements for Lake Erie can be partly attributed to the relative difficulty of measuring internal loading (compared to external loading) especially at lake or basin-wide scales, and partly because internal loading is relatively small compared to the external inputs. Extrapolation of the available results to estimate a basin-wide flux was attempted for the western basin for which the most data are available and aerobic conditions for the majority of the year can be assumed. Comparison of the 95% confidence interval of the mean aerobic internal flux to western Lake Erie obtained here with the Annex 4-recommended 6,000 metric tons of P per year loading shows that the internal loading is expected to contribute between 6% and 25% of the target load (combined range of the 95% confidence intervals based on two temperature corrections). While it does not appear that internal loading is sufficient to cause the algal blooms by itself, the contribution of the internal loading up to a quarter of the total target load should not be neglected. For the central basin, sparse data and mixed oxic-anoxic conditions make such extrapolation impossible.

Currently available data do not provide input into changes of internal loading into Lake Erie over time, as all available data were collected in a short span of time (2012–2014). Also, while most estimates presented in this chapter are a snapshot of the flux at a particular time, the real flux will vary throughout the year and is a function of the amount of new organic material deposited to the sediment, sediment resuspension and mixing, and temperature. It is important to consider that shifts in Lake Erie's hydrology and ecosystem dynamics are likely to occur in the future as a result of predicted climate change. For example, climate change is likely to cause more extreme precipitation events, particularly in the winter and spring (Cousino et al. 2015 and literature there cited), which would cause higher P delivery through runoff from farming land and lead to more and extended anoxic events (Gibbons 2015). Most studies agree that more intense springtime precipitation translates to increased intensity of the algal bloom during summer months (Stumpf et al. 2012; Obenour et al. 2014; Ho and Michalak 2017). On the other hand, the model designed by Cousino et al. (2015) showed that while a moderate ambient temperature increase could increase delivery of P to Lake Erie from the Maumee River through predicted higher evapotranspiration, a strong increase in the ambient temperature could actually cause a reduction in the P input from the Maumee River because of a relatively high increase in total precipitation compared to the moderate scenarios. In addition, an increase in temperature may lead to increased hypolimnion anoxia, increased P flux from sediments (Gibbons 2015), and an increase in the diffusion of SRP from the sediment by increasing the diffusion coefficient. Under hot, calm conditions, waters in the western basin of Lake Erie can become anoxic within a few days, and even though anoxic episodes in western Lake Erie are short-lived (modeled duration of four to five days) and occur only occasionally, as little as four days of anoxic P flux could deliver more P to the water column than the average spring load from the Maumee River (Bridgeman 2016). Hence, climate change-related water-quality problems in Lake Erie may be hard to predict and may result in strong year-to-year variability (Rucinski et al. 2016). In the central basin, an increased extent and duration of hypoxia can provide a positive feedback mechanism for P delivery by allowing release of P immobilized in sediment iron oxyhydroxides. Changes in primary production, light penetration depths, and oxygenation can lead to further ecological and nutrient cycling changes in the higher trophic levels, which can be hard to predict. For example, introduction of dreissenid mussels (zebra and quagga) in the Great Lakes in the 1990s resulted in initial improvement of water quality because the mussels are capable of filtering large amounts of algae from the water column. However, within the next several years dramatic food-web changes occurred, shifting the bulk of primary production from phytoplankton to bottom-dwelling algae and plants (Lake Erie LaMP 2011). Comparison of the positive impact of dreissenids on water quality through algae filtration with the negative effects due to nutrient excretion conducted using a two-dimensional ecological model of Lake Erie (Zhang et al. 2011) led to the conclusion that the presence of dreissenids results in increased internal loading of P.

Recently developed three-dimensional models such as WLEEM (Verhamme et al. 2016b), which allow incorporation of hydrodynamics, sediment and nutrient transport, and hydrometeorological conditions and then connect them with ecosystem-based nutrient cycling and phytoplankton dynamics may provide the most comprehensive tools to predict water-quality changes under different P input, environmental, and ecosystem structure scenarios. Reliable models can provide invaluable information required to prepare water quality management plans and set appropriate P reduction goals for controlling Lake Erie eutrophication in the decades to come. For example, WLEEM has been used to assess the likely effects of reduction in loadings from various sources (Detroit River, Maumee River, internal loading) to western Lake Erie (Verhamme et al. 2016a). Collection of additional data to supplement the somewhat sparse internal loading dataset for Lake Erie could improve model calibration and validation. To date, the most effective P reduction strategy for Lake Erie was achieved through controlling point sources (wastewater treatment plants), whereas attempts to decrease P input from nonpoint sources (e.g., through changes in fertilizer use and till practices on farmlands) have been less successful. Addressing this issue is becoming an important focus for future external P loading reduction plans (Lake Erie LaMP 2011). Runoff of P from farmland provides a qualitatively important source of P delivery because the P from fertilizers is readily bioavailable.

19.5 CONCLUSIONS

Lake Erie has been experiencing severe water-quality problems related to eutrophication for many decades. High inputs of P from agriculture and wastewater treatment plants increased primary productivity, leading to development of HABs in the western and central basins and increased the extent of anoxia in the central basin. P reduction programs implemented in the 1970s were initially heralded to be a great environmental success, but about two decades later the water quality improvements have reversed. While no clear answer exists as to what caused the return to eutrophic conditions, changes in ecosystem dynamics, changes in the form of P from external loading sources, weather and climate-related shifts in nutrient speciation and transport, and release of P from the sediments remain on the suspect list. Currently available internal loading data for Lake Erie are spatially and temporally limited because all the data were collected within the 2012–2014 time range, and a majority of the data were collected in the western basin, with only a few measurements from the central basin. Calculated basin-wide internal loading of P to the western basin suggests that the expected contribution of internal loading is between 6% and 25% of the target 6,000 mta recommended by the Annex 4 of the 2012 GLWQA (IJC 2012). While it is not possible to control internal loading and the planned water-quality improvement strategies focus on external loading reductions, contributions of the internal loading should not be neglected. To date, most P reduction plans for Lake Erie focused on point sources (improvement in wastewater treatment plants and decrease of P concentration in detergents) and on decreasing sediment-bound P from agricultural runoff. Strategies for achievement of Annex 4 target loadings focus mostly on further reducing the nonpoint sources, including SRP from agricultural land. Climate change-related increases in temperature, intensity of the winter and spring precipitation, and possible ecosystem dynamics shifts are expected to continue affecting the P cycling mechanisms in Lake Erie, driving the need for new experimental data and improved models to aid the design of efficient water-quality programs for this important economical and recreational asset of the Midwestern United States.

19.6 ACKNOWLEDGMENTS

This work was supported by an EPA Great Lakes Restoration Initiative project GL-00E01284 to the Ohio Lake Erie Commission. E. Verhamme and J. Bratton of LimnoTech (Ann Arbor, MI) are thanked for communicating the WLEEM modeling results that are cited in this work.

19.7 REFERENCES

Allan, JD; Smith, SDP; McIntyre, PB; Joseph, CA; Dickinson, CE; Marino, AL; Reuben, GB; Olson, JC; Doran, PJ; Rutherford, ES; et al. 2015. Using cultural ecosystem services to inform restoration priorities in the Laurentian Great Lakes. Frontiers Ecol Environ. 13(8):418–424.

Apell, JN; Tcaciuc, AP; and Gschwend, PM. 2015. Understanding the rates of nonpolar organic chemical accumulation into passive samplers deployed in the environment: guidance for passive sampler deployments. Integr Environ Assess Manag. 12(3):486–492.

Boudreau, BP. 1996. The diffusive tortuosity of fine-grained unlithified sediments. Geochim Cosmochim Acta. 60(16):3139–3142.

Bridgeman, T. 2016. Phosphorus Sources in Western Lake Erie: How Important is sediment P? Great Lake Restoration Presentation. Downloaded from https://ohioseagrant.osu.edu/p/9lttx on November 11, 2017.

Bridgeman, TB; Schloesser, DW; and Krause, AE. 2006. Recruitment of *Hexagenia* mayfly nymphs in western Lake Erie linked to environmental variability. Ecol Appl. 16:601–611.

Cousino, LK; Becker, RH; and Zmijewski, KA. 2015. Modeling the effects of climate change on water, sediment, and nutrient yields from the Maumee River watershed. J Hydrogeology: Regional Studies. (4):762–775.

Di Toro, DM. 2001. *Sediment Flux Modeling*. John Wiley & Sons. ISBN: 978-0-471-13535-7.

Ding, X; Behbahani, M; Gruden, C; and Seo, Y. 2015. Characterization and evaluation of phosphate microsensors to monitor internal phosphorus loading in Lake Erie sediments. J Environ Manag. 160:193–200.

Dittrich, M; Wehrli, B; and Reichert, P. 2009. Lake sediments during the transient eutrophication period: reactive-transport model and identifiability study. Ecol Model. 220:2751–2769.

Gibbons, KJ. 2015. Effect of Temperature on Phosphorus Release from Anoxic Western Lake Erie Sediments. M.S. Thesis, University of Toledo. p. 63.

Ho, JC and Michalak, AM. 2017. Phytoplankton blooms in Lake Erie impacted by both long-term and spring-time phosphorus loading. J Great Lakes Res. 43:221–228.

Holdren, GC and Armstrong DE. 1980. Factors affecting phosphorus release from intact lake sediment cores. Environ Sci Technol. 14:79–87.

IJC (International Joint Commission). 1978. Environmental management strategy for the Great Lakes system. Final report from the Pollution from Land Use Activities Reference Group (PLUARG). Windsor, Ontario, Canada, July 1978.

IJC (International Joint Commission). 2012. Great Lakes Water Quality Agreement 2012. Protocol amending the agreement between Canada and the United States of America on Great Lakes water quality. IJC: Windsor, Ontario, Canada, September 7, 2012.

IJC (International Joint Commission). 2014. A Balanced Diet for Lake Erie: Reducing Phosphorus Loadings and Harmful Algal Blooms. Report of the Lake Erie Ecosystem Priority. ISBN: 978-1-927336-07-6.

James, WF. 2012. Estimation of Internal Phosphorus Loading Contributions to the Lake of Woods. Minnesota, ERDC Report. p. 42.

Jézéquel, D; Brayner, R; Metzger, E; Viollier, E; Prévot, F; and Fiévet, F. 2007. Two-dimensional determination of dissolved iron and sulfur species in marine sediment pore-waters by thin-film based imaging. Thau lagoon (France). Estuar Coast Shelf Sci. 72:420–431.

Kaltenberg, EM. 2016. New Approaches in Measuring Sediment-Water-Macrobenthos Interactions. Ph.D. Dissertation Case Western Reserve University. Retrieved from https://etd.ohiolink.edu/

Lake Erie LaMP. 2011. Lake Erie Binational Nutrient Management Strategy: Protecting Lake Erie by Managing Phosphorus. Prepared by the Lake Erie LaMP Work Group Nutrient Management Task Group.

Li, Y and Gregory, S. 1974. Diffusion of ions in sea water and in deep-sea sediments. Geochim Cosmochim Acta. 38: 703–714.

Maccoux, MJ; Dove, A; Backus, SM; and Dolan, DM. 2016. Total and soluble reactive phosphorus loadings to Lake Erie. J. Great Lakes Res. 42(6): 1151–1165.

MacIsaac, HJ; Johannsson, OE; Ye, J; Sprules, WG; Leach, JH; and McCorquodale, JA. 1999. Filtering impacts of an introduced bivalve (*Dreissena polymorpha*) in a shallow lake: application of a hydrodynamic model. Ecosystems. 2:338–350.

MacIsaac, HJ; Sprules, WG; Johannsson, OE; and Leach, JH. 1992. Filtering impacts of larval and sessile zebra mussels (*Dreissena polymorpha*) in western Lake Erie. Oecologia. 92:30–39.

Matisoff, G and Ciborowski, JJH. 2005. Lake Erie trophic status collaborative study. J Great Lakes Res. 31:1–10.

Matisoff, G; Kaltenberg, EM; Steely, RL; Hummel, SK; Seo, J; Gibbons, KJ; Bridgeman, TB; Seo, Y; Behbahani, M; and James, WF. 2016. Internal loading of phosphorus to western Lake Erie. J Great Lakes Res. 42:775–788.

Mortimer, CH. 1941. The exchange of dissolved substances between mud and water. J Ecol. 29:280–329.

Myers, DN; Thomas, MA; Frey, JW; Rheaume, SJ; and Button, DT. 2000. Water quality in the Lake Erie-Lake Saint Clair drainages Michigan, Ohio, Indiana, New York, and Pennsylvania, 1996–98: U.S. Geological Survey Circular 1203, p. 35. online at https://pubs.water.usgs.gov/circ1203/

North, RL; Johansson, J; Vandergucht, D; Doig, LE; Liber, K; Lindenschmidt, K-E; Baulch, H; and Hudson, J.J. 2015. Evidence for internal phosphorus loading in a large prairie reservoir (Lake Diefenbaker, Saskatchewan). J. Great Lakes Res. 41(Suppl. 2):91–99.

Obenour, DR; Gronewold, AD; Stow, CA; and Scavia, D. 2014. Using a Bayesian hierarchical model to improve Lake Erie cyanobacteria bloom forecasts. Water Resour Res. 50:7847–7860.

Ohio Environmental Protection Agency (Ohio EPA) 2010. Ohio Lake Erie Phosphorus Task Force Final Report. (Columbus, OH, Available at) http://epa.ohio.gov/portals/35/lakeerie/ptaskforce/Task_Force_Final_Report_April_2010.pdf. (Last accessed July 15, 2015).

Orihel, DM; Schindler, DW; Ballard, NC; Graham, MD; O'Connell, DW; Wilson, LR; and Vinebrooke, RD. 2015. The "nutrient pump:" iron-poor sediments fuel low nitrogen-phosphorus ratios and cyanobacterial blooms in polymictic lakes. Limnol Oceanogr. 60(3):856–871.

Paytan, A; Roberts, K; Watson, S; Peek, S; Chuang, P-C; Defforey, D; and Kendall, C. 2017. Internal loading of phosphate in Lake Erie Central Basin. Sci Total Environ. 579:1356–1365.

Reichert, P. 1994. AQUASIM—a tool for simulation and data analysis of aquatic systems. Water Sci Technol. 30:21–30.

Richards, RP; Baker, DB; Crumrine, JP; and Stearns, AM. 2010. Unusually large loads in 2007 from the Maumee and Sanducky Rivers, tributaries to Lake Erie. J Soil Water Conserv. 65(6):450–462.

Rucinski, DK; DePinto, JV; Beletsky, D; and Scavia, D. 2016. Modeling hypoxia in the central basin of Lake Erie under potential phosphorus reduction scenarios. J Great Lakes Res. 42:1206–1211.

Scavia, D; Allan, JD; Arend, KK; Bartell, S; Beletsky, D; Bosch, NS; Brandt, SB; Briland, RD; Daloglu, I; DePinto, JV; et al. 2014. Assessing and addressing the re-eutrophication of Lake Erie: central basin hypoxia. J Great Lakes Res. 40:226–246.

Søndergaard, M; Jensen, JP; and Jeppesen, E. 1999. Internal phosphorus loading in shallow Danish lakes. Hydrobiologia. 408/409:145–152.

Strickland, JDH and Parsons, TR. 1972. A practical handbook of seawater analysis, bulletin 167 (second edition). Fisheries Research Board of Canada. Ottawa.

Stumpf, RP; Wynne, TT; Baker, DB; and Fahnenstiel, GL. 2012. Interannual variability of cyanobacterial blooms in Lake Erie. PLoS ONE. 7(8):e42444.

Tenberg, A; Stahl, H; Gust, G; Mueller, V; Arning, U; Andersson, H; and Hall, POJ. 2004. Intercalibration of benthic flux chambers I. Accuracy of flux measurements and influence of chamber hydrodynamics. Prog Oceanogr. 60:1–28.

Testa, JM; Damien, BC; DiToro, DM; Boyton, WR; Cornwell, JC; and Kemp, WM. 2013. Sediment flux modeling: Simulating nitrogen, phosphorus, and silica cycles. Estuar Coast Shelf Sci. 131:252–263.

United States Environmental Protection Agency (US EPA) and Government of Canada. 1995. The Great Lakes. An Environmental Atlas and Resource Book. Third Edition. ISBN 0-662-23441-3.

Verhamme, EM; Bratton, J; DePinto, J; Redder, T; and Schlea, D. 2016a. Western Lake Erie Ecosystem Model (WLEEM) Phosphorus Mass Balance and Sediment Flux. Great Lakes National Program Office, April 26, 2016.

Verhamme, EM; Redder, TM; Schlea, DA; Grush, J; Bratton, JF; and DePinto, JV. 2016b. Development of the Western Lake Erie Ecosystem Model (WLEEM): Application to connect phosphorus loads to cyanobacteria biomass. J Great Lakes Res. 42:1193–1205.

Zhang, H; Culver, DA; and Boegman, L. 2011. Dreissenids in Lake Erie: an algal filter of a fertilizer? Aquat Invasions. 6:175–194.

CHAPTER 20

INTERNAL PHOSPHORUS LOADING IN SUB-TROPICAL LAKE KINNERET, ISRAEL, UNDER EXTREME WATER LEVEL FLUCTUATION[1]

Werner Eckert[2], Yaron Beeri-Shlevin, and Aminadav Nishri

Abstract

The water column of Lake Kinneret (LK), a deep, warm, monomictic freshwater lake in northern Israel, has been studied intensively since 1969 by means of a weekly monitoring program. Prior to that, the mesotrophic-to-eutrophic lake had undergone severe man-made perturbations including impoundment, high allochthonous nutrient loads due to swamp drainage, and excessive pumping for drinking water. Extreme lake-level fluctuation that was enhanced by climate warming was one of the results. This chapter makes use of the phosphorus (P) dataset from the central lake station to investigate long-term changes in the internal P loading as expressed by the seasonal increase of the integrated phosphate concentrations during the stratified season. Calculations revealed a highly variable annual internal P load that showed an increasing trend during the 1970s and a gradual decline starting from the early 1980s. The latter can be interpreted as a response to management measures in the watershed (sewage diversion and partial reflooding of a former swamp area) that caused an overall decrease in the riverine nutrient load. Through investigating the critical control parameters for internal P loading in LK, we discovered its close dependency on the phytoplankton biomass during the stratified season and on the redox conditions at the sediment-water interface. Based on these findings, we successfully developed a conceptual model aimed at the quantification of the internal P load in LK with nitrate concentrations, thermocline depth, and hypolimnion temperature as additional drivers.

Key Words: Monomictic lake, sediment water interface, microbial activity, benthic boundary layer, hypolimnetic uptake, Jordan River

---

[1] Dedicated to our colleague Alon Rimmer who passed away tragically in April 2018

[2] Israel Oceanographic and Limnological Research, The Yigal Allon Kinneret Limnological Laboratory, P.O.Box 447, Migdal 14950, Israel. E-mail: werner@ocean.org.il

20.1   INTRODUCTION

Lake Kinneret (LK), commonly known as the Sea of Galilee, is a mesotrophic-to-eutrophic, warm, monomictic lake in Israel, located at 32°50′N, 35°35′E in the northern part of the Syro–African Rift Valley (see Figure 20.1a). It is the only natural freshwater lake in the Middle East and, until 2012, provided 25–30% of Israel's annual water consumption. At −210 m below Mediterranean Sea level, LK is the lowest freshwater lake on earth. The 168.7 km² lake is 20 km long and up to 12 km wide with mean and maximum depths of 24 m and 43 m, respectively. LK is thermally stratified from March until December/January, with temperatures ranging between 31–32°C in the epilimnion (summer) and 14–16°C during winter mixis. With the onset of stratification in March, the hypolimnion turns gradually anoxic, followed by the depletion of nitrate and the accumulation of sulfide, methane, and the mineralization products, ammonium and phosphate (Eckert and Conrad 2007; Nishri et al. 2000). The most relevant tributary of the lake is the Jordan River (JR) that accounts for ~70% of the recorded inflows. It drains a watershed of 2,730 km² that includes four different hydrological units (see Figure 20.1b): the Jurassic karst of Mt. Hermon; the basalt plateau of the Golan Heights; the carbonaceous karst of the Eastern Galilee Mountains; and the flat alluvial Hula Valley (Rimmer and Givati 2014). The JR also discharges more than 80% and 90% of the riverine

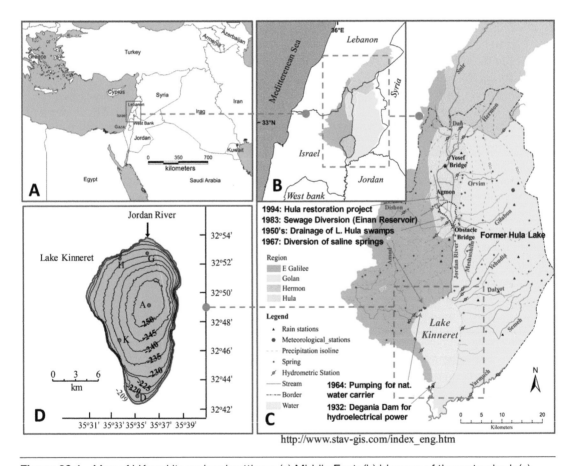

http://www.stav-gis.com/index_eng.htm

Figure 20.1 Map of LK and its regional settings: (a) Middle East; (b) blow up of the watershed; (c) detailed map of LK and its watershed including a history of management measures and sites; and (d) map of LK with depth contours and sampling sites.

phosphorus (P) and nitrogen (N) loads, respectively. The salinity of LK fluctuates between 200 and 300 mg L^{-1} Cl$^-$ and is significantly higher than that of the lake's headwaters due to submersed saline springs along the coast of the lake (Rimmer 2000). To prevent an increase in LK's salinity, some offshore saline springs were diverted around the lake (see Figure 20.1c) starting from 1967, causing chloride levels to drop from 400 to 190 mg L^{-1} by 1969 (Nishri et al. 1999). Until the mid-1990s, the phytoplankton biomass of LK was dominated by a spring dinoflagellate bloom, with chlorophytes, diatoms, and cyanobacteria having secondary contributions (Zohary et al. 2014). Then the proportions changed; the relative contribution of dinoflagellates has declined, while the shares of diatoms, chlorophytes, and especially cyanobacteria have increased.

Until the construction of the Degania Dam in 1932 (see Figure 20.1c), LK was a pristine lake with minimal anthropogenic influence and seasonal water-level fluctuations of up to 1.5 m (see Figure 20.2). During the following 85 years, it underwent a transition from a natural lake to a controlled hydrological regime. The regulated outflow for hydropower between 1932 and 1948 doubled seasonal water-level fluctuations. After the completion of the National Water Carrier (NWC) in 1964 (see Figure 20.1c), pumping activity of potentially up to 450×10^6 m^3 per annum (a^{-1}), in combination with an ongoing decline of the annual water discharge of the Jordan River by -5.8×10^6 m^3 a^{-1} (see Figure 20.3), caused the water level to drop to more than 5 m below maximum (see Figure 20.2). The growing influence of climate change in the region became evident in the 1990s when successive drought years were followed by record floods, such as during the winters of 1992/93, 2002/03, and 2012/13 (see Figure 20.3).

A reliable assessment of the relevance of these changes for the LK ecosystem is challenging due to the fact that water-quality monitoring began only in 1969 and we have no record of the lake's ecological and hydrochemical historic conditions (Nishri 2011). In the case of phosphorus (P) cycling, Hambright et al. (2004) concluded from their results from the sedimentary record that P accumulation rates have increased significantly because of engineering activities within the lake's catchment. These activities include the drainage of the Hula swamps in the 1950s (see Figure 20.1c) when a large—albeit unknown—amount of suspended solids was flushed into the lake. As a result,

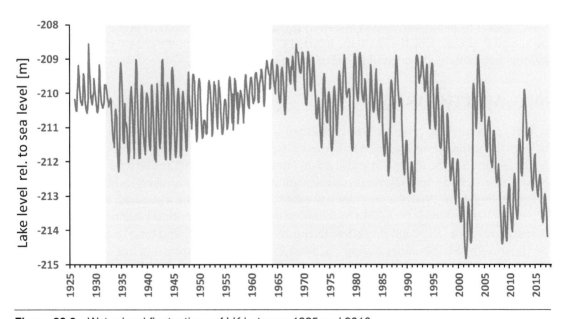

Figure 20.2   Water-level fluctuations of LK between 1925 and 2016.

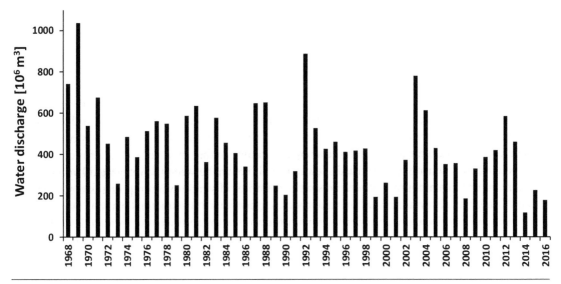

Figure 20.3 Annual water discharge of the JR from 1968 until 2016.

LK received an immense P load and at the same time lost an important natural pre-filter for the JR water before entering the lake (Hambright and Zohary 1998). Using the available dataset, Eckert and Nishri (2014) showed that between 1969 and 1981, the annual average total P (TP) inventory in the water column of LK increased from 50 to 100 metric tons, at an average rate of 3.7 ton a^{-1} followed by a gradual decline by -1.0 ton a^{-1} during the following years. When their calculation of the multiannual trends in the allochthonous P load with 3.6 and -1.0 tons a^{-1}, respectively, matched those of the P inventories, the authors could relate this decline to water management measures in the watershed. Most critical in this aspect were the construction of the Einan reservoir in 1983 for sewage and peat water diversion, and the reflooding of Lake Agmon to increase ground-water levels in the formerly drained swamp area (see Figure 20.1).

Key questions to be addressed in our study relate to the long-term effect of the gradually declining annual P discharge by the JR on the internal P load in LK and to the critical parameters for the seasonal variability in the internal P load of the lake.

20.2 METHODS

In 1969, a monitoring program was established to assess the water quality of LK. The program includes five stations (see Figure 20.1d), where discrete-depth water samples are collected at a high-depth resolution in the upper 7 m and at a 5-m interval from 10 m to the bottom sediments. During the stratified period (April–November), three additional metalimnion samples are collected at the central Station A. In 1970, the program was expanded to include the watershed, with the objective to quantify nutrient loads to the lake by following daily changes in critical parameters: that is, water discharge, chloride, all N and P species, alkalinity, sulfate, calcium, and total suspended matter in the JR, and smaller tributaries during discharge events. Sample processing has been according to Standard Methods (e.g., APHA 1992) and the data sets are stored in the LK Data Base (LKDB). P inventories for LK were calculated from the time series of the weekly-to-biweekly point measurements of TP, total dissolved P (TDP), and soluble reactive P (SRP) concentrations from fixed depths in the water column. Hereby SRP defines the P fraction that passes through a 0.45 μm filter and reacts with molybdate to form a color complex and as such, represents orthophosphate. By interpolating concentrations linearly between point measurements, we calculated the P content for each m³. Thus, the sum of the vertical profile represents the integrated P content per m². The annual allochthonous P load was quantified from daily measurements of SRP, TDP, and TP concentrations in the JR by the MEKOROT water company and from the parallel water-discharge data. Based on the findings by Eckert and Nishri (2014), we estimated the potentially bioavailable P (BAP) fraction in the JR load as: TDP + 22% of the particulate P. Internal P loading in LK corresponds to the amount of SRP accumulating in the hypolimnion during the stratified period. It is calculated by multiplying the SRP concentration for each depth meter by the volume of that specific layer as a function of the basin bathymetry. During the stratified season, the difference between the integrated hypolimnetic SRP content and the amount that was present at the onset of stratification represents the actual internal P load. For our multiannual trend analysis, we compared the seasonal maxima as described by Eckert and Nishri (2014). In order to explain the annual variability in the internal P loading we have developed a conceptual model based on driving variables defined later in the discussion of the obtained results.

20.3   RESULTS

Monthly averaged near-bottom concentrations of phosphate and ammonium in LK between 1974 and 2016 show a highly synchronized annual pattern for both solutes (see Figure 20.4). Concentration minima are typically measured during winter and early spring when the lake is fully mixed followed by a continuous increase during the stratified period. Concentration maxima in fall display a highly variable amplitude between 10 µg P L⁻¹ to 180 µg P L⁻¹, and 0.7 mg N L⁻¹ to 2.5 mg N L⁻¹. This strict periodic pattern was interrupted only once during the 1983–1984 cycle when the water column of LK did not fully mix.

A detailed account of the sequence of events during the redox transition from oxic to anoxic and to sulfide-enriched conditions in the near-bottom zone of LK is given by the time series of the critical solutes in the hypolimnion during the years 1992–1993 (see Figure 20.5). The winter of 1991/92

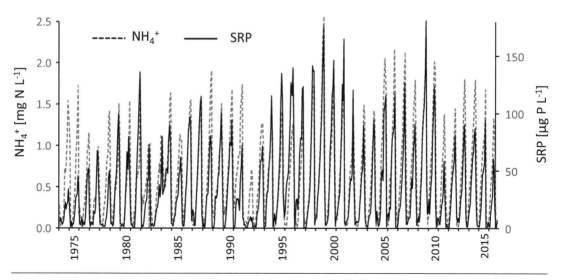

Figure 20.4   Time change of near-bottom ammonium and SRP concentrations at the central lake station of LK between 1974 and 2016.

brought a record flood (> 800×10^6 m³) to the lake with an exceptionally high external nutrient input while 1992/93, with a JR discharge of 500×10^6 m³, was regarded as average at that time (see Figure 20.3).

In winter and early spring when the whole water column is mixed, SRP (purple) and NH_4^+ (green) concentrations declined to near-zero concentrations (see Figure 20.5), while dissolved oxygen (DO) (blue) and nitrate (yellow) both increased dramatically to maximum concentrations reached at the end of March. With the onset of thermal stratification in spring, hypolimnetic DO concentrations began to decline and anoxic conditions were established in the near-bottom zone by the end of April (1993), a process that in the case of extreme cold winters (such as 1992), can extend into June. At DO levels below 1 mg L^{-1}, ammonium started to increase gradually while SRP still remained close to the detection limit. The depletion of DO was succeeded by that of nitrate with a delay of one month in 1992 and a few days in 1993. Only after the total exhaustion of the residual nitrate did hypolimnetic SRP concentrations increase—a process accompanied by the accumulation of total sulfide (red). Maximum SRP levels in 1992 and 1993 were 15 and 60 µg L^{-1}, respectively.

While the seasonal pattern of near-bottom concentrations of SRP and ammonium was nearly identical (see Figure 20.4), the vertical distribution of P in the hypolimnion differs significantly from that of ammonium as exemplified by two representative profiles from November 2003 and September 2014 (see Figure 20.6). With their maxima in the near bottom zone, NH_4^+ concentrations decrease more or less linearly throughout the hypolimnion and metalimnion, whereas elevated P concentrations are restricted to the zone below 25 m. Hereby, SRP represents nearly 100% of the TDP fraction and more than 90% of the TP. Taking into account this vertical hypolimnetic distribution, we interpolated our weekly point measurements of SRP to obtain an integrated SRP concentration per square meter. Its increased rate during the stratified period (April–October) rendered the internal P loading estimates for the years 1974–2016 (see Figure 20.7). The average release rate was 3.13 ± 1.58 mg P m^{-2} d^{-1}, with rates varying by one order of magnitude from 0.6 mg P m^{-2} d^{-1} in 1974 and 1992, to 6 mg P m^{-2} d^{-1} in 1986 and 1987.

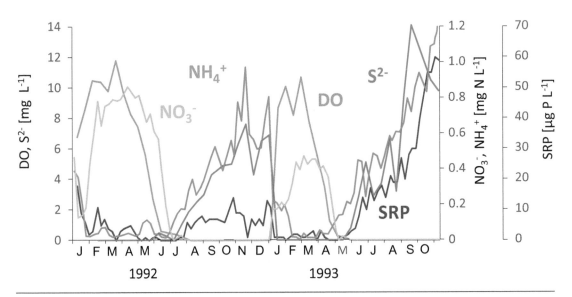

Figure 20.5 Weekly changes of near-bottom concentrations of dissolved oxygen (DO—blue), nitrate N (NO$_3^-$—yellow), ammonium (NH$_4^+$—green), phosphate (SRP—purple) and sulfide (S^{2-}—red) at the central lake station of LK.

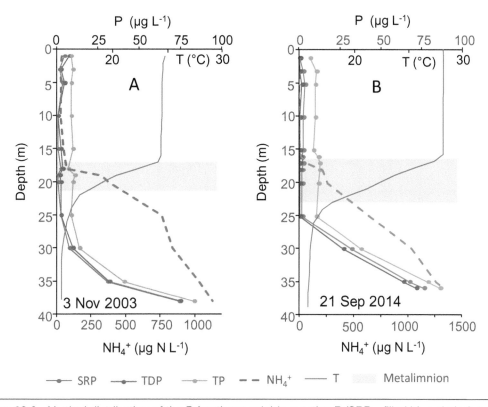

Figure 20.6   Vertical distribution of the P fractions: soluble reactive P (SRP—filled blue circles), total dissolved P (TDP—filled red circles), total P (TP—filled green circles), ammonium N (NH₄⁺—green dashed line) and temperature (purple line) in the water column of LK at the central lake station. Grey shaded area delineates the metalimnion. (A) data from November 3, 2003; and (B) data from September 21, 2014.

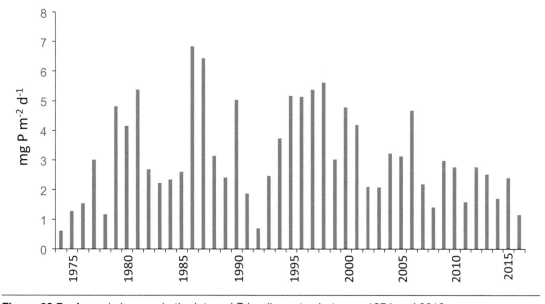

Figure 20.7   Annual changes in the internal P loading rates between 1974 and 2016.

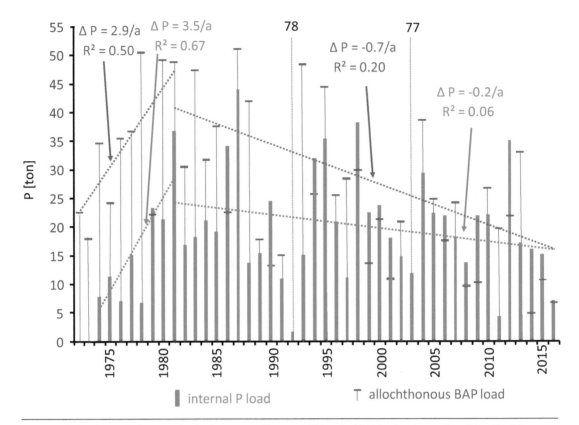

Figure 20.8   Annual changes in the integrated internal P load of LK (red columns) and in allochtho-nous BAP estimates from the JR (blue bars) from 1974 until 2016.

To compare the annual internal P load with the allochthonous riverine load of the JR, we cal-culated the total hypolimnetic SRP accumulation based on the time change in the measured SRP profiles, basin morphometry, and thermocline development (see Figure 20.8). Since we began SRP measurements in LK in 1974, the internal P load has varied between 1.5 metric tons (1992) and 44 tons (1987). During the time period 1974–1981, hypolimnetic SRP content increased linearly at a rate of 3.5 ton P a^{-1} ($r^2 = 0.67$) followed by a gradual long-term decline of 0.2 tons a^{-1} with great vari-ability ($r^2 = 0.04$). During the same time interval the allochthonous winter loads of BAP to the lake varied between 5 and 50 tons, apart from the two extreme flood years 1992/93 and 2003/04 when they were more than 75 tons. Up until ~1981, the external P input followed a similar increasing slope as that of the internal load at an average annual increase of 3.5 ton a^{-1} ($r^2 = 0.67$). Following 1981, JR loads turned highly variable while displaying a relative steeper declining rate of 0.7 ton a^{-1} ($r^2 = 0.04$) compared to that of the internal load.

20.4   DISCUSSION

The time series of monthly averages of near-bottom concentrations of ammonium and SRP conveys the stringent seasonality of internal nutrient loading in LK (see Figure 20.4). Concentrations of both nutrients are highly correlated ($r^2 = 0.78$), suggesting their common origin from organic matter decomposition and their concomitant release from the sediment water interface (SWI) to the near-bottom zone due to microbially-mediated mineralization processes at the SWI or diffusion from the pore water (Eckert and Nishri 2000). The common origin of both nutrients is further evidenced during extreme events, such as the winter of 1983/84 when the water column did not fully mix and concentrations remained high throughout spring—or during 1991/92, an extreme flood year, with a record low in hypolimnetic nutrient accumulation. Considering the multiannual change of maximum seasonal SRP concentrations, our data in Figure 20.4 (without 1992) display a linear increasing trend from 1974 until 2000 ($r^2 = 0.6$) from 50 µg L^{-1} to 170 µg L^{-1} followed by an average decline of -4 µg L^{-1} a^{-1} ($r^2 = 0.4$).

In order to understand the mechanisms that drive the P and N release, the hydrochemical changes at the SWI have to be followed at a higher temporal resolution. The data from weekly near-bottom measurements of solute concentrations at the central lake station between 1992–1993 show that NH$_4^+$ and SRP do not increase instantaneously under anoxic conditions (see Figure 20.5). While the former starts to increase at DO levels of 0.5 mg L^{-1}, the latter remains at zero until the depletion of both DO and nitrate. As such, SRP accumulation starts in parallel with the build-up of sulfide. The implication of this response for internal P loading in LK is that P release into the water column depends strictly on the succession and the dynamics of heterotrophic processes at the SWI and, consequently, on the availability of organic matter. This observation confirms the results obtained from a mesocom study with intact sediment cores: when in the presence of microbial activity, P release occurred only after the transition from oxic to anoxic conditions (Eckert et al. 1997). However, inhibition of the microbial activity in the overlaying water and at the SWI caused a similar P release under oxic and anoxic conditions, confirming microbial uptake as the major sink for upward diffusing P. The authors could confirm this hypothesis further by adding glucose, as an easily degradable energy source, to an active anoxic mesocosm when SRP concentrations dropped to the detection limit within a few hours.

These findings corroborate the critical—while ambiguous—role of microorganisms for the internal P loading process. As long as microorganisms experience carbon limitation, they will enhance the P flux at the SWI. The importance of microorganisms for the sedimentary P release process is further emphasized by Eckert and Nishri (2000) who compared measured P accumulation rates with theoretical rates based on diffusion from the pore water. They found that the P release from microbial mineralization at the SWI doubled the calculated diffusive P flux. The role of organic matter supply for the redox evolution of the hypolimnion and for nutrient accumulation is well known (Müller et al. 2012). In the case of LK, Hadas and Pinkas (1995) linked the timing of sulfide release by heterotrophic sulfate-reducing bacteria to the progress of thermal stratification of the water column at the time of the breakdown of the spring bloom of dinoflagellates. One of their take-home messages was that the stronger the stratification, the less the amount of organic matter that reaches the hypolimnion. This relates directly to the difference in the sequence of biogeochemical events at the SWI of LK during 1992 and 1993 (see Figure 20.5). The exceptionally cold and rainy winter of 1991/92 caused mixing temperatures to drop to a record low—below 13°C—and a late dinoflagellate bloom of which only a minor portion reached the sediments due to mineralization in the water column and out-flushing. The consequence was reduced heterotrophic activity in the hypolimnion that remained oxic until the end of June; thus, it took another month to deplete the nitrate. Only then did SRP start to accumulate in parallel with sulfide, but the apparent lack of organic matter kept P concentrations well below average levels.

Once released from the SWI, the vertical transport of solutes into the anoxic hypolimnion was shown to be controlled by turbulent diffusion that yields a linear concentration gradient from the bottom to the metalimnion (Rimmer et al. 2006). However, while this conclusion holds well for ammonium, SRP seems to follow different rules (see Figure 20.6). The phenomenon of high near-bottom P concentrations compared to a nearly P-free upper hypolimnion was shown by Nishri et al.

(2000) who pinpointed the importance of the benthic boundary layer for P accumulation. Eckert and Nishri (2014) suggested a combined hypolimnetic uptake and removal mechanism whereby upward-diffusing SRP is taken up by particle-associated microorganisms along their settling path to the lake bottom. This hypothesis is supported as well by the fact that there is no particulate P increase in the hypolimnion, as SRP at all times makes up more than 90% of the TP (see Figure 20.6). We estimated the effect of this SRP sink on internal loading by integrating the measured hypolimnetic SRP in Figure 20.6: 0.3 g m^{-2} (A) and 0.42 g m^{-2} (B) and by comparing it to the theoretical integrals of 1.00 g m^{-2} (A) and 0.88 g m^{-2} (B), which were calculated assuming a similar diffusive transport mechanism as in the case of ammonium. Accordingly, more than 50% of the sedimentary SRP that is released was removed from the water column by this anaerobic microbial P uptake.

The multiannual average of the rates of internal loading at the central lake station of LK is 3.13 mg P m^{-2} d^{-1} ± 1.58 mg P m^{-2} d^{-1} (see Figure 20.7). This relatively large variability (CV = 51%) is greater than that of the annual water discharge of the JR (see Figure 20.3) with CV = 41% (423 m^3 a^{-1} ± 172 × 10^6 m^3 a^{-1}); a major goal of our research has been to find an explanation for this annual variability. In a first approach, we investigated how far the annual changes in total internal P load might relate to the riverine BAP load of the JR (see Figure 20.8). Regression analysis rendered an insignificant correlation. Nevertheless, both loads display increasing trends during the 1970s, with 2.9 ton P a^{-1} and 3.5 ton P a^{-1}, which are similar to the 3.7 ton a^{-1} in the TP inventory of LK found by Eckert and Nishri (2014). Similarly, the declining trend in the allochthonous BAP load of −0.7 ton a^{-1} approaches the −1.0 ton a^{-1} drop of TP in LK after 1981. For the same period, we observe a decreasing trend in the internal P load of about −0.2 ton a^{-1} but with far greater variability as indicated by the low r^2 of 0.06. In summary, we can conclude for the comparison between both P loads that until the mid-1980s, the relative contribution of the internal load to the P inventory was on average 50% of the external load, whereas during recent years both sources have contributed roughly equally to the lake's P budget.

We tested possible links between the annual variability of the internal P load and that of other water-column parameters including phytoplankton biomass, nitrate concentration, lake level, thermocline depth, and hypolimnion temperatures; only phytoplankton biomass exhibited a significant positive correlation with an r^2 = 0.36 (not shown). Under the assumptions that organic matter decomposition is the major source for internal P loading in LK and that the process itself occurs in the absence of DO and nitrate, hypolimnetic SRP accumulation will be a function of the amount of organic matter reaching the sediment and of the redox buffer capacity of the hypolimnion. In other words, internal loading should be affected by the strength of stratification as expressed by both the temperature gradient across the thermocline and the thermocline depth, as well as by the hypolimnetic nitrate concentration at the onset of thermal stratification. We have combined these considerations into the following conceptual internal P loading model:

$$IPL = (B \times 0.11 - N_H) \times (T_H \times 0.01) \times (D_T \times 0.01) \qquad \text{[Eq. 20.1]}$$

where: IPL = internal P load [ton], B = average phytoplankton biomass between April and October [ton], N$_H$ = hypolimnetic nitrate inventory in June [ton], T$_H$ = hypolimnion temperature in May [°C], and T$_D$ = thermocline depth in April [m]. The factors (in sequence of appearance [ton], [°C^{-1}], [m^{-1}]) are the result of model optimization. As shown in Figure 20.9, our IPL model follows closely the measured loads between 1981 and 1999 and in the aftermath of 2007. At an overall correlation of r^2 = 0.58 (see Figure 20.10), the latter improves to 0.80 when omitting the 1999 through 2007 period.

The degree to which our model is applicable to obtain internal loading estimates in other stratified lakes remains to be shown. Our goal for the future is to find the missing parameter(s) that may

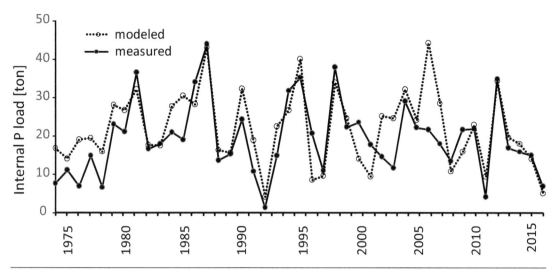

Figure 20.9   Time change of measured (filled circles) and modeled (empty circles) internal P loads.

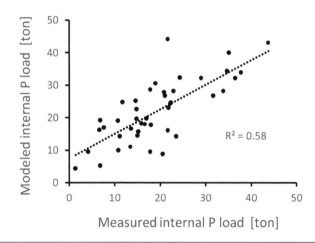

Figure 20.10   Linear correlation between modeled and measured annual internal P loads.

help to explain the discrepancy found during the early 2000s. Our overall approach however, may stimulate others to parametrize our model with data sets from their specific lakes.

20.5   CONCLUSIONS

We show that internal P loading in LK is critically affected by the phytoplankton biomass that reaches the lake bottom and by the hydrochemical conditions in the hypolimnion during the stratified season. Hereby the settling flux of organic matter is controlled by the physical conditions in the water column. The stronger the stratification, the less organic matter reaches the sediment-water interface. The further role of this seston for internal P loading depends on the redox buffer capacity in the near-bottom zone. In the presence of oxygen and/or nitrate, microbial uptake effectively reduces any sedimentary P release in Lake Kinneret. We combined these findings into a P loading model with phytoplankton biomass, hypolimnetic nitrate concentration, temperature, and thermocline depth as drivers. With an overall correlation of $r^2 = 0.58$, our model is in relatively good agreement with the measured loads. The cause for the large overestimations in 2006 and 2007 is still unknown. We intend to investigate this issue in the future and to develop an improved model. Regarding the long-term trend in the internal P load under the effect of management activities in and around Lake Kinneret, the increasing loads until the early 1980s are likely driven by the increasing input of allochthonous P from the JR. The subsequent decline has to be seen as a co-effect of controlled nutrient reduction in the watershed and a decline in the tributary P load due to the climate warming-related negative trend in the water discharge of the JR.

20.6 ACKNOWLEDGMENTS

The authors are grateful to Tamar Zohary for allowing the use of the phytoplankton biomass data and to the MEKOROT water company for the P concentration data from the Jordan River.

20.7 REFERENCES

American Public Health Association (APHA). 1992. *Standard Methods for the Examination of Water and Wastewater*, 18th ed.

Eckert, W and Conrad, R. 2007. Sulfide and methane evolution in the hypolimnion of a subtropical lake: a three-year study. Biogeochemistry. 82:67–76.

Eckert, W and Nishri, A. 2000. Sedimentary phosphorus flux in Lake Kinneret: precipitation vs. release. Adv Limnol. 55:397–411.

Eckert, W and Nishri, A. 2014. Phosphorus cycle. In: Zohary T, Sukenik A, Berman T, Nishri A, (Eds.). Lake Kinneret: Ecology and Management. Dordrecht (NL) Springer, p. 347–363.

Eckert, W; Nishri, A; and Parparova, R. 1997. Factors regulating the flux of phosphate at the sediment-water interface of a subtropical calcareous lake: a simulation study with intact sediment cores. Water Air Soil Poll. 99:401–409.

Hadas, O and Pinkas, R. 1995. Sulfate reduction in the hypolimnion and sediments of Lake Kinneret, Israel. Freshwater Biol. 33: 63–72.

Hambright, KD; Eckert, W; Leavitt, PR; and Schelske, CL. 2004. Effects of historical lake level and land use on sediment and phosphorus accumulation rates in Lake Kinneret. Environ Sci Technol. 38:6460–6467.

Hambright, KD and Zohary, T. 1998. Lakes Hula and Agmon: destruction and creation of wetland ecosystems in northern Israel. Wetl Ecol Manag. 6:83–89.

Müller, B; Bryant, LD; Matzinger, A; and Wüest, A. 2012. Hypolimnetic oxygen depletion in eutrophic lakes. Environ Sci Technol. 46:9964–9971.

Nishri, A. 2011. Long-term impacts of draining a watershed wetland on a downstream lake, Lake Kinneret, Israel. Air Soil Wat Res. 4:57–70.

Nishri, A; Imberger, J; Eckert, W; Ostrovsky, I; and Geifman, Y. 2000. The physical regime and the respective biogeochemical processes in Lake Kinneret lower water mass. Limnol Oceanogr. 45:972–981.

Nishri, A; Stiller, M; Rimmer, A; Geifman, Y; and Krom, MD. 1999. Lake Kinneret (The Sea of Galilee): the effects of diversion of external salinity sources and the probable chemical composition of the internal salinity sources. Chem Geol. 158:37–52.

Rimmer, A. 2000. The influence of lake level on the discharge of the Kinneret saline springs. Adv Limnol. 55:55–67.

Rimmer, A; Eckert, W; Nishri, A; and Agnon, Y. 2006. Evaluating hypolimnetic diffusion parameters in thermally stratified lakes. Limnol Oceanogr. 51:1906–1914.

Rimmer, A and Givati, A. 2014. Hydrology. In: Zohary, T, Sukenik, A, Berman, and T, Nishri, A, (Eds.). *Lake Kinneret: Ecology and Management*. Dordrecht (NL) Springer, pp. 97–111.

Zohary, T; Yacobi, YZ; Alster, A; Fishbein, T; Lippman, S; and Tibor, G. 2014. Phytoplankton. In: Zohary, T; Sukenik, A; Berman, T; and Nishri, A, (Eds.). *Lake Kinneret: Ecology and Management*. Dordrecht (NL) Springer, pp. 161–190.

CHAPTER 21

EXTERNAL AND INTERNAL PHOSPHORUS LOADS TO A COASTAL URBAN LAGOON, JACAREPAGUÁ LAGOON, RIO DE JANEIRO, BRAZIL

Marcelo Manzi Marinho[1], Natália Pessoa Noyma[1], Leonardo de Magalhães[1], Jônatas de Souza Mercedes[1], Vera Huszar[2], and Miquel Lürling[3,4]

Abstract

Anthropogenic activities have led to major degradation of coastal urban lagoons worldwide, where eutrophication is the most important water quality problem, often promoting the occurrence of harmful cyanobacterial blooms. Reducing the inflow of nutrients is a straightforward mitigation measure. However, external load reduction is not a guarantee for fast recovery because the internal loading of nutrients can delay recovery for many years. In this chapter, we evaluated the external and internal loads of phosphorus (P) to a coastal urban lagoon (Jacarepaguá Lagoon, Rio de Janeiro, Brazil) located in a region that has undergone a disordered process of urban and industrial growth over the last four decades, resulting in nutrient enrichment and development of harmful cyanobacterial blooms. Jacarepaguá Lagoon suffers from a high P load (45.1 mg m^{-2} day^{-1}) generated by both external and internal loading. The main tributaries deliver a high external annual P input of about 55 metric tons yr^{-1}. Jacarepaguá also has a rather continuous flux of P from the sediment, yielding an annual internal load of 6.2 tons P yr^{-1}, which is around 11% of

continued

[1] Laboratory of Ecology and Physiology of Phytoplankton, Department of Plant Biology, University of Rio de Janeiro State, Rua São Francisco Xavier 524—PHLC Sala 511a, 20550-900, Rio de Janeiro, Brazil. Corresponding author (M.M. Manzi): manzi.uerj@gmail.com.

[2] Museu Nacional, Federal University of Rio de Janeiro, 20940-040, Rio de Janeiro, Brazil.

[3] Aquatic Ecology & Water Quality Management Group, Department of Environmental Sciences, Wageningen University, P.O. Box 47, 6700 AA, Wageningen, The Netherlands.

[4] Department of Aquatic Ecology, Netherlands Institute of Ecology (NIOO-KNAW), P.O. Box 50, 6700 AB, Wageningen, The Netherlands.

the total P load. The estimated average internal load of 4.6 mg m^{-2} day^{-1} can be considered high when compared to other eutrophic systems and should not be neglected. Reduction of the external P load from the rivers is an absolute necessity for restoration of the lagoon, but sediment P release still might be sufficient to fuel phytoplankton blooms. Restoring the lagoon needs more than external load control—and *in situ* geo-engineering techniques seem far more cost-effective than standard dredging procedures.

Key-words: Cyanobacteria, eutrophication control, lake restoration, sediment P release

21.1 INTRODUCTION

Coastal lagoons are shallow aquatic ecosystems that develop at the interface between coastal terrestrial and marine ecosystems, and can be permanently open or intermittently closed-off from the adjacent sea by depositional barriers (Kjerfve 1994). Neotropical coastal lagoons are typically located within densely populated areas, and are probably among the most human-dominated and threatened ecosystems on earth (Berkes and Seixas 2005). Hence, cultural eutrophication is viewed as the most common problem affecting Neotropical coastal lagoons (Esteves et al. 2008). Urban coastal lagoons with relatively long water residence times are particularly vulnerable to nutrient enrichment (Kennish et al. 2014), which may lead to the development of harmful algal blooms (Heisler et al. 2008).

The focus of this study is Jacarepaguá Lagoon, situated in the Municipality of Rio de Janeiro (see Figure 21.1). It is part of an urban system that comprises three lagoons (Jacarepaguá, Camorim, and Tijuca) connected to each other in a sequence (see Figure 21.1). Camorim Lagoon constitutes, in fact, a channel between the two others, and the lagoon system is connected to the sea via Tijuca Lagoon.

This morphology results in a strangulated system in which the seawater inflow is constrained, leading to limited inflow from the sea and inefficient water exchange. Outflow is not accompanied by the sediment, which is preferably retained within the lagoons (Fernandes et al. 1994). The tidal range is in the order of 1 m in the sea inlet area and a few centimeters (not more than 10 cm) in the innermost lagoon. Jacarepaguá Lagoon has the larger watershed of the lagoon system (see Table 21.1), where the main tributaries (Marinho River, Camorim River, Caçambé River, and Pavuna River) flow mainly through urban settlements.

The Jacarepaguá Lagoon is located in a region of intense and disorganized urban occupation, where the strong stress generated by the growing and continuous socio-economic-environmental

Table 21.1 Main morphometric and hydrological features of Jacarepaguá lagoon (Benedetti 2011)

Area (km^2)	3.7
Volume (10^6 m^3)	12.3
Maximum depth (m)	13.0
Mean depth (m)	3.0
Residence time (days)	176
Watershed area (km^2)	102.8

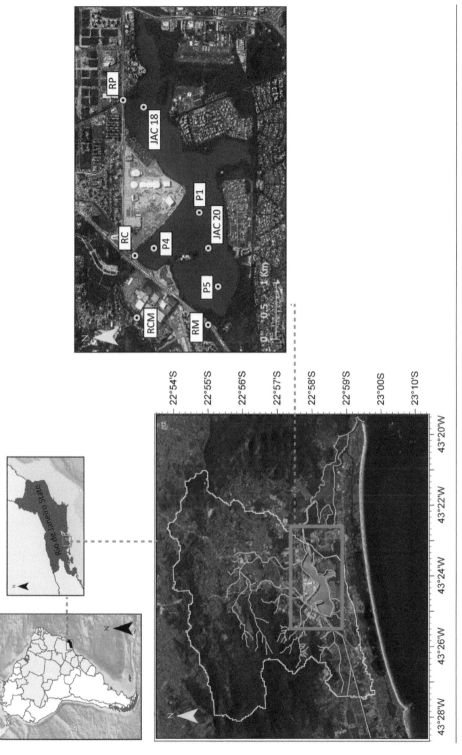

Figure 21.1 Location of Jacarepaguá Lagoon. *Upper left*: continental location (South America); blow up of continental region (Rio de Janeiro state). *Lower left*: lagoonal system, with the drainage basin (yellow outline) and tributary rivers of the Jacarepaguá Lagoon within urban region in the west part of Rio de Janeiro Municipality. *Right*: Jacarepaguá Lagoon (lagoon stations: JAC 18 and JAC 20—water column and sediment sampling stations; P1, P2, and P4—sediment sampling stations; tributaries stations: RP—Pavuna river, RC—Caçambé river, RCM—Camorim River, and RM—Marinho).

demands over the last four decades have resulted in a rapid and continuous degradation of the environment. The lagoons of this system serve basically to receive the disposal of domestic effluents from 16 districts with about 700,000 inhabitants, in addition to industrial effluents from various chemical, pharmaceutical, metallurgical, food, and electronic industries. This population density, coupled with lack of sanitation, releases 45 tons per day^{-1} of organic biological oxygen demand (BOD) load and 80 tons per day^{-1} of garbage (COHIDRO 2006). A recent study showed that this pollution promoted the proliferation of not only the indigenous microbial community, but also opportunistic pathogenic bacteria (Salloto et al. 2012).

The artisanal fishing activity had great economic importance in the region until the mid-2000s, mainly trading tilapia (*Oreochromis niloticus*), twait shad (*Brevoortia pectinata*), and mullet (*Mugil* spp.). However, pollution from untreated effluent discharge resulted in local extinction of most of the 89 species recorded in the early 1990s, with frequent fish kills being observed. In addition, the occurrence of toxic blooms of *Microcystis aeruginosa* in the Jacarepaguá Lagoon has resulted in contamination of fish with microcystins (Magalhães et al. 2001).

Eutrophication symptoms in the lagoon system have been registered since the 1970s (Semeraro and Costa 1972). Blooms of cyanobacteria were first reported in the beginning of the 1990s (Fernandes 1993) and have become more frequent since the middle 1990s (Gomes et al. 2009). Currently, Jacarepaguá Lagoon has been classified as hypereutrophic and suffers constant cyanobacterial blooms, mainly *M. aeruginosa*, as a result of the input of high nutrient concentrations (Gomes et al. 2009; de Magalhães et al. 2017).

The key factors challenging the health of Jacarepaguá Lagoon are the untreated industrial and domestic effluent discharges. These inflows contribute to high loads of nutrients and organic matter that have accelerated eutrophication, resulting in the occurrence of toxic blooms of cyanobacteria and opportunistic pathogens, posing serious risks to the environment and to public health.

Nutrient loading derived from anthropogenic activities in the drainage basin—agriculture, domestic, and industrial sewage—represents the highest contribution to eutrophic waterbodies (Paerl and Paul 2012). Reducing or stopping the nutrient inflows may mitigate eutrophication symptoms in systems that have a short residence time and no major P loading from the sediment (Edmondson 1970). In cases where the sediment has been loaded with nutrients because of decades of external inputs, simply implementing external load reduction measures will not yield immediate relief and the recovery can take decades due to recycling of phosphorus (P) from the P-rich sediments (Søndergaard et al. 1999, 2001; Gulati and van Donk 2002; Cooke et al. 2005).

In Jacarepaguá Lagoon, it is expected that the high external load of nutrients (primarily P) entering the system during the last four decades has resulted in P accumulation in the sediment. We expect this persistent external load to have resulted in a high internal P load with the potential to maintain the hypereutrophic condition of the lagoon following external load reduction. Our hypothesis is that a successful rehabilitation of the lagoon to prevent harmful cyanobacterial blooms requires drastic external load reduction, as well as the control/management of internal P loading.

21.2   METHODS

21.2.1   Sampling (Periodicity, Sampling Stations)

Samples from Jacarepaguá Lagoon were collected monthly at two sampling stations (JAC 18—S22°59′14″ W43°24′10″ and JAC 20—S22°58′37″ W43°22′49″) from November 2014 to December 2017 (see Figure 21.1). These two stations have been shown to be representative of the lagoon (Ferrão-Filho et al. 2002; Gomes et al. 2009). Water samples from the main tributaries of Jacarepaguá

Lagoon were collected in June 2017 and September 2017, and then monthly from November 2017 to July 2018 (Caçambé River—S22°58′29″ W43°24′13″, Arroio Pavuna River—S22°58′24″ W43°22′44″, Camorim River—S22°58′31″ W43°24′48″, and Marinho River—S22°59′14″ W43°24′53″).

21.2.2   Abiotic and Biotic Variables—Sample Analysis

Water temperature, dissolved oxygen (DO) concentration, pH, and salinity were measured with a multiparameter sonde (YSI model 600 QS) at the top, 0.5 m depth, and bottom of the sampling stations in the lagoon. Water transparency was measured by Secchi disk, and the euphotic zone (Z_{eu}) was estimated at 2.7 × the Secchi disk depth (Cole 1994). Dissolved and total P (TP), chlorophyll a (Chl a), and quantitative phytoplankton (fixed with Lugol's solution) samples were collected with a tube (4.5 cm diameter) integrating 1 m of the water column. TP was measured in unfiltered water samples, while soluble reactive P (SRP) was determined in filtered water samples (1.2 μm glass fiber filters, GF-3 Macherey-Nagel). The analyses were done with a FIAlab 2500 flow injection analyzer (FIALab Instruments Inc., Seattle, Washington), which was run according to manufacturer instructions. Chl a concentrations were measured using a PHYTO-PAM, phytoplankton analyzer (Heinz Walz GmbH, Effeltrich, Germany). Phytoplankton populations were enumerated according to the settling technique (Utermöhl 1958) in random fields (Uehlinger 1964) using an inverted microscope (Olympus, CKX41). Phytoplankton biovolume ($mm^3 L^{-1}$) was estimated by multiplying the density of each species by the mean volume of its cells (Hillebrand et al. 1999).

21.2.3   Sediment P Content

The sediment from Jacarepaguá Lagoon was collected in two different campaigns using a gravity Uwitec core sampler (Uwitec, Mondsee, Austria). The first campaign was done in September 2015 when five sediment cores from JAC 20 were obtained. At that time, the water-column pH was 9.9, salinity was 5.2 ppt, and alkalinity was 3.6 mEq L^{-1} ± 0.13 mEq L^{-1}. The second campaign took place in July 2018 when sediment cores were collected from five stations. These sediment sampling sites were JAC 18 (S22°59′14.1″, W43°24′9.6″), P1 (S22°59′09″, W43°23′46″) located at the mouth of Camorim Canal; P4 (S22°58′42″, W43°24′09″) close to the mouth of both Camorim and Caçambé Rivers; and P5 (S22°59′12″, W48°24′29″) in front of the fisherman's village. In this campaign, the water of Jacarepaguá Lagoon had a pH of 8.5 ± 0.08, salinity of 7.2 ppt ± 0.57 ppt, and water alkalinity of 3.5 mEq L^{-1} ± 0.15 mEq L^{-1}.

The pool of potentially releasable sediment P (RSP) in the sediment was determined by a sequential extraction protocol modified from Paludan and Jensen (1995), and described in Cavalcante et al. (2018). The RSP was extracted from the top 10 cm of sediment from one core from each station, in both sampling campaigns. P present in the interstitial pore water was extracted by washing 1 g of wet sediment 2 times with 25 mL demineralized and deoxygenated water. The anoxic water was prepared by bubbling with N_2. After centrifugation, the supernatant was collected, acidified with 0.5 mL 2M H_2SO_4, and filtered through a 0.8 μm membrane filter before TP and SRP analysis. The sediment pellet was subsequently extracted with a strong reducing agent (bicarbonate/sodium dithionite, BD reagent) under N_2 addition to obtain P adsorbed onto the surface of oxidized Fe and Mn (BD-P). After acidification of the supernatant with 3 mL 2M H_2SO_4, the remaining dithionite was oxidized by aeration and the supernatant subsequently filtered through a 0.8 μm membrane filter before TP and SRP analysis. The sediment pellet was further extracted with 2 × 25 mL 0.1M NaOH and 23.5 mL demineralized water to obtain organic P and P adsorbed to clay minerals and aluminum-oxides. The three combined supernatants (73.5 mL) were filtered as before and acidified with 1.5 mL 2M H_2SO_4, (Psenner et al. 1984; Jensen and Thamdrup 1993). Nonreactive P (NaOH-NRP) was calculated as the difference between SRP and TP. The pool of RSP was estimated from the content of the water-P, BD-P, and NaOH-NRP fractions (Schauser et al. 2006).

21.2.4   Internal P Fluxes Estimated from Sediment

The internal P flux from the sediment was determined using intact cores sampled with an Uwitec core sampler (Uwitec, Mondsee, Austria). Four sediments cores from each station of the first and second campaigns were transported to the laboratory and used as replicates to determine the P flux from the sediment into the overlying water.

In the first campaign, the cores were left untouched and the overlying lagoon water reached anoxia (DO < 1 mg L^{-1}). In the second campaign, the lagoon water was removed by siphoning and replaced with demineralized water that was gently added to minimize disturbance of sediment; the oxygen concentration decreased by about 50% but did not reach anoxia. In both cases, the sediment cores were closed with a rubber stopper and placed in the laboratory at 25°C in the dark. The tubes contained black sediment between 18 cm and 29 cm deep with 20 cm to 43 cm of over-lying water. In the first campaign, 10 mL samples were taken for SRP analysis at the start and then after 1, 3, 15, 22, 29, 35, 64, and 96 days. In the second campaign, 10 mL samples were taken for SRP analysis at the start and then after 1, 7, and 14 days. SRP samples were filtered and stored at $-20°C$ before being analyzed with the flow injection analyzer. The differences between initial and final SRP concentrations were used to estimate the SRP fluxes (mg P m^{-2} d^{-1}) using the formula: $\{(P_{final} - P_{start}) \times$ water height$\}/\Delta t$, with P in mg m^{-3}, water height in m, and Δt in days (d).

21.2.5   External P Loading from Rivers/Tributaries

The external loading from rivers was calculated from the TP concentrations measured in water samples collected in the five tributaries. The P loading (kg day^{-1}) was estimated for each tributary using the formula: $P_{load} = (TP \times discharge \times 86,400) / 1000$, with TP in g m^{-3} and river discharge in m^3 s^{-1}. We used the most recent data of river discharge (Masterplan 2015). Because river flow information for the Jacarepaguá Lagoon tributaries is scarce, and no monthly or even seasonal measurements exist, river discharges were considered permanent (constant in time) for the estimates of P load.

21.3   RESULTS

21.3.1   Physical and Chemical Regime

Jacarepaguá Lagoon is a brackish, shallow, turbid waterbody. On average, the depth at the sampling stations was 1.2 m (see Figure 21.2). However, light rarely reaches the bottom, and the euphotic zone extends to 0.6 m on average (see Figure 21.2). During the study, the lagoon showed oligo-mesohaline conditions, with salinity ranging from 1.3 ppt (0.02 ppt in one case) to 13.1 ppt, and showed no variation through the water column, except for some sampling dates when salinity was higher near the bottom (see Figure 21.2). We observed no significant differences among sampling stations or seasons ($p > 0.05$).

Temperature was typical of tropical aquatic systems (20.9°C–35.0°C) with significantly ($p < 0.001$) higher values in summer (average 28.9°C $\pm$ 2.1°C) and lower during fall/winter (average 23.9°C $\pm$ 1.9°C). As expected for a large, shallow system, the water column showed no indications of temperature stratification, but oxygen evidently declined through the water column and was lowest near the sediment (see Figure 21.2).

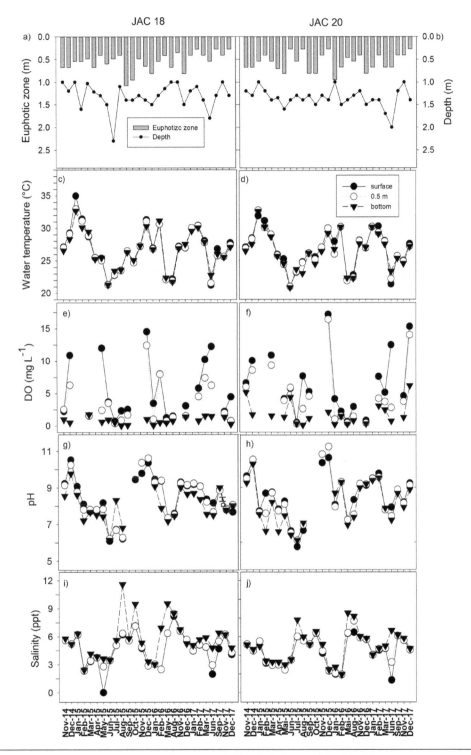

Figure 21.2   Euphotic zone (grey bars) and depth (filled circles and line) (a, b), water temperature (c, d), DO (e, f), pH (g, h), and salinity (i, j) in the sampling stations JAC 18 (left panels) and JAC 20 (right panels) at the surface (filled circles), 0.5 m depth (open circles), and at the bottom (inverted closed triangles) during the monitoring period (November 2014 to December 2017) in Jacarepaguá Lagoon.

Jacarepaguá Lagoon has alkaline waters (average pH 8.5) most of the time, although neutral or acidic values (5.8–6.7) were registered in winter (see Figure 21.2). No differences were observed between the stations or through the water column. However, lower pH values were often observed near the lagoon bottom. Warmer months had higher pH (p < 0.001 – average 9.1 ± 0.9) than the coldest period (average 8.5 ± 1.2).

The water column presented a clinograde profile of DO, with hypoxia at the bottom most of the time (see Figure 21.2). In the euphotic zone, average DO was 4.8 mg L^{-1}. However, supersaturation occurred in many months, especially during warmer periods.

21.3.2   Occurrence of Cyanobacteria

Jacarepaguá Lagoon had high Chl a concentrations (average 178 µg L^{-1}) and based on the Chl a trophic state index, the system is classified as a hypertrophic waterbody (see Figure 21.3). A clear

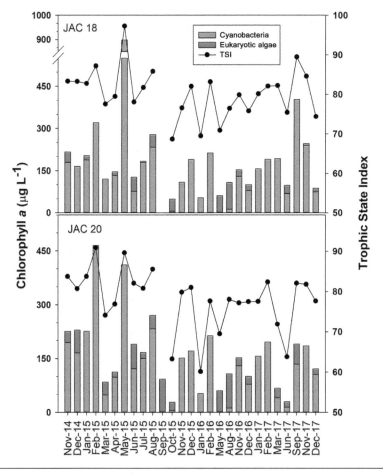

Figure 21.3   Chl a concentration (µg L^{-1}) of cyanobacteria (cyan stacks), eukaryotic algae (green stacks), and Trophic State Index (TSI) (filled circles) in the sampling stations JAC 18 and JAC 20 during the monitoring period (November 2014 to December 2017) in Jacarepaguá Lagoon.

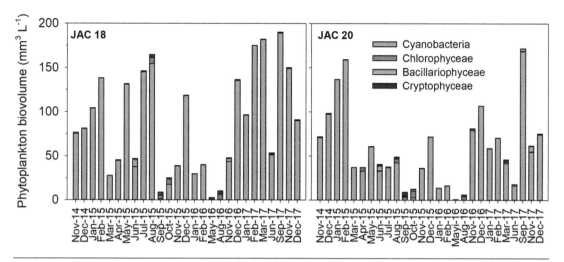

Figure 21.4   Contribution of phytoplanktonic groups to the phytoplankton biomass (mm³ L⁻¹) in the sampling stations JAC 18 and JAC 20 during the monitoring period (November 2014 to December 2017) in Jacarepaguá Lagoon.

dominance of cyanobacteria was observed almost all the time, except during some months in the cold/dry season.

Phytoplankton biomass was dominated by the cyanobacterium *M. aeruginosa*. Sampling stations showed significant differences, but higher phytoplankton biomass was found during the warmer periods (p < 0.05), when the *M. aeruginosa* contribution was > 80% of total biovolume (see Figure. 21.4).

21.3.3   Water-Column and Sediment P Content

P in the water of Jacarepaguá Lagoon was high during the study (see Figure 21.5). Total P concentration was similar at both sampling stations, ranging from 0.7 mg L⁻¹ to 1.8 mg L⁻¹. On average, TP was 10% lower (p < 0.05) during the rainy season (spring/summer), but there was no clear seasonal pattern. Dissolved inorganic P (DIP) concentrations ranged from 0.2 mg L⁻¹ to 0.9 mg L⁻¹ and showed no spatial or seasonal differences (see Figure 21.5).

The RSP content is high in Jacarepaguá Lagoon, but not uniformly distributed over the sediment (see Figure 21.6). RSP contents varied among sampling stations, where the highest values were found near river mouths (JAC 18 and P4). Most of the RSP is in the redox-sensitive bound fraction (BD-P, 50–80%). The organic-bound fraction (NaOH–NRP) also was important, especially at JAC 18 and JAC 20 (see Figure 21.6).

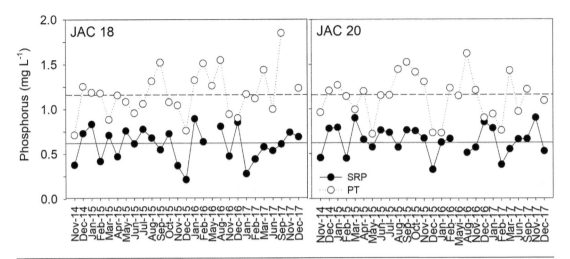

Figure 21.5 TP (open circles) and DIP (filled circles) concentration (mg L^{-1}) in the sampling stations JAC 18 and JAC 20 during the monitoring period (November 2014 to December 2017) in Jacarepaguá Lagoon.

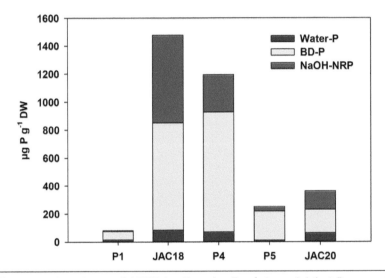

Figure 21.6 Concentration of potential RSP fractions (μg P g^{-1} dry weight) at five sampling stations in Jacarepaguá Lagoon.

21.3.4 P Input from Rivers/Tributaries

P measured in tributary rivers of Jacarepaguá Lagoon was high and almost constant, except for November 2017, when the highest values were observed in all tributaries (see Figure 21.7). This peak of TP cannot be attributed to rainfall, since the accumulated rain in the 96 hours prior to sampling was only 1.2 mm.

In general, TP measured in the lagoon (JAC 18 and JAC 20) reflected the TP input from the tributaries, except for the Camorim River, which showed lower TP values compared to other tributaries and the lagoon (H$_6$ = 21.294; p < 0.05).

During the study, the P load from tributary rivers to Jacarepaguá Lagoon was high (> 40 kg P d^{-1}), ranging from 42.5 kg P d^{-1} to 208 kg P d^{-1} during the year, except for the huge load estimated for November 2017 (see Figure 21.7). The P load entering the lagoon averages 152.1 kg P d^{-1} (SD = 159.9 kg P d^{-1}, n = 11) when the November extreme high is included but declines to 104.4 kg P d^{-1} (SD = 55.6 kg P d^{-1}, n = 10) when the November load is excluded. The Arroio Pavuna River was the main source of external P from the watershed, but the Caçambé River also was an important contributor (see Figure 21.8).

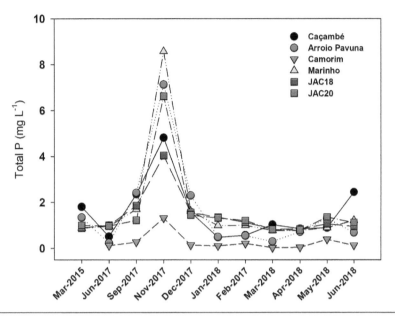

Figure 21.7 TP concentration (mg L^{-1}) in the tributary rivers (Caçambé, Arroio Pavuna, Camorim, and Marinho) and in sampling stations JAC 18 and JAC 20 in Jacarepaguá Lagoon during the monitoring period (March 2015 to June 2018).

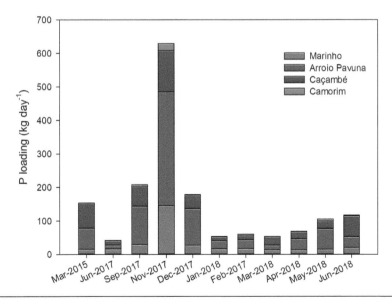

Figure 21.8 Estimated external P loading (kg d^{-1}) from the Tributary Rivers (Caçambé, Arroio Pavuna, Camorim, and Marinho) during the monitoring period (March 2015 to June 2018).

21.3.5   Internal P Fluxes Estimated from Sediment Cores

Estimates of internal fluxes of P from the sediments of Jacarepaguá Lagoon were calculated from incubation experiments with sediment cores collected in 2015 at one station (JAC 20) and in July 2018 at four stations (P1, JAC 18, P4, and P5). In both experiments, the increase in SRP concentration was linear, even in the long-term incubation core of 2015 (see Figure 21.9).

In general, the estimated SRP fluxes at the sampling stations were < 5 mg P m^{-2} d^{-1}, except at JAC 20 (see Figure 21.10). The estimated fluxes were not different from sampling stations in 2018

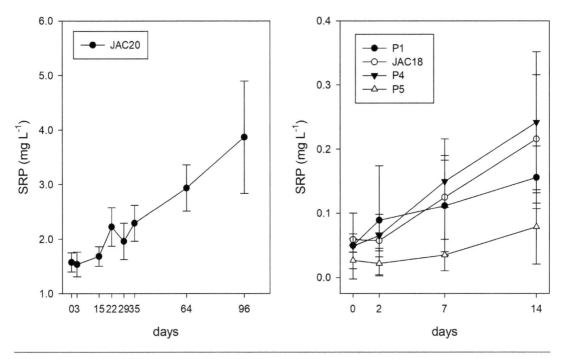

Figure 21.9   Course of the SRP concentrations in sediment cores collected in 2015 (JAC 20) and 2018 (P1, JAC 18, P4, and P5) (n = 4).

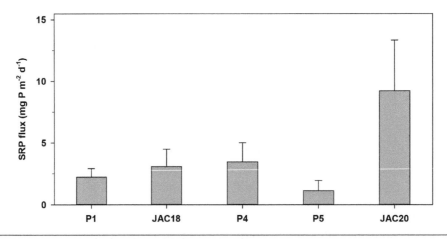

Figure 21.10   Estimated internal fluxes of P (mg P m^{-2} d^{-1}) from five sampling stations in Jacarepaguá Lagoon.

($F_{3,15} = 3.02$; $p > 0.05$) with an average of 2.5 (±1.4) mg m^{-2} day^{-1}. However, P flux for JAC 20 (9.2 mg m^{-2} day^{-1} ± 4.1 mg m^{-2} day^{-1}) in 2015 was significantly higher than in 2018 ($F_{4,19} = 11.0$; $p < 0.001$).

21.4 DISCUSSION

Jacarepaguá Lagoon is a shallow system and mixing of the water column is expected on a daily basis. In fact, no stratification was observed in the study period; however, hypoxic or anoxic conditions near the sediments prevailed frequently. This chemical stratification was probably caused by decomposition of the high amounts of organic matter from sewage effluent entering the lagoon from its inflowing tributaries. So, we can assume that the anoxic condition of the sediment occurs throughout the year.

Roy et al. (2012) considered the bulk of the surface sediments to be permanently anaerobic when calculating the annual internal load of SRP to Lake Pontchartrain (USA). The authors argued that even under an oxygenated water column, subjected to repeated wind mixing, the fine-grained sediments in Lake Pontchartrain largely limit the diffusion of oxygen.

Since most of the potential RSP is in the redox-sensitive bound fraction, we can expect a continuous flux of P from the sediment in Jacarepaguá Lagoon. However, the internal loading of P calculated for Jacarepaguá Lagoon varied in time when we compare the internal P fluxes estimated for 2015 (6.3 mg m^{-2} day^{-1}) versus 2018 (2.5 mg m^{-2} day^{-1}). These estimates yield an annual load of 8.5 tons P yr^{-1} and 3.3 tons P yr^{-1}, for 2015 and 2018, respectively. Considering the grand mean (4.6 mg m^{-2} day^{-1}), we calculate an annual internal load of 6.2 tons P yr^{-1}.

The observed inter-annual variation may be related to both the conditions used to estimate P fluxes in the experiments and the differences in the water quality. One of the experiments conducted by de Magalhães et al. (2018) to estimate internal P fluxes was run under anoxic conditions and gave the highest flux (9.9 mg m^{-2} day^{-1}). On the other hand, the estimates with cores sampled in 2018 were made under oxic conditions (DO > 3 mg L^{-1}). Another aspect to be considered is the difference in the water quality of the lagoon during the sampling. DO and salinity were higher in 2018 (3.3 mg L^{-1} and 7.6 ppt, respectively) than in 2015 (0.6 mg L^{-1} and 5.3 ppt, respectively).

In addition, an alkaline pH was observed for the overlying water in 2015 (> 8.0) while it was circumneutral (7.2) in 2018. Both pH and alkalinity play an important role in sediment P binding (Orihel et al. 2017). At higher pH values, the capacity of P binding to oxidized iron decreases, which in turn, increases the P availability into the water column. For instance, Boers (1991) observed a 10-fold increase in sediment P release rate at pH 9.5 compared to the rate at pH 8.3. Boers (1991) achieved the higher pH by adding NaOH, but hydroxyl ion concentrations may drastically increase due to photosynthesis, favoring ligand exchange reactions (Lijklema 1977). Moreover, iron-P and aluminum-P complexes may dissolve at higher pH (Olila and Reddy 1995). In Lake Volvi (Greece), P desorption at low redox potential occurred, but ion-exchange with OH$^-$ at higher pH seemed the major mechanism behind internal P loading (Christophoridis and Fytianos 2006). In Jacarepaguá Lagoon, both mechanisms seem to play a role in sediment P release.

An additional factor in Jacarepaguá Lagoon is elevated salinity. SRP efflux from sediments is enhanced at higher salinities because sulfides may interfere with iron-bound P, but this salinity effect can be offset by high pH promoting SRP release at lower salinities (Hartzell and Jordan 2012). In Jacarepaguá Lagoon, the SRP efflux in sediment cores from campaign one (lower salinity, lower oxygen, higher pH) was clearly higher than the SRP efflux in sediment cores from campaign two (higher salinity, higher oxygen, lower pH). These differences underpin the importance of regular measurement of sediment SRP fluxes to determine the influence of the different environmental factors and to get more insight into the seasonal variability in SRP fluxes.

21.4.1 Internal Versus External Loading in Jacarepaguá Lagoon

Our results indicate that the eutrophication of the Jacarepaguá Lagoon is promoted by high external sources of P. Considering the nutrient load from the main inflows, we estimate an annual input of about 54.7 tons yr^{-1}. This equates to an external loading of 40.5 mg m^{-2} day^{-1}, which falls between the range measured in other freshwater eutrophic Brazilian lakes that receive sewer effluents (13.4 mg P m^{-2} day^{-1}: Henry et al. 2004; 72.3 mg P m^{-2} day^{-1}: Torres et al. 2007). The authorities are planning river treatment units, which have already been implemented in the Arroio Fundo River that feeds into the Tijuca Lagoon, located downstream of Jacarepaguá. This treatment unit uses aluminum sulfate and polymers at a cost of approximately R$550,000 (Brazilian *real*) or $140,000 (United States *dollars*) per month. Reducing the external nutrient load is an absolute necessity for restoration of the lagoon, but it might not be enough.

High sediment P release rates, averaging 9.2 mg P m^{-2} d^{-1}, were observed in JAC 20 sediment cores when they became anoxic during the incubation. Likewise, in oligohaline Lake Pontchartrain, the sediment P flux of 1.06 mg P m^{-2} d^{-1} in anoxic incubations was clearly higher than the 0.48 mg P m^{-2} d^{-1} in oxic conditions (Roy et al. 2012). In general, sediment P release is greater under anoxic than oxic conditions, and thus the results from the sediment incubations are in line with the redox sensitivity of iron/manganese-bound P. However, our field data in Jacarepaguá Lagoon show that the oxygen concentrations close to the sediment are anoxic ($<$ 1 mg L^{-1}) almost the whole year. Consequently, the sediment P release rates retrieved during the second campaign might be an underestimate of *in situ* release.

Despite the significant influence of low temperature in temperate lakes, we did not expect the P fluxes calculated here to be less in winter than in summer. Our lower temperatures are close to 25°C and sufficient to stimulate high bacterial production. Despite the marine influence, interference from salinity is not expected since the sea connection via canals and other lagoons is limited, which is reflected in only slight salinity variations in Jacarepaguá Lagoon. The effect of salinity on P release from sediment is contradictory in the literature, as some studies have shown a decrease in dissolved P concentration due to higher salinity (e.g., Baldwin et al. 2006; Weston et al. 2006), while others have shown increases in P concentration (Caraco et al. 1990; Zhang and Huang 2011). Kim et al. (2017) showed no difference in P release from sediment when salinity varied from 10.8 psu to 38.3 psu in anoxic sediment of a brackish water lake in South Korea. In the case of Jacarepaguá Lagoon, the morphometry and location of the lagoon make it less suitable to high salinity variation; hence, a significant influence of salinity on P flux from sediment is not expected. However, considering the effect of salinity on SRP efflux from sediment (Hartzell and Jordan 2012), we recommend further examination of interactions between redox and salinity on sediment P processes in Jacarepaguá Lagoon.

Estimations of external and internal P loading for coastal lagoons are scarce. Vadrucci et al. (2004) estimated the contribution from external and internal P inputs for Lake Alimini Grande, a salt-marsh ecosystem (Southern Italy) which, despite the considerable load of nutrients from the catchment area, is classified as meso-oligotrophic (De Donno et al. 2017). The estimates of external and internal P loading for Lake Alimini Grande averaged 0.34 and 0.3 mg m^{-2} day^{-1}, respectively (Vadrucci et al. 2004; data recalculated), which is two and one orders of magnitude lower, respectively, than rates estimated for Jacarepaguá Lagoon.

The presence of dense urbanization without wastewater treatment, together with the high concentration of P in the sediment, has led to the high level of eutrophication in Jacarepaguá Lagoon. This study revealed a huge P load (45.1 mg m^{-2} day^{-1}) generated by the external and internal loading to the water column of Jacarepaguá Lagoon. The external load of P is much higher than the internal load, which represents around 11% of the P load to the water column. However, internal load from sediments can be considered high when compared to other eutrophic systems and should not be neglected. We conclude that in-lake measures are necessary even if the external loading is reduced.

Even after successful reduction of the external P load from the rivers, the sediment P release might be sufficient to fuel phytoplankton blooms. The turbid water and organic matter production may remain for many years—or even decades—since a true flushing of the system, given its morphometry and water flow, seems unlikely. Cracking the cycle of phytoplankton biomass production and bacterial organic matter remineralization near the sediment is a prerequisite to accelerated recovery. With a highest potential releasable sediment P content of $1-1.4$ mg g^{-1}, the high sediment Fe content in Jacarepaguá (on average = 28 mg g^{-1}; Júnior et al. 2012) is clearly exceeding the molar Fe:P ratio of 1:1 below which phosphate can be released from the sediment, but above which phosphate can be effectively trapped by iron(hydr)oxides (Smolders et al. 2001). An oxidizing environment at the sediment-water interface is a prerequisite for effective control of sediment P release by iron (Smolders et al. 2006). This can be achieved artificially through aeration or oxygenation (see Chapter 5).

Another approach for increasing oxygen conditions near the sediment is to limit phytoplankton production by reducing the availability of nutrients and thereby limiting autochthonous organic matter production. A recent study with sediment from Jacarepaguá Lagoon showed that a low dose of poly-aluminum chloride combined with the lanthanum-modified clay Phoslock® cleared the water column of cyanobacteria and drastically reduced the phosphate concentrations by stripping the water column and hampering the release from sediment stored P (de Magalhães et al. 2018). If such intervention is applied in the lagoon, it would need a total budget of around $10 million, which is only 4–5% of the estimated dredging costs (de Magalhães et al. 2018). Restoring the lagoon needs more than external load control; *in situ* geo-engineering techniques seem far more cost-effective than standard dredging procedures.

21.5 CONCLUSIONS

Jacarepaguá Lagoon is a shallow hypereutrophic system that presents hypoxic or anoxic conditions near the sediment, almost permanently, resulting in a continuous flux of P from the sediment with an estimated annual load of 6.2 tons P yr^{-1}.

The estimated internal loading of P varied between the study years and can be related to environmental factors (e.g., pH, salinity), but more measurements are needed for a better understanding of how environmental factors influence the release of P from the lagoon sediment.

Eutrophication of the Jacarepaguá Lagoon is promoted by a high external load, with an annual input of about 55 tons P yr^{-1} from the main tributaries. The estimated internal load represents around 11% of the total P load to the water column, and can be considered high when compared to other eutrophic systems.

A successful rehabilitation of the lagoon to prevent harmful cyanobacterial blooms requires drastic external load reduction, as well as the control/management of internal P loading. Laboratory studies indicate that in-situ geo-engineering P-capping techniques seem very promising and also are more cost-effective than standard dredging.

21.6 REFERENCES

Baldwin, DS; Rees, GN; Mitchell, AM; Watson, G; and Williams, J. 2006. The short-term effects of salinization on anaerobic nutrient cycling and microbial community structure in sediment from a freshwater wetland. Wetlands. 26(2):455–464.

Benedetti, PE. 2011. Caracterizaçã o geoambiental dos sedimentos da Lagoa de Jacarepaguá. [Masters Dissertation]. Pontifícia Universidade Católica do Rio de Janeiro.

Berkes, F and Seixas, CS. 2005. Building resilience in lagoon social–ecological systems: a local-level perspective. Ecosystems. 8(8):967–974.

Boers, PCM. 1991. The influence of pH on phosphate release from lake sediments. Water Res. 25(3):309–311.

Caraco, N; Cole, J; and Likens, GE. 1990. A comparison of phosphorus immobilization in sediments of freshwater and coastal marine systems. Biogeochemistry. 9(3):277–290.

Cavalcante, H; Araújo, F; Noyma, NP; and Becker, V. 2018. Phosphorus fractionation in sediments of tropical semiarid reservoirs. Sci Total Environ. 619:1022–1029.

Christophoridis, C and Fytianos, K. 2006. Conditions affecting the release of phosphorus from surface lake sediments. J Environ Qual. 35(4):1181–1192.

COHIDRO. 2006. Estudo de Impacto Ambiental para Estabilizaçã o da Barra do Canal de Sernambetiba e sua Interligação com as Lagoas de Jacarepaguá, Tijuca e Marapendi. SEMADUR/SERLA.

Cole, GA. 1994. *Textbook of Limnology*. Project Heights (IL): Waveland Press Inc.

Cooke, GD; Welch, EB; Peterson, S; and Nichols, AS. 2005. *Restoration and Management of Lakes and Reservoirs*. Boca Raton (FL): CRC press.

De Donno, A; Bagordo, F; Serio, F; Grass, T; Devoti, G; and Guido, M. 2017. Environmental quality and hygienic safety of the Alimini Lakes (Puglia, Italy): 20 years of monitoring (1995–2014). Rend Lincei Sci Fis. 28(2):317–328.

de Magalhães, L; Noyma, NP; Furtado, LL; Drummond, E; Leite, VBG; Mucci, M; van Oosterhout, F; Huszar, VL; Lürling, M; and Marinho, MM. 2018. Managing eutrophication in a tropical brackish water lagoon: testing lanthanum-modified clay and coagulant for internal load reduction and cyanobacteria bloom removal. Estuaries Coast. https://doi.org/10.1007/s12237-018-0474-8.

de Magalhães, L; Noyma, NP; Furtado, LL; Mucci, M; van Oosterhout, F; Huszar, VL; Marinho, MM; and Lürling, M. 2017. Efficacy of coagulants and ballast compounds in removal of cyanobacteria (*Microcystis*) from water of the Tropical Lagoon Jacarepaguá (Rio de Janeiro, Brazil). Estuaries Coast. 40(1):121–133.

Edmondson, WT. 1970. Phosphorus, nitrogen, and algae in Lake Washington after diversion of sewage. Science. 169(3946):690–691.

Esteves, FA; Caliman, A; Santangelo, JM; Guariento, RD; Farjalla, VF; and Bozelli, RL. 2008. Neotropical coastal lagoons: an appraisal of their biodiversity, functioning, threats and conservation management. Braz J Biol. 68(4):967–981.

Fernandes, HM; Bidone, ED; Veiga, LHS; and Patchineelam, SR. 1994. Heavy-metal pollution assessment in the coastal lagoons of Jacarepaguá, Rio de Janeiro, Brazil. Environ Pollut. 85(3): 259–264.

Fernandes, VO. 1993. Estudos limnológicos na Lagoa de Jacarepaguá (RJ): Variáveis abióticas e mudanças na estrutura e dinâmica da comunidade perifítica em *Typha domingensis* Pers [dissertation]. [São Carlos (SP)]: UfsCar.

Ferrão-Filho, AS; Domingos, P; and Azevedo, SM. 2002. Influences of a *Microcystis aeruginosa* Kützing bloom on zooplankton populations in Jacarepaguá Lagoon (Rio de Janeiro, Brazil). Limnologica. 32(4):295–308.

Gomes, AMA; Sampaio, PL; da Silva Ferrão-Filho, A; de Freitas Magalhaes, V; Manzi Marinho, M; de Oliveira, ACP; dos Santos, VB; Domingos, P; and de Oliveira e Azevedo, SMF. 2009. Floraçõ es de cianobactérias tóxicas em uma lagoa costeira hipereutrófica do Rio de Janeiro/RJ (Brasil) e suas consequências para saúde humana. Oecol Bras. 13(2):329–345.

Gulati, RD and Van Donk, E. 2002. Lakes in the Netherlands, their origin, eutrophication and restoration: state-of-the-art review. In: *Ecological Restoration of Aquatic and Semi-Aquatic Ecosystems in the Netherlands (NW Europe)*. Dordrecht: Springer. pp. 73–106.

Hartzell, JL and Jordan, TE. 2012. Shifts in the relative availability of phosphorus and nitrogen along estuarine salinity gradients. Biogeochemistry. 107(1–3):489–500.

Heisler, J; Glibert, PM; Burkholder, JM; Anderson, DM; Cochlan, W; Dennison, WC; Gobler, C; Dortch, Q; Heil, C; Humphries, E; et al. 2008. Eutrophication and harmful algal blooms: a scientific consensus. Harmful Algae. 8(1):3–13.

Henry, R; Do Carmo, CF; and Bicudo, DC. 2004. Trophic status of a Brazilian urban reservoir and prognosis about the recovery of water quality. Acta Limnol Bras. (Online). 16(3):251–262.

Hillebrand, H; Dürselen, CD; Kirschtel, D; Pollingher, U; and Zohary, T. 1999. Biovolume calculation for pelagic and benthic microalgae. J Phycol. 35(2):403–424.

Jensen, HS and Thamdrup, B. 1993. Iron-bound phosphorus in marine sediments as measured by bicarbonate-dithionite extraction. In: *Proceedings of the Third International Workshop on Phosphorus in Sediments.* Dordrecht: Springer. p. 47–59.

Júnior, JTA; Benedett, PA; Pires, PJM; and Almeida, RFR. 2012. Sediments Quality Assessment of Jacarepaguá Lagoon: The Venue of the 2011 Rock in Rio. Clean—Soil, Air, Water. 40 (9), 906–910.

Kennish, MJ; Brush, MJ; and Moore, KA. 2014. Drivers of Change in Shallow Coastal Photic Systems: an Introduction to a Special Issue. Estuaries Coasts. 37(1):3–19.

Kim, TH; Kang, JH; Kim, SH; Choi, I; Chang, KH; Oh, JM; and Kim, KH. 2017. Impact of salinity change on water quality variables from the sediment of an artificial lake under anaerobic conditions. Sustainability. 9(8):1429.

Kjerfve, B. 1994. Coastal lagoon processes. Elsevier Oceano Series. 60: 1–8. doi.org/10.1016/S0422-9894(08)70006-0.

Lijklema, L. 1977. Role of iron in the exchange of phosphate between water and sediments. In: Interactions Between Sediments and Fresh Water; Proceedings of an International Symposium.

Magalhães, VF; Soares, RM; and Azevedo, SM. 2001. Microcystin contamination in fish from the Jacarepaguá Lagoon (Rio de Janeiro, Brazil): ecological implication and human health risk. Toxicon. 39(7): 1077–1085.

MASTERPLAN. 2015. Estudo de impacto ambiental das obras de prolongamento do enrocamento (molhe) existente na entrada do Canal da Joatinga e as melhorias da circulaçã o hídrica do Complexo Lagunar de Jacarepaguá. p. 4120.

Olila, OG and Reddy, KR. 1995. Influence of pH on phosphorus retention in oxidized lake sediments. Soil Sci Soc Am J. 59:946–959.

Orihel, DM; Baulch, HM; Casson, NJ; North, RL; Parsons, CT; Seckar, DC; and Venkiteswaran, JJ. 2017. Internal phosphorus loading in Canadian fresh waters: a critical review and data analysis. Can J Fish Aquat Sci. 74(12):2005–2029.

Paerl, HW and Paul, VJ. 2012. Climate change: links to global expansion of harmful cyanobacteria. Water Res. 46(5):1349–1363.

Paludan, C and Jensen, HS. 1995. Sequential extraction of phosphorus in freshwater wetland and lake sediment: significance of humic acids. Wetlands. 15(4):365–373.

Psenner, R; Pucsko, R; and Sage, M. 1984. Fractionation of organic and inorganic phosphorus compounds in lake sediments, an attempt to characterize ecologically important fractions (Die Fraktionierung Organischer und Anorganischer Phosphorverbindungen von Sedimenten, Versuch einer Definition Okologisch Wichtiger Fraktionen). Arch Hydrobiol. 1(1).

Roy, ED; Nguyen, NT; Bargu, S; and White, JR. 2012. Internal loading of phosphorus from sediments of Lake Pontchartrain (Louisiana, USA) with implications for eutrophication. Hydrobiologia. 684(1):69–82.

Salloto, GR; Cardoso, AM; Coutinho, FH; Pinto, LH; Vieira, RP; Chaia, C; Lima, JL; Albano, RM; Martins, OB; and Clementino, MM. 2012. Pollution impacts on bacterioplankton diversity in a tropical urban coastal lagoon system. PloS One. 7(11):e51175.

Schauser, I; Chorus, I; and Lewandowski, J. 2006 Effects of nitrate on phosphorus release: comparison of two Berlin lakes. Acta Hydrochim Hydrobiol. 34(4):325–332.

Semeraro, J and Ferreira da Costa, A. 1972. Plâncton e a poluiçao nas lagoas da Tijuca, Camorin e Jacarepagua. In: IES Publicaçao (Vol 73,). Instituto de Engenharia Sanitária.

Smolders, AJP; Lamers, LPM; Moonen, M; Zwaga, Z; and Roelofs, JGM. 2001. Controlling phosphate release from phosphate-enriched sediments by adding various iron compounds. Biogeochemistry. 54:219–228.

Smolders, AJP; Lamers, LPM; Lucassen, ECHET; Van Der Velde, G; and Roelofs, JGM. 2006. Internal eutrophication: How it works and what to do about it—a review. Chem Ecol. 22(2):93–111.

Søndergaard, M; Jensen, JP; and Jeppesen, E. 1999. Internal phosphorus loading in shallow Danish lakes. In: *Shallow Lakes' 98.* Dordrecht: Springer. pp. 145–152.

Søndergaard, M; Jensen, PJ; and Jeppesen, E. 2001. Retention and internal loading of phosphorus in shallow, eutrophic lakes. Sci World J. 1:427–442.

Torres, IC; Resck, RP; and Pinto-Coelho, RM. 2007. Mass balance estimation of nitrogen, carbon, phosphorus and total suspended solids in the urban eutrophic, Pampulha reservoir, Brazil. Acta Limnol Bras. (Online). 19(1):79–91.

Uehlinger, V. 1964. Etude statistique des méthodes de dénombrement planctonique [Doctoral dissertation]. Kundig.

Utermöhl, H. 1958. Zur Vervollkommnung der quantitativen Phytoplankton-Methodik. Mitt Int Ver Theor Angew Limnol. 9(1):1–38.

Vadrucci, MR; Semeraro, A; Zaccarelli, N; and Basset, A. 2004. Nutrient loading and spatial–temporal dynamics of phytoplankton guilds in a Southern Italian coastal lagoon (Lake Alimini Grande, Otranto, Italy). Chem Ecol. 20(sup1):285–301.

Weston, NB; Dixon, RE; and Joye, SB. 2006. Ramifications of increased salinity in tidal freshwater sediments: Geochemistry and microbial pathways of organic matter mineralization. J Geophys Res Biogeosci. 111(G1).

Zhang, JZ and Huang, XL. 2011. Effect of temperature and salinity on phosphate sorption on marine sediments. Environ Sci Technol. 45(16):6831–6837.

CHAPTER **22**

INPUTS, OUTPUTS, AND INTERNAL CYCLING OF PHOSPHORUS IN TROPICAL LAKE MALAWI, AFRICA

Harvey A. Bootsma[1] and Robert E. Hecky[2]

Abstract

While phosphorus (P) budgets have been constructed for many tropical lakes, there are virtually no measurements of internal P cycling rates in these systems. Measurement of various P fluxes in meromictic Lake Malawi reveals that the upward flux of dissolved P from the hypolimnion and metalimnion to the epilimnion is ~1.4 times greater than external loading from rivers and the atmosphere. Much of the P recycling within the hypolimnion and metalimnion appears to occur in the sediment, possibly due to rapid sinking of the river P load, which is dominated by particulate P. There is evidence that the redoxcline near the top of the hypolimnion may be a zone of dissolved P loss, due to microbial uptake and/or complexing with iron oxyhydroxides, similar to the redox shuttle that exists in some marine systems. While external P loading appears to have some influence on the seasonal dynamics of plankton, both short-term and long-term plankton dynamics are strongly regulated by internal P loading, which in turn is driven by vertical mixing rates. Much of this mixing occurs as upwelling at the southern end of the lake, where plankton and fish production are greatest. Due to the long residence time and slow vertical exchange rates in Lake Malawi, the response of the plankton and fish community may lag several decades behind changes in external P loading. The measurements of internal P fluxes presented here can be used to parameterize mechanistic models that will help to predict this response and guide nutrient management strategies.

Key Words: Lake Malawi, phosphorus, meromictic, tropical, anoxic

[1] University of Wisconsin-Milwaukee, School of Freshwater Sciences, 600 E. Greenfield Ave., Milwaukee, WI 53204; hbootsma@uwm.edu

[2] Large Lakes Observatory, University of Minnesota-Duluth, Duluth, MN 55812; rehecky@gmail.com

22.1 INTRODUCTION

Large, tropical lakes have several features that distinguish them from many temperate systems, and that have important implications for internal phosphorus (P) dynamics. Kilham and Kilham (1990) pointed out that, while large temperate lakes tend to be oligotrophic, large tropical lakes are highly productive relative to their external nutrient loads; this is due to efficient internal recycling of P, which is ultimately the result of year-round warm temperatures and deep mixed layers. Efficient internal P recycling may also be promoted by anoxic or hypoxic conditions (Hecky et al. 1996, 2010; North et al. 2014), which are common in large tropical lakes because they tend to be meromictic or oligomictic and their warmer temperatures result in lower saturation concentrations of dissolved oxygen (DO) and higher respiration rates (Lewis 1987; Hecky 2000).

While the importance of internal nutrient cycling has been inferred for many tropical lakes, few studies have directly quantified internal nutrient dynamics in these systems. P load has been determined for numerous tropical systems (e.g., Salas and Martino 1991), but with a few exceptions (Kalff 1983), internal processes such as phytoplankton P uptake, P sedimentation, and P recycling have rarely been quantified. This dearth of data makes it difficult to parameterize mechanistic models, which are needed to predict lake response to external stressors such as land use change, fishing, and climate change (e.g., Lam et al. 2002; Naithani et al. 2007). This need is especially acute for the African Great Lakes. Most notably, Lake Victoria has experienced severe eutrophication over the past half-century, with consequences for ecosystem structure, function, and biodiversity (Bootsma and Hecky 1993; Seehausen et al. 1997; Hecky et al. 2010). While changes in the deeper Lake Tanganyika and Lake Malawi/Nyasa (hereafter referred to as Lake Malawi) have not been as dramatic, there is evidence that these lakes are also being altered by changes in climate (Vollmer et al. 2005; Cohen et al. 2016) and land use (Hecky et al. 2003; Otu et al. 2011). Therefore, it is critical that management of these systems be guided by predictive models, which in turn, rely on an understanding of how physical, biological, and chemical dynamics interact in these systems. We report here the results of a multiyear study to quantify external P loading and various internal P fluxes in Lake Malawi.

22.1.1 The Lake Malawi Ecosystem

Lake Malawi (known as Lake Nyasa in Tanzania and Lake Niassa in Mozambique) is the 4th deepest lake in the world, the 15th largest by surface area, and the 5th largest by volume (rankings include the Caspian Sea; Herdendorf 1982) (see Figure 22.1 and Table 22.1). The lake is located at the southern end of the western branch of the East African Rift System. Reflecting its tectonic origins, the lake basin consists of a series of half-grabens, resulting in a shoreline that alternates between steep mountains and more gently sloping plains. The basin contained a lake as early as the late Miocene (~8.6 Ma) but deep-water conditions likely did not exist until the early Pliocene (~4.5 Ma), with large fluctuations in lake depth since then (Delvaux 1995). Over the past century, lake level has fluctuated by ~2 m. The lake drains via the Shire River, which is a tributary to the Zambezi River. However, since the early 1800s there have been several periods when lake level was low enough that there was no outflow (Sene et al. 2017). The deepest parts of the lake are underlain by > 4 km of sediment, which provides a valuable paleorecord of climatological, geological, biological, and limnological conditions (Scholz et al. 2011). Because many of the proxies used to reconstruct past conditions are influenced by in-lake physical and biogeochemical processes, interpretation of the sedimentary record is facilitated by the study of neolimnological conditions.

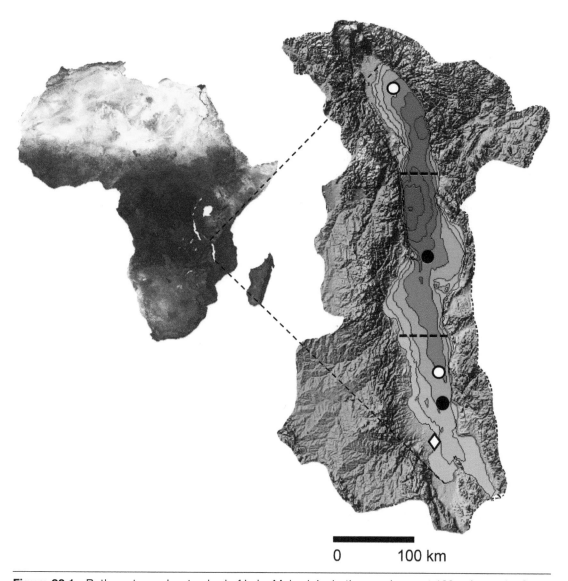

Figure 22.1   Bathymetry and watershed of Lake Malawi. Isobaths are shown at 100 m intervals. Open circles are locations of sediment cores, filled circles are locations of sediment traps, and the open diamond is the location of atmospheric deposition measurements. Horizontal dashed lines separate the southern, central, and northern sections of the lake that were used to assess spatial distribution of river P loads. Images of continent and Lake Malawi watershed topography from NASA.

Surface temperatures of Lake Malawi fluctuate seasonally between ~23°C and > 29°C (Bootsma 1993a; Ngochera and Bootsma 2011). The hypolimnetic temperature is relatively constant on an annual basis, but over the past century it has steadily warmed from 22°C to 23°C (Vollmer et al. 2005; Li et al. 2018). This has been accompanied by warming of the surface mixed layer (Vollmer et al. 2005), and so the lake has remained meromictic. During the warmest months (December to March), a primary thermocline exists between 40 m and 60 m. During the cool, windy season (June–August), a weak thermal gradient persists between the surface and ~250 m, with temperatures below 250 m being relatively constant throughout the year. As a result of the permanent stratification,

Table 22.1 Physical and chemical properties of Lake Malawi. All values except pH, chlorophyll *a*, and photosynthesis are from Bootsma and Hecky (2003) and sources cited therein.

Surface area	29,500 km^2
Max. depth	700 m
Avg. depth	264 m
Volume (V)	7,775 km^3
Altitude	474 m
River inflow	29 km^3 yr^{-1}
River outflow	12 km^3 yr^{-1}
Direct rainfall	39 km^3 yr^{-1}
Flushing time (V/I)*	114 yr
Specific conductivity	230 μS cm^{-1}
Alkalinity	2,300 μeq L^{-1}
pH (epilimnion)	8.0–8.9
pH (hypolimnion)	7.5–8.0
Chl *a***	0.3–0.6 mg m^{-3}
Photosynthesis***	500–740 mg C m^{-2} yr^{-1}

* I = inflow, which includes river inflow and direct rainfall. Note that if flushing time is calculated as volume/outflow, it is much longer (648 yr).
** 25th to 75th percentile chlorophyll *a* for epilimnion.
*** Based on measurements reported by Bootsma (1993) and Guildford et al. (2007).

the hypolimnion is anoxic, with relatively high concentrations of dissolved nutrients (Bootsma and Hecky 1993; Li et al. 2018). Concentrations of dissolved nitrogen (N) and P are low in the epilimnion (usually < 0.2 μM; Bootsma and Hecky 1993). Chlorophyll *a* concentrations are also low (usually < 1 μg L^{-1}), but phytoplankton photosynthesis rates are relatively high, ranging from ~0.4 g C m^{-2} d^{-1} to > 1 g C m^{-2} d^{-1} (Bootsma 1993a; Guildford et al. 2007).

Unlike the Laurentian Great Lakes, which are strongly P limited, the African Great Lakes appear to be balanced between P and N limitation, which is due in part to high rates of denitrification in these systems (Hecky et al. 1996; Bootsma and Hecky 2003). Although Lake Malawi phytoplankton can at times show signs of strong N limitation (Guildford et al. 2003), nutrient assay experiments and seston stoichiometry indicate that P limitation is also common, and because N deficits can be compensated for by N fixation (Gondwe et al. 2008), ultimately P is the element that regulates phytoplankton productivity (Hecky et al. 1996; Ngochera and Bootsma 2018).

Lake Malawi is perhaps best known for the diversity of its fish community. The lake is home to between 500 and 1,000 fish species that belong to 11 families, although most of the diversity is within the family Cichlidae (Snoeks and Konings 2004). Diversity is particularly high within the nearshore rocky zones, where as many as 500 individuals and 22 species can exist in a 50 m^2 area (Ribbink et al. 1983). From a scientific perspective, the fish community provides fascinating opportunities to study fish ecology and speciation. For the human populations living around the lake, the fish are a critical food source, providing most of the dietary animal protein for many people (Bland and Donda 1995). Estimates of annual fish catch range from 30,000 to 80,000 metric tons (Jamu et al. 2011).

More complete descriptions of Lake Malawi's biota, limnology, meteorology, and geological setting are provided by Tiercelin and Lezzar (2002), Bootsma and Hecky (1993, 2003), and Hamblin et al. (2003).

22.1.2   Challenges/Stressors

Several recent publications have highlighted the key challenges to sustainable management of the Lake Malawi ecosystem (Bootsma and Jorgensen 2004; Weyl et al. 2010; Jamu et al. 2011). Primary threats include overfishing, increased sediment and nutrient loads due to deforestation and poor agricultural practices, loss of biodiversity, climate change, and changes in water levels, which have implications for both fish habitat and hydroelectric power generation. Several of these threats will affect, or be affected by, P dynamics in the lake. While total fish catches have not significantly declined over the past several decades, the small, pelagic cyprinid, *Engraulicypris sardella*, has become a major component of the fishery (Weyl et al. 2010). Recruitment and growth of this species appears to be strongly regulated by plankton production (Thompson and Irvine 1997), which in turn, is dependent on P supply.

Measurements of river nutrient loads (Hecky et al. 2003) and burial in sediments (Hecky et al. 1999; Otu et al. 2011) indicate that P loading to Lake Malawi has increased since the early 1960s, especially in the southern half of the lake. There are also indications that P loading from the atmosphere is exceptionally high (Bootsma et al. 1996). While there is limited data with which to determine the ecosystem response to this loading, there is concern that increases in P loading that are not accompanied by proportional increases in N loading may lead to greater prevalence of N_2-fixing cyanobacteria (Hecky et al. 2003), with negative consequences for water quality, fish production, and human health. The eutrophication of Lake Victoria, another African Great Lake, over the past half-century has provided an object lesson in the consequences of changing land use and increasing nutrient loads, which include increased abundance of cyanobacteria, deoxygenation, and loss of biodiversity (Seehausen et al. 1997; Hecky et al. 2010). The great depths of Lake Malawi and Lake Tanganyika have likely buffered their response to increased nutrient loads to some extent, but recent changes in phytoplankton composition in Lake Malawi (Otu et al. 2011) suggest that this lake is responding to increased nutrient loads, especially in the southern region.

Like many of the world's lakes, Lake Malawi is getting warmer (Vollmer et al. 2005; Li et al. 2018). In another deep, rift valley lake, Lake Tanganyika, there is evidence that such warming has led to stronger vertical stratification, reduced nutrient supply to the surface mixed layer, and reduced plankton and fish production (Cohen et al. 2016). It remains uncertain whether the same phenomenon is occurring in Lake Malawi because there is insufficient data to determine whether the epilimnion is warming faster or slower than the hypolimnion. Several massive fish kills have occurred in the lake over the past two decades, usually at the end of the cool, mixing season, which suggests the possibility that stratification may have weakened rather than strengthened, resulting in the upward mixing of anoxic water.

22.1.3   Key Questions

Management of the Lake Malawi ecosystem has focused primarily on fisheries, relying primarily on fishery catch and effort data. However, there is increasing recognition of the need to expand to an ecosystem-based management strategy (Bootsma and Jorgensen 2004; Ngochera et al. 2014). Such a strategy requires that the fishery be viewed in a context that acknowledges the influence of lower food web and biogeochemical processes, which in turn, are regulated by ecosystem scale drivers such as climate and land use. A critical question is how the lake and its biota will respond to changes in nutrient loading, which requires an accurate quantification of nutrient loads as well as internal nutrient dynamics. This is especially important considering the long temporal lags that may exist between increased nutrient loads, plankton production and composition, and response to any mitigation efforts (Bootsma and Hecky 1993; Sharpley et al. 2013).

In light of the expected trajectory of global air temperatures, Lake Malawi will continue to warm for at least the next few decades. The implications of this warming for the biota of Lake Malawi will depend to a large degree on how it affects the internal dynamics of P, N, carbon, and DO. Predicting these effects requires an understanding of how these processes are linked to the physical processes of stratification, vertical diffusion, and upwelling because these physical processes are directly influenced by meteorological conditions.

Land use within the Lake Malawi watershed has also changed (Hecky et al. 2003; Otu et al. 2011) and may continue to change in the near future due to deforestation and intensive agriculture. The consequences of the increased loading of organic carbon and nutrients that accompany these changes depend on the transport and fate of these inputs within the lake and their relative contribution to annual loads to the euphotic zone. Specific questions that need to be addressed include:

1. How are river nutrient loads partitioned between the various density strata in the lake? Because vertical mixing in the lake is slow, this will determine the time scales over which the plankton community responds to external nutrient loads and to any mitigation efforts.
2. Related to the previous question: What fraction of river-loaded P is recycled within the lake and what fraction is buried without entering the lake's internal cycle?
3. How will changes in land use affect the bioavailable N:P loading ratio? This ratio is an important determinant of phytoplankton species composition, which in turn, affects food quality for higher trophic levels.

22.2 METHODS

Much of the data for the analysis presented here was acquired during an intensive field-sampling program from 1997 to 1999. One of the goals of this program was to quantify P loads to the lake and various P fluxes within the lake, including sinking, burial, and upward vertical fluxes of dissolved P. Specific details of the measurement of nutrient pools and fluxes are presented elsewhere (Bootsma et al. 2003; Hecky et al. 2003). A brief summary of the methods used for measurement of each flux is presented here.

External loading was quantified for both river and atmospheric inputs. Major tributaries were sampled near their mouths monthly throughout the year. In the rainy season (December–March), rivers in the southern portion of the lake's watershed were sampled one to three times per week (see Bootsma et al. 2003 for further details). The P pools that were measured included soluble reactive P (SRP), total dissolved P (TDP), and particulate P (PP). Total annual river P loads were estimated following an approach similar to that used by Dolan and Chapra (2012) for the Laurentian Great Lakes. For all P pools, no significant relationship was observed between river discharge and P concentration, and so concentrations between sampling dates were estimated by linear interpolation. Daily discharge measurements for major tributaries were provided by the Malawi Water Department. For portions of the watershed that were not monitored for discharge, daily river discharge was determined by averaging the runoff of adjacent river basins, and river P concentrations were set as the monthly mean concentration of nearby river basins with similar physiography (Kingdon et al. 1999). To determine the magnitude of potential error using this approach, 95% confidence intervals for P concentrations for adjacent rivers were applied to derive a range of loading estimates.

Loss of P via the Shire River outflow was determined as the mean concentration of total P (TP) in samples collected ~1 km downstream from where the river exits the lake (38.1 ± 18.5 mg m^{-3}) times the total annual discharge (5.84 km^3). While this approach does not account for seasonal variability of TP concentration or discharge, any error introduced has a negligible effect on the lake P budget, as P outflow is a minor component of the budget (discussed below).

Atmospheric wet deposition was determined by collecting rain with duplicate sampling containers at a lakeshore station in Senga Bay. Containers were deployed immediately before each rain event and retrieved within six hours following a rain event. Dry deposition was estimated by deploying duplicate containers, each of which contained 5 L of filtered lake water (GF/F filters, 0.7 μm pore size), for a 24-hour period on stands immediately next to the lakeshore in Senga Bay. Deposition was determined as the change in mass of the various P pools in the collection containers over the 24-hour period. Between October 1996 and March 1998, wet atmospheric deposition was measured on 90 dates. Dry deposition was measured on 21 dates over one annual cycle, April 1997 to March 1998.

Within the lake, P sinking rates were determined by deploying sediment traps (10.4 cm diameter, 1.14 m tall) at two locations—one near the center of the lake (bottom depth = 400 m) and one in the southern part of the lake (bottom depth = 200 m) (see Figure 22.1). At the location near the center of the lake, triplicate traps were deployed at each of three depths (100 m, 180 m, and 300 m), and at the southern station traps were deployed in triplicate at depths of 100 m, 140 m, and 180 m. Traps were deployed for six consecutive periods between February 1997 and January 1999. At the time of deployment, 500 mL of a 7 g L^{-1} solution of KCl and 5 mL of chloroform were added to each trap to minimize mixing and to preserve trap contents during the deployment period.

Estimates of sediment P burial rates are derived from five dated sediment cores from the southern, central, and northern regions of the lake (see Figure 22.1). Details of coring methods, as well as results of biogenic silica measurements for the northern core and P measurements for other cores, are presented in Johnson et al. (2001) and Otu et al. (2011). P burial was determined as the product of P concentration per sediment dry weight times sediment dry-weight burial rate, which was determined based on ^{210}Pb-dating at 1-cm intervals, assuming a constant rate of ^{210}Pb supply to the sediment (Johnson et al. 2001).

We did not directly measure sediment-water fluxes of dissolved P. In the P budget presented below, we use data reported by Li et al. (2018), who measured SRP profiles in sediments from four locations in Lake Malawi and estimated sediment-water flux rates using a classic Fickian diffusion model, which assumes that flux is dominated by molecular diffusion.

Vertical fluxes of dissolved P in the water column were determined as:

$$F = k \frac{V_2}{V_1}(C_2 - C_1)\frac{V_1}{A} \qquad \text{[Eq. 22.1]}$$

where: F = flux (mg m^{-2} yr^{-1}), k is the annual vertical exchange coefficient that quantifies the vertical mixing between strata, V_1 and V_2 are volumes of adjacent strata (m^3; upper and lower, respectively), A is lake surface area (m^2), and C_1 and C_2 are the volume-weighted average concentrations of dissolved P in the adjacent strata (mg m^{-3}). C_1 and C_2 are calculated by:

$$C = \frac{\sum(\Delta Z_i A_i C_i)}{\sum(\Delta Z_i A_i)} \qquad \text{[Eq. 22.2]}$$

where: ΔZ_i is the thickness of a depth layer within a stratum (set at 1 m), A_i is the horizontal area of a layer, and C_i is the dissolved P concentration within the layer. Values for C were determined based on several deep vertical profiles of dissolved P measured over a 16-month period, which were accompanied by conductivity, temperature, and depth (CTD) profiles to measure the vertical distribution of temperature, conductivity, and DO. Vertical exchange coefficients (k) were reported by Vollmer et al. (2002), who were able to determine mean annual mixing rates based on the vertical distribution of chlorofluorocarbon-12 (CFC-12) in Lake Malawi. CFC-12 is an anthropogenic gas that has existed in the atmosphere for less than a century and which steadily increased in concentration from the 1930s to the late 1990s. Therefore, it can be used as a tracer of vertical mixing in aquatic systems where mixing occurs on time scales of years or longer. Based on the vertical distribution of CFC-12, DO, and water density in Lake Malawi, Vollmer et al. (2002) identified three distinct strata (which we refer to as epilimnion, metalimnion, and hypolimnion) separated by inflection points at 105 m and 220 m. Using a model of CFC-12 dynamics that accounts for the atmospheric history of CFC-12 concentration and CFC-12 degradation within the anoxic hypolimnion, Vollmer et al. estimated the epilimnion-metalimnion exchange time to be 3.4 years (k = 0.294) and the metalimnion-hypolimnion exchange time to be 15.9 years (k = 0.063).

All chemical measurements were made following the methods of Stainton et al. (1977).

22.3 RESULTS

22.3.1 River Loading and Output

Total river discharge during 1997 was 32.9 km^3, which is slightly higher than the long-term average discharge of 28.6 km^3 yr^{-1} (Kidd 1983). P input from rivers is highly seasonal, due to the seasonality of rainfall in the Lake Malawi region (see Figure 22.2). Highest concentrations of TP were measured

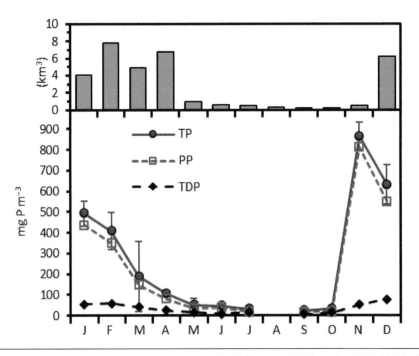

Figure 22.2 Monthly total river discharge and mean river P concentrations in 1997. Vertical bars represent 95% confidence intervals. Discharge for river basins that were not monitored was determined by averaging the areal runoff for adjacent monitored basins.

in the early rainy season (November–December). Although total river discharge remained relatively constant from January to April, TP concentrations steadily decreased as the rainy season progressed. P load was dominated by PP, which averaged 77% ± 17% of TP for all samples collected. Of the dissolved P, the SRP fraction was variable (6% – 100%), but usually high (mean = 77% ± 31%).

The spatial distribution of river P load was assessed by dividing the lake into three regions from south to north (see Figure 22.1). Loading to the central portion of the lake is much less than that to the southern and northern sections (see Figure 22.3). This is due in part to the larger size of the watershed near the lake's southern and northern sections. More intensive agricultural activity in the southern section and in some river basins in the northern section (Hecky et al. 2003) also contribute to this pattern, as does higher annual rainfall in the northern section. Over the full annual cycle, total river P load is estimated to be between 266 mg m^{-2} yr^{-1} and 398 mg m^{-2} yr^{-1} (95% C.I.).

The effect of riverine P inputs on phytoplankton production depends in part on the depth at which river plumes enter the lake. P entering the epilimnion may have an immediate effect on production, whereas P that sinks or is advected to deeper layers may be either permanently buried and have no effect on phytoplankton production or be recycled, in which case the phytoplankton response may be dampened and expressed over longer time scales. To assess the potential importance of river plume sinking, we determined the density of river water following the approach of McCullough et al. (2007) and compared it with the density structure within the lake, which was determined using the equations of state of Chen and Millero (1986) and salinity derived from CTD specific conductivity measurements (Wüest et al. 1996). The depth of equivalent density was determined for each sampling date for 11 of the major tributaries. The proportion of annual inflow that has a density greater than that at 100 m ranged from 11% to 70% (mean = 38%), with northern tributaries contributing a higher proportion than southern tributaries due to the higher average elevation (and hence, cooler temperatures) of the northern river basins.

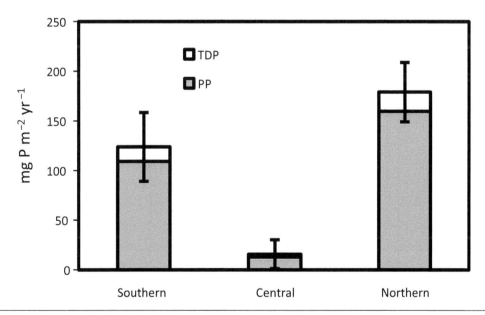

Figure 22.3 River P loads to the southern, central, and northern regions of the lake in 1997 (see Figure 22.1 for delineation of regions). Vertical bars represent 95% confidence intervals for total P loads.

22.3.2 Atmospheric Deposition

The volume-weighted mean TDP concentrations in rainfall was 8.1 mg m^{-3}, and wet deposition rates in the two rainy seasons were similar (see Figure 22.4). Total P tended to be dominated by SRP, which on average, made up 79% of TP. In the atmosphere, virtually all P is in particulate form, and thus the dominance of SRP in rain samples suggests that the atmospheric particulate forms are highly soluble and bioavailable. Rain TP concentration was inversely related to rain amount per event, and as a result, there was not a significant relationship between rainfall amount and deposition per event. Therefore, total annual wet P deposition was determined as the product of the mean P deposition per rain event for the two rainy seasons sampled between 1996 and 1998 (0.16 mg P m^{-2} event^{-1}, 95% C.I. = 0.12 mg P m^{-2} yr^{-1} – 0.19 mg P m^{-2} event^{-1}), and the total number of rain events in 1997 (81). This results in an estimated load of 9.9 mg P m^{-2} yr^{-1} – 15.8 mg P m^{-2} yr^{-1}.

The mean dry deposition rate for 21 days sampled over one annual cycle was 0.43 mg P m^{-2} d^{-1} (95% C.I. = 0.26 mg P m^{-2} yr^{-1} – 0.61 mg P m^{-2} d^{-1}). The highest dry deposition rates were observed near the end of the dry season (see Figure 22.4). However, this temporal trend was not consistent enough to apply season-specific deposition rates when calculating total annual dry deposition, and so annual deposition was determined as the mean daily rate times the number of dry days in 1997 (284), resulting in an estimated load of 124 mg P m^{-2} yr^{-1} (95% C.I. = 74 mg P m^{-2} yr^{-1} – 174 mg P m^{-2} yr^{-1}). It is difficult to determine the short-term solubility of PP in dry deposition with the method applied in this study. On several occasions the TDP concentration in collection basins decreased over a 24-hour period, even though the TP concentration increased. This was likely due to adsorption of dissolved P onto particles, or assimilation of dissolved P by bacteria or phytoplankton in the lake water that may have passed through the GF/F filters.

Atmospheric deposition to the lake surface can be influenced by proximity to land (Cole et al. 1990), and so the deposition rates reported here may be greater than those occurring over the open lake. During the study period, prevailing winds were from the south, with wind direction being between 150° and 250° on 93% of the days sampled. On 45% of the dates sampled, wind direction was

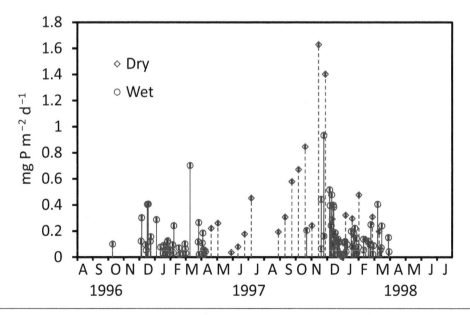

Figure 22.4   Dry and wet atmospheric deposition of phosphorus over two rainy seasons and one dry season.

off the lake (< 180°), and so deposition on those days is assumed to reflect over-lake conditions. A comparison of deposition rates with wind direction revealed no significant trend, and for both wet and dry P deposition, mean deposition rate for days when wind direction was < 180° was not significantly different from that for days when wind direction was > 180° (t-test, $\alpha = 0.01$). Therefore, for the purposes of the P budget presented here, we assume that the deposition rates measured at the lakeshore station are representative of over-lake rates.

22.3.3   Vertical Fluxes and Burial

Depth-specific sinking rates derived from sediment traps are presented in Figure 22.5. Several features of these data stand out. Mean sinking rates increased slightly between 110 m and 180 m. Similar patterns have been observed in other large lakes (Eadie et al. 1984; Baker et al. 1991). The most likely mechanisms that would produce this pattern are horizontal transport of particulate material at depth and resuspension of bottom material. Sediment traps were located >20 km from shore, and so it is possible that much of the river-borne particulate material sinks out of surface waters before reaching these offshore sites, either to disperse at depths of greater water density as interflows (McCullough et al. 2007) or to settle on the lake bottom followed by horizontal transport at depth during resuspension events. Previous profiles of suspended solids in Lake Malawi have revealed large peaks in concentration at depths >100 m (Halfman and Scholz 1993).

Sinking rates at 300 m were lower than at shallower depths. The sinking rates shown in Figure 22.5 are normalized to lake surface area, and so the lower sinking rate at 300 m is due in part to the decrease in lake bottom area with depth. Without normalization, the P sinking rate at 300 m was 60% of that at 180 m, likely reflecting the decomposition of particulate material as it sinks through the anoxic hypolimnion.

Whole-lake burial rates are difficult to determine based on the small number of cores, due both to spatial variation and temporal variation as reflected in core profiles. Cores span time intervals

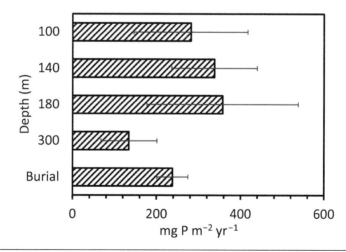

Figure 22.5 Sinking rate of particulate P measured in sediment traps deployed at four depths. Measurements made at each depth are as follows: 100 m = the mean of measurements at both trap locations between September 1997 and September 1998; 140 m = the mean of measurements at the southern trap from May 1997 to January 1999; 180 m = mean of measurements at southern trap May 1997 to January 1999 and northern trap May 1997 to May 1998; 300 m = mean of measurements at northern trap May 1997 to May 1998. Horizontal lines represent 95% confidence intervals.

from the early 1900s to the late 1990s. Near the north end of the lake, burial rates have been relatively constant for most of the past century, but burial rates at the southern core sites have increased over the past half-century, fluctuating between ~150 mg m^{-2} yr^{-1} and ~740 mg m^{-2} yr^{-1}. For the budget presented here, we use the mean burial rate for cores post-1960, which is 212 mg m^{-2} yr^{-1}. Like Alin and Johnson (2007), we assume that permanent burial occurs only at depths >100 m, which results in a burial rate of 372 mg m^{-2} yr^{-1} (95% C.I. = 245 mg m^{-2} yr^{-1} – 496 mg m^{-2} yr^{-1}) when normalized to lake surface area (see Figure 22.5).

Vertical profiles of temperature, DO, and TDP are presented in Figure 22.6. In the upper 200 m, temperatures fluctuated seasonally, but they were nearly constant at 22.77°C – 23.0°C below 200 m. Total dissolved P concentrations increased gradually from the surface down to the oxic-anoxic interface, which fluctuated between 182 m and 221 m, and increased rapidly below that depth, to a maximum of ~120 mg m^{-3}. Volume-weighted TDP concentrations for each of the three layers, along with estimates of vertical fluxes of TDP, are presented in Table 22.2. A comparison of the upward flux rates within the water column with the sediment-water flux rates reported by Li et al. (2018) suggests that approximately half of the P recycling within the metalimnion (105 m – 220 m) occurs at the sediment-water interface, with the remainder occurring in the water column. In the hypolimnion, recycling at the sediment water interface appears to be even more important, accounting

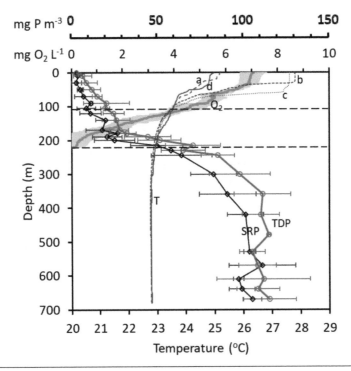

Figure 22.6   Vertical profiles of temperature (T: red), DO (O$_2$: green), soluble reactive phosphorus (SRP: black), and total dissolved phosphorus (TDP: blue). The shaded area around the DO profile and the horizontal lines on the TDP points represent 95% confidence intervals. TDP and oxygen concentrations are the means of profile measurement on four dates: (a) September 13, 1997; (b) December 18, 1997; (c) May 2, 1998; and (d) September 17, 1998. The dashed horizontal lines indicate the operational boundaries between the epilimnion, metalimnion, and hypolimnion. Below 200 m, SRP concentrations were determined as the sum of measured SRP and PP, to correct for precipitation of dissolved P due to oxidation during the filtration process; likewise for TDP.

Table 22.2   Vertical distribution of dissolved P concentrations (±95% confidence interval), vertical exchange coefficients, and dissolved P fluxes. The k value and TDP flux for each stratum represent the exchange rate and upward flux between that stratum and the overlying stratum. Flux rates are normalized to lake surface area. TDP concentrations are the means of measurements on four dates between September 13, 1997 and September 17, 1998. For each date, the mean TDP concentration in each stratum was determined as the area-weighted mean of concentrations at multiple depths within that stratum. Sediment fluxes represent the mean of measurements reported by Li et al. (2018) (1.15 ± 0.43 mg P m^{-2} d^{-1}) for four locations ranging from 104 to 650 m, normalized to lake surface area. TDP flux from sediment within the epilimnion is likely high, but cannot be determined with available data.

Stratum	Volume (km³)	Lake Bottom Area (km²)	k (yr⁻¹)	TDP (mg m⁻³)	TDP Flux (mg m⁻² yr⁻¹)
Epilimnion (0–105 m)	2,754	5,924		12.4 ± 4.3	
Metalimnion (105–220 m)	2,310	6,900	0.294	41.2 ± 5.6	667 ± 56
Hypolimnion (220–700 m)	2,711	16,436	0.063	92.7 ± 5.3	301 ± 47
Metalimnion Sediment					335 ± 127
Hypolimnion Sediment					236 ± 90

for ~80% of total recycling. The flux from the metalimnion to the epilimnion is approximately two times that from the hypolimnion to the metalimnion. The greater flux at the shallower interface is expected since this flux will include P transported from the hypolimnion to the metalimnion, as well as any P recycled within the metalimnion.

22.4   DISCUSSION

Virtually all the P that enters Lake Malawi remains in the lake. Total P load is between 388 mg m^{-2} yr^{-1} and 549 mg m^{-2} yr^{-1}, and less than 2% of that leaves the lake via the Shire River outflow (see Figure 22.7). At a whole-lake scale, the P budget is not balanced. The mean P burial rate of 237 mg m^{-2} yr^{-1} is less than half of the estimated total loading rate. Of these two rates, the burial rate is likely the more uncertain because it relies on only five cores that were not well distributed across the lake, and even this small number of cores indicates high spatial variability of burial rates. Rainfall and river discharge during the study period were similar to the long-term average for the region, so there is no reason to suspect that P loading is overestimated. The apparent disparity between total load and burial may also indicate that the lake is not at steady state with regard to P. There is evidence that P loads to Lake Malawi have increased over the past half-century. River basins with high population densities and intensive agriculture have much higher P and sediment export coefficients than those in less inhabited parts of the watershed (Hecky et al. 2003; Otu et al. 2011), and sediment cores indicate the P burial rate has more than doubled in the southern part of the lake since 1960 (Hecky et al. 1999; Otu et al. 2011). Our analysis assumes that none of this P is permanently buried at depths <100 m, but additional coring is needed to confirm this as high burial rates at these shallower depths may account for the missing P sink in our budget.

The effects of increased P load on the lake's phytoplankton and other biota will depend to a large degree on the transport pathways of riverine P immediately after it enters the lake. If river P is retained within the epilimion, the response will likely be immediate and large in magnitude. Conversely, if river P is injected into the metalimnion and hypolimnion, a significant fraction may be permanently buried without contributing to the lake's internal P cycle; the fraction that remains in the water column will be made available to phytoplankton in the epilimnion over time scales of years to decades due to the low vertical mixing rates and the long hypolimnetic P residence time

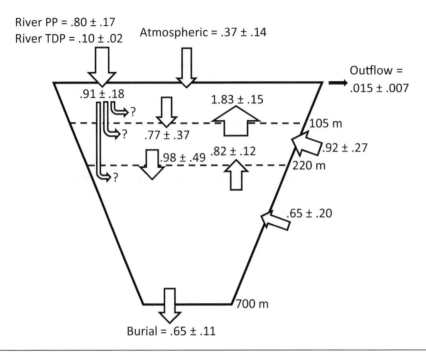

Figure 22.7 Estimate of P loading and internal fluxes for Lake Malawi. Loading estimates are based on measurements of river input and atmospheric deposition in 1997, while sinking and upward flux estimates are based on sediment trap and nutrient profile measurements in 1997 and 1998. All values are mg P m^{-2} d^{-1} ± 95% confidence intervals. The dashed horizontal lines indicate the operational boundaries between the epilimnion, metalimnion, and hypolimnion. Arrow width is proportional to flux rate.

(~24 years). Such a temporal lag poses a challenge to lake management for two reasons: (1) the temporal disconnect between land use and lake response would likely result in a delay of any mitigation efforts; and (2) the lake's response to mitigation efforts would be delayed as it continues to recycle legacy P loaded in previous years. The response of P burial rate will likely exhibit a similar time lag, and so monitoring of external loads, rather than burial rates or in-lake concentrations of P, is the best means of acquiring timely information on whether nutrient loads to the lake are changing.

The comparison of river-water density with density profiles in the lake likely results in an overestimate of the sinking depth of river loads in the southern part of the lake because the gently sloping bottom in this region allows for loss of sediment load and mixing of river water with lake water before river plumes reach great depths (McCullough et al. 2007). Likewise, CTD profiles offshore of river mouths in the central and northern regions of the lake suggest that much of the river inflow is entrained in the metalimnion (Halfman 1993) because of the steeper bottom slopes and the cooler river temperatures in those regions. A comparison of various P fluxes within the epilimnion and metalimnion further highlights the potential importance of transport to, and recycling within, the metalimnion. Total loading to the epilimnion, which includes river inputs, atmospheric deposition, and upward flux from the metalimnion, is greater than 930 mg P m^{-2} yr^{-1}, yet sedimentation as measured in pelagic sediment traps at the base of the epilimnion is only ~280 mg P m^{-2} yr^{-1} (see Figure 22.7). This suggests that significant amounts of P are being transported to the metalimnion in the nearshore zone, possibly as turbidity currents along the lake bottom (Soreghan et al. 1999). Much of this P appears to be recycled within the metalimnion, resulting in a P flux from the metalimnion to the epilimnion (667 mg P m^{-2} yr^{-1}) that is more than twice that of the hypolimnion to the metalimion (301 mg P m^{-2} yr^{-1}) (see Figure 22.7). Much of the difference between these two upward fluxes appears to be accounted for by recycling in metalimnetic sediment which, according to the measurements of Li et al. (2018), approximates 335 mg P m^{-2} yr^{-1}. The importance of the metalimnetic sediment as a site for P recycling may be due to the dominance of the particulate fraction in the riverine P load, which likely reaches the lake bottom much more quickly than dissolved P that is assimilated and recycled within the water column.

Near-bottom advection may be a mechanism by which P is also transported to deep sediments. The deep burial rate, normalized to lake surface area, is estimated as 372 ± 171 mmol P m^{-2} yr^{-1}, which is nearly three times greater than the sedimentation rate measured in sediment traps at 300 m. Particle transport to the deep sediments may be due to direct sinking of sediment-laden turbidity currents originating from river mouths or to resuspension and downslope focusing of epilimnetic and metalimnetic sediments. The presence of subaqueous channels in the central and norther regions (Johnson et al. 1995), and the existence of unstratified homogenites within the sedimentary record (Barry et al. 2002) provide evidence for both of these mechanisms. In Lake Tanganyika, another rift valley lake, there is also evidence for direct loading of river-borne P into the deep hypolimnion of the north basin, resulting in anomalously high concentrations of phosphate and ammonia in the deep water (Hecky et al. 1991).

The large difference between P sedimentation at 180 m and at 300 m (see Figure 22.5) is surprising. The only mechanisms that can account for this difference are recycling of particles as they sink between these two depths, or lateral transport of particles at some depth between these two layers, followed by deposition in other parts of the lake. Using a sequencing sediment trap, Pilskaln (2004) recorded a mean carbon sedimentation rate of 2.92 ± 1.62 mg C m^{-2} d^{-1} at depths of 300 m – 350 m, which is even lower than the 300 m carbon sedimentation rate measured in this study (60.8 ± 1.7 mg C m^{-2} d^{-1}, not normalized to lake surface area). Pilskaln (2004) suggested that resuspension may account for the higher sedimentation rates in shallower traps. It appears that both resuspension and lateral advection result in sedimentation rates that are highly variable in space, both horizontally and vertically. Obviously, the transport and fate of PP in Lake Malawi remain poorly understood, and longer-term measurements with greater spatial coverage are needed to better understand these dynamics.

The important role of internal P loading is highlighted by a comparison of internal fluxes with external load. Vertical flux from the metalimnion to the epilimnion (667 ± 56 mg m^{-2} yr^{-1}) is ~1.4 times the total external load. The estimates of upward fluxes of dissolved P are based on mixing coefficients derived from reverse modeling of CFC-12 profiles (Vollmer et al. 2002). This mixing may be due to a number of mechanisms, including turbulent mixing, entrainment resulting from seasonal cooling and deepening of the thermocline, and upwelling. Upwelling is especially prominent at the lake's southern end, where the energy associated with large internal seiches is translated into high current speeds and turbulence. Hamblin et al. (2003) used a time series of temperature and nutrient profiles to model nutrient flux due to turbulence and upwelling in the southern part of the lake (south of the southern sediment trap location; see Figure 22.1). Using data for three separate dates between March and May 1997, they estimated upwelling flux rates at the base of the euphotic zone (~60 m) ranging from 6 mg P m^{-2} d^{-1} to 120 mg P m^{-2} day^{-1}. The area over which upwelling occurs is variable, but temperature profiles and satellite imagery indicate that the upwelling-influenced region can extend up to 200 km from the lake's southern end (Hamblin et al. 2003). If upwelling is assumed to occur in the region south of the southern sediment trap location (114 km from the lake's southern end) for three months of the year (the southeast trade winds are strongest between June and September), then the previously mentioned P upwelling rates are equivalent to 93 mg P m^{-2} yr^{-1} to 1,767 mg P m^{-2} yr^{-1} when normalized to the entire lake surface area. By comparison, the total upward flux into the epilimnion based on vertical TDP gradients and CFC-12-derived mixing rates is 667 ± 56 mg m^{-2} yr^{-1}. A more precise estimate of upwelling P flux will require measurements with greater temporal resolution, but the aforementioned analysis suggests that a large portion of the internal loading of P to the epilimnion occurs during upwelling at the lake's southern end. This mechanism effectively scavenges much of the P load from the lake's entire watershed and eventually funnels a large proportion of bioavailable P into the southern region of the epilimnion. As a result, plankton production and fishery catches are much greater in the southeast and southwest arms of the lake than in other regions (Bootsma 1993a, b; Weyl et al. 2010).

The use of CFC-12-derived mixing rates and dissolved nutrient profiles to determine vertical nutrient fluxes assumes that dissolved nutrients behave conservatively within the hypolimnion and metalimnion. This approach cannot be used to determine dissolved N flux from the hypolimnion to the metalimnion because dissolved N is lost to coupled nitrification-denitrification near the oxic-anoxic interface (Bootsma and Hecky 1993; Hecky et al. 1996). There is some evidence that this interface may also be a partial barrier to vertical P flux. Within the anoxic hypolimnion, dissolved iron is present in its reduced form (Fe^{2+}). As this mixes near the oxic-anoxic interface, it oxidizes and precipitates as iron oxyhydroxides, which can adsorb and/or co-precipitate phosphate. This precipitate will sink, and re-dissolve on reaching a lower redox potential. This ferrous wheel has been observed in other lakes (Campbell and Torgersen 1980) and in marine systems with deep anoxic layers (Shaffer 1986; Dellwig et al. 2010; McParland et al. 2015). Two lines of evidence suggest this may occur in Lake Malawi. First, when water samples collected from below the oxic-anoxic interface are filtered immediately after collection, a brown-orange precipitate forms on the filter. Analysis of this precipitate indicates that it has a high Fe and P content, with an Fe:P ratio of ~4.0 (see Figure 22.8). This is formed by oxidation during the filtering process; the amount of precipitate formed increases with the length of time samples are exposed to ambient air. Second, on most dates when deep profiling was conducted, an SRP minimum was observed near the oxic-anoxic interface (see Figure 22.9).

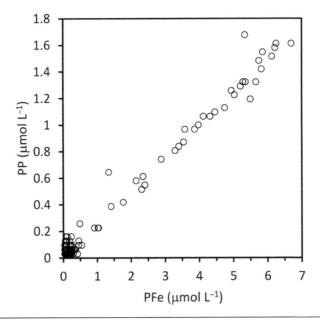

Figure 22.8   Particulate P and Fe concentrations measured in precipitate that accumulated on filters through which water samples from 165–700 m were filtered. No trend of particulate P or Fe with depth was observed, with most values of particulate Fe below 200 m being between 4.0 and 6.5 µmol L^{-1}.

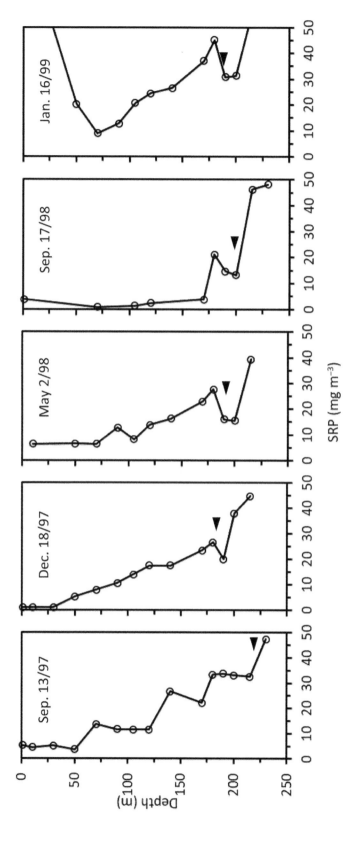

Figure 22.9   SRP profiles near mid-lake on five dates listed at the top of each graph. Measurements are plotted to the depth of the nitrate minimum, below which it is assumed that Fe exists in the reduced form (resulting in precipitation of SRP during filtration). Black arrows indicated the depth of the oxic-anoxic interface on each date.

This minimum may be due in part to P uptake by chemoautotrophic bacteria, as appears to occur in Lake Kivu (Haberyan and Hecky 1987), but the presence of a Fe-P precipitate on filters suggests that abiotic precipitation is also occurring. Currently, the rate of SRP loss at the oxic-anoxic interface and the significance of this loss within the lake's P cycle remain uncertain. Unlike dissolved inorganic N concentrations, which approach zero just below the oxic-anoxic interface (Bootsma and Hecky 1993), SRP concentrations at this depth remain well above those usually observed in the epilimnion, and thus, vertical mixing of hypolimnetic water remains an important internal loading mechanism.

The findings presented here offer several major insights into P dynamics in Lake Malawi:

1. P supply available to phytoplankton is dominated by internal loading. Annual flux from the metalimnion to the epilimnion is approximately twice as great as the total external load from rivers and the atmosphere.

2. A large fraction of external load appears to directly enter depths below the epilimnion. Hence the internal load of P from the metalimnion to the epilimnion likely meets a much larger fraction of phytoplankton P uptake than inferred from a comparison with total external load.

3. Much of the internal P recycling occurs within the sediment. In the hypolimnion, the sediment-water P flux accounts for > 70% of the flux from the hypolimnion to the metalimnion. Likewise, in the metalimnion, the sediment-water flux of 0.92 ± 0.27 mg P m^{-2} d^{-1} accounts for most of the difference between the hypolimnion-metalimnion flux (0.82 ± 0.12 mg P m^{-2} d^{-1}) and the metalimnion-epilimnion flux (1.83 ± 0.15 mg P m^{-2} d^{-1}). The relatively high proportion of recycling that occurs within the sediments may be due to the dominance of PP in the river P load.

4. Atmospheric deposition accounts for about one third of the total external P load. However, the effect of atmospheric P deposition on plankton production is likely greater than inferred from a direct comparison with river load because much of the atmospheric P appears to be bioavailable and, unlike much of the river load, it is retained in the epilimnion for immediate uptake by phytoplankton. Both the atmospheric and river loads of P are greater than earlier estimates that were based on sparser data sets (Bootsma et al. 1996), but both the earlier and current estimates highlight the significance of atmospheric deposition as a P source to Lake Malawi. Similar observations have been made for Lake Tanganyika (Langenberg et al. 2003) and Lake Victoria (Tamatamah et al. 2005).

Our results also highlight several knowledge gaps that need to be addressed before we can confidently predict the response of Lake Malawi to external stressors such as climate change and land-use change. If warming is leading to stronger stratification, decreases in vertical nutrient flux and declines in plankton and fish production can be expected—as appears to be occurring in Lake Tanganyika (Cohen et al. 2016). However, Lake Malawi differs from Lake Tanganyika in some significant ways, including a smaller but higher-altitude drainage basin (Bootsma and Hecky 1993), greater catchment rainfall, less cloud cover, and greater evaporation (Nicholson and Yin 2002). While there is strong evidence that Lake Malawi is warming (Vollmer et al. 2005), the data are insufficient to indicate whether the hypolimnion is warming faster or slower than the surface waters. In contrast to observations for Lake Tanganyika, there is some evidence that warming is weakening the stratification in other African lakes (Marshall et al. 2013; Mahere et al. 2014). If this occurs in Lake Malawi, internal P loading and plankton production might be expected to increase. The magnitude of plankton and fish community response to any changes in internal P loading requires a more complete understanding of vertical and lateral mixing mechanisms (turbulent diffusion, entrainment, upwelling, currents) and their relation to meteorological conditions.

Temporal fluctuations in plankton abundance (Bootsma 1993a, b) and plankton carbon isotopes (Ngochera and Bootsma 2011) indicate that there is a seasonal response to both internal nutrient loading and river loading, with the response to internal loading being more pronounced. While internal P loading is the primary mechanism that regulates plankton production on seasonal and annual time scales, the lake's productivity is ultimately set by external P loads. Because much of the river P load appears to be injected to depths below the euphotic zone, the plankton response to river P load is modulated by internal cycling processes. To predict the lake's response to changes in external P load, there are several critical questions that need to be addressed:

1. What is the fate of river-loaded particulate P? Where is it transported, and what fraction is recycled versus permanently buried?
2. How might the transport of river-loaded P and the vertical flux of P within the lake be affected by continued climate warming? Specifically, will the P supply to phytoplankton within the epilimnion increase or decrease?
3. What is the time scale over which plankton production responds to river P load?

If a large portion of river P load is cycled through the metalimnion and hypolimion before being made available to phytoplankton, the lake's response to fluctuations in external loading will be dampened and the impact of any increased P load on phytoplankton production will take several decades to be fully expressed. Likewise, the lake's response to any mitigation efforts will also lag behind any decrease in river P loads, as highlighted by Bootsma and Hecky (2003). Continuous monitoring of river nutrients, along with the development and application of physical/biogeochemical models to predict lake response (e.g., Lam et al. 2002), are needed to preempt any long-term changes in land use that might lead to eutrophication.

Lake Malawi supports a fishery that serves as a critical food source for the region. Management of this fishery relies primarily on catch and effort data, but the acquisition of reliable data is difficult (Weyl et al. 2010). An important fishery species is the small, pelagic cyprinid *Engraulicypris sardella*, a planktivore whose abundance is affected little by fishing pressure. Rather, because it is r-selective, with a density-independent larval survival (Weyl et al. 2010), the recruitment of *E. sardella* is strongly linked to zooplankton abundance, with the abundance of both zooplankton and *E. sardella* larvae reaching maxima during the June–September mixing season in Lake Malawi (Thompson and Irvine 1997). Hence, short-term fluctuations in this component of the fishery appear to be strongly regulated by internal nutrient loading. Other longer-lived species that feed on *E. sardella* might also be expected to respond to changes in internal nutrient loading, albeit over longer time scales (Weyl et al. 2010), as has been observed for some marine fisheries (Iverson 1990). While there is a real need for better catch and effort data, fishery management may also benefit by accounting for the effect of both external and internal nutrient loading on the year-class strength of lower trophic-level fish species.

22.5 CONCLUSION

Lake Malawi is an immense natural resource that serves as a vital regional food source, contains a valuable sedimentary record, and is home to the world's most diverse array of freshwater fish species. Its large volume and long residence time, as well as minimal industrial development within its watershed, have buffered it to some degree against external stressors. However, the sedimentary record indicates that nutrient loading from parts of the watershed have increased over the past half-century and continues to increase, and long-term temperature records indicate the lake is warming. Both of these stressors can be expected to affect P dynamics within the lake, with consequences for water quality, fish production, and biodiversity. The data presented here highlight the importance of internal loading in the lake's P cycle and provide insight into the temporal and spatial scales over which the lake responds to external P loading. But several critical processes remain poorly understood. These include the effect of climate warming and changes in rainfall patterns on lake thermal structure, the transport of river-borne P after entering the lake, the source and magnitude of atmospheric P deposition, the fate of allochthonous particulate P after it enters the lake, and the role of a potential barrier to upward P flux at the oxic-anoxic interface. Because of the lake's slow response rate to changes in external P load, management of the lake and its watershed must be guided by monitoring of river nutrient loads, as well as by physical/biogeochemical models that can predict the effects of changes in nutrient loads and meteorological forcing.

22.6   ACKNOWLEDGMENTS

This research was conducted as part of the SADC/GEF Lake Malawi/Nyasa/Niassa Biodiversity Conservation Project, with funding provided by the Canadian International Development Agency (now merged with Global Affairs Canada), the Global Environmental Facility, and the UK Department for International Development. Analyses of particulate P and Fe samples were conducted at the Freshwater Institute (Fisheries and Oceans Canada) under the direction of M. Stainton. We are grateful to B. Mwichande, J. Mwita, M. Kingdon, S. Smith, R. Mollot, and S. Page for assistance with field sampling and chemical analyses in Malawi. We are also grateful for the support of Capt. M. Day and the rest of the crew of the R/V Usipa.

22.7   REFERENCES

Alin, SR and Johnson, TC. 2007. Carbon cycling in large lakes of the world: A synthesis of production, burial, and lake-atmosphere exchange estimates. Global Biogeochem Cycles. 21(3): GB3002, doi:10.1029/2006GB002881.

Baker, JE; Eisenreich, SJ; and Eadie, BJ. 1991. Sediment trap fluxes and benthic recycling of organic carbon, polycyclic aromatic hydrocarbons, and polychlorobiphenyl congeners in Lake Superior. Environ Sci Technol. 25(3):500–509.

Barry, SL; Filippi, ML; Talbot, MR; and Johnson, TC. 2002. Sedimentology and geochronology of late Pleistocene and Holocene sediments from northern Lake Malawi. In: Odada, EO and Olago, DO, (Eds.). *The East African Great Lakes: Limnology, Palaeolimnology and Biodiversity*. Kluwer, Dordrecht. p. 369–391.

Bland, SJ and Donda, SJ. 1995. Common property and poverty: fisheries co-management in Malawi. Fisheries Bulletin. 30:1–16.

Bootsma, HA. 1993a. Algal dynamics in an African Great Lake, and their relation to hydrographic and meteorological conditions [dissertation]. [Winnipeg (MB)]: University of Manitoba.

Bootsma, HA. 1993b. Spatio-temporal variation of phytoplankton biomass in Lake Malawi, Central Africa. Verh—Int Ver Theor Angew Limnol. 25(2):882–886.

Bootsma, HA; Bootsma, MJ; and Hecky, RE. 1996. The chemical composition of precipitation and its significance to the nutrient budget of Lake Malawi. In: Johnson, TC, Odada, EO, (Eds.). *The Limnology, Climatology and Paleoclimatology of the East African Lakes*. Amsterdam: Gordon and Breach Publishers. pp. 251–266.

Bootsma, HA and Hecky, RE. 1993. Conservation of the African Great Lakes: a limnological perspective. Cons Biol. 7(3):644–656.

Bootsma, HA and Hecky, RE. 2003. A comparative introduction to the biology and limnology of the African Great Lakes. J Great Lakes Res. 29:3–18.

Bootsma, HA; Hecky, RE; Johnson, TC; Kling, HJ; and Mwita, J. 2003. Inputs, outputs, and internal cycling of silica in a large, tropical lake. J Great Lakes Res. 29(Suppl. 2):121–138.

Bootsma, HA and Jorgensen, SE. 2004. Lake Malawi / Nyasa, p. 36. In: Nakamura M. (Ed.). Managing Lake Basins; Practical Approaches for Sustainable Use. International Lake Environment Committee (ILEC). pp. 259–276.

Campbell, P and Torgersen, T. 1980. Maintenance of iron meromixis by iron redeposition in a rapidly flushed monimolimnion. Can J Fish Aquat Sci. 37:1303–1313.

Chen, CTA and Millero, FJ. 1986. Thermodynamic properties for natural waters covering only the limnological range. Limnol Oceanogr. 31(3):657–662.

Cohen, AS; Gergurich, EL; Kraemer, BM; McGlue, MM; McIntyre, PB; Russell, JM; Simmons, JD; and Swarzenski, PW. 2016. Climate warming reduces fish production and benthic habitat in Lake Tanganyika, one of the most biodiverse freshwater ecosystems. Proc Natl Acad Sci. 113(34):9563–9568.

Cole, JJ; Caraco, NF; and Likens, GE. 1990. Short-range atmospheric transport: a significant source of phosphorus to an oligotrophic lake. Limnol Oceanogr. 35(6):1230–1237.

Dellwig, O; Leipe, T; März, C; Glockzin, M; Pollehne, F; Schnetger, B; Yakushev, EV; Böttcher, ME; and Brumsack, HJ. 2010. A new particulate Mn–Fe–P-shuttle at the redoxcline of anoxic basins. Geochim Cosmochim Acta. 74(24):7100–7115.

Delvaux, D. 1995. Age of Lake Malawi (Nyasa) and water level fluctuations. Mus R Afr Centr Tervuren (Belg) Dept Geol Min Rapp Ann 1993 & 1994:99–108.

Dolan, DM and Chapra, SC. 2012. Great Lakes total phosphorus revisited: 1. Loading analysis and update (1994–2008). J Great Lakes Res. 38:730–740.

Eadie, BJ; Chambers, RL; Gardner, WS; and Bell, GL. 1984. Sediment trap studies in Lake Michigan: resuspension and chemical fluxes in the southern basin. J Great Lakes Res. 10(3):307–321.

Gondwe, MJ; Guildford, SJ; and Hecky, RE. 2008. Planktonic nitrogen fixation in Lake Malawi/Nyasa. Hydrobiologia. 596(1):251–267.

Guildford, SJ; Bootsma, HA; Taylor, WD; and Hecky, RE. 2007. High variability of phytoplankton photosynthesis in response to environmental forcing in oligotrophic Lake Malawi/Nyasa. J Great Lakes Res. 33:170–185.

Guildford, SJ; Hecky, RE; Taylor, WD; Mugidde, R; and Bootsma, HA. 2003. Nutrient enrichment experiments in tropical great lakes Malawi/Nyasa and Victoria. J Great Lakes Res. 29:89–106.

Haberyan, K and Hecky, RE. 1987. The late Pleistocene and Holocene stratigraphy and paleolimnology of Lakes Kivu and Tanganyika. Palaeogeogr Palaeoclimatol Palaeoecol. 61:169–197.

Halfman, JD. 1993. Water column characteristics from modern CTD data, Lake Malawi, Africa. J Great Lakes Res. 19(3):512–520.

Halfman, JD and Scholz, CA. 1993. Suspended sediments in Lake Malawi, Africa: a reconnaissance study. J Great Lakes Res. 19(3):499–511.

Hamblin, PF; Bootsma, HA; and Hecky, RE. 2003. Surface meteorological observations over Lake Malawi/Nyasa. J Great Lakes Res. 29(Suppl. 2):19–33.

Hecky, RE. 2000. A biogeochemical comparison of Lakes Superior and Malawi and the limnological consequences of endless summer. Aquat Ecosyst Health Manag. 3:23–33.

Hecky, R; Bootsma, HA; and Kingdon, M. 2003. Impact of land use on sediment and nutrient yields to Lake Malawi/Nyasa (Africa). J Great Lakes Res. 29 (Suppl. 2):139–158.

Hecky, RE; Bootsma, HA; Mugidde, R; and Bugenyi, FB. 1996. Phosphorus pumps, nitrogen sinks, and silicon drains: plumbing nutrients in the African Great Lakes. In: Johnson, TC and Odada, EO, (Eds.). *The Limnology, Climatology and Paleoclimatology of the East African Lakes*. Amsterdam: Gordon and Breach Publishers. pp. 205–223.

Hecky, RE; Coulter, GW; and Spigel, RH. 1991. The nutrient regime. In: Coulter, GW, (Ed.). *Lake Tanganyika and Its Life*. Oxford University Press. pp.76–89.

Hecky, RE; Kling, HJ; Johnson, TE; Bootsma, HA; and Wilkinson, P. 1999. Algal and sedimentary evidence for recent changes in the water quality and limnology of Lake Malawi/Nyasa. In: Bootsma, HA and Hecky, RE, (Eds.). Water quality report: Lake Malawi/Nyasa Biodiversity Conservation Project. Southern African Development Community/Global Environmental Facility (SADC/GEF). pp. 191–214.

Hecky, RE; Mugidde, R; Ramlal, PS; Talbot, MR; and Kling, GW. 2010. Multiple stressors cause rapid ecosystem change in Lake Victoria. Freshwater Biol. 55:19–42.

Herdendorf, CE. 1982. Large lakes of the world. J Great Lakes Res. 8(3):379–412.

Iverson, RL. 1990. Control of marine fish production. Limnol Oceanogr. 35(7):1593–1604.

Jamu, D; Banda, M; Njaya, F; and Hecky, RE. 2011. Challenges to sustainable management of the lakes of Malawi. J Great Lakes Res. 37:3–14.

Johnson, TC; Barry, SL; Chan, Y; and Wilkinson, P. 2001. Decadal record of climate variability spanning the past 700 yr in the Southern Tropics of East Africa. Geology. 29(1):83–86.

Johnson, TC; Wells, JD; and Scholz, CA. 1995. Deltaic sedimentation in a modern rift lake. Geol Soc Am Bull. 107(7):812–829.

Kalff, J. 1983. Phosphorus limitation in some tropical African lakes. Hydrobiologia. 100(1):101–112.

Kidd, CHR. 1983. A water resources evaluation of Lake Malawi and the Shire River. UNDP Project MLW/77/012, World Meteorological Organization, Geneva.

Kilham, P and Kilham, SS. 1990. OPINION Endless summer: internal loading processes dominate nutrient cycling in tropical lakes. Freshwater Biol. 23(2):379–389.

Kingdon, MF; Bootsma, HA; Mwita, J; Mwichande, B; and Hecky, RE. 1999. River discharge and water quality. In: Bootsma, HA and Hecky, RE, (Eds.). Water quality report: Lake Malawi/Nyasa Biodiversity Conservation Project. Southern African Development Community/Global Environmental Facility (SADC/GEF). pp. 29–69.

Lam, DCL; Leon, L; Hecky, R; Bootsma, H; and McCrimmon, RC. 2002. A modeling approach for Lake Malawi/Nyasa/Niassa: integrating hydrological and limnological data. In: Odada, EO and Olago, DO, (Eds.). *The East African Great Lakes: Limnology, Palaeolimnology and Biodiversity*. Kluwer, Dordrecht. pp. 189–208.

Langenberg, VT; Nyamushahu, S; Roijackers, R; and Koelmans, AA. 2003. External nutrient sources for Lake Tanganyika. J Great Lakes Res. 29 (Suppl. 2):169–180.

Lewis Jr., WM. 1987. Tropical limnology. Annu Rev Ecol Syst. 18(1):159–184.

Li, J; Brown, ET; Crowe, SA; and Katsev S. 2018. Sediment geochemistry and contributions to carbon and nutrient cycling in a deep meromictic tropical lake: Lake Malawi (East Africa). J Great Lakes Res. 44(6):1221–1234.

Mahere, TS; Mtsambiwa, MZ; Chifamba, PC; and Nhiwatiwa, T. 2014. Climate change impact on the limnology of Lake Kariba, Zambia–Zimbabwe. Afr J Aquat Sci. 39(2):215–221.

Marshall, BE; Ezekiel, CN; Gichuki, J; Mkumbo, OC; Sitoki, L; and Wanda, F. 2013. Has climate change disrupted stratification patterns in Lake Victoria, East Africa? Afr J Aquat Sci. 38:(3):249–253.

McCullough, GK; Barber, D; and Cooley, PM. 2007. The vertical distribution of runoff and its suspended load in Lake Malawi. J Great Lakes Res. 33(2):449–465.

McParland, E; Benitez-Nelson, CR; Taylor, GT; Thunell, R; Rollings, A; and Lorenzoni, L. 2015. Cycling of suspended particulate phosphorus in the redoxcline of the Cariaco Basin. Mar Chem. 176:64–74.

Naithani, J; Darchambeau, F; Deleersnijder, E; Descy, JP; and Wolanski, E. 2007. Study of the nutrient and plankton dynamics in Lake Tanganyika using a reduced-gravity model. Ecol Model. 200:225–233.

Ngochera, M; Donda, S; Mafaniso, H; and Berge, E. 2014. Discusson and recommendations for defragmentation of resource management on the Southeast Arm of Lake Malawi. In: Donda, S; Hara, M; Ngochera, M; and Berge, E, (Eds.). *Fragmentation of Resource Management on the South East Arm of Lake Malawi: Dynamics around Fisheries* (Vol. 3). LIT Verlag Münster. pp. 147–173.

Ngochera, MJ and Bootsma, HA. 2011. Temporal trends of phytoplankton and zooplankton stable isotope composition in tropical Lake Malawi. J Great Lakes Res. 37:45–53.

Ngochera, MJ and Bootsma, HA. 2018. Carbon, nitrogen and phosphorus content of seston and zooplankton in tropical Lake Malawi: Implications for zooplankton nutrient cycling. Aquat Ecosyst Health Manag. 21(2):185–192.

Nicholson, SE and Yin, X. 2002. Mesoscale patterns of rainfall, cloudiness and evaporation over the Great Lakes of East Africa. In: Odada, EO and Olago, DO, (Eds.). *The East African Great Lakes: Limnology, Palaeolimnology and Biodiversity*. Kluwer, Dordrecht. pp. 93–119.

North, RP; North, RL; Livingstone, DM; Köster, O; and Kipfer, R. 2014. Long-term changes in hypoxia and soluble reactive phosphorus in the hypolimnion of a large temperate lake: consequences of a climate regime shift. Glob Change Biol. 20:811–823.

Otu, M; Ramlal, P; Wilkinson, P; Hall, RI; and Hecky, RE. 2011 Paleolimnological evidence of the effects of recent cultural eutrophication during the last 200 years in Lake Malawi, East Africa. J Great Lakes Res. 37:61–74.

Pilskaln, CH. 2004. Seasonal and interannual particle export in an African rift valley lake: A 5-yr record from Lake Malawi, southern East Africa. Limnol Oceanogr. 49(4):964–977.

Ribbink, AJ; Marsh, BA; Marsh, AC; Ribbink, AC; and Sharp, BJ. 1983. A preliminary survey of the cichlid fishes of rocky habitats in Lake Malawi. S Afr J Zool. 18:147–310.

Salas, HJ and Martino, P. 1991. A simplified phosphorus trophic state model for warm-water tropical lakes. Water Res. 25(3):341–350.

Scholz, CA; Cohen, AS; Johnson, TC; King, J; Talbot, MR; and Brown, ET. 2011. Scientific drilling in the Great Rift Valley: the 2005 Lake Malawi Scientific Drilling Project—an overview of the past 145,000 years of climate variability in southern hemisphere East Africa. Palaeogeogr Palaeoclimatol Palaeoecol. 303:3–19.

Seehausen, O; Van Alphen, JJ; and Witte, F. 1997. Cichlid fish diversity threatened by eutrophication that curbs sexual selection. Science. 277(5333):1808–1811.

Sene, K; Piper, B; Wykeham, D; Mcsweeney, RT; Tych, W; and Beven, K. 2017. Long-term variations in the net inflow record for Lake Malawi. Hydrol Res. 48(3):851–866.

Shaffer, G. 1986. Phosphate pumps and shuttles in the Black Sea. Nature. 321:515–517.

Sharpley, A; Jarvie, HP; Buda, A; May, L; Spears, B; and Kleinman, P. 2013. Phosphorus legacy: overcoming the effects of past management practices to mitigate future water quality impairment. J Environ Qual. 42:1308–1326.

Snoeks, J and Konings, A. 2004. The Cichlid diversity of Lake Malawi/Nyasa/Niassa. Cichlid Press.

Soreghan, MJ; Scholz, CA; and Wells, JT. 1999. Coarse-grained, deep-water sedimentation along a border fault margin of Lake Malawi, Africa: seismic stratigraphic analysis. J Sediment Res. 69(4):832–846.

Stainton, MP; Capel, MJ; and Armstrong, FAJ. 1977. *The Chemical Analysis of Fresh Water*, 2nd ed. Fish Mar Serv Misc Spec Publ. 25:166.

Tamatamah, RA; Hecky, RE; and Duthie, H. 2005. The atmospheric deposition of phosphorus in Lake Victoria (East Africa). Biogeochemistry. 73(2):325–344.

Thompson, AB and Irvine, K. 1997. Diet-shifts and food-dependent survival in *Engraulicypris sardella* (Cyprinidae) larvae from Lake Malawi, Africa. J Plankton Res. 19(3):287–301.

Tiercelin, J-J and Lezzar, K-E. 2002. A 300 million years history of rift lakes in Central and East Africa: an updated broad review. In: Odada, EO and Olago, DO, (Eds.). *The East African Great Lakes: Limnology, Palaeolimnology and Biodiversity*. Kluwer, Dordrecht. pp. 3–60.

Vollmer, MK; Bootsma, HA; Hecky, RE; Patterson, G; Halfman, JD; Edmond, JM; Eccles, DH; and Weiss, RF. 2005. Deep-water warming trend in Lake Malawi, East Africa. Limnol Oceanogr. 50(2):727–732.

Vollmer, MK; Weiss, RF; and Bootsma, HA. 2002. Ventilation of Lake Malawi/Nyasa. In: Odada, EO and Olago, DO, (Eds.). *The East African Great Lakes: Limnology, Palaeolimnology and Biodiversity*. Kluwer, Dordrecht. pp. 209–233.

Weyl, OL; Ribbink, AJ; and Tweddle, D. 2010. Lake Malawi: fishes, fisheries, biodiversity, health and habitat. Aquat Ecosyst Health Manag. 13(3):241–254.

Wüest, A; Piepke, G; and Halfman, JD. 1996. Combined effects of dissolved solids and temperature on the density stratification of Lake Malawi. In: Johnson, TC and Odada, EO, (Eds.). *The Limnology, Climatology and Paleoclimatology of the East African Lakes*. Amsterdam: Gordon and Breach Publishers. p. 183–202.

Section III

Integration and Synthesis

CHAPTER 23

FACTORS INFLUENCING INTERNAL PHOSPHORUS LOADING: A META-ANALYSIS

Emily Kindervater, Nicole Hahn, and Alan D. Steinman[1]

Abstract

Greater attention is being devoted to internal phosphorus (P) loading in water bodies around the world, with the growing recognition that this release of P from sediments can delay aquatic ecosystem recovery for years or decades. Studies have identified various factors as important in driving internal P loading—such as temperature, dissolved oxygen (DO) concentration, wind-wave action, and labile sediment P. We examined over 1,350 studies from the peer-reviewed literature that studied internal P loading from freshwater lakes and reservoirs, and retained 27 studies from 29 lakes for analysis. Criteria for retention included studies that: measured internal loading with sediment core incubations; were not undergoing any type of in-lake restoration; included sediment cores that were not manipulated prior to or during incubation; and reported original data. Both total P (TP) and filtered P release rates required four principal components to explain ~75% of the variance. Maximum lake depth (positive), lake surface area (positive), DO (negative), and incubation temperature (positive) were strongly related to TP release rate. Surface area, maximum depth, sampling duration, and DO level were all negatively related to filtered P release rate. The most explanatory multiple linear regression models identified three factors consistent in both TP and filtered release rates: DO, incubation temperature, and latitude. Despite the considerable variability found in our analyses, it appears that DO levels and temperature are master factors influencing internal P loading. We provide recommendations for future studies, which will facilitate comparisons among lakes.

Key words: Internal phosphorus loading; meta-analysis; dissolved oxygen; temperature

[1] Annis Water Resources Institute, Grand Valley State University, 740 West Shoreline Drive, Muskegon, MI 49441, USA.
E-mail: steinmaa@gvsu.edu

23.1 INTRODUCTION

Phosphorus (P) loading can be generically considered as all physical, chemical, and biological processes by which P is mobilized and translocated from the benthic environment to the water column (see Chapter 1). The role of internal P loading in the eutrophication of water bodies, as well as in delaying their recovery following watershed restoration actions, has gained increasing attention throughout the world over the past few decades (Sas 1989; Søndergaard et al. 2003; Jeppesen et al. 2005; see Chapters 6–22). Despite this increasing recognition as a potential source of P to water bodies and its attendant effect on stimulating algal blooms, very few studies have attempted to formally examine all published data for common and emerging trends. Prior studies have implicated temperature, DO, trophic status, and lake morphometry, among others, as potentially important factors (Orihel et al. 2017), but to the best of our knowledge, this is the first formal analysis to examine these factors and others on a broad scale. Carey and Rydin (2011) examined how burial patterns of sediment P relate to lake trophic state, but did not explicitly address release rates.

Meta-analysis is a common approach to utilize formal statistical techniques for summarizing the results of independent experiments (Hedges et al. 1999). Although response ratio (the ratio of the mean outcome in the experimental group compared to that in the control group) is often used as an indicator of how large an effect is occurring, for our analysis there was no true control group. Hence, we used principal component analysis (PCA) and multiple linear regression to identify factors that best explain patterns of internal P loading in lakes with a focus on sediment P flux measurements. This type of information can be extremely valuable when it comes to management and control of internal P loading.

23.2 METHODS

23.2.1 Approach

Our meta-analysis began with a search of the Web of Science database using the search term *phosphorus* with each of the following in turn: *internal load, internal release,* or *internal flux*. These searches returned a total of 1,357 studies through August 2018. Abstracts were used to eliminate any paper not related to aquatic science. Then, more detailed readings of each paper were required to determine if each met the following criteria. Papers were deemed usable, if:

1. Full published papers were accessible through Grand Valley State University Library or Web of Science database;
2. The study sampled freshwater lakes, reservoirs, or ponds (hereafter all termed *lakes*);
3. The study included incubations of sediment to measure a P release rate;
4. No in-lake restoration or treatment such as alum, Phoslock®, hypolimnetic pumping, O_2 injection, destratification via machine, dredging, etc., was applied prior to core sampling;
5. No homogenization, drying, freezing, or resuspension of core sediments occurred prior to or during incubation;
6. The study reported original data;
7. The methods and reporting were clear enough for accurate data extraction.

The studies included in our analysis were incubations (including benthic chambers and continuous flow-through incubations) of undisturbed sediment for measurement of P release rates (either originally reported in mg P m^{-2} d^{-1} or converted from original units). All physical, chemical, and biological data thought to possibly influence internal P loading were recorded. After data mining was concluded, data were cleaned to ensure consistent units within each factor. The data were separated by P type:

1. TP;
2. Filtered P—consisting of soluble reactive P (SRP), dissolved reactive P (DRP), or dissolved inorganic P (DIP).

A flow chart (see Figure 23.1) shows our approach for processing the studies, which resulted in a reduction from the original 1,357 publications down to the final 27. The factors included in the final dataset, based on availability in all studies were: waterbody type (lake or reservoir), latitude, longitude, lake surface area, maximum lake depth, trophic level, sampling year, sampling season, core length (depth of sediment in core), incubation condition (level of overlying water disturbance during incubation), DO level, incubation temperature, sampling days (incubation duration), and P release rate. All data were quantitative or ordinal except for waterbody type. We used the following rules when finalizing the data for analysis:

1. For release rates below detection levels, half the detection limit value was used in analysis;
2. When a range of values was reported (e.g., multiple observations in time), the median value was used in analysis;
3. Time frames reported as one month were considered to be 30 days;
4. For morphometric characteristics that were not reported, data from other published papers on the same lake were used as a proxy;
5. If latitude and longitude were not reported and could not be found within another published study, the lake was located (if possible) on Google Maps and the latitude and longitude recorded from that source;
6. Seasons were categorized as following: Spring (March–May); Summer (June–August); Fall (September–November); and Winter (December–February) (for northern hemisphere);
7. DO concentration was grouped into levels: < 2 mg L^{-1} = Hypoxic, 2–4 mg L^{-1} = Low DO, > 4 mg L^{-1} = High DO;
8. Graphed data were extracted using the online software WebPlotDigitizer, version 4.1 (Sabater et al. 2018);
9. Sediment cores taken from multiple sites within a single lake were averaged to reduce the influence of heavily sampled lakes, as long as they were incubated under similar conditions.

23.2.2 Data Analysis

Each analysis was conducted separately but with identical methods for TP and filtered P datasets. Initial correlations and histograms of all remaining factor pairs were used to identify patterns and potential problems with each dataset. PCA was conducted in scaling = 2 (i.e., graphing distances are based on the position of the factors) to estimate the amount of variation explained by the factors: latitude, surface area, maximum depth, trophic level, sampling season, core length, incubation condition, DO level, incubation temperature, sampling days, and P release rate. Longitude and sampling year were not included in our analyses because any part of variance explained would be artificial; variance would be due to lake sampling availability and human choice, not because of direct impacts of time and climate on release rate. Since the ordinal factors were semi-quantitative, they were included in this analysis (Borcard et al. 2011).

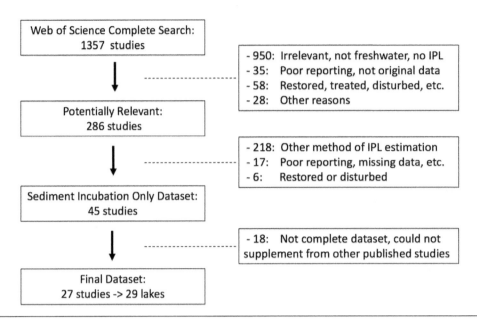

Figure 23.1   Flowchart of data compiling and exclusion. IPL = internal P loading.

Multiple linear regression was conducted on each data set for all factors: latitude, surface area, maximum depth, trophic level, sampling season, core length, incubation condition, DO level, incubation temperature, sampling days, and P release rate. Variance inflation factors, Q-Q norm plots, and Bonferonni outlier tests were used to identify multicollinearity and outliers. Release rate, surface area, maximum depth, and sampling days (for TP only) were log-transformed. After transformation and removal of problematic data and factors, the linear model was re-executed and all potential models were compared using Akaike's Information Criterion (AIC). All analyses were conducted in R (3.4.2).

23.3   RESULTS AND DISCUSSION

The final data set included 29 lakes in 27 studies: 10 studies represented 10 lakes for TP release and 17 studies represented 19 lakes for filtered P release (see Figure 23.2). Although this dataset represents a small fraction of all the internal P loading work being conducted around the globe, by comparing studies that employed the same methodology, we remove the uncertainty and bias that is present in the complete dataset (see Table 23.1). In addition, we selected the methodology (sediment core incubations) that was employed most often to measure internal P loading, in an intentional attempt to boost the number of observations for our analysis. Unfortunately, constraining ourselves to sediment core incubations eliminated the relatively few studies that have taken place in the southern hemisphere (see Table 23.2).

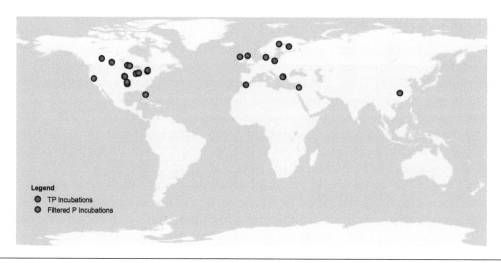

Figure 23.2 Map of included incubation study lakes

Table 23.1 Number of internal P loading (IPL) studies by each IPL method

Measurement Method	Number of Studies
Incubation and Benthic Chambers	79
Mass Balance	47
Modeling	41
Fractions & Isotopes	29
Peeper and Diffusive Gradients in Thin Films	14
Sedimentation & Resuspension	14
Water Column	8
Pore Water and Sediment, Fick's Law	8
Redox	2
Multiple	36

Table 23.2 Number of internal P loading studies per continent. Those numbers in parentheses represent studies that occurred on continent boundaries or contained lakes from multiple continents. All eight African studies were conducted in either Morocco or Egypt. Four of the six Australian studies were conducted in New Zealand. The studies in South America were conducted in Brazil.

Continent	Number of Studies
Europe	114 (+6)
North America	91 (+1)
Asia	50 (+5)
Africa	8
Australia	6
South America	3
Total Studies	278

23.3.1 PCA Analyses

The PCA analyses for TP resulted in complex ordination plots with four main principal components (PC). These four components accounted for 75.6% of the variation, with the graphed PC1, PC2, and PC3 accounting for 27.4%, 20.4%, and 18.3%, respectively (see Figure 23.3). In the first biplot (see Figure 23.3a, PC1 versus PC2), TP release rate is closely aligned with lake surface area and maximum depth along the PC1 axis, and negatively correlated mostly with DO level. The PC2 versus PC3 biplot (see Figure 23.3b) shows release rate aligned closely to trophic level and incubation temperature. Most of the reservoirs aggregated in similar ordination space, with one outlier in both biplots, although there was no clear relationship to any particular environmental factor.

The PC1 versus PC2 biplot (see Figure 23.3a) explained almost half the total variance of TP; therefore, the close relationship between TP release rate and both lake surface area and maximum lake depth is worth exploring. Deeper lakes are more likely to stratify, and therefore experience low DO conditions, promoting internal loading in those systems where P release is influenced by redox-sensitive reactions (Mortimer 1941). We speculate the positive relationship between TP release rate and lake surface area is indirect, and a function of lake depth and surface area being correlated (Hayes 1957). The negative relationship between release rate and DO level is consistent with the influence of desorption of P from redox-sensitive minerals. The PC2 versus PC3 ordination (see Figure 23.3b) also accounted for a substantial amount of the TP release-rate variance (38.7%). The relationship with trophic level is consistent with the analysis by Orihel et al. (2017), who found trophic state, along with oxygen and pH, to be the main drivers of internal loading variation in Canadian lakes. The positive relationship with incubation temperature likely reflects increased metabolic rates of heterotrophs in the sediment, utilizing ferric iron as an electron acceptor in respiration, leading to greater pools of available P as the P desorbs from iron (oxy)hydroxides. Field and laboratory studies have shown that internal loading rates are highest during warmer months (Steinman et al. 2009; Spears et al. 2011; James 2017; see Chapters 10, 13–15), consistent with this finding.

The PCA results for filtered P exhibited both similarities and differences to those for TP. Again, there were four main principal components that explained most of the variance: 73.5% in total (see Figure 23.4a, b, graphing only PC1-PC3), with approximately half the variation explained by the first two components. In the PC1 versus PC2 biplot (see Figure 23.4a), the P release rate appears to be negatively correlated mostly with lake surface area, maximum depth, sampling duration, and DO level. Some of these relationships may be an artifact of the incubation and not related to real-world circumstances. For example, the negative relation of release rate with sampling duration may be due to the accumulation of P in the water column over time, which decreases the magnitude of the P concentration gradient between sediment and water, resulting in lower P release. The negative relation of filtered P release rate with lake surface area and maximum depth contrasts with their positive relation with TP release rate (see Figure 23.3a, b); this may be due to filtered P being more bioavailable than TP, resulting in lower water-column concentrations, and an apparently lower P release rate. Unfortunately, too few studies reported whether the incubation water was filtered or not, so we were unable to determine if algal uptake might have played a role. The PC2 versus PC3 biplot (see Figure 23.4b) accounted for 30.6% of the variance, with TP release rate negatively related to DO level; as previously discussed, this is consistent with stratified lakes, when low DO levels influence the dissolution of iron (oxy)hydroxides, thus releasing bound P (Mortimer 1941).

Even when collapsing our database of studies by selecting for those that measured internal loading via sediment core incubations, the PCA nonetheless resulted in four important principal components. This suggests there is considerable variability among waterbodies and that no single factor, or set of factors, is exhibiting undue influence on P release rates.

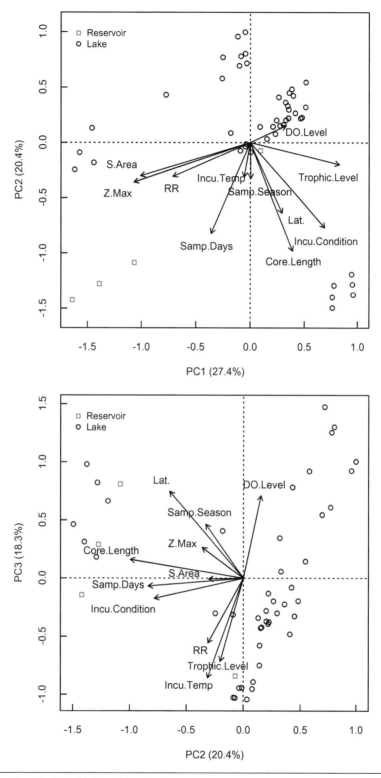

Figure 23.3a,b Total P PCA biplot with principal component percent variation explained. (a) PC1 vs. PC2; (b) PC2 vs. PC3; RR = release rate; Z.Max = maximum waterbody depth.

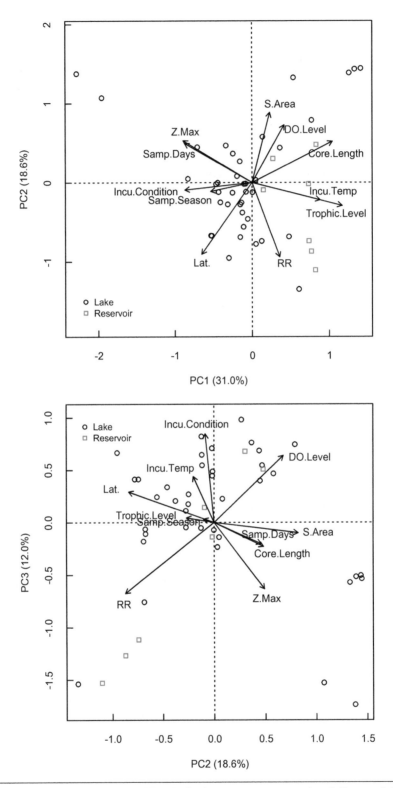

Figures 23.4a,b Filtered P PCA biplot with principal component percent variation explained. (a) PC1 vs. PC2; (b) PC2 vs. PC3; RR = release rate; Z.Max = maximum waterbody depth.

23.3.2 Multiple Linear Regression Analyses

Multiple linear regression was used to determine general trends between the TP release rate and the other lake and incubation factors (see Table 23.3). After testing for multicollinearity, maximum depth was removed from the regression. This regression model identified DO level, latitude, incubation temperature, surface area, and core length as significant factors. Using forward-backward AIC, the best-fit model with the fewest factors was [logRR.4 ~ DO.Level + Lat. + logS.Area + logLength + Incu.Temp] with an AIC value of −89.44. The next best model also included incubation condition with an AIC of −88.45 (see Table 23.3).

The starting regression for filtered-P identified latitude, DO level, incubation condition, and maximum depth as significant factors (see Table 23.4). Using forward-backward AIC, the best fit model with the fewest factors was [logFiRR.4 ~ Trophic.Level + DO.Level + Incu.Condition + Lat. + logFiZ.Max + Samp.Days + Incu.Temp] with an AIC value of −79.48. The next-best model also included surface area with an AIC of −78.3 (see Table 23.4).

Table 23.3 Total P multiple linear regression models ordered by AIC. Significant factors from full model were DO level, latitude, surface area, core length, and incubation temperature (p = < 0.0001, 0.010, 0.012, 0.015, 0.024, respectively.)

Model	Adj. R²	F-statistic	d.f.	p	AIC
DO + Lat. + logS.Area + logCore.Length + Incu. Temp	0.531	11.85	5, 43	< 0.0001	−89.44
DO + Lat. + logS.Area + logCore.Length + Incu. Temp + Incu.Condition	0.529	9.994	6, 42	< 0.0001	-88.45
DO + Lat. + logS.Area + logCore.Length + Incu. Temp + Incu.Condition + logSamp.Days	0.532	8.787	7, 41	< 0.0001	−87.89
DO + Lat. + logS.Area + logCore.Length + Incu. Temp + Incu.Condition + logSamp.Days + Trophic. Level	0.531	7.779	8, 40	< 0.0001	−86.97
DO + Lat. + logS.Area + logCore.Length + Incu. Temp + Incu.Condition + logSamp.Days + Trophic. Level + Season	0.523	6.845	9, 39	< 0.0001	−85.43

Table 23.4 Filtered P multiple linear regression models ordered by AIC. Significant factors from full model were latitude, DO level, incubation condition, and maximum depth (p = 0.0001, 0.003, 0.016, 0.030, respectively.)

Model	Adj. R²	F-statistic	d.f.	p	AIC
Trophic.Level + DO + Incu.Condition + Lat. + logMax.Depth + Samp.Days + Incu.Temp	0.556	9.055	7, 38	< 0.0001	−79.48
Trophic.Level + DO + Incu.Condition + Lat. + logMax.Depth + Samp.Days + Incu.Temp + logS. Area	0.556	8.054	8, 37	< 0.0001	−78.73
Trophic.Level + DO + Incu.Condition + Lat. + logMax.Depth + Samp.Days + Incu.Temp + logS. Area + Core.Length	0.545	6.986	9, 36	< 0.0001	−76.81
Trophic.Level + DO + Incu.Condition + Lat. + logMax.Depth + Samp.Days + Incu.Temp + logS. Area + Core.Length + Samp.Season	0.534	6.152	10, 35	< 0.0001	−75.00

The final multiple regression models for TP and filtered P both included incubation temperature, DO, and latitude. As previously noted, the mechanisms explaining why internal P loading is influenced by both DO (negatively) and temperature (positively) are well-known. The positive relation of latitude was weak for TP (slope = 0.055) but stronger for filtered P (slope = 0.368), although there was considerable variability in release rate. The lakes with the highest release rates tended to occur in middle latitudes (42° to 55°N), but even at those latitudes, there were many lakes with low release rates. The paucity of lakes below 30° and above 70°N, as well as any systems from the southern hemisphere, preclude any robust conclusions regarding latitude.

23.3.3   Caveats

We caveat this meta-analysis because our dataset is not fully representative of internal P-loading measurements across the world or of all the factors that can potentially impact P release from sediments. Our meta-analysis identified several trends from past studies that limited our ability to robustly generalize our findings:

1. P release and uptake rates that are measured to be negligible or negative are less likely to be reported than positive release rates;
2. Conductivity and pH, despite being important to sediment P release (Christophoridis and Fytianos 2006; Orihel et al. 2017), are infrequently measured and even less frequently reported;
3. The bubbling of CO_2 to decrease DO levels in sediment core tubes results in a decrease in the pH of overlying water when performed without a buffer. While an expected outcome, it is another reason to report incubation pH since reducing DO levels with this method impacts more than just redox-dependent P release (Christophoridis and Fytianos 2006);
4. Many studies stated that a variety of parameters were measured, both *in situ* and during the incubations, but did not report the results, even as supplementary information;
5. Reporting of sediment core handling prior to and during incubations was inconsistent and often vague; any disturbance, homogenization, drying, or freezing of sediments will affect the sediment release rate in comparison to *in situ* conditions (Kelton and Chow-Fraser 2005).

We restricted our analysis to sediment flux incubations, which provided sufficient consistency in methodology to allow reasonable interpretation of patterns. However, laboratory-based sediment incubations do not provide a whole-system understanding of internal P loading. Ideally, researchers would implement a combination of approaches that include experimental manipulations to test for P release potential under differing environmental conditions (e.g., temperature, DO, light, physical disturbance) (cf. Steinman et al. 2006) and *in situ* measurements (see Chapters 3, 5) that more accurately reflect lake P dynamics.

23.4   CONCLUSIONS AND RECOMMENDATIONS

There was considerable variability in which factors best explained internal release from sediment core incubations; however, it appears that DO levels and temperature are master factors. This finding is consistent with prior studies conducted on individual lakes. However, our dataset was limited in geographic scope, so there is a need for more studies in tropical, subtropical, and polar lakes, especially in the southern hemisphere, in order to obtain a more comprehensive global picture.

In addition to the need for a broader geographic array of lakes, the variation and lack of consistency in the type of factors reported from the sediment core incubation studies limited our analysis.

Some important factors are difficult to measure and understandably are not often reported. However, for sediment core incubation experiments, the measurement and reporting of DO, bottom water TP, sediment chemistry (including P and metals), conductivity, pH, and temperature are strongly recommended, as all of these factors have been shown to influence P release from sediments (Christophoridis and Fytianos 2006; Jiang et al. 2008; Orihel et al. 2017).

We also recommend the formation of an international committee to identify the common factors to be measured for future internal P loading studies, including measurement methodology. A consistent set of standards will facilitate comparisons across the planet and allow for a more robust analysis of trends, identification of global drivers, and recommendations for management options.

23.5 ACKNOWLEDGMENTS

We are very grateful to Jim McNair and Meg Woller-Skar for their advice on statistical analyses and Bryan Spears for his editorial review, which greatly improved the paper. Clark Meston assisted with literature review. The full dataset is available upon request from the authors.

23.6 REFERENCES

Borcard, D; Gillet, F; and Legendre, P. 2011. *Unconstrained Ordination.* "Numerical Ecology with R." New York (NY): Springer.

Carey, CC and Rydin, E. 2011. Lake trophic status can be determined by the depth distribution of sediment phosphorus. Limnol Oceanogr. 56:2051–63.

Christophoridis, C and Fytianos, K. 2006. Conditions affecting the release of phosphorus from surface lake sediments. J Environ Qual. 35:1181–1192.

Hayes, FR. 1957. On the variation in bottom fauna and fish yield in relation to trophic level and lake dimensions. J Fish Res Bd Can. 14:1–32.

Hedges, LV; Gurevitch, J; and Curtis, PS. 1999. The meta-analysis of response ratios in experimental ecology. Ecology. 80:1150–1156.

James, WF. 2017. Diffusive phosphorus fluxes in relation to the sediment phosphorus profile in Big Traverse Bay, Lake of the Woods. Lake Reserv Manage. 33:360–368.

Jeppesen, E; Søndergaard, M; Jensen, JP; Havens, KE; Anneville, O; Carvalho, L; Coveney, MF; Deneke, R; Dokulil, MT; Foy, BO; et al. 2005. Lake responses to reduced nutrient loading–an analysis of contemporary long-term data from 35 case studies. Freshwater Biol. 50:1747–71.

Jiang, X; Jin, X; Yao, Y; Li, L; and Wu, F. 2008. Effects of biological activity, light, temperature and oxygen on phosphorus release processes at the sediment and water interface of Taihu Lake, China. Water Res. 42:2251–2259.

Kelton, N and Chow-Fraser, P. 2005. A simplified assessment for factors controlling phosphorus loading from oxygenated sediments in a very shallow eutrophic lake. Lake Reserv Manage. 21:223–230.

Mortimer, CH. 1941. The exchange of dissolved substances between mud and water in lakes. J Ecol. 29:280–329.

Orihel, DM; Baulch, HM; Casson, NJ; North, RL; Parsons, CT; Seckar, DC; and Venkiteswaran, JJ. 2017. Internal phosphorus loading in Canadian fresh waters: a critical review and data analysis. Can J Fish Aquat Sci. 74:2005–2029.

Sabater, S; Bregoli, F; Acuna, V; Barcelo, D; Elosegi, A; Ginebreda, A; Marce, R; Munoz, I; Sabater-Liesa, L; and Ferreira, V. 2018. Effects of human-driven water stress on river ecosystems: a meta-analysis. Nature: Scientific Reports. 8:11463.

Sas, H. 1989. Lake restoration by reduction of nutrient loading. Expectations, experiences, extrapolation. St. Augustin: Academic Verlag.

Søndergaard, M; Jensen, JP; and Jeppesen, E. 2003. Role of sediment and internal loading of phosphorus in shallow lakes. Hydrobiologia. 506–509:135–145.

Spears, BM; Carvalho, L; Perkins, R; Kirika, A; and Paterson, DM. 2011. Long-term variation and regulation of internal phosphorus loading in Loch Leven. Hydrobiologia. 681:23–33.

Steinman, A; Chu, X; and Ogdahl, M. 2009. Spatial and temporal variability of internal and external phosphorus loads in Mona Lake, Michigan. Aquat Ecol. 43:1–8.

Steinman, AD; Nemeth, L; Nemeth, E; and Rediske, R. 2006. Factors influencing internal phosphorus loading in a western Michigan, drowned river-mouth lake. J N Am Benthol Soc. 25:304–312.

CHAPTER 24

SYNTHESIS, IMPLICATIONS, AND RECOMMENDATIONS

Bryan M. Spears[1] and Alan D. Steinman[2]

24.1 INTRODUCTION

Our goal in this book is to present a comprehensive assessment of internal phosphorus (P) loading in lakes, drawing on peer-reviewed literature, as well as data and expertise from case studies. Most importantly, our coauthors have imparted hundreds (collectively) of years of expertise in measurement, modeling, and management of internal loading in lakes, producing a comprehensive evidence base with which to inform future practitioners in this field. This expertise spans 16 countries from all continents (with the exception of Antarctica). The introductory chapters provide a blueprint for the study of internal loading in lakes.

Our case studies offer impressive data from some of the world's longest limnological long-term monitoring programs. Analysis and interpretation of these data are presented to quantify the causes of long-term variation in internal loading in 18 lakes, spanning 16 countries, including three transboundary lakes. Cumulatively, these lakes represent 64,678 km^2 of lake surface area draining more than 256,673 km^2 of catchment surface area (see Table 24.1). These case studies offer detailed process understanding with which to examine the responses of internal loading to different elements of environmental change. For example, authors highlight the importance of large-scale social change as drivers of legacy P and internal loading in lakes, as well as how agricultural intensification and urbanization following World War II started the accumulation of legacy P in many lake-bed sediments, which today, drives internal loading.

The authors also explore potential future changes in internal loading processes associated with climate change, in shallow and deep lakes, utilizing empirical and process modeling approaches. This body of work offers new insights into important processes that are common in small water bodies, as well as some of the world's largest lakes, which support millions of people through the provision of drinking water (Lake Erie, USA/Canada: see Chapter 19), energy (Lake Kinneret, Israel: see Chapter 20), and food (Lake Peipsi, Estonia/Russia: see Chapter 10). These lakes are increasingly being viewed as sources of income to neighboring communities, such as through eco-tourism (Esthwaite Water, UK: see Chapter 14). Our authors have confirmed that internal loading is causing large-scale social impacts; one common example is the promotion of cyanobacteria dominance leading to

[1] UK Centre for Ecology & Hydrology, Bush Estate, Penicuik, Midlothian, EH26 0QB. E-mail: spear@ceh.ac.uk.

[2] Annis Water Resources Institute, Grand Valley State University, 740 West Shoreline Drive, Muskegon, MI 49441, USA. E-mail: steinmaa@gvsu.edu.

Table 24.1 Summary of the lakes featured in the case studies within this book. Both catchment total P load (CLoad) and internal total P load (ILoad) estimates reported here were calculated using various mass-balance approaches, described in detail in each chapter. Comparisons between CLoad and ILoad should be made with caution given that CLoad was typically calculated using annual mean conditions, whereas ILoad was calculated either as annual mean or as the mean of the period during which ILoad was observed to occur (typically summer or summer-autumn means), as outlined in each chapter. To simplify comparisons, we report here daily fluxes. In the few cases where data on lake typology was not reported in the chapters, peer-reviewed literature sources were used. ILoad for Malawi as metalimnion to epilimnion upward flux. ND = no data available.

Lead Author	Lake Name	Catchment Area (km²)	Surface Area (km²)	Max Depth (m)	Mean Depth (m)	Residence Time (yr)	Does It Stratify?	CLoad (mg TP m⁻² d⁻¹)	ILoad (mg TP m⁻² d⁻¹)
Nürnberg	Simcoe, CA	1,097	722	15.30	41.00	16.00	Y	0.32	76.00
Reddy	Apopka, USA	485	125	5.40	1.60	7.50	N	0.35	0.96
Reddy	Okeechobee, USA	14,000	1,732	3.90	2.70	3.50	N	0.79	0.18
Nogaro	Grand Lake St. Marys, USA	241	52	4.90	1.50	1.60	N	3.15	ND
Tammeorg	Peipsi, EE/RUS	18,460	3,555	15.30	7.10	2.00	N	0.94	2.30
Xie	Taihu, CH	36,500	2,338	2.60	1.90	0.77	N	2.55	24.61
Spears	Leven, UK	158	13.30	25.30	3.90	0.50	Y	2.02	1.15
Phillips	Barton, UK	109	0.70		1.30	0.04	N	5.50	4.10
Mackay	Esthwaite, UK	17	1.00	16.00	6.90	0.27	Y	1.33	ND
Søndergaard	Soebygaard, DK	12	0.40	1.90	1.00	0.08	N	10.41	ND
Huser	Hjälmaren, SE	3,800	483	24.00	6.10	3.40	Y	0.30	1.50
Noordhuis	Eemmeer, NL	ND	15	2.00	1.70	0.09	N	20.40	1.00
Ventalla	Pyhäjärvi, FI	431	155	26.00	5.50	5.00	N	0.30	2.00
Kaltenberg	Erie, USA/CA	78,000	25,700	64.00	19.00	2.60	Y	0.12	0.95
Eckert	Kinneret, IL	2,730	168.70	43.00	24.00	10.00	Y	0.32	3.13
Marinho	Jacarepaguá, BRL	103	12	9.90	3.30	0.33	N	12.56	4.60
Bootsma	Malawi, MWI/TZA/MOZ	100,500	29,600	706	292	114	Y	1.31	1.87
Hupfer	Arendsee, DE	30	5.14	49	29	56	Y	0.30	7.75

contaminated drinking water (Lake Taihu, China: see Chapter 11) and food (Jacarepaguá Lagoon, Brazil: see Chapter 21). Importantly, many of our case studies include sites of international importance for freshwater biodiversity (Lake Malawi, Malawi, Mozambique, and Tanzania: see Chapter 22). In these lakes, internal loading is causing ecosystem scale extinction of rare species and pushing whole ecosystems to the brink (and beyond in some cases), aided by extrinsic drivers, including extreme weather events (Lake Okeechobee, USA: see Chapter 7).

A common concern raised in most chapters is that internal loading is likely to degrade the ecology of lakes for decades, even after improved sanitation and agricultural practices are implemented in catchments. The direct control of internal loading may offer, at least temporarily, relief from eutrophication symptoms and represents a major contemporary field, as outlined by Lürling et al. (see Chapter 5) and considered for large-scale application by Huser et al. (see Chapter 16), but critically evaluated by Nogaro et al. (see Chapter 8).

A common rationale for all case studies was to deliver water-quality improvements to meet a multitude of societal needs. The relationship between water quality and economic value is well-accepted (cf. Pretty et al. 2003; Dodds et al. 2009; Isely et al. 2018), so managing internal loading provides both environmental and economic benefits. Hundreds of millions of dollars have been invested across our case studies to improve water quality through catchment management and, commonly, internal P loading was reported as a key confounding factor in achieving water-quality management goals.

Clearly, a comprehensive review of this material would be unpalatable for the reader. Instead, in this final chapter, we focus on presenting novel developments and future challenges associated with the study and control of internal loading in lakes. We frame this discussion to assess the societal need to reconcile economic growth with ecological and water-quality management in lakes, in the face of unprecedented environmental change.

24.2 THE NEED FOR STANDARD METHODOLOGIES FOR MEASURING INTERNAL P LOADING IN LAKES

Impressive advances have been made in methodological development toward understanding interannual and spatial variability in internal loading within lakes. This work has led to an improved understanding of the processes that are driving internal loading as reviewed by Søndergaard and Jeppesen (see Chapter 4), although the extent to which these processes operate across lakes is less well-understood. Hupfer et al. (see Chapter 2) provide a detailed analysis of the methods available for quantifying internal P loading, and our case studies demonstrate the application of these approaches. Despite efforts to standardize approaches, such as the analysis of sediment P composition and content (Hupfer et al. 1995), there remains no single standard approach for the measurement of internal loading or of its component processes. This limits our ability to assess commonalities in internal loading responses to environmental change, and in turn, the assessment of management approaches that may deliver wider benefits.

A few notable studies have demonstrated the power of multi-lake assessments when data from comparable methods are available. We argue that much is to be gained from these studies in providing a better understanding of the likely responses in lake ecosystems to climate and land-use change (discussed further later on). For example, Carey and Rydin (2011) reported a relationship between sediment total P (TP) profiles and water column TP concentrations using multi-lake analyses. Dithmer et al. (2016) used sediment P fractionation and nuclear magnetic resonance (NMR) techniques to demonstrate differences in multiple lake responses in sediment chemistry following applications of La-bentonite to control internal loading. Nürnberg (1988) demonstrated the power of multi-lake analyses to produce relationships with which internal loading in deep lakes can be predicted. Kinder-vater et al. (see Chapter 23) conducted a meta-analysis, and found that while previously accepted significant factors such as dissolved oxygen (DO) and temperature had explanatory value in models of sediment P release, there was still considerable unexplained variance suggesting additional study is needed to develop robust predictive models. Long-term water quality data, often produced to meet regulatory requirements at national and regional scales, provide an impressive resource with which to conduct such analyses, potentially with global reach. However, the prerequisite comparable data is often hard to assemble; large scale, multi-lake data from multiple sources (e.g., different re-search institutions, countries, or continental regions), is often not comparable due to inconsistent methodologies, reporting, or language barriers. Indeed, 35 of the 68 potentially relevant papers on sediment incubation estimations of internal P loading evaluated for the meta-analysis (see Chapter 23) could not be used due to poor reporting, unclear methodologies, or incomplete datasets.

This lack of methodological consistency in quantifying internal loading at the ecosystem scale across lakes was apparent also in our collection of case studies. The most consistent approach em-ployed to estimate internal loading was the ecosystem scale mass-balance approach, although a number of methods were used to calculate these estimates. In some cases, process studies were con-ducted at small scales (i.e., cm or meter) in the laboratory (e.g., sediment core incubations) or *in situ* (e.g., benthic chambers) to calculate short-term (i.e., daily to weekly) flux, which was then scaled up to ecosystem and seasonal or annual scales. In other cases, ecosystem scale mass-balance approaches utilizing monthly water-column TP variation were employed to estimate internal loading over lon-ger time periods (i.e., seasonal to annual frequency). Finally, more comprehensive process-based modeling estimates were conducted utilizing direct measurements from the study sites to quantify specific transformation and translocation processes providing net-effects estimates of whole-lake internal loading. There is a pressing need to comprehensively assess the relationships between these methods and to produce internationally accepted methods for wide-scale application. Hupfer et al. (see Chapter 2) draw on their impressive methodological expertise to provide a compelling concep-tual model on the definition of *internal loading* with which to frame such a comparison, and Kinder-vater et al. (see Chapter 23) recommend a consistent set of standards to facilitate comparisons of internal loading studies that would allow for a more robust analysis of trends, identification of global drivers, and recommendations for management options.

24.3 THE NEED TO BETTER UNDERSTAND THE DRIVERS OF VARIATION ACROSS SCALES

Søndergaard and Jeppesen (see Chapter 4) introduce the factors known to control internal loading in lakes, drawing on evidence from the literature and from their experiences of working in many lakes around the world. Their analysis and interpretation of long-term monitoring data from Danish lakes has provided unparalleled insights into the generality of these processes. These and other studies, for example, have identified strong seasonality in flux intensity when, in temperate lakes during summer periods, sediments act as a net source of P to the water column and are characterized by higher diffu-sion rates and lower redox conditions. In these lakes, the impact of P released from the sediment on epilimnetic water quality may increase with decreasing depth and decreasing flushing rate.

The importance of the geochemical composition of the sediments is also an important factor in determining the likelihood of sediment P release. Ostrofsky and Marcbach (2019) utilize sediment P sequential extraction data coupled with sediment P flux estimates to demonstrate that in stratify-ing lakes, only sediments with Al:Fe molar ratios of less than three may be expected to exhibit P release (but see Parsons et al. 2017). Further, the magnitude of the flux where Al:Fe is less than three was strongly correlated with the reductant-extractable sediment P content. Ostrofsky and Marcbach present one of the first decision trees, based on empirical data, whereby the effects of climate change and land-use change on internal loading can be considered in the context of water quality management. Finally we turn to very large lakes, where our case studies have demonstrated the need to understand the influence of hydrologic connectivity, and the physical limnological processes driving it, on P dynamics. In Lake Malawi, for example, there exists a P "conveyor belt" where hydrological processes transform and transport P, resulting in heterogeneous ecological responses, far from the point of pollutant entry (see Chapter 22). In very deep lakes, where the transport time of organic P from the epilimnion to the lake bed is long, remineralization can occur on descent, mimicking the effects of sediment P release that is detectable as hypolimnetic P accumulation (see Chapter 22). Similar complexity exists in Lake Hjälmaren where the three main basins perform different P processes, depending upon their morphometry, which drive P connectivity and ecological responses at the whole-lake scale (see Chapter 16).

24.4   UNDERSTANDING LEGACY P IN THE WATERSHED

The importance of legacy processes is now becoming clear at the global scale. Losses of P from land to freshwaters have increased globally from 5 to 9 Tg P yr^{-1} (equivalent to $5-9 \times 10^6$ metric tons (t) yr^{-1}), about half of which is retained by fresh waters (Beusen et al. 2016). P in runoff has resulted in widespread eutrophication and represents a significant and continuous source of legacy P to lakes (Sharpley et al. 2012). Goyette et al. (2018) report that P losses via runoff from land to freshwaters will occur when soil P content exceeds 0.03 t km^{-2} – 8.7 t km^{-2} and that losses are likely to proceed for between 100 to 2,000 years following land-based abatement measures. They call for new strategies to address long-term losses of legacy P from soils to fresh waters in P-replete watersheds. The control of internal loading in lakes represents an opportunity to address, directly, legacy P within receiving waterbodies to deliver improvements during the transient recovery period associated with land-based measures.

Following reductions in catchment (external) load associated with eutrophication management strategies, lake bed sediments can switch from being a sink of P to being a source (Søndergaard 2007; Sharpley et al. 2012). It is likely that the variation in recovery time is influenced by lake morphology and extrinsic stressors, such as hydrodynamic forcing. Reddy et al. (see Chapter 7) highlight the importance of legacy P relinquishment from drainage basins to the receiving water bodies of Lakes Apopka and Okeechobee, Florida, USA. In both lakes, recovery was confounded by internal loading following catchment management. In recent decades, our understanding of legacy P processes has improved (Sharpley et al. 2012). This research has acknowledged the issues associated with legacy P as it travels slowly through land and aquatic ecosystems to receiving water bodies. It is important that opportunities to deliver improvements in the short term are considered more widely, as these early wins can generate broad public support, and can be incorporated within the context of integrated management approaches.

In lakes, measures to control legacy P and to mitigate the effects of climate change have been developed and when applied correctly, can provide rapid water quality improvements over years to decades (Huser et al. 2016). Lürling et al. (see Chapter 5) present a much needed critical assessment of internal loading-control demonstration studies and Hickey and Gibbs (2008) present comprehensive evidence on internal loading-control measures, including decision support and risk assessment. Measures include direct interventions designed to remove or immobilize P, such as sediment dredging (Oldenborg and Steinman 2019) and P inactivation (Lürling et al. 2016), respectively, as well as indirect measures designed to force desirable ecological responses, such as manipulating the fish community to restrict the yield of phytoplankton per unit of P. The common bond of these measures is that they rarely deliver long-term benefits while nutrient loading from the catchment remains elevated, and in some cases, may not even offer temporary relief when applied inappropriately (Grand Lake St. Mary's: see Chapter 8).

The control of internal loading may represent an attractive approach to climate-change mitigation given the advances made in this field in recent years (see Chapter 5). This approach is highly innovative and acknowledges the importance of scale with respect to the multiple drivers acting on our lake ecosystems. For example, drivers such as climate change operate at scales beyond the scope of a water manager's influence, yet the resulting increase in weather variability limits their capacity to deliver improvements or prevent further degradation of a lake and its catchment (Friberg et al. 2017).

24.5 UNDERSTANDING MULTIPLE AND INTERACTING STRESSORS OF INTERNAL LOADING

We observed in our case studies, significant inter-annual variation in internal P loading, in some cases following catchment load reduction. What is less clear is whether or not this variability increases as the dominance of external loading recedes or varies with lake morphology, which would be intuitive and predictable. For example, internal loading was reported to *shut down* from one year to the next, in the decades following reductions in catchment P loading to Loch Leven (see Chapter 12). This study, at first glance, indicates the operation of nonlinear, multiple stressor-response relationships responsible for the regulation of internal loading in this shallow lake. In contrast, more gradual reductions in internal loading following catchment load reduction have been reported in other lakes (e.g., Barton Broad: see Chapter 13; Lake Søbygaard: see Chapter 15), where longer recovery times extending to many decades have been associated with high legacy P content of the bed sediments. A better understanding of this inter-annual variability would provide a higher level of confidence in the predictions of future water quality. This is particularly important for lakes where internal loading plays a key role in summer or annual water quality, as is the case in many of the lakes included in our case studies. In addition, where climate change-related stressors are important, it would provide a measure of the potential future water-quality envelope in relation to internal loading under future change scenarios. Regardless of what drives the inter-annual variation, the results from Lake Simcoe (Canada: see Chapter 6) indicate that modeled predictions of internal P loading are more meaningful when assessed over clusters of years as opposed to individual years, presumably due to the reduced influence of extreme years.

Søndergaard and Jeppesen (see Chapter 4) conclude that ". . . internal phosphorus loading is a highly complex process that involves a number of biogeochemical mechanisms operating on different spatial and temporal scales" and that ". . . the importance of internal loading depends on a number of interacting mechanisms. . . ." The data presented in the case studies allow us to explore catchment and lake typology as drivers of internal loading at the ecosystem scale. However, we acknowledge that the wide range of methodologies used to calculate internal loading, as evidenced in our case studies, limits direct comparisons among lakes and with catchment load. In addition, variation in depth, geology, and land use within the same lake can result in variable internal loading estimates (cf. Lake Simcoe: see Chapter 6; Lake Taihu: see Chapter 11; Lake Hjälmaren: see Chapter 16; Lake Erie: see Chapter 19), potentially confounding attempts to derive patterns across lakes. Nevertheless, we used the data from our case studies to explore general trends in lake typology and estimates of internal loading, using mass-balance estimates of whole lake internal loading as the most commonly reported value. Using Spearman Rho analysis, we identified catchment P load and lake residence time as the two factors most strongly correlated (positively) with internal loading. We then constructed a multiple linear regression model using these factors, the results of which were significant (see Figure 24.1).

We expected internal loading to be negatively related to depth and to be positively related to retention time because lakes that flush quickly leave less time for hypoxia to form, limiting redox-driven P release, and deeper lakes that stratify will be more prone to high sediment P release rates under anoxia. Carey and Rydin (2011), and work by others reviewed by Nurnberg (see Chapter 3), suggest that sediment P release rates should increase with increasing catchment P load or lake trophic status and our model is in general agreement with this conclusion. In addition, it is likely that retention time and depth co-vary and thus cannot be considered independent predictors. Utilizing the case-study evidence, further general principles can be drawn with respect to the drivers of internal P loading at the whole-lake scale.

Weather played a key role in driving internal loading in some case studies, suggesting the importance of climate change as a future driver of internal loading. For example, in Esthwaite Water, UK (see Chapter 14), hypolimnetic P flux was found to be a key factor controlling whole-lake P budgets and this, in turn, was regulated strongly by changes in anoxia and the strength of stratification, driven mainly by warmer temperatures. Ventelä et al. (see Chapter 18), Tammeorg et al. (see Chapter 10), and Reddy et al. (see Chapter 7) demonstrated the importance of wind-induced disturbance of bed sediments as a driver of internal loading in large shallow lakes suggesting that an increase in storminess will potentially increase internal loading in large shallow lakes. Ventelä et al. (see Chapter 18) highlighted the importance of the duration of ice cover and precipitation as drivers of internal P loading in Lake Pyhäjärvi. Spears et al. (see Chapter 12) reported on the combined effects

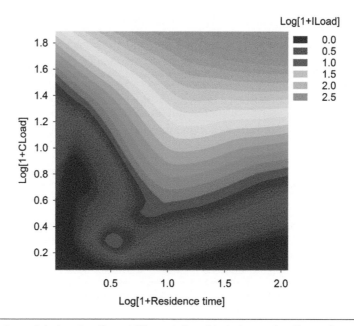

Figure 24.1 Contour plot showing the additive relationship between log [1+ . . .] transformed CLoad and hydraulic residence time estimates and responses in mass-balance estimates for ILoad from 15 lakes in Table 24.1, with available data. The relationship was confirmed using multiple linear regression in Sigmaplot, Version 13. ILoad = −0.128 + (0.88 × CLoad) + (0.42 × Residence time). p = 0.001; R^2 = 0.67; CLoad: p = < 0.001; Residence Time: p = 0.022; Total DF 14; F = 12.4. The data passed the Shapiro-Wilk Normality Test (p = 0.924) and the Constant Variance Test (p = 0.171).

of multiple climate change-related stressors where internal loading increased with increasing wind speed and decreasing precipitation, with the latter most likely operating through flushing rate and dilution of high P concentrations in lake water with low P water from the catchment. Hydrologic residence time can also influence lag times of internal loading where, in very large lakes, increased precipitation may decrease the lag times between the deposition of allochthonous P to the bed and release back to the water column, as noted in Lake Malawi (see Chapter 22).

Biology also can play a critical role in internal loading (see Chapter 1). Noordhuis et al. (see Chapter 17) present a trophic cascade with unexpected results: reducing P from the catchment to Lake Eemmeer (The Netherlands) can result in less growth of dreissenid mussels, which otherwise keep phytoplankton populations in check due to their filtration. Hence, in this case, reducing P may actually increase phytoplankton growth due to reduced filtration by a higher trophic level. In turn, this phytoplankton community eventually settles on the lake bottom and becomes an important contributor to future internal loading, as shown in Lake Kinneret (see Chapter 20).

As we gain a better understanding of the factors influencing internal loading, it is clear from our case studies and syntheses that much remains unresolved. In addition to warming, other trends facing lakes around the world include over-withdrawal, brownification (Leech et al. 2018), and salinization (Dugan et al. 2017). Over-withdrawal is taking a toll on the ecology and water quality of Lake Kinneret (see Chapter 20) while salinization has not prevented internal loading from being a major impairment in Brazil's Jacarepaguá Lagoon (see Chapter 21) or freshwater systems in North America (Kaushal et al. 2018).

The evidence outlined previously confirms the importance of climate change as a key driver of internal loading with effects being manifest through changes in temperature, precipitation, and wind in both shallow polymictic and deep monomictic lakes. We combine these stressors with others to present an overarching conceptual model of internal loading processes in stratifying and mixed waterbodies (see Figure 24.2). In regions where precipitation will decrease, especially in summer, and where summer temperatures and wind speeds will increase, we expect more prolonged and intense internal loading leading to further degradation of lake water quality. However, key research gaps are evident with respect to potential future hydroclimatic forcing of internal loading.

The regional effects of climate change in relation to climate predictions should be further assessed utilizing short- and long-term weather predictions. It should be possible to produce forecasts of internal loading using this approach, although admittedly, there will be considerable uncertainty

---

Figure 24.2   (See next page)   The P cycle in a eutrophic lake with a shallow area and a deep thermally stratified area, with a focus on internal P load release mechanisms from chemical (redox-related reactions during anoxic conditions), biological (bioturbation such as Chironomidae larvae, disturbance by benthic fish, respiring microbes), and physical (wind/wave action) release processes. The main external stressors to exacerbate internal loading are climate change (changes in precipitation, wind, increasing temperatures, and increased atmospheric pollution deposition), and increased catchment loading from increased population growth and land use change. P can enter the system through inflows, the atmosphere, groundwater exchange, and sediment release. The loss of P from a lake system includes outflows through groundwater exchange and sedimentation. P is tied up in biomass in the lake (bacteria, phytoplankton, zooplankton, macrophytes, macro-invertebrates (not shown), fish, and water birds) and passed through the food web via predation/bacterivory or released back into the system through biomass degradation and excretion as organic matter (OM). The mineralization of OM releases dissolved organic P (DOP), dissolved inorganic P (DIP), and particulate organic P (POP). Phytoplankton and heterotrophic prokaryotes uptake DOP and DIP and can transform POP to DOP and release DIP into the system. In the sediment, bacteria-mediated reactions transform POP to DOP to DIP and inorganic P (PIP) to DIP. Diagram created by Kate Waters, Centre for Ecology & Hydrology, UK.

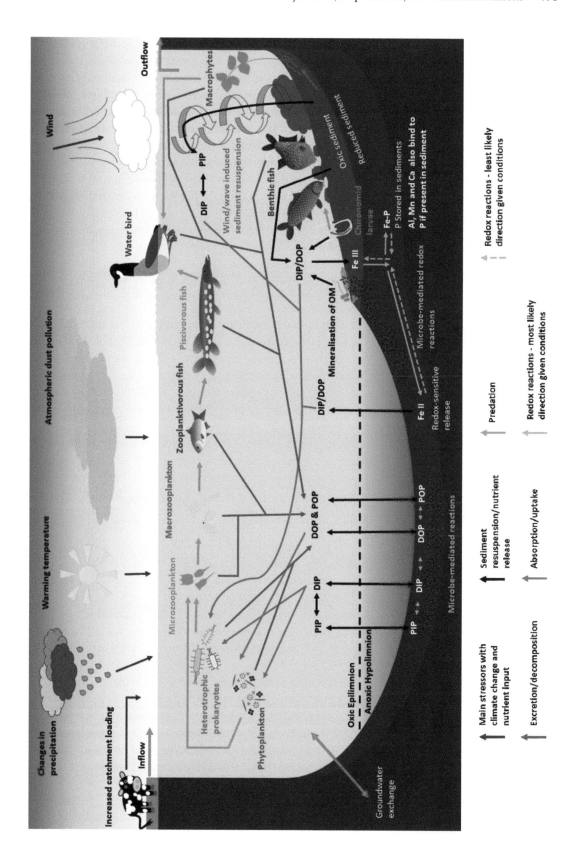

at first. The significant correlations between hydroclimatic indices and internal loading (Lake Taihu: see Chapter 11 and Barton Broad: see Chapter 13) suggest that teleconnection processes should lead to coherence in inter-annual and long-term variation in internal loading across lakes of the same districts or regions. The importance of lake type and other local factors, such as land use and lithology, in mediating the effects of large-scale hydroclimatic forcing should be quantified to produce information on climate sensitivities with which to inform future water quality management and conservation initiatives. The extent to which the advances in internal loading control can be used to mitigate the effects of climate change need further exploration (see Chapter 16).

24.6   RECONCILING THE NEED FOR ECONOMIC DEVELOPMENT WITH THE NEED TO PROVIDE ECOLOGICAL INTEGRITY

It is well recognized that lakes provide a number of critical ecosystem services, including use values such as drinking water, fishing, recreational boating, pollution reduction, flood control, and navigation, as well as non-use values such as cultural experiences, spiritual connection, and psychological well-being (Steinman et al. 2017). As a consequence, restoring impaired habitat or water quality of lakes often results in an increase in the economic value of the water body (Pretty et al. 2003; Austin et al. 2007; Braden et al. 2008; Isely et al. 2018)—and this was commonly the driver of restoration efforts reported in our case studies. In some cases, wider water quality directives were important, such as the European Water Framework Directive, although these focused predominantly on restoring for biodiversity and conservation and less on the delivery of multiple ecosystem services (Carvalho et al. 2019).

In lakes where internal loading is a key driver, its control can result in net economic benefits to local communities, although few empirical relationships have been developed to underpin cause and effect between water quality and ecological responses to management and ecosystem service delivery (Spears et al. 2019). The extent to which ecosystem services will be improved through the control of internal P loading will be strongly influenced by each system's unique cultural and environmental setting. For example, the concerns about P enrichment affecting food web interactions and fishery yield is much more profound in Lake Malawi than in Grand Lake St. Mary's, where locals don't depend on fish for their primary food supply. Benefits can take many different forms, including: generating employment and profits to businesses (BenDor et al. 2015; Blignaut et al. 2014); greater access to blue spaces for leisure and recreational activities; greater positive impacts on individuals' perceived physical, emotional, and spiritual well-being (Bell et al. 2015); an enhanced sense of place (Völker and Kistemann 2013; Poe et al. 2016); increased community cohesion (Weinstein et al. 2015); cultural renewal (Kittinger et al. 2016); and improved collective efficacy, thereby reducing crime (Weinstein et al. 2015).

The positive values associated with restoration must be considered in the broader context of the cost of P control or management. Huser et al. (2016), drawing on experiences in the Minneapolis (MN, USA) Chain of Lakes, report on the complexity of assessing the cost effectiveness of internal P loading control (cost range: \$14–\$95 kg P^{-1} removed) in combination with catchment based interventions (cost range: \$1,187–\$17,996 kg P^{-1}). In this study, cost effectiveness of internal P load control interventions (with alum, in this case) increased, relative to catchment load reduction measures, as urbanization increased, making catchment measures more costly to implement.

Assessing the economic trade-off of restoration is not straightforward, as some ecosystem services do not flow through traditional markets, so they may be undervalued when using standard economic metrics. In addition, the choice of a discount value is often subjective and will influence the long-term valuation. Conversely, estimating the cost of restoration can be challenging, especially when factoring in operation and maintenance costs for long-term projects and considering economies of scale. Promising new approaches, such as landscape planning, are already being developed to address this problem (Mueller et al. 2016, 2019). These strategies are relevant for the valuation of internal P loading control measures. Even relatively short-term (less than one year) approaches, such as chemical inactivation or sediment dredging, may need to be repeated within 5–10 years, whereas aeration may be continuous. Hence, deciding on a management strategy using economics alone is short-sighted, although economics can serve to demonstrate the scale of the problem and the associated costs. For example, in the United Kingdom, the Office for National Statistics has estimated the value of freshwaters at £39.5 billion (approximately $48 billion USD) (ONS 2015). Excess P is a major stressor impacting UK lakes and rivers, thereby representing both a national scale challenge and a threat to this economic asset (Leaf 2019).

In the European Union, little progress has been made in the last ten years in reducing P concentrations in lakes, despite the implementation of the Water Framework Directive (EEA 2015); approximately 31,819 km^2 of EU lake surface area (ca. 40% of monitored lakes by number) are deemed to fall short of ecological quality targets, many of which are impacted by P pollution. A further 37,153 km^2 of EU lake surface area are deemed to have *acceptable* levels of degradation. In China, monitoring of 862 lakes indicated that P load reduction measures implemented since 2000, predominantly through improvements in sanitation and agriculture, resulted in a reduction in P concentrations in 60% of monitored lakes (Tong et al. 2017). However, monitoring data also indicated that current conditions may not yet support ecological recovery and that the biggest initial responses were achieved in the most polluted sites. In the United States, P concentrations in 72% of rivers and 79% of lakes exceed background levels as a result of human activity (Dodds et al. 2009). At the ecosystem scale, the costs associated with this level of degradation are staggering. Losses associated with our cases studies, for a single algal bloom, include $30 million in Lake Taihu in 1990 (Le et al. 2010), and in Loch Leven in the 1990s at $1.3 million. National scale estimates for losses are sparse but include these estimated costs for responding to eutrophication: (1) USA: 2.2 billion USD yr^{-1} (Dodds et al. 2009); (2) UK: 160 million USD yr^{-1} (Pretty et al. 2003); and (3) Australia: 154 million USD yr^{-1} (LWRRDC 1998), as reviewed by Douglas et al. (2016).

24.7 CONCLUSION

The control of internal loading that utilizes established or novel interventions is likely to become the norm in the future as the need to make societal gains and minimize economic losses from eutrophic waters; this is especially true in urban lakes where the cost effectiveness of catchment management is limited. The information provided in this book reveals our current state of knowledge regarding internal P loading, including management and controls, but also identifies significant knowledge gaps and opportunities. We strongly urge the lake research and management communities to work together to develop standard approaches for the measurement and mitigation of internal loading. Our ultimate goal should be to develop and demonstrate adaptive management approaches that address both catchment-focused and in-lake interventions, with the potential to deliver significant ecological and socio-economic benefits globally, and counteract the predicted negative effects of climate change.

24.8   REFERENCES

Austin, JC; Anderson, ST; Courant, PN; and Litan, RE. 2007. Healthy Waters, Strong Economy: the Benefits of Restoring the Great Lakes Ecosystem. Washington, DC: Brookings Institution.

Bell, SL; Phoenix, C; Lovell, R; and Wheeler, BW. 2015. Seeking everyday wellbeing: the coast as a therapeutic landscape. Social Sci Med. 142:56–67.

BenDor, T; Lester, TW; Livengood, A; Davis, A; and Yonavjak, L. 2015. Estimating the size and impact of the ecological restoration economy. PloS One. 10(6): e0128339.

Beusen, AHW; Bouwman, AF; Van Beek, LPH; Mogollon, JM; and Middelburg, JJ. 2016. Global riverine N and P transport to ocean increased during the 20th century despite increased retention along the aquatic continuum. Biogeosciences. 13:2441–2451.

Blignaut, J; Aronson, J; and de Groot, R. 2014. Restoration of natural capital: a key strategy on the path to sustainability. Ecol Engin. 65:54–61.

Braden, JB; Taylor, LO; Won, D; Mays, N; Cangelosi, A; and Patunru, AA. 2008. Economic benefits of remediating the Buffalo River, New York area of concern. J Great Lakes Res. 34:649–660.

Carey, CC and Rydin, E. 2011. Lake trophic status can be determined by the depth distribution of sediment phosphorus. Limnol Oceanogr. 56:2051–2063.

Carvalho, L; Mackay, E; Cardoso, AC; Baatrtrup-Pedersen, A; Birk, S; Blackstock, KL; Borics, G; Borja, A; Feld, CK; Ferreira, MT; et al. 2019. Protecting and restoring Europe's waters: an analysis of the future development needs of the Water Framework Directive. Sci Tot Environ. 658:1228–1238.

Dithmer, L; Nielsen, UG; Lurling, M; Spears, BM; Yasseri, S; Lundberg, D; Moore, A; Jensen, ND; and Reitzel, K. 2016. Responses in sediment phosphorus and lanthanum concentrations and composition across 10 lakes following applications of lanthanum modified bentonite. Water Res. 97:101–110.

Dodds, WK; Bouska, WW; Eitzmann, JL; Pilger, TJ; Pitts, KL; Riley, AJ; Schloesser, JT; and Thornbrugh, DJ. 2009. Eutrophication of U.S. freshwaters: analysis of potential economic damages. Environ Sci Technol. 43:12–19.

Douglas, G; Hamilton, DP; Robb, MS; Pan, G; Spears, BM; and Lurling, M. 2016. Guiding principles for the development and application of solid-phase phosphorus adsorbents for freshwater ecosystems. Aquat Ecol. 50:385–405.

Dugan, HA; Bartlett, SL; Burke, SM; Doubek, JP; Krivak-Tetley, FE; Skaff, NK; Summers, JC; Farrell, KJ; McCullough, IM; Morales-Williams, AM; et al. 2017. Salting our freshwater lakes. Proc Natl Acad Sci. 114:4453–4458.

EEA. 2015. The European environment—state and outlook 2015: synthesis report. European Environment Agency, Copenhagen. DOI: 10.2800/944899

Friberg, N; Buijse, T; Carter, C; Hering, D; Spears, BM; Verdonschot, P; and Moe, TF. 2017. Effective restoration of aquatic ecosystems: scaling the barriers. WIREs Water. 4:e1190.

Goyette, JO; Bennett, EM; and Maranger, R. 2018. Low buffering capacity and slow recovery of anthropogenic phosphorus pollution in watersheds. Nature Geoscience. 11:921–925.

Hickey, CW and Gibbs, MM. 2008. Lake sediment phosphorus release management—decision support and risk assessment framework. NZ J Mar Fresh Res. 43:819–856.

Hupfer, M; Gachter, R; and Giovanoli, R. 1995. Transformation of phosphorus species in settling seston during early sediment diagenesis. Aquat Sci. 57:305–324.

Huser, BJ; Egemose, S; Harper, H; Hupfer, M; Jensen, H; Pilgrim, KH; Reitzel, K; Rydin, E; and Futter, M. 2016. Longevity and effectiveness of aluminum addition to reduce sediment phosphorus release and restore lake water quality. Water Res. 97:122–132.

Huser, BJ; Futter, M; Lee, JT; and Perniel, M. 2016. In-lake measures for phosphorus control: the most feasible and cost-effective solution for long-term management of water quality in urban lakes. Water Res. 97:142–452.

Isely, P; Sterrett Isely, E; Hause, C; and Steinman, AD. 2018. A socioeconomic analysis of habitat restoration in the Muskegon Lake area of concern. J Great Lakes Res. 44:330–339.

Kaushal, SS; Likens, GE; Pace, ML; Utz, RM; Haq, S; Gorman, J; and Grese, M. 2018. Freshwater salinization syndrome on a continental scale. Proc Natl Acad Sci. 115(4):E574–83.

Kittinger, JN; Bambico, TM; Minton, D; Miller, A; Mejia, M; Kalei, N; Wong, B; and Glazier, EW. 2016. Restoring ecosystems, restoring community: socioeconomic and cultural dimensions of a community-based coral reef restoration project. Regional Environ Change. 16(2):301–313.

Le, C; Zha, Y; Li, Y; Sun, D; Lu, H; and Yin, B. 2010. Eutrophication of lake waters in China: cost, causes, and control. Environ Manage. 4:662–668.

Leaf, S. 2016. Taking the P out of pollution: an English perspective on phosphorus stewardship and the Water Framework Directive. Water Environ J. 32:4–8.

Leech, DM; Pollard, AI; Labou, SG; and Hampton, SE. 2018. Fewer blue lakes and more murky lakes across the continental US: implications for planktonic food webs. Limnol Oceanogr. 63(6):2661–2680.

Lürling, M; Mackay, E; Reitzel, K; and Spears BM. 2016. A critical perspective on geo-engineering for eutrophication management in lakes. Water Res. 97:1–10.

LWRRDC. 1998. Evaluation of the impact of research projects relating to Australia's natural resources. Occasional Paper Land and Water Resources Research and Development Corporation, Canberra, Australia. No. IR 01/99.

Mueller, H; Hamilton, DP; and Doole, GJ. 2016. Evaluating services and damage costs of degradation of a major lake ecosystem. Ecosys Serv. 22: 370–380.

Mueller, H; Hamilton, D; Doole, G; Abell, J; and McBride, C. 2019. Economic and ecosystem costs and benefits of alternative land use and management scenarios in the Lake Rotorua, New Zealand, catchment. Global Environ Change. 54:102–112.

Nürnberg, GK. 1988. Prediction of phosphorus release rates from total and reductant-soluble phosphorus in anoxic sediments. Can J Fish Aquat Sci. 45:453–462.

Oldenborg, K and Steinman, AD. 2019. Impact of sediment dredging on sediment phosphorus flux in a restored riparian wetland. Sci Total Environ. 650:1696–1979.

ONS. 2015. UK natural capital—freshwater ecosystem assets and services accounts. Office for National Statistics. https://www.ons.gov.uk/economy/environmentalaccounts/articles/uknaturalcapitalfreshwaterecosystem assetsandservicesaccounts/2015-03-20.

Ostrofsky, ML and Marbach, RM. 2019. Predicting internal phosphorus loading in stratified lakes. Aquatic Sci. 81:18–27.

Parsons, CT; Rezanezhad, F; O'Connell, DW; and Van Cappellen, P. 2017. Sediment phosphorus speciation and mobility under dynamic redox conditions. Biogeosciences. 14:3585–3602.

Poe, MR; Donatuto, J; and Satterfield, T. 2016. "Sense of Place": human wellbeing considerations for ecological restoration in Puget Sound. Coastal Manage. 44(5):409–426.

Pretty, JN; Mason, CF; Nedwell, DB; Hine, RE; Leaf, S; and Dils, R. 2003. Environmental costs of freshwater eutrophication in England and Wales. Environ Sci Technol. 37:201–208.

Sharpley, A; Jarvie, HP; Buda, A; May, L; Spears, B; and Kleinman, P. 2013. Phosphorus legacy: overcoming the effects of past management practices to mitigate future water quality impairment. J Environ Qual. 42:1308–1326.

Søndergaard, M. 2007. Nutrient dynamics in lakes—with emphasis on phosphorus, sediment and lake restorations. Doctor's dissertation (DSc). National Environmental Research Institute, University of Aarhus, Denmark. p. 276.

Spears, BM; Hamilton, DP; Pan, Y; and May, L. Accepted 2019. Lake management: is prevention better than the cure? Inland Waters.

Steinman, AD; Cardinale, BJ; Munns, WR; Ogdahl, ME; Allan, JD; Angadi, T; Bartlett, S; Brauman, K; Byappanahalli, M; Doss, M; et al. 2017. Ecosystem services in the Great Lakes. J Great Lakes Res. 43:161–168.

Tong, Y; Zhang, W; Wang, X; Couture, RM; Larssen, T; Zhao, Y; Li, J; Liang, H; Liu, X; Bu, X; et al. 2017. Decline in Chinese lake phosphorus concentration accomplished by shift in sources since 2006. Nature Geoscience. 10:507–512.

Völker, S and Kistemann, T. 2013. "I'm always entirely happy when I'm here!" Urban blue enhancing human health and well-being in Cologne and Düsseldorf, Germany. Social Sci Med. 78:113–124.

Weinstein, N; Balmford, A; DeHaan, CR; Gladwell, V; Bradbury, RB; and Amano, T. 2015. Seeing community for the trees: the links among contact with natural environments, community cohesion, and crime. BioScience. 65:1141–1153.

INDEX